Advances in Experimental Medicine and Biology

Volume 1087

More information about this series at http://www.springer.com/series/5584

Junjie Xiao
Editor

Circular RNAs

Biogenesis and Functions

Springer

Editor
Junjie Xiao
School of Life Science,
Institute of Cardiovascular Sciences
Shanghai University
Shanghai, China

ISSN 0065-2598 ISSN 2214-8019 (electronic)
Advances in Experimental Medicine and Biology
ISBN 978-981-13-1425-4 ISBN 978-981-13-1426-1 (eBook)
https://doi.org/10.1007/978-981-13-1426-1

Library of Congress Control Number: 2018956322

This Springer imprint is published by the registered company Springer Nature Singapore Pte Ltd.
The registered company address is: 152 Beach Road, #21-01/04 Gateway East, Singapore 189721, Singapore

Contents

Part I

Overview

1 An Overview of Circular RNAs

Rajendra Awasthi, Anurag Kumar Singh, Gaurav Mishra, Anand Maurya, Dinesh Kumar Chellappan, Gaurav Gupta, Philip Michael Hansbro, and Kamal Dua

Abstract

Circular RNAs (cirRNAs) are long, noncoding endogenous RNA molecules and covalently closed continuous loop without 5′–3′ polarity and polyadenylated tail which are largely concentrated in the nucleus. CirRNA regulates gene expression by modulating microRNAs and functions as potential biomarker. CirRNAs can translate in vivo to link between their expression and disease. They are resistant to RNA exonuclease and can convert to the linear RNA by microRNA which can then act as competitor to endogenous RNA. This chapter summarizes the evolutionary conservation and expression of cirRNAs, their identification, highlighting various computational approaches on cirRNA, and translation with a focus on the breakthroughs and the challenges in this new field.

Keywords

cirRNA · Circular RNAs · Gene expression · Translation

1 Introduction

In 1976 Sanger and coworkers proposed that the viroids are single-stranded structures covalently bound to circular RNAs (cirRNAs). These are pathogenic to certain plants of higher class. It was primitively reported as a viroid, consisting of a covalently closed cirRNA molecule, and pathogenic to particular higher plants [1].

CirRNAs, a class of noncoding endogenous RNA, regulate gene expression in mammals at the transcriptional or posttranscriptional level by

R. Awasthi (✉)
Amity Institute of Pharmacy, Amity University, Noida, Uttar Pradesh, India

A. K. Singh
Centre of Experimental Medicine & Surgery, Institute of Medical Sciences, Banaras Hindu University, Varanasi, Uttar Pradesh, India

G. Mishra · A. Maurya
NKBR College of Pharmacy and Research Centre, Meerut, Uttar Pradesh, India

D. K. Chellappan
Department of Life Sciences, School of Pharmacy, International Medical University, Kuala Lumpur, Malaysia

G. Gupta
School of Pharmaceutical Sciences, Jaipur National University, Jaipur, India

P. M. Hansbro · K. Dua (✉)
School of Biomedical Sciences and Pharmacy, University of Newcastle, Hunter Medical Research Institute, Newcastle, Australia

Discipline of Pharmacy, Graduate School of Health, University of Technology Sydney, Sydney, NSW, Australia

J. Xiao (ed.), *Circular RNAs*, Advances in Experimental Medicine and Biology 1087,
https://doi.org/10.1007/978-981-13-1426-1_1

interacting with microRNAs [2–6]. For many years, cirRNAs were overlooked as rare isoforms that result from splicing artifacts or gene rearrangements [7]. These rediscovered RNA molecules mainly arise from exon circularization or intron circularization and covalently joined 3′ and 5′ ends of a single-stranded RNA molecule by backsplice events (an upstream splice acceptor is joined to a downstream splice donor), thus presenting as covalently closed continuous loops [1, 7–9].

CirRNAs are misinterpreted as splicing errors. Recently, cirRNAs have shown to be widespread and diverse in eukaryotic cells [10]. CirRNAs are relatively stable in the cytoplasm [5]. These are produced by a backsplicing process, wherein downstream exons are spliced to upstream exons in reverse order [2]. CirRNAs are more stable than linear RNA isoforms due to the lack of accessible ends, which resist exonucleases. However, the mechanism of cirRNA formation and their cellular function are still unclear. Relating to the function of cirRNA, it is hypothesized that these molecules are epigenetic microRNA sponges [7]. Human CDR1as/ciRS-7 are examples of functional exonic cirRNAs which have been experimentally validated to function as miRNA sponges and involve in gene expression regulation [10]. However, it is not clear whether all cirRNA molecules work as miRNA sponges or not [7]. The stable nature of cirRNAs makes these moieties intriguing candidates as functional molecules in circulating body fluid [11]. However, currently, there is no systematic approach available for identifying exonic cirRNAs in the human transcriptome [10].

Various challenges associated with the detection of cirRNA include exclusion of sequencing errors, unfair treatment between exonic cirRNAs, and other types of RNAs (e.g., trans-spliced RNAs and genetic rearrangements) on the basis of prejudice, adjustment of errors, in vitro artifacts, and the reconciliation of heterogeneous results [10].

CirRNAs are specific to certain diseases such as neuronal disorders and atherosclerosis [4, 5]. Our insensitivity about cirRNAs is due to an insufficiency of available sequencing data for cirRNA detection [12]. CirRNAs have great potential as clinical diagnostic markers and new therapeutic molecules for the disease therapy. Till today few reports have been published on cirRNAs due to low expression level. Originally these molecules were considered as by-products of alternative splicing and were named as a genetic accident or experimental errors [1].

2 Identification and Appropriate Validation of cirRNA

The cirRNA was firstly recognized in the early 1990s. The recognition on a large scale was not focused in the early stages because of its tedious traditional method of study and due to the lack of useful information. Therefore, the developed recent method of study and identification brings a very precise way to explore cirRNA. Moreover, a key element of cirRNA is out of-order arrangement of exons kenned as a backsplice (described beneath) is not one of a kind to cirRNAs. An early RNA-seq mapping algorithm filtered out such sequences. These issues have been tended to through the improvement of exonuclease-based enhancement approaches. Novel bioinformatic devices such as sequencing with longer reads and higher throughput and sequencing of ribosomal RNA (rRNA)-depleted RNA libraries (as opposed to poly(A)-advanced libraries) make easy to separate cirRNAs from other RNAs and also maintain its circularity [13].

The identification of cirRNAs is exceptionally valuable for understanding the regulatory mechanisms and for potential ramifications for remedial applications, for instance, working as miRNA sponges for oncogenic miRNAs. lncRNA is effectively recognized from other little ncRNA, such as miRNA, siRNA, and snoRNA, by utilizing straightforward property transcript size. However, for cirRNA identification from different lncRNAs, it has been nearly unrealistic to distinguish them just on simple features. cirRNA has shown some extraordinary succession attributes from different lncRNAs, for example, GT-AG match of sanctioned graft locales, combined Alu rehash, and backsplice [14]. Sequence features cumulating with machine learning are accounted to be puissant to prognosticate gene regulation, splicing sites, and chromatin 18. They promote sequence-based strategy possibly used to recognize cirRNA from different lncRNAs efficaciously [15].

Discovery of cirRNA articulation can be accomplished utilizing various techniques such as polymerase chain reaction (PCR) of the Northern blot, two-dimensional gel electrophoresis, gel trap electrophoresis, in situ hybridization, and RNase degradation assay [16].

2.1 Identification and Validation of cirRNA by PCR

PCR is the speediest and most effortless technique to distinguish the expression of cirRNAs. Primers are utilized as a part of PCR for recognition of protein-coding or noncoding RNAs. These are essentially planned and focused to permit enhancement of the primer-flanked nucleic acid region. The utilization of different oriented primer sets is fundamental for the recognition of cirRNA articulation utilizing PCR [17]. Sanger sequencing is the fundamental technique to identify various circular transcripts by semiquantitative or quantitative PCR. Sanger sequencing is also useful for further refinement of the PCR product to validate backsplice site. Backsplice sequence information can be generated from RNA sequencing data or publicly accessible sets of non-poly(A)-culled RNA sequencing data from the National Center for Biotechnology Information – Gene Expression Omnibus (NCBI – GEO) database. Primers to categorically detect cirRNAs by PCR should be planned divergently which can be straightforwardly achieved utilizing free online implements such as Primer3.

Identification of circular RNA by PCR is done in the following steps:

a. *Primer design*: For the determination of chromosome position of the terminuses presaged to pair for backsplicing, we need to determine which exons/introns are to be included in the backsplicing utilizing the genome browser. To get a general summary of the required exons in cirRNA, an entire backsplice sequence is inserted in the BLAT implement by using https://genome.ucsc.edu/cgi-bin/hgBlat. The corresponding exon sequences are fetched to the respective gene and species using www.ensembl.org. The exon order is reversed, keeping 5′→3′ orientation of both exons. The sequence has to be pasted into the corresponding box using http://primer3.ut.ee/. The box is changed from a product size in the range of 70–150 bp and cull pick primers. Cull primer pair ascertains amplified region covering the backsplice site and controls the presaged primer tm to 60 °C. It is suggested that the primers should not overlap the backsplice site. The primer sequence is examined by UCSC in silico PCR implement to check the amplification in genome assembly and the UCSC-annotated genes; no presaged amplification is expected.
b. *Semiquantitative PCR*: The accompanying convention is depicted for the enhancement of cirRNAs. For reference, articulation of the straight RNA of the quality of intrigue ought to be evaluated. Briefly, the buffer concentrate is defrosted, dNTP is mixed, and random hexamer-primed cDNA is kept on the ice. The quantity of responses is computed, and 1–2 extra responses are incorporated to make up for inevitable misfortune by pipetting out. PCR Master is mixed maintaining Taq Reaction Buffer (10×) 2.5 μL, 1 μL of Taq polymerase (1 U/μL), 0.5 μL of 10 mM dNTPs, forward and reverse primers (1 μL each), and 14 μL of RNase/DNase-free water. PCR Master Mix (20 μL) is distributed for each reaction, and 5 μL of random hexamer-primed cDNA (an RNA/cDNA equivalent of >10 ng per reaction is recommended) is added. Control PCR Master (15 μL) containing H_2O and 5 μL of RNase-/DNase-free H_2O is added and mixed. PCR is carried out at 95 °C for 2 min, 95 °C for 10 s, 60 °C for 20 s at 30–35 PCR cycles, and 72 °C for 15 s. The time and temperature are subject to the individual polymerase and the item estimate. PCR items are

broken down by gel electrophoresis utilizing 2% agarose gels.

c. *The accompanying convention is portrayed utilizing the SYBR Green Master Mix for a standard 96-well qPCR*: Heumuller and Boeckel described specificity of the PCR examine for cirRNA discovery utilizing semi-quantitative PCR and gel electrophoresis preceding qPCR. Besides, the qPCR item ought to dependably be prepared by liquefying bend investigation and in any event once by consequent gel electrophoresis. Dissolve bend investigation is not vital when utilizing hydrolysis test-based qPCR. In this situation, a hydrolysis test ace blend is utilized rather than the SYBR Green Master Mix (SYBR-GMM) in the accompanying convention. SYBR-GMM is defrosted, and irregular hexamer cDNA is prepared on ice. The quantity of responses is figured out, and 1–2 extra responses are incorporated to make up for inevitable misfortune by pipetting. This is followed by planning qPCR Master Mix. The qPCR Master Mix contains 10 μL of SYBR-GMM, 3 μL of water, and 1 μL of the forward and switch 10 μM groundwork stock. To perform hydrolysis test-based qPCR, it is recommended to utilize 10 μL of hydrolysis test ace blend, 2 μL of water, 1 μL of the forward and turnaround 10 μM preliminary stock each, and 1 μL of the hydrolysis test. PCR ace blend (15 μL) is circulated for every response, and 5 μL irregular hexamer-prepared cDNA is incorporated. This is followed by the incorporation of H_2O control comprising of 15 μL of the PCR ace blend and includes 5 μL H_2O. It is ensured to liquefy bend for every groundwork. The information is broken down utilizing the 2-CT strategy or the 2-ΔCT technique when a housekeeping quality (e.g., the mRNA of RPLP0) has been estimated.

d. *PCR items increasing the back-grafted area ought to be filtered utilizing phenol/chloroform/isoamyl liquor precipitation*: In this way, Sanger sequencing (PCR sequencing) is utilized to approve the presence of the back-join site and to control the specificity of the differently oriented preliminaries utilized as a part of the PCR. An equivalent volume of phenol/chloroform/isoamyl liquor (25:24:1 (v/v/v)) is added to the PCR item in a 1.5 mL response tube mixed and centrifuged for 5 min at 12,000 × g at room temperature (25 °C). The tube is handled deliberately and abstained from irritating stage partition-exchange, the upper (watery) stage to another 1.5 mL response tube. It is recommended not to aggravate the lower (natural) stage. Tainting with the lower stage can bring about diminished extraction productivity. The tube containing the lower stage (phenol squander) is disposed of. The product was blended with 2.5 mL of ice-cold ethanol and mixed for 15 min at 4 °C to hasten the DNA. The supernatant is evacuated using a pipette. The pellet is dried on a warm obstruct with open cover at 37 °C for 0.5–2 min. The pellets are resuspended in 10 μL TE cushion. DNA sum ought to be resolved, and the test is sent to PCR sequencing utilizing the disparate forward and invert preliminary.

2.2 Identifying cirRNAs by RNA Fluorescence In Situ Hybridization (FISH)

FISH permits the representation of various RNA species inside the cell. This section contains an all-around appropriate technique to recognize cirRNA through a junction specific test. CirRNAs are set apart by making a beeline for a tail-ligated intersection that is not found in some other RNAs. Till today, this convention is very strong and delicate. Numerous tests marked by an alternate fluor can be taken into consideration to achieve synchronous identification of different targets [18]. RNA FISH depends on the straightforward idea of uncovering settled cells or tissues to short DNA oligonucleotides in adequately high fixations to enable blending with corresponding RNA molecules to frame stable DNA-RNA half-breeds. The tests comprise of a pooled set of ~32–48 DNA oligos of various arrangements, every 20 nucleotides in length and named with a solitary fluorophore at its 3′ end. The convention can recognize single RNA molecule with high specificity (a couple of false positives) and high affectability (a couple of false negatives) and does not require flag intensification steps, which tend to render single-atom identification approaches less quantitatively [19].

2.3 Northern Blot Analysis of cirRNAs

Northern smear hybridization makes the strategy for the decision to convincingly show round setup of putative cirRNAs. CirRNA identification can be proficient by short tests spreading over the round graft intersection or by longer tests covering as much as a whole circularized exon. This alternative winds up significantly if the specificity for roundabout isoform is not fundamental (for instance, if the straight structures do not enter the gel, if both direct and roundabout isoforms ought to be identified in parallel, or if there should be an occurrence of solely roundabout RNAs). Northern blots are thus basic part of any cirRNA portrayal, because of their incredible flexibility. To start with the decision of test districts (round or straight joint intersection or exonic areas) and identification standard (digoxigenin or 32P-named tests) decides the specificity for roundabout versus direct isoforms. The decision of gel network includes greater adaptability in northern smudge examine. Agarose gels are reasonable for cirRNAs from 0.2 kb up to a few kb. In agarose gels, round and straight RNAs of a similar size cannot be recognized by their running conduct. Actually, in denaturing polyacrylamide gels, direct RNA keeps running at the normal size, while cirRNAs have a lower evident versatility with respect to straight markers; this hindrance impact is upgraded by expanding acrylamide fixations [20]. Due to this impediment, cirRNAs up to 1 kb can be examined by polyacrylamide gel electrophoresis. Thus, at any rate for a far-reaching investigation of one or a couple of putative cirRNAs, not for a medium- to high-throughput screening endeavors, Northern blot tests give an extremely profitable and exceedingly useful approach.

2.4 Portrayal of cirRNA Concatemers

The model on cirRNA biogenesis suggests that the rearranged rehashes take part in base blending, accordingly situating the two splice sites in nearness. Wang and his colleagues outlined the embodiment of exon 2 from beta-globin (HBB) in the middle of modified components. As far as anyone is concerned, this exon is not creating cirRNA in its normal setting. However, when flanked with transformed rehashes, the exon produces one particular cirRNA, as well as a step of cirRNA-like items (cirRNA concatemers) [21]. The identification and profiling of cirRNA are normally done by cutting-edge sequencing (NGS) or by qRT-PCR [22]. The cirRNA in the first place contained exon rehashes or whether the monotony was presented by the RT chemical. Barrett et al. presented a blueprint of basic biochemical tests projected by northern smearing to ponder the idea of these cirRNAs and demonstrated that they are made out of exon rehashes (cirRNA concatemers). To recognize concatemers and interwoven cirRNAs (topologically bolted single exon cirRNAs), three particular examinations such as (1) RNase R absorption to approve the roundabout structure of the cirRNA species, (2) RNase H absorption to decide the structure of exons by crumbling the cirRNAs into their exon units, and (3) a soluble treatment to tenderly scratch the cirRNA into a relating direct RNA have been suggested [23].

3 Computational Approaches on cirRNA

The hereditary data streams of life, in which DNA and protein are considered as primary on-screen characters of cell life, retain RNA as basic part of protein synthesis. However, this perspective of the organic part of RNA experienced various challenges [24]. Computational approaches to deal with RNA tertiary structure expectation are based on the examination of RNA tertiary themes, and diagram hypothesis for RNA and RNA endeavors plan to enhance the in vitro test choice for aptamer outline. Thus, the examination of RNA basic correlation is important thought of root-mean-square deviation (RMSD) since the forecasts are not exact for RNA during this phase [25]. There are numerous different zones of advancement in RNA bioinformatics, for instance, auxiliary structure forecasts [26].

Current discoveries in the field of noncoding RNA are focused on cirRNA [27] which are produced by nonlinear backsplicing linked to a downstream splice donor and upstream splice acceptor. CirRNA is present in all eukaryotic clades, including insects, fungi, and plants, and it also exists in humans to establish several thousand different cirRNAs. In the immense majority, the function of cirRNA is not clear. A small subset of cirRNAs has been reported to act as steerers for miRNAs [28, 29] or to bind and regulate protein function [30, 31]. The diversity of cirRNAs can be explicated based on the gene fraction and antisense strand of some genes and from intergenic regions [32–34]. The length of cirRNA ranges from 100 bp to 4 kb [35]. cirRNAs may hold multiple or single exon [36] and are present in different cell lines, tissues, and extracellular exosomes. Biogenesis of cirRNA is based on lariat-driven, intron-pairing-driven, and RNA-binding protein-driven circularization mechanisms [37, 38]. It has been proposed that cirRNA acts as microRNA sponges and regulates multiple gene expressions. The source quality, mode of exon creation, biogenesis, and capacity make them different than other RNAs. Comment-free recognition calculations can be utilized as a part of an extensive variety of living beings. However, it requires more careful systems to guarantee unwavering quality.

The majority of the location techniques are upgraded for their assigned aligners, and these can be additionally partitioned into joint mindful aligners and adaptable read mappers. Paired-end sequencing gives more data to diminish false positives for discovery techniques that receive sifting in light of paired-end mapping. In view of identification calculations, promising computational techniques have been developed for the downstream investigations of cirRNAs. However, new computational techniques to remake full length of cirRNAs and measure their demeanor are critically required. Late examinations have shown that cirRNAs are universal and have different capacities and systems of biogenesis. In such investigations, computational profiling of cirRNAs has been pervasively utilized as an irreplaceable strategy to give high-throughput ways to deal with identifying and breaking down of cirRNAs. In any case, without a general comprehension of the basic methodologies, these computational techniques may not be exactly chosen or utilized for a particular research reason, and a few misguided judgments may bring about predispositions in the examinations. Gao and Zhao reviewed the key advances and abridged trade-off of various systems, covering every single prominent calculation for cirRNA discovery and different downstream investigations [39].

The computational approach plays an important role in high-throughput RNA-seq data examination and in expression of cirRNA profiling. Till today about 11 computational approaches for cirRNA detection have been reported. CIRI, CIRCexplorer [40], and KNIFE [41] are more functional than other approaches. All the reported computational approaches have their own advantages and essential point sensitivity, precision, and computational cost (Tables 1.1 and 1.2). Downstream computational approaches are significant due to their primary detection results. These approaches are linked to the quantification and differential expression analysis (Table 1.3) [42, 43].

3.1 Detection of cirRNA Using Annotation, Genome Reference, and GT-AG Splicing Signals

Genomes are essential for algorithm sensing and can be used in the detection workflows. It mainly works for the direct alignment of sequencing reads against the standard genome. UROBORUS [44],

Table 1.1 Compilation of 11 cirRNA detection methods

Category	Method	Mapper(s)	Characteristics	Mapper type
Split-alignment-based	CIRI	BWA-MEM	Filtering stringent PEM Restore of unbalanced BSJ read multiple seed matching	Versatile
Split-alignment-based	CIRCexplorer	TopHat/STAR	Noncollinearity detection	Splice-aware
Split-alignment-based	DCC	STAR	GT-AG splice sites	Splice-aware
Split-alignment-based	cirRNA_finder	STAR	GT-AG splice sites	Splice-aware
Split-alignment-based	MapSplice	Bowtie	Embedded in algorithm to detect cirRNA	Versatile
Pseudoreference-based	KNIFE	Bowtie, Bowtie 2	De novo detection as remedy	Versatile
Split-alignment-based	Find circ	Bowtie 2	No PEM filtering Two 20bp anchors for noncollinearity detection	Versatile
Split-alignment-based	segemehl	Per se	Few of the filters adopted	Versatile
Pseudoreference-based	NCLscan	BWA, BLAT, Novoalign	Trans-spliced transcript detection in addition to cirRNA detection	Mixed
Split-alignment-based	UROBORUS	TopHat	Storage of unbalance BSJ reads	Mixed
Pseudoreference-based	PTESFinder	Bowtie, Bowtie 2	No PEM filtering	Versatile

Table 1.2 Performance comparison among 11 cirRNA detection methods by third-party evaluation

Method	Hs68 true positive	Hs68 precision (%)	HeLa true positive	HeLa precision (%)
KNIFE	2359	66.53	2055	44.26
MapSplice	1854	76.33	1766	54.11
DCC	2107	63.08	1760	45.22
UROBORUS	279	19.73	761	31.00
CIRI	3400	69.49	3210	54.20
PTESFinder	2474	63.29	2054	35.65
Segemehl	3094	8.74	2506	14.32
NCLscan	892	64.73	954	45.06
Find_circ	2377	59.75	2092	39.99

Table 1.3 Summary of computational methods for downstream analysis of cirRNAs

Method	Language	Input requirement	Function
FUCHS	Python	BAM/SAM formatted alignment	miRNA seed analysis
CIRI-AS	Perl	cirRNA, references of SAM formatted	Detection of internal structure
CircView	Java	CirRNA list	Visualization
Sailfish-cir	Python	GTF format annotation	Very close quantification
CirPro	Perl	FASTQ-formatted sequencing reads	Protein-coding potential estimation
CircTest	R	Parental gene with read-count	Differential expression test

Fig. 1.1 Pseudoreference-based approach for the cirRNA detection
The reference genome is combined with the corresponding genome annotation to build pseudo-sequence

CIRCexplorer [45], find circ [46], and CIRI [47] are the examples of detection algorithms. The circularity pathway of cirRNA is different from other categories of RNAs, and thus an evident feature can be captured from the circle junction alignment which is used as a backsplice junction (BSJ). By contrast, forward-spliced junction (FSJ) in mRNA, the developed sequencing reads that are collinearly aligned on the genome, interprets spanning BSJs are divided into segment and are aligned to the address/reference sequence in reverse order. Hence, detection algorithms in this category can be termed as split-alignment-based approaches.

For different algorithms, such as NCLscan [48] and KNIFE [49], the reference genome is combined with the corresponding genome annotation to build pseudo-sequence around putative BSJs in the first few steps (Fig. 1.1). Consequent steps are centered on the complete alignment of sequencing reads against such pseudo-sequences to identify BSJ reads. In addition to a BSJ pseudo-sequence database, KNIFE also constructed a FSJ sequence information according to the annotation to remove candidate reads with high-score alignment in both databases. In this category, detection algorithms may be termed pseudoreference-based approaches.

The application of annotation is much useful. As an example, a comprehensive evaluation of RNA-seq aligners which actively addressed that annotation can help to increase the sensitivity for junction reorganization versus de novo detection [50].

3.2 Various Read Mappers or Specified Splice-Aware Aligner

BSJ reads such as split-alignment and pseudoreference approaches are identified with the help of alignment of transcriptomic read. Many algorithms, for example, KNIFE and CIRI, prefer read mappers which are generally applicable in reference-based RNA/DNA sequence studies. The simplest way is splice-aware aligners which were developed for RNA-seq reads across intron-sized gaps on genome references such as Novoalign, STAR [51], and TopHat [52]. Detection algorithms, such as CIRCexplorer and DCC [53], depend on this type of aligner. An evident advantage is that the aligners are optimized according to eukaryotic transcription which is easier than various read mappers. The flexibility of splice-aware aligners is less than the versatile mappers, which have developed both end-to-end and local alignment. In other condition, nearly all splice-aware aligners are based on versatile read mappers.

4 Translation of cirRNA

The introduction of nearly all cirRNAs from exons and their confinement in cytoplasm increase the probability of cirRNA translation [54]. Translation of several viral proteins in many organisms, including humans, depends on internal ribosome entry site (IRES) and cellular IRESs. However, their mechanism of action remains controversial [55]. Based on this theory, artificial cirRNAs with IRES have been translated [56, 57]. It has also been noticed that principal cirRNAs can be translated in vitro and in vivo [58]. Translation of cirRNAs in living human cells is based on rolling circle amplification mechanism. The elements including IRES are not required for the translation of cirRNA in eukaryotic translation system [59]. The cirRNAs have been endogenously translated and indirectly tested [59].

Binding of open pre-initiation complex containing small ribosome is the beginning step for canonical translation process in eukaryotes [59]. The interaction between the cap-binding protein

(CBP) poly(A) regions leads to the circularization of mRNA competent for translation [60]. Small ribosomal subunits and mRNAs are further scanned by small ribosomal subunits for the start codon. After that the 60S ribosomal subunit is required. Also, the internal start codons can recirculate the ribosomes internally by an IRES-dependent mechanism.

4.1 Translation of cirMbl

Till date, the investigations are limited to the in vivo translation of endogenous cirRNAs. Ribo-cirRNAs have specific ribosome profiling and are denoted as cirRNAs. Ribosome profiling datasets of *Drosophila* have shown presence of cirRNA-specific junctions [61]. Expression of cirRNA in *Drosophila* S2 cells depends on the presence of intron-exon-intron minigenes. The minigenes have been reported to express V5-tagged proteins. V5-tagged proteins have been reported from the cells transfected with circCdiV5, circPde8V5, or circMblV5. This was not observed with the cells which are not transfected with minigenes of cir-cHaspinV5 or circCamKIV5. The protein of desired size has been observed using antiMBL antibody. For cirMbl, cirRNA has been established as the main source of detected protein. Transgenic flies containing MBL-immunoreactive bands have been originated to express in vivo cirMbl minigene [62]. In ribosome footprinting (RFP) reads of fly heads about 122 ribo-cirRNAs have been identified. Protein domains have been identified in many ribo-cirRNA-encoded proteins. Protein expression of V5-tagged cirRNA minigenes was not affected by co-expression of 4E-BP that inhibits cap-dependent translation. circMbls are translated in a cap-independent manner, and the untranslated region (UTR) sequence of circMbl is capable of facilitating cirRNA translation [61].

4.2 Translation of circ-ZNF609

Circ-ZNF609 codon starts with a linear transcript and terminates at a stop codon. It contains 753-nt ORF [62]. A substantial percentage of circ-ZNF609 was indicated to sediment with heavy polysome. Treatment with puromycin disrupted active translation of ribosomes which shifted circ-ZNF609 to lighter polysomes. p-circ3XF containing 3XFLAG-coding sequence of stop codon has been tested to express circ-ZNF609 protein-coding ability. The study resulted in two flagged isoforms. On the other side, p-lin3XF containing circ-ZNF609 ORF was also produced and expressed same proteins more efficiently. The RNA amount was normalized in both cir-c3XF and p-lin3XF. The results suggested that the translation efficiency of p-circ3XF was lower than that of p-lin3XF. CRISPR/Cas9 has been utilized to introduce 3XFLAG-code in endogenous ZNF609 gene which translated circZNF609 from the chromosomal gene. Positive clone with alleles (1) holding expected flag and (2) holding a deletion that prevents circ-ZNF609 production has been reported. However, only clone carrying flagged allele can develop circ-ZNF609.

In an investigation the positive clone cell lysates were immunoprecipitated with anti-FLAG antibody and subject to mass spectrometry. One peptide mapping to the cir-ORF was observed in the positive clone cell, whereas several peptides were present in the protein from cells overexpressing p-circ3XF and p-lin3XF. This resulted in the formation of lower cir-RNAs from the chromosomal gene and lower translational capacity. The heat shock resulted to increased translation of circZNF609. In this process no cap structure was present in cirRNA. The study proposed the possibility of translation through sequences with internal ribosome entry site (IRES) activity [62].

5 m6 A Modification in cirRNA Translation

Yang and coworkers discovered sequences to induce translation of cirRNA [56]. RRACH fragment involves in the methylation of N6 position of adenosine (m6 A). m6 Higher peak density of m6 A has been observed in cirRNA when compared to mRNAs [63, 64] suggesting its role in mRNA translation [65, 66]. These findings support the hypothesis that m6 A plays an important role in cirRNA translation. To validate this hypothesis, a short fragment containing m6 A motif was introduced before the start codon in cirRNA reporter. The level of GFP protein output was assessed in all the transfected cells [57], and it was observed that cirRNA containing m6 A motif was translated expeditiously. The methylation of cirRNAs with m6 A motif is associated with RNA-immunoprecipitation (RNA-IP). Negative effect of m6 ademethylase FTO co-expression on the amount of immunoprecipitated RSV-containing cirRNA and GFP translation from cirRNA has been observed. eIF4G2, a non-canonical protein which recognizes IRES, also plays an important role in cirRNA translation and partakes in cap-independent translation and m6 A reader protein YTHDF3 [56, 67]. N6-methylation requires the translation of mRNA and leads to GFP protein translation from m6 A-containing cirRNA by heat shock stress [56, 65, 67].

6 Conclusions

It is evident that the number of cirRNAs with known functions is expanding during the last few years. However, the function of various cirRNAs remains unknown, and very little information is available about the control of backsplicing process of cirRNA generation. This could be due to the availability of limited and challengeable methods to detect and characterize cirRNAs. A detailed study of cirRNA biogenesis and in vivo research may allow testing for functional consequences of cirRNA expression and will provide novel insights into cellular development human disease.

Competing Financial Interests The authors declare no competing financial interests.

References

1. Liu L, Wang J, Khanabdali R et al (2017) Circular RNAs: isolation, characterization and their potential role in diseases. RNA Biol 14(12):1715–1721
2. Li J, Yang J, Zhou P, et al (2015a) Circular RNAs in cancer: novel insights into origins, properties, functions and implications. Am J Cancer Res 5(2):472
3. Li P, Chen S, Chen H et al (2015b) Using circular RNA as a novel type of biomarker in the screening of gastric cancer. Clin Chim Acta 444:132–136
4. Greene J, Baird AM, Brady L et al (2017) Circular RNAs: biogenesis, function and role in human diseases. Front Mol Biosci 4:38
5. Hansen TB, Kjems J, Damgaard CK (2013) Circular RNA and miR-7 in cancer. Cancer Res 73(18):5609–5612
6. Bachmayr-Heyda A, Reiner AT, Auer K et al (2015) Correlation of circular RNA abundance with proliferation–exemplified with colorectal and ovarian cancer, idiopathic lung fibrosis, and normal human tissues. Sci Rep 5:8057
7. Barrett SP, Salzman J (2016) Circular RNAs: analysis, expression and potential functions. Development 143(11):1838–1847
8. Lasda E, Parker R (2014) Circular RNAs: diversity of form and function. RNA 20(12):1829–1842
9. Chen I, Chen CY, Chuang TJ (2015) Biogenesis, identification, and function of exonic circular RNAs. Wiley Interdiscip Rev RNA 6(5):563–579
10. Bahn JH, Zhang Q, Li F et al (2015) The landscape of microRNA, Piwi-interacting RNA, and circular RNA in human saliva. Clin Chem 61(1):221–230
11. Gao Y, Wang J, Zhao F (2015) CIRI: an efficient and unbiased algorithm for de novo circular RNA identification. Genome Biol 16(1):4
12. Jeck WR, Sharpless NE (2014) Detecting and characterizing circular RNAs. Nat Biotechnol 32(5):453
13. Jeck WR, Sorrentino JA, Wang K et al (2013) Circular RNAs are abundant, conserved, and associated with ALU repeats. RNA 19(2):141–157
14. Xiong HY, Alipanahi B, Lee LJ et al (2015) The human splicing code reveals new insights into the genetic determinants of disease. Science 347(6218):1254806
15. Hansen TB (2018) Characterization of circular RNA concatemers. In: Circular RNAs. Humana Press, New York, pp 143–157
16. Hansen TB, Wiklund ED, Bramsen JB (2011) miRNA-dependent gene silencing involving Ago2-mediated cleavage of a circular antisense RNA. EMBO J 30(21):4414–4422
17. Gingeras TR (2009) Implications of chimaeric non-co-linear transcripts. Nature 461(7261):206
18. Itzkovitz S, Van Oudenaarden A (2011) Validating transcripts with probes and imaging technology. Nature Methods 8(4s):S12
19. Günzl A, Palfi Z, Bindereif A (2002) Analysis of RNA–protein complexes by oligonucleotide-targeted RNase H digestion. Methods 26(2):162–169
20. Tabak HF, Van der Horst G, Smit J (1988) Discrimination between RNA circles, interlocked RNA circles and lariats using two-dimensional polyacrylamide gel electrophoresis. Nucleic Acids Res 16(14):6597–6605

21. Wang PL, Bao Y, Yee MC et al (2014) Circular RNA is expressed across the eukaryotic tree of life. PloS one 9(3):e90859
22. You X, Vlatkovic I, Babic A (2015) Neural circular RNAs are derived from synaptic genes and regulated by development and plasticity. Nat Neurosci 18(4):603
23. Barrett SP, Wang PL, Salzman J (2015) Circular RNA biogenesis can proceed through an exon-containing lariat precursor. Elife 4
24. Jakobi T, Dieterich C (2018) Deep computational circular RNA analytics from RNA-seq data. Methods Mol Biol 1724:9–25
25. Reznichenko A (2012) Translational renal genetics. University Library Groningen Host. Humana Press, New York
26. Shapiro BA, Yingling YG, Kasprzak W et al (2007) Bridging the gap in RNA structure prediction. Curr Opin Struct Biol 17(2):157–165
27. Ebbesen KK, Kjems J, Hansen TB (2016) Circular RNAs: identification, biogenesis and function. Biochim Biophys Acta 1859:163–168
28. Hansen TB, Jensen TI, Clausen BH et al (2013) Natural RNA circles function as efficient microRNA sponges. Nature 495:384–388
29. Memczak S, Jens M, Elefsinioti A et al (2013) Circular RNAs are a large class of animal RNAs with regulatory potency. Nature 495:333–338
30. Zheng Q, Bao C, Guo W et al (2016) Circular RNA profiling reveals an abundant circHIPK3 that regulates cell growth by sponging multiple miRNAs. Nat Commun 7:11215
31. Ashwal-Fluss R, Meyer M, Pamudurti NR et al (2014) circRNA biogenesis competes with pre-mRNA splicing. Mol Cell 56:55–66
32. Du WW, Yang W, Liu E et al (2016) Foxo3 circular RNA retards cell cycle progression via forming ternary complexes with p21 and CDK2. Nucleic Acids Res 44:2846–2858
33. Gao Y, Wang J, Zhao F (2015) CIRI: an efficient and unbiased algorithm for de novo circular RNA identification. Genome Biol 16(1):4
34. Memczak S, Jens M, Elefsinioti A et al (2013) Circular RNAs are a large class of animal RNAs with regulatory potency. Nature 495(7441):333
35. Salzman J, Chen RE, Olsen MN et al (2013) Cell-type specific features of circular RNA expression. PLoS Genet 9(9):e1003777
36. Lasda E, Parker R (2014) CircularRNAs: diversity of form and function. RNA 20:1829–1842
37. Li Z, Huang C, Bao C et al (2015) Exon-intron circular RNAs regulate transcription in the nucleus. Nat Struct Mol Biol 22(3):256
38. Conn SJ, Pillman KA, Toubia J (2015) The RNA binding protein quaking regulates formation of circRNAs. Cell 1(6):1125–1134
39. Hansen TB, Jensen TI, Clausen BH et al (2013) Natural RNA circles function as efficient microRNA sponges. Nature 495(7441):384
40. Gao Y, Zhao F (2018) Computational strategies for exploring circular RNAs. Trends Genet 34(5):389–400
41. Zhang XO, Wang HB, Zhang Y et al (2014) Complementary sequence-mediated exon circularization. Cell 159(1):134–147
42. Szabo L, Morey R, Palpant NJ et al (2015) Statistically based splicing detection reveals neural enrichment and tissue-specific induction of circular RNA during human fetal development. Genome Biol 16(1):126
43. Glažar P, Papavasileiou P, Rajewsky N (2014) circBase: a database for circular RNAs. RNA 20(11):1666–1670
44. Meng X, Chen Q, Zhang P et al (2017) CircPro: anintegrated tool for the identification of circRNAs with protein-coding potential. Bioinformatics 33:3314–3316
45. Memczak S, Jens M, Elefsinioti A (2013) Circular RNAs are a large class of animal RNAs with regulatory potency. Nature 495(7441):333
46. Zhang XO, Wang HB, Zhang Y et al (2014) Complementary sequence-mediated exon circularization. Cell 159(1):134–147
47. Gao Y, Zhang J, Zhao F (2017) Circular RNA identification based on multiple seed matching. Brief Bioinform. https://doi.org/10.1093/bib/bbx014
48. Song X, Zhang N, Han P et al (2016) Circular RNA profile in gliomas revealed by identification tool UROBORUS. Nucleic Acids Res 44(9):e87–e87
49. Chuang TJ, Wu CS, Chen CY et al (2015) NCLscan: accurate identification of non-co-linear transcripts (fusion, trans-splicing and circular RNA) with a good balance between sensitivity and precision. Nucleic Acids Res 44(3):e29–e29
50. Szabo L, Morey R, Palpant NJ et al (2015) Statistically based splicing detection reveals neural enrichment and tissue-specific induction of circular RNA during human fetal development. Genome Biol 16(1):126
51. Baruzzo G, Hayer KE, Kim EJ et al (2017) Simulation-based comprehensive benchmarking of RNA-seq aligners. Nat Methods 14(2):135
52. Dobin A, Davis CA, Schlesinger F et al (2013) STAR: ultrafast universal RNA-seq aligner. Bioinformatics 29(1):15–21
53. Trapnell C, Pachter L, Salzberg SL (2009) TopHat: discovering splice junctions with RNA-Seq. Bioinformatics 25(9):1105–1111
54. Cheng J, Metge F, Dieterich C (2015) Specific identification and quantification of circular RNAs from sequencing data. Bioinformatics 32(7):1094–1096
55. Jackson RJ (2013) The current status of vertebrate cellular mRNA IRESs. Cold Spring Harb Perspect Biol 5:a011569
56. Yang Y, Fan X, Mao M (2017) Extensive translation of circular RNAs driven by N-6-methyladenosine. Cell Res 27:626–641

57. Wang Y, Wang Z (2015) Efficient backsplicing produces translatable circular mRNAs, RNA-Publ. RNA Soc 21:172–179
58. Chen CY, Sarnow P (1995) Initiation of protein synthesis by the eukaryotic translational apparatus on circular RNAs. Science 268:415–417
59. Abe N, Matsumoto K, Nishihara M (2015) Rolling circle translation of circular RNA in living human cells. Sci Rep 5:16435
60. Guo JU, Agarwal V, Guo H et al (2014) Expanded identification and characterization of mammalian circular RNAs. Genome Biol 15:409
61. Aitken CE, Lorsch JR (2012) A mechanistic overview of translation initiation in eukaryotes. Nat Struct Mol Biol 19:568–576
62. Pamudurti NR, Bartok O, Jens M et al (2017) Translation of CircRNAs. Mol Cell 66:9–21
63. Legnini I, Di Timoteo G, Rossi F et al (2017) CircZNF609 is a circular RNA that can be translated and functions in myogenesis. Mol Cell 66:22–37
64. Dominissini D, Moshitch-Moshkovitz S, Schwartz S (2012) Topology of the human and mouse m(6) A RNA methylomes revealed by m(6)A-seq. Nature 485:201–284
65. Meyer KD, SaletoreY ZP et al (2012) Comprehensive analysis of mRNA methylation reveals enrichment in 3' UTRs and near stop codons. Cell 149:1635–1646
66. Meyer KD, PatilDP ZJ et al (2015) 5′ UTR m(6) A promotes cap-independent translation. Cell 163:999–1010
67. Fu Y, Dominissini D, Rechavi G (2014) Gene expression regulation mediated through reversible m(6)A RNA methylation. Nat Rev Genet 15:293–306

Part II

Bioinformatics for Circular RNAs

RNA sequencing and Prediction Tools for Circular RNAs Analysis

2

Elena López-Jiménez, Ana M. Rojas, and Eduardo Andrés-León

Abstract

Circular RNAs (circRNAs) are noncoding and single-stranded RNA transcripts able to form covalently circular-closed structures. They are generated through alternative splicing events and widely expressed from human to viruses. CircRNAs have been appointed as potential regulators of microRNAs (miRNAs), RNA-binding proteins (RPBs), and lineal protein-coding transcripts. Although their mechanism of action remains unclear, the deregulation of circular RNAs has been confirmed in different diseases such as Alzheimer or cancer.

The introduction of high-throughput next-generation sequencing (NGS) technology provides millions of short RNA sequences at single-nucleotide level, allowing an accurate and proficient method to measure circular RNAs. Novel protocols based on non-polyadenylated RNAs, rRNA-depleted, and RNA exonuclease-based enrichment approaches (RNase R) have taken even further the possibility of detecting circRNAs.

Besides, the identification of circRNAs presence requires the development of specific bioinformatics tools to detect junction-spanning sequences from transcriptome deep-sequencing samples. Thus, recently established bioinformatics' approaches have permitted the discovery of an elevated number of different circRNAs in diverse organisms. In that sense, recent studies have compared different methods and advocate the simultaneous use of more than one prediction tool. For that reason, we want to highlight pipelines such as miARma-Seq that is able to execute various circular RNA identification algorithms in an easy way, without the tedious installation of third-party prerequisites.

Keywords

CircRNAs · CircRNA RNA-seq · CircRNA prediction tools

E. López-Jiménez
Imperial College London, London, UK

A. M. Rojas
Computational Biology and Bioinformatics Group, Institute of Biomedicine of Seville, Seville, Spain

E. Andrés-León (✉)
Bioinformatics Unit, Instituto de Parasitología y Biomedicina "López-Neyra", Consejo Superior de Investigaciones Científicas (IPBLN-CSIC), Granada, Spain
e-mail: eandres@ipb.csic.es

J. Xiao (ed.), *Circular RNAs*, Advances in Experimental Medicine and Biology 1087,
https://doi.org/10.1007/978-981-13-1426-1_2

1 Introduction

Circular RNAs (circRNAs) are a class of single-stranded RNA transcripts able to form covalently circular-closed structures. This type of RNA molecule has been considered as noncoding RNA because no protein product is expressed, although recent studies point out in an opposite direction. The existence of circular transcripts was asserted almost three decades ago [1], but they were considered as RNA splicing artifacts [2]. Nevertheless, current next-generation sequencing (NGS) techniques of non-polyadenylated RNAs have revealed large numbers of widespread circRNAs highly and stably expressed in cells and tissues [3]. Successive studies had revealed that the expression level of circRNAs is rigorously controlled and specific among the different tissues and even between the different cell types. For instance, in humans they are remarkably expressed in the brain, exosomes, and peripheral blood [4]. Moreover, circRNAs are present in numerous organisms throughout evolution [5] such as viruses, bacteria, and plants [6].

Functional studies unveil that circular RNAs can control the translation of lineal RNA protein transcripts and microRNAs (miRNAs) and consequently encompass a key role in gene expression regulation [7]. During the last decade, several studies related with this class of noncoding RNAs have emerged, pointing out possible relationships with relevant diseases such as cancer, neurodegenerative diseases, and cardiovascular disorders [7–9].

1.1 Basic Characteristics of Circular RNAs

CircRNAs are comprised mainly by sequences coming exonic or intronic regions having 5′ and 3′ ends covalently closed as a result of a backsplicing event [10]. This event occurs between a splice donor site followed by a preceding splice acceptor site, whereas in a conventional linear splicing, this happens preferably to a rearward acceptor.

Due to their circular shape, circRNAs are molecules missing the 3′ polyadenylated tail and hence, resistant to RNA-degrading enzymes, which increases their cellular half-life to approximately 48 h, while linear RNAs have a roughly half-life of 10 h [7]. However, circRNAs are not stable in circulating serum exhibiting a short half-life lower than 15 s, apparently due to RNA endonucleases [11]. The amount of circular RNA molecules is frequently small, constituting between 5% and 10% of their linear counterparts.

1.2 Biogenesis

CircRNAs, like the majority of RNAs, are transcribed by the RNA polymerase II (Pol II) enzyme [12], as a pre-messenger RNA (pre-mRNA). These pre-mRNAs are the principal product of transcription, which frequently undertake a splicing process to harvest linear mRNAs. In the case of circRNAs, they suffer from a backsplice event, promoting the circularization process. Consequently, there is a frequently reduced expression of linear mRNAs when they are circularized, which produces an inverse correlation between the number of regular and circular RNA molecules [12]. Two different processes have been suggested for mammalian exonic circRNA circularization by the spliceosome machinery [13]. The first proposed mechanism ensues if a downstream splice donor pair having a non-spliced upstream splice acceptor and the contributing RNA are covalently closed. The second one, named "exon skipping" mechanism, involves a splicing event within loop structures (lariats) formed from the process that consists of avoiding exons [14].

Introns close to backsplice sites tend to be larger than regular introns; nevertheless, contiguous introns can be lesser than average [15]. Furthermore, the size of an exon appears to be related also with the circularization process as it was previously described that the average size of exons which constitute a circRNA for their own has a mean size three times longer in comparison with all expressed exons [3]. It has been also described the existence of paired Alu repeats (repetitive sequences typical in the human DNA and specific for primates) localized close to back-splice regions, where there is an elevated occurrence of human exonic circRNA generation [16].

In such a way, longer exons than average, surrounded by small introns having reversed Alu repeats, appear to be main features present in the RNA circularization mechanism.

1.3 Putative Roles of CircRNAs

The function of circular RNAs still remains unclear. Various analyses have identified numerous exonic circRNAs having conserved circularization sites in orthologous exons [15, 16]. This evolutionary conservation suggests the execution of important roles in an organism; hence, numerous possible functions have been suggested. Among these plausible functions, we can highlight ongoing functional studies indicating that circular RNAs are able to regulate the expression of linear mRNA transcripts.

1.3.1 MiRNA Sponges

The vast majority of known circRNAs are enclosed in the cellular cytoplasm [3, 16] as they are transported outside the nucleus. It has been demonstrated that circular RNAs contain abundant miRNA binding sites to interact with. This fact has allowed them to be called "miRNAs sponges" as they bind miRNAs, preventing them from executing their regulatory roles [17]. A well-studied example comes from the Sry gene, discovered in 1993, and belonging to the sex-responsible region Y. In specific conditions when miR-138 is overexpressed, it coprecipitates with argonaute 2 (AGO2) and with the Sry circular transcript due to the presence of 16 binding sites for miR-138 within the circRNA sequence. Besides, in mouse cells, the expression level of miR-138 is negatively correlated with Sry; therefore, miR-138 expression is reduced while Sry expression is increased [17]. *ANRIL*, an antisense RNA from the tumor suppressor INK4 locus, contributes in transcription inhibition [18]. The expression of circular *ANRIL* RNA is directly linked with *INK4/ARF* expression levels. This fact has been described as a risk factor for atherosclerosis disease [19]. Finally, the circular RNA ciRS-7 which functions as a sponge, is expressed in neuronal tissues and disposes more than 70 binding sites for miR-7 [10]. This microRNA is drastically repressed in patients exhibiting sporadic Alzheimer disease [9].

1.3.2 Gene Expression Regulation

Nowadays, most of the identified circRNAs is demonstrated to be derived from exons, even though there are some intron-containing circular RNAs (named ciRNAs). They are characterized by their constrained expression in the nucleus [20], as most linear RNAs having retained introns are usually confined in the nucleus of the cells [21]. Several evidences suggest that these ciRNAs allow the transcription regulation of genes in *cis*. Particularly, they can promote the RNA polymerase II (Pol II) transcription activity of their host genes. Nevertheless, the causal mechanism remains unclear [15, 22].

Moreover, the generation of circRNAs via circularization of exons has been suggested to be a process that competes with the splicing machinery as they perform their function on the same splice-sequence sites. This fact was observed in neural tissue, in which an inverse expression level was described, being the circRNAs more abundantly expressed than their linear counterparts [22]. Similarly, in brain tissue during the aging process, there is a raised expression of circRNAs opposite to the low levels of linear RNAs [8]. This high level of circRNAs in certain tissues sustains the idea that RNA circularization can control gene expression by displacing the canonical splicing of linear RNAs [23].

1.3.3 Interaction with RNA-Binding Proteins

In a similar manner to some other no protein-coding linear RNA transcripts, circRNAs are able to interact with RNA-binding proteins, for instance, AGO [10]. It has been also suggested that they could serve as "scaffolding" for RNA-binding proteins interacting with numerous proteins that increase the stability of the circRNA transcript [13]. Another example of interaction has been shown in Foxo3, a tumor suppressor gene [24]. Circ-Foxo3 has been implicated in cell cycle due to its interactions with some proteins involved on that pathway, regulating and preventing an abnormal proliferation. In detail, the cell division protein kinase 2 (CDK2) and cyclin-dependent kinase inhibitor 1 (p21) interact with circ-Foxo3 to establish a RNA-protein complex that reduces cell cycle progression. Consequently, it produces a cell cycle arrest as a consequence of CDK2 and p21 proteins depletion, and the cell is retained in the G1/S phase.

1.4 Losing the Identity: Could CircRNAs be still considered as Noncoding RNAs?

The possibility of the translation of the circRNAs emerged more than two decades ago as a consequence of the existence of an internal ribosome entry site (IRES) that could allow it [25]. Theoretically, if a circRNA owns an IRES and an ATG sequence, it would be able to be translated. Chen *et al.* corroborated that idea using the hepatitis δ agent by a noncanonical mechanism, and it was thought to have probably been specific for some viral agents. After that, Jerk et al. and other authors took into consideration the protein-coding features of numerous ATG-containing exonic circRNAs, but they could not identify any naturally protein produced from a circRNA [15, 16].

More recent studies experimentally demonstrated the translation into protein of some circRNAs [26, 27]. Pamudurti et al. described a group of circRNAs that were translated *in vivo* in *Drosophila melanogaster*. They showed that these circRNAs commonly presented the main features related with the translation process, for instance, they encode proteins having specific domains although are translated in a non-5′ cap mode. Besides, they shared the start codon with the hosting RNA. In this work, none of the processed sequences or the results that they obtained could separately support the existence of circRNA translation. But in a combinatorial way and with the additional experimental validation that they perform, they were able to provide a strong evidence of the presence of translation associated with circular RNAs. Their conclusions pointed out to the presence of a particular sequence to allow the translation process in a regular endogenous framework. Even more, they reported strong evidences of the fact that the translation of this subset of circRNAs is not by chance and presented these results suggesting a specific and regulated effect. Simultaneously, Legnini et al. were able to describe an example of a eukaryotic protein-coding circRNA called circ-ZNF609. In this work, they also conclude that the translation of this RNA is splicing-related and 5′ cap-independent. After that, Yang et al. finally revealed that a single m^6A motif is sufficient to lead translation initiation in human cells. Furthermore, he also discovered that circRNAs have an elevated number of m^6A sites [26, 28]. Besides, these studies reported strong evidences showing that the translation of these circRNAs is not a random effect and presented their results suggesting that it is a specific fact.

This discovery could allow us to have a deeper knowledge of the possible regulatory functions over gene expression levels that circRNAs could perform. However, more information is needed about the mechanisms of translation of circRNAs to establish this fact as other layer of control for genomic regulation. Therefore, with the rapid advance of current molecular and sequencing techniques and the development of new bioinformatic methodologies, novel functions will be discovered, and some of the actual unresolved questions will be determined in the near future.

2 Experimental Methodologies for CircRNA Discovery and Characterization

2.1 Sample Treatment

The presence of circular RNAs is not easy to detect and distinguish from other small RNAs and miRNAs due to their size or mobility properties. Nowadays, the most frequently used methodology requires destroying the circularity of these RNA species that could allow to their identification, because of the amplification and/or fragmentation steps performed.

Some techniques, such as "rapid amplification of cDNA ends" (RACE) or poly(A) enrichment of the samples for NGS transcriptome studies, cannot be effective in this case due to circRNAs not having neither a defined end nor a free 3′ or 5′ that could be modified for allowing to the detection. Furthermore, one of the main features of circRNAs, their generation by a "backsplice" process, is not exclusive of these species of small RNAs, and initial RNA-seq aligners tools eliminated those sequences.

Recently, with the development of new methodology that improves the selection of circRNAs during the processing of the samples like exonuclease-enrichment approaches, as well as sequencing of ribosomal RNA (rRNA)-depleted libraries instead of poly(A)-enriched libraries with longer reads and higher coverage and the generation of novel bioinformatic tools, this problem has been sorted out.

In the study described by Jeck et al. [13], a new biochemical protocol called Circle-Seq was introduced. This methodology involves the treatment of RNA samples with an exonuclease enzyme (RNAse R). In this way, linear RNAs are processed leaving the circRNAs intact. Nevertheless, it has recently been claimed that resistance to this enzyme alone is not sufficient to determine the circularity of a RNA transcript, due to the fact that some circRNAs were sensitive to this enzyme. This strategy then could interfere in the global selection of the circRNAs within a sample, generating a bias, as well as not be able to completely eliminate other RNA species still resistant to this process. Other studies suggest the employment of additional biochemical procedures for the isolation of circRNAs, like the use of a 2D (two-dimensional) denaturing polyacrylamide gel electrophoresis or ribosomal RNA (rRNA) depletion and poly(A)-depletion for increasing the amount of circRNAs in sequencing samples [15].

2.2 Microarrays from CircRNA Identification

The only commercially available circRNA microarray for human has been developed by Arraystar company to facilitate the analysis of circular RNA data. It is also available for mouse and rat. This platform contains a total of 13,617 different human probes, matching the circRNA-specific junctions, and distributed in an 8*15K format platform. These probes were selected from a total of six different recent studies describing the datasets [10, 15, 16, 29–31].

They offer a highly sensitive and specific platform for circRNA discovery, providing a service with circular junction sequences, linear RNA digestion by RNase R enzyme (pre-treatment of the RNA samples), and an efficient circRNA labeling system.

In comparison with the RNA sequencing for detecting circRNAs, they point out some of the specific features of the circular RNA, arguing that (1) junction-spanning sequences are only a small portion of the circular RNA compared with linear RNA at the similar expression level, so it could be not detected by the conventional RNA-seq methods [31]; (2) in order to perform a differential expression analysis in this type of data, not enough numbers of sequences from circular RNA are achieved; and (3) for the detection of the presence of circRNAs, only few reads are needed, whereas for a reliable quantification, a greater number of read counts are required.

Moreover, the protocol for preparing the RNA samples adds a group of exogenous RNA controls developed by the External RNA Controls Consortium (ERCC) as spike-in controls. Including this in the protocol, RNA amplification, labeling, or hybridization procedural effects can be corrected for obtaining an accurate and reliable result across the samples.

One of the disadvantages of using this platform is the high input of total RNA needed for preprocessing the samples. Depending on the field of the study, it could be a limiting condition due to the extremely low amount of material available for each sample (e.g., human patient samples, early embryonic stages, etc.). However, Arraystar also offers the use of an amplification step in the preparation of samples with a low-input material, which unfortunately could increase the noise on the results and the cost of the process.

In summary, the use of this platform is highly recommended in those cases in which a candidate-based approach could be applied (accepting the bias that this kind of platform could introduce in the data) for a faster and reliable acquisition of the data. The use of the RNA sequencing is recommended for the novel discovery of circular molecules since only a small number of read counts are required, but it is inappropriate for an accurate analysis of differential expression or specific quantification of circRNAs, taking into account the current methodological procedures.

2.3 RNA Sequencing (RNA-Seq)

2.3.1 Genomic Detection and Isolation Methods

In the last two decades, important changes have occurred in the scenario of the circular RNA genome-wide studies. The methodology for discriminating the different RNA species has been improved since the discovery of the intrinsic characteristics of circRNAs (circularity, absent of 3′ or 5′ ends, non-polyadenylated 3′, cytoplasmic location, etc.) in several previous studies [3, 13, 16, 32, 33].

Based on these features, recent high-throughput studies have been performed using a deeper sequencing with longer reads strategy that allows the detection of circular RNAs. Moreover, there was an improvement in the algorithms used for mapping these reads appropriately, and ribosomal RNA depletion was used in order to enhance the sequencing of non-polyadenylated RNA species.

Focusing in the library preparation for sequencing, researchers have developed and compared different strategies for enriching the libraries in circRNAs, trying to avoid the noise of the presence of other similar RNA species or the bias that the elimination of some of them could provoke. These library preparation strategies have been combined with two main different approaches to identify the precise candidate junction for every circRNA: (1) a large candidate junction-based approach, based on existing transcript models, and (2) the identification of these junctions searching for those sequences able to target directly to the genome (as it was previously applied for spliced alignment algorithms). These methods have been able to identify circRNAs that were experimentally confirmed by sequencing, RNase R protocol, and other different techniques.

Analyzing in depth the combination of these methodologies, we can observe some different aspects: the first option, using a large candidate junction-based approach has obtained reliable results and is the fastest one for applying in ribosomal RNA-depleted libraries. This selection generates a bias in the results that is adding the inconvenient of being unable to detect novel circRNAs present in the samples. Also, this strategy doesn't provide evidences of circularity of the species detected. Within this option, there are two different ways to perform this approach: (a) applying an RNase R enrichment that eliminates the linear RNA species and (b) without RNase R pre-treatment, in which a 75 nt paired-end sequencing has to be performed and the pairs of reads, containing one of the read the splice junction, will be divided in a group in which the paired read without the subsequent splice was aligned to a coding exonic region amid back-spliced exons (which can arise and be explained by circRNAs) and a group in which the pair mate is located in an exon, separated of the backspliced exons (which are considered as artifacts of sequencing). After the sequencing, statistics has to be performed with these two groups of reads in order to calculate an accurate score for every single junction detected. It has the advantage of providing a false discovery rate (FDR) cutoff instead of a random read depth-based threshold.

A combined approach employing a qPCR assay and the addition of RNase exonuclease was used to validate those newly discovered circRNAs. This first methodology showed the resistance to RNAse R enzyme of the transcripts, and other properties of backsplice-containing linear RNA were missed.

The second option consists in using rRNA-depleted and RNase R-treated libraries for high-throughput sequencing, and after that, mapping reads directly to de novo genomic positions and discovers backspliced reads in specific sequences. This method avoids the bias of a candidate-based approach, allowing the identification of novel circRNAs. This method consists selecting the reads that could not be directly aligned to the genome and take the two terminations of a single read and map them separately, based on the backsplice properties of the sequences (has to be flanked by GT/AG splice site in the genome context). This method is less accurate than a candidate-based approach but allows the detection of unannotated splice junctions.

This methodological system, which combines the biochemical properties of the circRNAs for improving the library preparation with the enrichment on circular RNA species, and posterior sequencing, was named as Circle-Seq for Jeck et al. [13], and it was described in archaea studies [33] and in mammals [16]. This library preparation is followed by the application of a mapping algorithm called MapSplice [34] that is able to identify apparent backsplice sequences.

In this strategy, they performed an RNase R treatment of the samples before the library preparation. For mammalian samples, the rRNA depletion step is required, but not in archaea. This technique is based on the use of two exonic circRNA features for the identification: (1) the inclusion of the backsplice junction reads, by means of a segmented mapping approach; and (2) the samples have been pre-treated with a step of RNase R, eliminating the linear RNA species and enriching in circular ones (in comparison with the mock-treated control). An example of the data obtained using this methodology was exhibited in the study that described cANRIL [16], the circular RNA from the ANRIL gene.

The use of the RNase R treatment doesn't avoid the presence of lariat RNAs (circRNAs mostly intronic that were formed during the canonical RNA splicing and possess a 2′–5′ carbon linkage at the splicing fork site). Although these lariat RNAs are easily distinguishable from circRNAs, their branch region sequence resembles backsplice read in these parts of the sequence, and besides they are also disordered in comparison with their genomic annotation.

Although the Circle-Seq protocol could generate a high depth of circular and lariat products, it has some limitations, for instance, as we commented before regarding the microarrays platform, for performing Circle-Seq, a higher amount of total RNA than in a regular sequencing protocol is needed, without any enrichment, which is more prone to suffer from endonuclease contamination. One important point to keep in mind is that this process can generate a bias on the results due to the possible elimination of longer circRNA products, as a single nicking event would confer exonuclease sensitivity.

However, for detecting backsplicing alternative events in circRNAs, non-poly(A) (without RNase R) or rRNA-depleted (without RNase R) samples are recommended. Both of these methods have discovered different circRNAs that were later validated by other alternative detection methods, as an example sequencing or RNAse R testing. This is a key step for the detection of putative circRNAs by bioinformatic algorithms. Depleting highly expressed RNA species with different splicing patterns such as ribosomal RNAs or linear mRNAs favors the identification and quantification of the expression of circular RNAs (Fig. 2.1).

2.4 CircRNA Validation

The identification and validation of circRNAs are required from several specific methods based on biochemical approaches. One of the most basic tools that can be used for validating them is the reverse transcription PCR (RT-qPCR). Following these assays, as the cDNA is going to arise from the circRNA, the sequence should comprehend the "exon junctional" region which is not present in the canonical spliced mRNA. To achieve this goal, primers have to be designed to detect and amplify this indicative region. The specific design of these primers called inverse or outward-facing

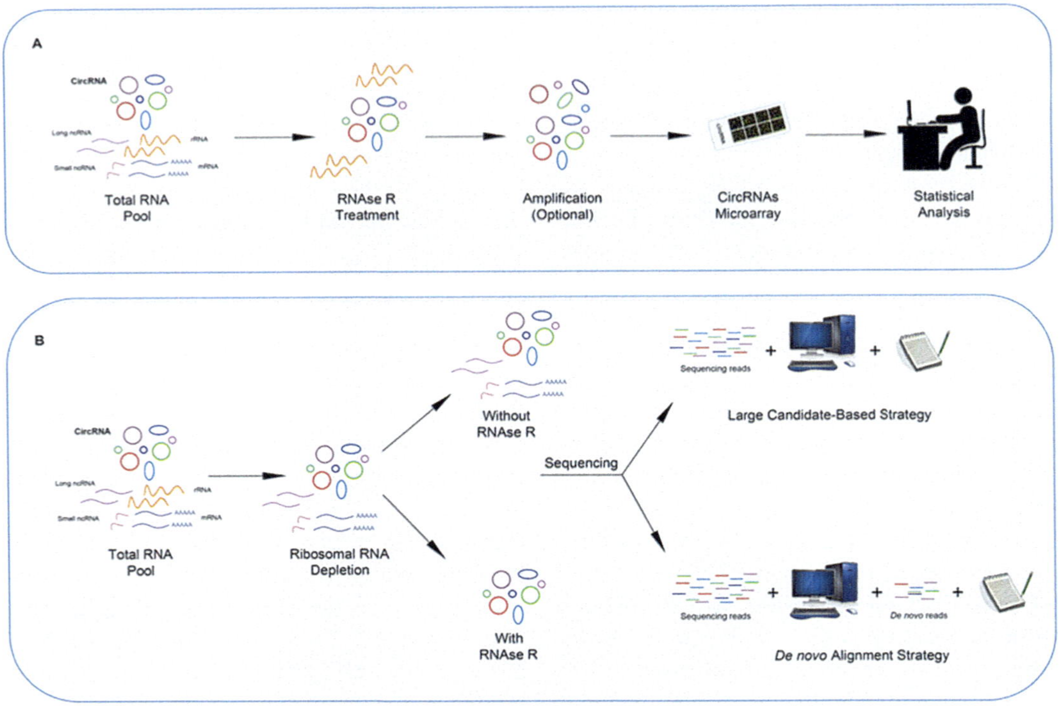

Fig. 2.1 Schematic workflow for sample preparation. (**a**) Schematic workflow of the process for a microarray circular RNA samples processing and data acquisition. (**b**) Schedule of different options for total RNA sample preparation and enrichment of circular RNAs (with or without RNAse R treatment) for sequencing processing and subsequently data analysis

primers prevents the alignment and amplification of other RNA (such as mRNA) species or DNA containing the diagnostic sequence. This is a quantitative approach that can be used in order to obtain the relative abundance of circRNAs in a biological context. RT-qPCR could also have artifacts and biases [35], so the result should be confirmed by means of sequencing the PCR products to check the presence of the junctional sequence [36]. Even though, it is possible to detect other species in the sequencing that make us aware of the level of noise of the experimental procedure for such specific condition.

Other important molecular technique for the validation is Northern blot [37]. The probes have to be designed for targeting the circular sequence or the specific junction sequence, and additional probes for the same circular RNA can be used in individual blottings for ensuring the presence of a specific circRNA. This is a very simple and specific procedure (because it is based on the mobility of the different species) to confirm the results in a qualitative manner.

In situ hybridization (ISH) techniques allow to confirm the presence of circRNAs in addition to identify the specific expression patterns of both cells and explicit tissues. To do this, specific probes are included, which are able to bind back-spliced junction sites [38]. In contrast, specific exonic sequences only existent in mRNAs are used as a control of the presence of the counterpartying linear transcripts expression, belonging to the same locus as circRNA.

Less frequent system to check the circularity and validate circRNA presence involves the use of RNase H (an endoribonuclease protein able to cleave RNA and RNA-DNA double strands) [37]. This method is based on the different binding patterns that two short DNA probes generate when they bind with the RNA of interest in the presence of each probe separately. In addition, a different migratory pattern can be observed depending if the species are circular or linear, because they exhibit a different behavior in a polyacrylamide gel.

Two-dimensional denaturing polyacrylamide gel electrophoresis (2D gel) can be also used to discriminate among linear and circular RNAs due to the different migratory patterns of both types of molecules. The specific pattern in 2D gel of the linear RNA is along the diagonal trajectory in the gel (and depending on the size), and circular RNA exhibits an accurate pattern. The gel trap method is other possibility for being used in RNAse R-treated samples [39]. Gel-trapping technique holds the pool enriched for circRNAs in the well of an electrophoresis gel, and at the same time, linear RNAs migrate away. The result could be directly extracted and sequenced using NSG technology [40].

The emerging results related to the circular transcriptome due to the revolution of the sequencing techniques in the last years come to light the need of an effective method for validating *in silico* results and predictions. None of the strategies explained before alone have the complete accuracy for ensuring the validation of the results. In that sense, a combinatorial validation strategy should be taken in consideration.

3 Computational Predictions of CircRNAs

As it was mentioned, circRNAs are distinguished by a "backspliced" process that occurs between a splice donor site and an upstream splice acceptor site. Therefore, the identification of circRNA presence requires the development of specific bioinformatic tools to detect junction-spanning sequences that reveal this backspliced from transcriptome deep-sequencing samples [10, 16]. The introduction of this high-throughput next-generation sequencing technology and an accurate protocol to reduce lineal mRNA (RNase R or non-polyadenylated procedures) has improved the description of numerous circular RNAs in different organisms.

Therefore, several algorithms have been already designed; thus, a vast amount of circRNAs resulting from exonic, intergenic, intronic, and UTR has been identified [3, 10, 41] and deposited in specialized databases such as circBase [42] or CIRCpedia [43].

Currently, there are several algorithms able to process RNA-seq samples in order to identify circular RNAs. Most of them have been benchmarked recently [44, 45], and the results are quite comparable. Besides, they conclude that although particular methods perform better than others, the highest percentage of true positives is obtained when results are combined from various methods and removing those circRNAs that do not appear in at least two different methods [44]. Because of this relevant conclusion and the rapid development of new prediction tools, we will present diverse softwares and discuss in detail their advantages and disadvantages (Table 2.1).

The majority of algorithms responsible for the identification of circular RNAs are divided in two different types with regard to the implemented methodology to discover circRNAs. The first group of programs is based on a "pseudo-reference" approach; briefly, they build a putative circRNA sequence reference from a gene annotation repository, to subsequently identify junction-spanning reads. The other strategy is called "segmented-based" and relies on the identification of backsplicing junctions from the mapping information provided by aligning reads to the reference genome or transcriptome (Fig. 2.2).

Among the pseudo-reference algorithms, we highlight KNIFE [46] and PTESFinder [47].

3.1 KNIFE

KNIFE [46] starts by mapping independently each paired-end read to the genome, ribosomal RNA sequences, lineal or scrambled exon-exon junction indexes, using Bowtie2 [48]. It rejects potential backspliced junction sequences if they also align with abnormal scores to lineal and scramble junction sequences. Those reads are

Table 2.1 CircRNA prediction tools

Tool name	Category	Aligners	References
KNIFE	Pseudo-reference	Bowtie 1 and Bowtie 2	[46]
PTESFinder	Pseudo-reference	Bowtie 1 and Bowtie 2	[47]
MapSplice	Segmented-based	Bowtie 1	[34]
CIRCexplorer	Segmented-based	TopHat and TopHat-fusion (Bowtie 1 and Bowtie 2), STAR	[30, 43]
CIRI	Segmented-based	BWA-MEM	[50]
Acfs	Pseudo and segmented-based	BWA-MEM	[51]

List of well-recognized circRNA prediction tools. They are organized according to a category (pseudo-reference, junction-spanning reads from potential circular RNAs are used to build a putative circRNA sequence which will be used as a reference; segmented-base, reads are aligned against a reference genome/transcriptome, and short segments from those reads are inspected to find backsplicing junctions and appropriate mapping signals supporting circRNA structures). The table also shows the mapper utility needed for each prediction tool and the research article that includes further information.

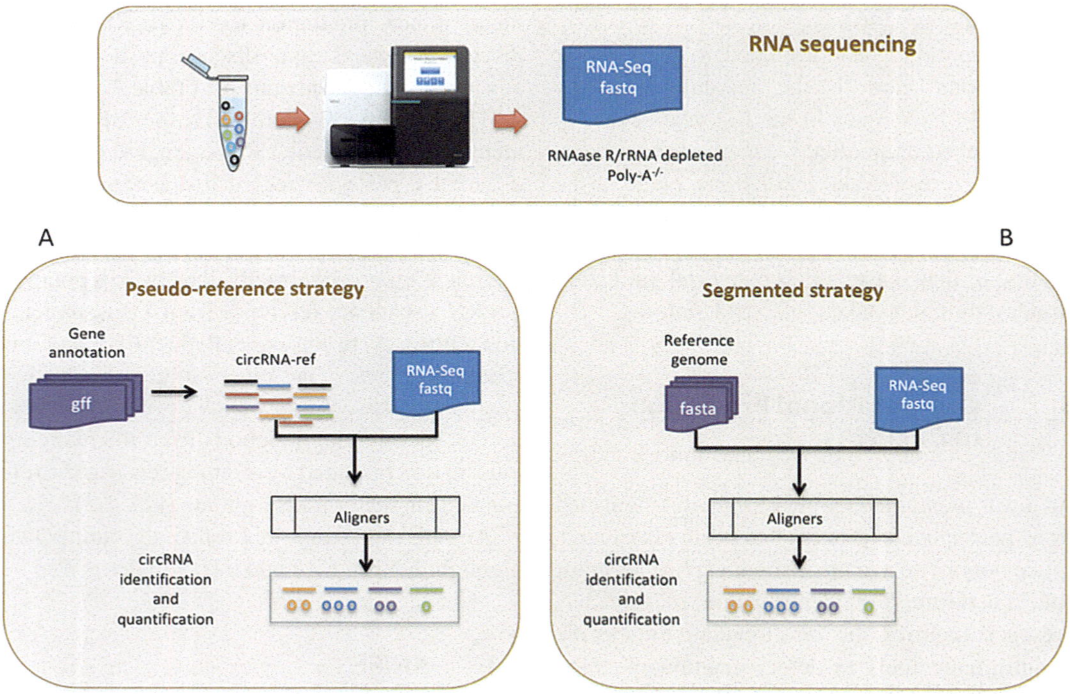

Fig. 2.2 Classification of circular RNA prediction tools according to the strategy. RNA samples of interest are sequenced. Those RNA-seq results can be processed for different methods; these can be classified into two groups: (**a**) the first group is composed of methods that rely on a "pseudo-reference" approach based on the generation of a putative circRNA sequence reference using gene annotation information. Reads are studied under this reference to identify junction-spanning reads. (**b**) The other strategy is called "segmented-based" and relies on the identification of backsplicing junctions from the mapping information provided by aligning reads to the reference genome or transcriptome

considered false positives and used to model all false positives according to two classes: real alignments (mapped to lineal mRNAs) or artifacts, when the paired-end reads alignment orientations are not coherent with neither a linear nor a circular RNA (named "decoy" alignments). The statistic model is based in a generalized model (GLM) focused on alignments scores, mapping quality, and offset position (category 1 or 2, "decoy" alignment).

One of the advantages of this algorithm is that they compute a posterior probability for each junction consistent or not, with decoy reads. This approach highlights those reads belonging to a circular RNAs.

Finally, unmapped reads are incorporated in a de novo algorithm aiming to discover unannotated splice sites responsible of RNA circularization.

3.2 PTESFinder

PTESFinder is a software that identifies putative posttranscriptional exon shuffling (PTES) structures from RNA-seq reads [47]. This tool is based in the assumption that circRNAs are transcripts characterized by the existence of exonic junction reads having an incongruous orientation according to their location in the genome. This program is divided into three consecutive phases: a discovery phase, an evaluation phase, and a filtering phase. In the first step, short sequences (20 base pairs by default) of each read end is aligned against the reference transcriptome using Bowtie [49]. A pair of sequences from the same read that map to the same gene but in reverse positions from their original order in the read sequence is potentially recognized as PTES. In the following phase, all initial reads are realigned to newly classified PTES structures using Bowtie2 [48]. Accordingly, it permits to obtain mapping scores from PTES in order to compare with those scores coming from lineal transcriptomic alignments. This information is used in the filtering stage to remove presumably false positives having higher scores using genomic or transcriptomic alignments rather than PTES mapping.

An important improvement included in this method is that it allows a "guided" evaluation by providing previously discovered PTES structures, hence avoiding in this way to perform the finding phase again. However, one of the weaknesses is that it does not use the information obtained from paired-end mapping (PE) since it interferes with the filtering phase affecting the specificity.

Among the "segmented-based" strategies, we will emphasize MapSplice [34], CIRCexplorer [30], CIRI [50], and Acfs [51] (which also uses a "pseudo-reference" approach as a filtering step).

3.3 MapSplice

MapSplice [34] is an algorithm for the detection of backsplice junction sequences which is independent of the splice-site information. This approach allows to discover novel splicing events along with noncanonical junctions in any transcriptome sample. Even more, it also identifies canonical junctions. MapSplice is splitted into two phases, the "tag alignment" step, where mRNA tags are mapped to a reference genome. The identification of candidate tag alignments is performed, in turn, in three parts: first, tags are subdivided into successive shorter pieces, which are aligned to the reference. In the next stage, sections lacking from an exonic alignment are considered for a spliced alignment method using a splice junction exploration procedure which includes adjacent pieces previously aligned. Finally, in the final step, tag alignments are combined to trace global candidate alignments for every single tag. Although, tags including splice junctions should involve a gapped alignment that ought to correspond to a removed intron by the splicing machinery in the transcription step. The second step, called the "splice inference phase," examines splice junctions that appear in the alignments of each tag to infer a splice significance value based on the quality and variety of the alignments. The goal of this step is to help with the selection process of the most reliable alignments for each tag, based on a mixture of quality alignment values and implication of the splicing event and based on that criteria, and discard spurious sequences.

3.4 CIRCexplorer

CIRCexplorer [30, 43] is a tool capable of identifying alternative backsplice and canonical splice junctions from single and paired-end reads. This software uses TopHat [52] coupled with TopHat-Fusion [53], although it optionally supports multiple circular RNA aligners such as STAR [54], BWA, [55] or segemehl [56]. Their approach is a two-step mapping strategy, where reads are first mapped against the reference sequence genome and later, nonaligned sequences are remapped using the TopHat-Fusion utility. New reads extracted from the fusion alignment in a noncollinear order coming from the same chromosome are remapped to a combination of (novel or known) gene annotation to conclude the exact location of backsplice sites from reliable circRNA structures. In the upgraded CIRCexplorer version [43], sequences aligned to the reference genome and collinear exon junction reads are now studied further in a de novo assembly, which permits to track down novel exons and therefore new splicing processes. Interestingly, in accordance with the authors and in comparison with their linear equivalent RNAs, circRNAs seem to present different alternative splicing and backsplicing patterns.

This program has many advantages, for instance, it is capable of working with single and paired-end reads. Besides, it also permits the usage of different kinds of aligners which allows a very detailed study of circular RNAs based on different mappers. In addition, it is able to predict circRNAs with high reliability while allowing the study of splicing patterns to identify new transcripts and exons. Among the disadvantages we can indicate that since each aligner requires different indexes, specific parameters and inputs files, it is a package indicated for researchers with high knowledge in the field.

3.5 CIRI

CIRI [50] uses the underlying strategy employed by BWA-MEM [55] which incorporates a local alignment with an affine-gap algorithm by seeding with a maximal exact match, typical of spanning junction reads in circular RNAs. This tool uses a method which relies on paired chiastic clipping (PCC) signal recognition in BWA aligned files, combined with various filtering phases aimed to eliminate false positives. Therefore, it collects PCC signals from aligned files which support the backsplicing junctions and the proper paired-end mapping (PEM) profiles coherent with circRNA structures. CIRI employs PEM information if existing, for an initial filter of spurious PCC signals. According to the authors, as two segments of a reliable junction sequence theoretically point out to the ends where all circRNA reads aligned, a probable junction indicates a circRNA if its paired mate read is mapped within the putative circular RNA region when supplied by the fragments of the junction site. Subsequently, it searches for those junction sites with known GT/AG splice signals. Finally, it clusters the unbalanced junction reads by employing a dynamic alignment methodology to remove putative false-positive junctions resulting from repetitive or homologous regions. Furthermore, if an accurate gene annotation is available for the organism of interest, it could be of interest to expand the search to other possible contiguous exon boundaries splice-site signals.

3.6 ACFS

Acfs [51], similar to CIRI, employs the BWA-MEM mapper to identify and quantify the abundance of circRNAs from single or paired-ended reads, although this tool is mainly designed to determine backsplice junctions from single-end transcriptome data. Acfs methodology is divided in three different steps: preprocessing phase, identification phase, and quantification phase. In the first part, reads are mapped to the genome.

If paired-end reads are included, they are processed and treated as single-end sequences. In the second step, potentially originated reads from the backsplice junctions are scrutinized by selecting those that align in a genomic position on the same chromosome and strand. Subsequently, Acfs inspects the strength of each backsplice junction alignment using a maximum entropy model [57] in order to identify the exact genomic position. In the latest phase, Acfs implements a supplementary alignment approach to precisely measure the abundance of the inferred circular RNAs. So, for each potential circRNA, a pseudo-circular reference is generated; then, the tool aligns each sequence to the genome, and before reporting the circRNA expression level, it examines alignments spanning the backsplice junctions.

This software is also capable of identifying circRNAs originating from noncanonical gene structure such as fusion genes, besides from the detection of circRNAs derived from regular genes. In that sense, as the splicing machinery is involved in the generation of circular RNAs, it would be probable that fusion gene loci could also produce these circular molecules.

Most of the programs discussed here, with the exception of Acfs, have recently been benchmarked using RNase R and poly(A)-depleted samples to measure the level of false positives [44, 45]. However, it has been claimed that resistance to the RNase R enzyme as a unique factor cannot be employed to conclude whether an mRNA is circular or not, since it was observed that some circRNAs were susceptible to exonuclease degradation [16, 46]. Nevertheless, KNIFE, CIRI, and CIRCexplorer achieved higher values of precision and sensitivity compared to other tools. As these studies highlight, given that the highest percentage of true positives is achieved from the intersection of predictions resulting from two different methods, here we present the new version of the miARma-seq pipeline, which is, as far as we know, the only one framework that includes several circRNA prediction tools in a single bundle.

3.7 MiARma-Seq

MiARma-Seq [58] (*miRNA-Seq And RNA-Seq Multiprocess Analysis*) is a tool designed to find mRNAs, miRNAs, and circRNAs, as well as for differential expression, target prediction (using the miRGate database [59] and its application programming interface [60]), and functional analysis in transcriptome samples. This software intends to reduce some of the principal difficulties that researchers may face when analyzing next-generation sequencing data such as (1) easy installation, removing third-party requisites that make the installation and configuration hard for researchers with little experience in the field; (2) speed of execution, allowing analysis in a standard computer or in a high-performance computing infrastructure (taking advantage of its parallelization); and (3) consistency, as the pipeline includes most of the standard tools available, in order to perform all calculations.

MiARma asks for a configuration file with general information about the experiment, and upon request, it can perform a quality step using FastQC [61] and a trimming phase using cutadapt [62] or kraken [63]. Subsequently, the user can perform an identification study of circRNAs based on different methods that miARma-seq includes, such as CIRI version 1.x and 2.x [50], KNIFE [46], CIRCexplorer [30, 43], and PTESFinder [47]. The possibility of using up to four different circRNA prediction methods in a single package without any prerequisite installation makes miARma one of the easiest tools in the identification of circular RNAs field. Interestingly, this pipeline offers the possibility of carrying out a differential expression analysis (using edgeR [64] or NOISeq [65]) among two conditions. In that sense, it could be possible, for instance, to compare transcriptome data from healthy controls and patient samples and correlate the presence/absence of circular RNAs with the onset or the prognosis of the disease. But also, it could be useful to compare RNase R-treated samples against untreated samples, given that this approach would allow to assess the

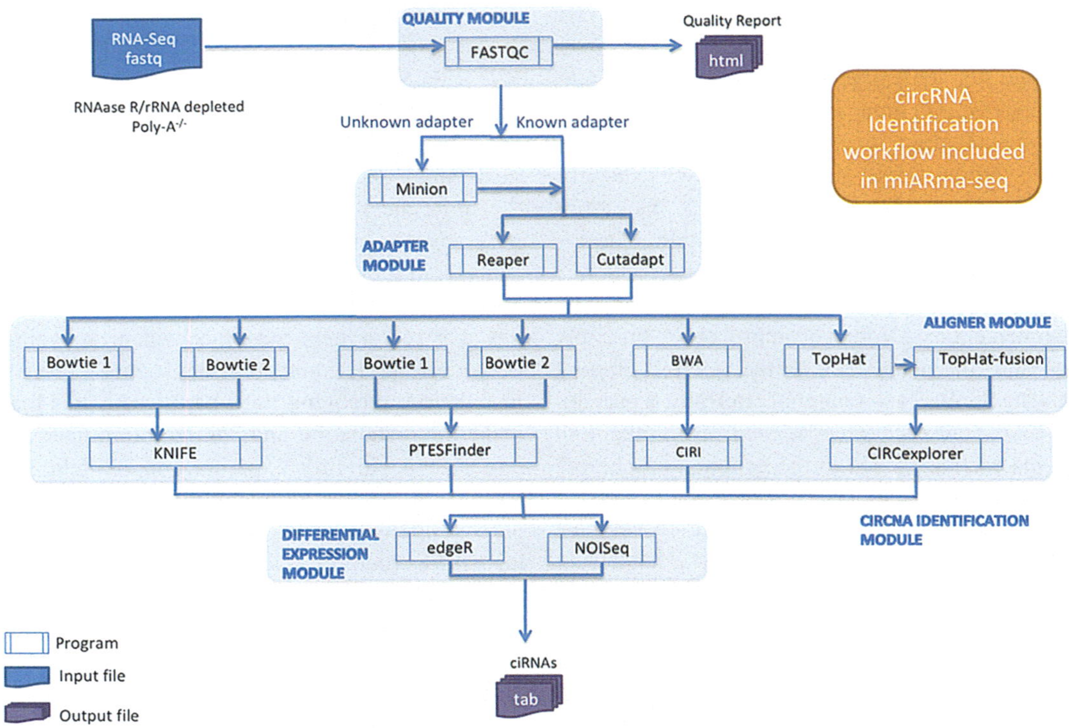

Fig. 2.3 Overview of the integrated tools available in miARma for circRNA identification and quantification. This workflow illustrates all software included in miARma-seq (only related to circRNA study) and all available modules. Hence, users can perform a complete analysis, from raw reads to circRNA identification, quantification, and differential expression

capacity of the method to accurately identify circRNAs. An overview of the integrated tools and the all-possible workflows available in miARma for circRNA identification is shown in detail in Fig. 2.3.

miARma-Seq is freely available at *http://miarmaseq.com* along with a complete documentation and diverse examples of usage.

3.8 CircRNA Databases (CircBase and CIRCpedia)

To date, numerous researchers have reported an elevated number of circular RNAs (circRNAs) expressed in diverse organisms and under different conditions. All information coming from these RNAs along with the identified alternative splicing or backsplicing events, and newly discovered exons, are available in the specialized databases such as CIRCpedia [43] or circBase [42]. The existence of this circRNA information, many of them experimentally validated, has allowed the collection of a bona fide dataset to facilitate the development of new stringent and reliable bioinformatics tools.

Currently, CIRCpedia contains circRNA backsplicing and alternative splicing data from 13 human cell lines, tissue, and species samples. This database provides query support by gene names and includes a helpful table with genomic coordinates, circRNA accession names, host gene names, relative expression values, and alternative (back)-splicing sequences from circRNAs along with exon identity. Links are also offered to download all the information for additional studies.

On the contrary, circBase [42] contains identified circRNAs in six organisms: human, mouse, *C. elegans*, *D. melanogaster*, *L. chalumnae*, and

L. menadoensis. For each of these species, the repository stores accurate material from different tissues, organs, and cell lines and allows exploring public circRNA datasets and downloading the scripts needed to discover and annotate your own circRNAs.

4 Future Perspectives

The massive generation of high-throughput data by means of the new sequencing technologies unveils the lack of an efficient and accurate methodology for the isolation and validation of circRNAs. Here we highlight that any of the existent methodologies is enough to assort the presence of this RNA species; they should be used in a combinatorial way to analyze them. It could be essential to standardize the procedure for detecting the circRNA using only one method that allows us to isolate and enrich only the circular RNA fraction within the RNA, in a simple and accurate way, avoiding the bias that other techniques could generate in different databases in order to be able to compare them. Thus, novel RNA-seq protocols and newly developed bioinformatics methodologies have emerged as a source of discovery for thousands of circRNAs in diverse organisms. Although dozen of projects based on circRNAs are published daily, several questions related to biogenesis and function of circRNA remain uncovered. Some research in cancer has revealed how the identification of upregulated and downregulated circRNA levels can be employed as a biomarker for specific types of cancer like in the case of laryngeal cancer [66]. Furthermore, the content of exosomes, which are highly enriched in circular RNAs, is altered by the generation and progression of cancer, as well as neurodegenerative and infectious diseases [67]. The recent discovery of the translation of a subset of circRNAs in eukaryotes opens the door to a more direct and specific functionality of the circular RNAs mediated by protein production [26–28], and their possible implication in human diseases has been pointed out in several studies [68].

As a summary, the identification and characterization of new circRNAs, together with the modification of protocols, such as Circle-Seq, to increase the presence of these circular molecules versus their linear counterparts, can help the development of more reliable tools. Without any doubt, this will respond unanswered questions to allow the use of the circRNAs as disease biomarker.

Acknowledgments We wish to thank the No Surrender Cancer Trust for supporting the position and projects of ELJ at Imperial College London.

Competing Financial Interests The authors declare no competing financial interests.

References

1. Nigro JM, Cho KR, Fearon ER et al (1991) Scrambled exons. Cell 64(3):607–613
2. Cocquerelle C, Mascrez B, Hetuin D et al (1993) Mis-splicing yields circular RNA molecules. FASEB J 7(1):155–160
3. Salzman J, Gawad C, Wang PL et al (2012) Circular RNAs are the predominant transcript isoform from hundreds of human genes in diverse cell types. PLoS One 7(2):e30733
4. Abu N, Jamal R (2016) Circular RNAs as promising biomarkers: a mini-review. Front Physiol 7:355
5. Morris KV, Mattick JS (2014) The rise of regulatory RNA. Nat Rev Genet 15(6):423–437
6. Lasda E, Parker R (2014) Circular RNAs: diversity of form and function. RNA 20(12):1829–1842
7. Chen LL (2016) The biogenesis and emerging roles of circular RNAs. Nat Rev Mol Cell Biol 17(4):205–211
8. Gruner H, Cortes-Lopez M, Cooper DA et al (2016) CircRNA accumulation in the aging mouse brain. Sci Rep 6:38907
9. Lukiw WJ (2013) Circular RNA (circRNA) in Alzheimer's disease (AD). Front Genet 4:307
10. Memczak S, Jens M, Elefsinioti A et al (2013) Circular RNAs are a large class of animal RNAs with regulatory potency. Nature 495(7441):333–338
11. Umekage S, Uehara T, Yoshinobu F et al (2012) In vivo circular RNA expression by the permuted intron-exon method. Innov Biotechnol 76:1
12. Ashwal-Fluss R, Meyer M, Pamudurti NR et al (2014) circRNA biogenesis competes with pre-mRNA splicing. Mol Cell 56(1):55–66
13. Jeck WR, Sharpless NE (2014) Detecting and characterizing circular RNAs. Nat Biotechnol 32(5):453–461

14. Zaphiropoulos PG (1996) Circular RNAs from transcripts of the rat cytochrome P450 2C24 gene: correlation with exon skipping. Proc Natl Acad Sci U S A 93(13):6536–6541
15. Salzman J, Chen RE, Olsen MN et al (2013) Cell-type specific features of circular RNA expression. PLoS Genet 9(9):e1003777
16. Jeck WR, Sorrentino JA, Wang K et al (2013) Circular RNAs are abundant, conserved, and associated with ALU repeats. RNA 19(2):141–157
17. Hansen TB, Jensen TI, Clausen BH et al (2013) Natural RNA circles function as efficient microRNA sponges. Nature 495(7441):384–388
18. Yap KL, Li S, Munoz-Cabello AM et al (2010) Molecular interplay of the noncoding RNA ANRIL and methylated histone H3 lysine 27 by polycomb CBX7 in transcriptional silencing of INK4a. Mol Cell 38(5):662–674
19. Burd CE, Jeck WR, Liu Y et al (2010) Expression of linear and novel circular forms of an INK4/ARF-associated non-coding RNA correlates with atherosclerosis risk. PLoS Genet 6(12):e1001233
20. Li Z, Huang C, Bao C et al (2015) Exon-intron circular RNAs regulate transcription in the nucleus. Nat Struct Mol Biol 22(3):256–264
21. Braunschweig U, Barbosa-Morais NL, Pan Q et al (2014) Widespread intron retention in mammals functionally tunes transcriptomes. Genome Res 24(11):1774–1786
22. Rybak-Wolf A, Stottmeister C, Glazar P et al (2015) Circular RNAs in the mammalian brain are highly abundant, conserved, and dynamically expressed. Mol Cell 58(5):870–885
23. Huang S, Yang B, Chen BJ et al (2017) The emerging role of circular RNAs in transcriptome regulation. Genomics 109(5–6):401–407
24. Du WW, Yang W, Liu E et al (2016) Foxo3 circular RNA retards cell cycle progression via forming ternary complexes with p21 and CDK2. Nucleic Acids Res 44(6):2846–2858
25. Chen CY, Sarnow P (1995) Initiation of protein synthesis by the eukaryotic translational apparatus on circular RNAs. Science 268(5209):415–417
26. Legnini I, Di Timoteo G, Rossi F et al (2017) Circ-ZNF609 is a circular RNA that can be translated and functions in myogenesis. Mol Cell 66(1):22–37 e29
27. Pamudurti NR, Bartok O, Jens M et al (2017) Translation of CircRNAs. Mol Cell 66(1):9–21 e27
28. Yang Y, Fan X, Mao M et al (2017) Extensive translation of circular RNAs driven by N(6)-methyladenosine. Cell Res 27(5):626–641
29. Guo JU, Agarwal V, Guo H et al (2014) Expanded identification and characterization of mammalian circular RNAs. Genome Biol 15(7):409
30. Zhang XO, Wang HB, Zhang Y et al (2014) Complementary sequence-mediated exon circularization. Cell 159(1):134–147
31. Labaj PP, Leparc GG, Linggi BE et al (2011) Characterization and improvement of RNA-Seq precision in quantitative transcript expression profiling. Bioinformatics 27(13):i383–i391
32. Suzuki H, Zuo Y, Wang J et al (2006) Characterization of RNase R-digested cellular RNA source that consists of lariat and circular RNAs from pre-mRNA splicing. Nucleic Acids Res 34(8):e63
33. Danan M, Schwartz S, Edelheit S et al (2012) Transcriptome-wide discovery of circular RNAs in Archaea. Nucleic Acids Res 40(7):3131–3142
34. Wang K, Singh D, Zeng Z et al (2010) MapSplice: accurate mapping of RNA-seq reads for splice junction discovery. Nucleic Acids Res 38(18):e178
35. Roy CK, Olson S, Graveley BR et al (2015) Assessing long-distance RNA sequence connectivity via RNA-templated DNA-DNA ligation. elife 4:e03700
36. Cocquet J, Chong A, Zhang G et al (2006) Reverse transcriptase template switching and false alternative transcripts. Genomics 88(1):127–131
37. Capel B, Swain A, Nicolis S et al (1993) Circular transcripts of the testis-determining gene Sry in adult mouse testis. Cell 73(5):1019–1030
38. Braissant O, Wahli W (1998) A simplified in situ hybridization protocol using non-radioactively labeled probes to detect abundant and rare mRNAs on tissue sections. Biochemica 1:6
39. Schindler CW, Krolewski JJ, Rush MG (1982) Selective trapping of circular double-stranded DNA molecules in solidifying agarose. Plasmid 7(3):263–270
40. Awan AR, Manfredo A, Pleiss JA (2013) Lariat sequencing in a unicellular yeast identifies regulated alternative splicing of exons that are evolutionarily conserved with humans. Proc Natl Acad Sci U S A 110(31):12762–12767
41. Chen BJ, Mills JD, Takenaka K et al (2016) Characterization of circular RNAs landscape in multiple system atrophy brain. J Neurochem 139(3):485–496
42. Glazar P, Papavasileiou P, Rajewsky N (2014) circBase: a database for circular RNAs. RNA 20(11):1666–1670
43. Zhang XO, Dong R, Zhang Y et al (2016) Diverse alternative back-splicing and alternative splicing landscape of circular RNAs. Genome Res 26(9):1277–1287
44. Hansen TB, Veno MT, Damgaard CK et al (2016) Comparison of circular RNA prediction tools. Nucleic Acids Res 44(6):e58
45. Zeng X, Lin W, Guo M et al (2017) A comprehensive overview and evaluation of circular RNA detection tools. PLoS Comput Biol 13(6):e1005420
46. Szabo L, Morey R, Palpant NJ et al (2015) Statistically based splicing detection reveals neural enrichment and tissue-specific induction of circular RNA during human fetal development. Genome Biol 16:126
47. Izuogu OG, Alhasan AA, Alafghani HM et al (2016) PTESFinder: a computational method to identify post-transcriptional exon shuffling (PTES) events. BMC Bioinform 17:31

48. Langmead B, Salzberg SL (2012) Fast gapped-read alignment with bowtie 2. Nat Methods 9(4):357–359
49. Langmead B, Trapnell C, Pop M et al (2009) Ultrafast and memory-efficient alignment of short DNA sequences to the human genome. Genome Biol 10(3):R25
50. Gao Y, Wang J, Zhao F (2015) CIRI: an efficient and unbiased algorithm for de novo circular RNA identification. Genome Biol 16:4
51. You X, Conrad TO (2016) Acfs: accurate circRNA identification and quantification from RNA-Seq data. Sci Rep 6:38820
52. Trapnell C, Pachter L, Salzberg SL (2009) TopHat: discovering splice junctions with RNA-Seq. Bioinformatics 25(9):1105–1111
53. Kim D, Salzberg SL (2011) TopHat-fusion: an algorithm for discovery of novel fusion transcripts. Genome Biol 12(8):R72
54. Dobin A, Davis CA, Schlesinger F et al (2013) STAR: ultrafast universal RNA-seq aligner. Bioinformatics 29(1):15–21
55. Li H, Durbin R (2009) Fast and accurate short read alignment with Burrows-Wheeler transform. Bioinformatics 25(14):1754–1760
56. Otto C, Stadler PF, Hoffmann S (2014) Lacking alignments? The next-generation sequencing mapper segemehl revisited. Bioinformatics 30(13):1837–1843
57. Yeo G, Burge CB (2004) Maximum entropy modeling of short sequence motifs with applications to RNA splicing signals. J Comput Biol 11(2–3):377–394
58. Andres-Leon E, Nunez-Torres R, Rojas AM (2016) miARma-Seq: a comprehensive tool for miRNA, mRNA and circRNA analysis. Sci Rep 6:25749
59. Andres-Leon E, Gonzalez Pena D, Gomez-Lopez G et al (2015) miRGate: a curated database of human, mouse and rat miRNA-mRNA targets. Database (Oxford) 2015:bav035
60. Andres-Leon E, Gomez-Lopez G, Pisano DG (2017) Prediction of miRNA-mRNA interactions using miRGate. Methods Mol Biol 1580:225–237
61. Anders S (2010) FastQC: a quality control tool for high throughput sequence data. Available online at: http://www.bioinformatics.babraham.ac.uk/projects/fastqc
62. Martin M (2011) Cutadapt removes adapter sequences from high-throughput sequencing reads. EMBnet J 17:1
63. Davis MP, van Dongen S, Abreu-Goodger C et al (2013) Kraken: a set of tools for quality control and analysis of high-throughput sequence data. Methods 63(1):41–49
64. Robinson MD, McCarthy DJ, Smyth GK (2010) edgeR: a Bioconductor package for differential expression analysis of digital gene expression data. Bioinformatics 26(1):139–140
65. Tarazona S, Garcia-Alcalde F, Dopazo J et al (2011) Differential expression in RNA-seq: a matter of depth. Genome Res 21(12):2213–2223
66. Xuan L, Qu L, Zhou H et al (2016) Circular RNA: a novel biomarker for progressive laryngeal cancer. Am J Transl Res 8(2):932–939
67. Lin J, Li J, Huang B et al (2015) Exosomes: novel biomarkers for clinical diagnosis. ScientificWorldJournal 2015:657086
68. Lyu D, Huang S (2017) The emerging role and clinical implication of human exonic circular RNA. RNA Biol 14(8):1000–1006

3 Online Databases and Circular RNAs

Seyed Hamid Aghaee-Bakhtiari

Abstract

circRNAs are a novel class of ncRNAs that unlike other ncRNAs are not linear and have a circular structure. These valuable ncRNAs have been detected in a wide range of organisms from plants to animals and in all cell lines and tissues. Commonly, circRNAs have several functions as gene expression regulation at transcriptional or posttranscriptional level, miRNA partnership, and splicing intercede. Currently, circRNAs are roughly remarked in a widespread collection of diseases, and circRNAs simply can be recognized in liquid samples for disease detection and progression assessment. Considering these features of circRNAs, these molecules are evolving the impeccable collection of original biomarkers for disease therapy and diagnosis. As the critical role of these molecules in different aspects medicine and biology, circRNAs are considered as key and critical class of ncRNA in the current ncRNA search field. To simplify the assessment of diverse features of circRNAs, several databases have been established such as circBase, CircInteractome, CircNet, Circ2Traits, CircR2Disease, TCSD, and CSCD. In this chapter, we have an overview on these main circRNA databases and introduce key features of each database.

Keywords

circRNA · Online databases · Web-accessible databases · Online resources

1 Introduction

Noncoding transcriptome comprises a collection of RNAs that have many controlling and fundamental functions [1]. MicroRNAs (miRNAs) and circRNAs are two classes of noncoding RNAs that their significance is completely demonstrated in numerous biological processes and several diseases [2, 3]. miRNAs are small endogenous RNAs that regulate gene expression, and numerous studies have established the important role of miRNAs in usual actions, such as cell propagation [4], cell differentiation [5], and cell cycle [6]. Furthermore, miRNAs have essential roles in human diseases and have a countless significance as biomarkers [7], therapeutic agents [8], and disease advancement assessment [9]. circRNAs are a group of newly discovered endogenous ncRNAs [10] and have been determined in many tissues and cell lines across most creatures [11].

S. H. Aghaee-Bakhtiari (✉)
Biotechnology Research Center, Mashhad University of Medical Sciences, Mashhad, Iran

Department of Medical Biotechnology, Faculty of Medicine, Mashhad University of Medical Sciences, Mashhad, Iran
e-mail: aghaeibh@mums.ac.ir

J. Xiao (ed.), *Circular RNAs*, Advances in Experimental Medicine and Biology 1087,
https://doi.org/10.1007/978-981-13-1426-1_3

Generally, circRNAs control gene expression transcriptionally or posttranscriptionally by transcription regulation, intervening with splicing, and collaboration with miRNAs [12]. Today, circRNAs are broadly noticed in an extensive range of diseases [13]. Furthermore, circRNAs have the cell or tissue specificity and stability and also easily can be detected in saliva [14] or blood samples [15]. Consequently, circRNAs are becoming the perfect group of novel biomarkers for disease diagnosis and therapy.

ncRNA online resources are essential tools for investigators to obtain data, and the number of them has been rapidly growing [16]. Considering the critical role of these molecules in molecular biology, circRNA has been converted to the hub in the current ncRNA exploration field. To facilitate the study of the different aspects of circRNAs, numerous databases have been developed such as circBase, CircInteractome, CircNet, Circ2Traits, CircR2Disease, TCSD, and CSCD.

2 circRNA Databases

circBase database covers circRNA data available up to 2013 and commonly becomes up to date with new published records [17]. circBase is established by Glažar P et al. and is accessible by the web server at http://www.circbase.org/. Now circBase presents data from human, mouse, *C. elegans*, and *Latimeria* organisms. The sequence and the supportive evidence of their expression of circRNAs can be retrieved, downloaded, and searched within the genomic context. Simple search, list search, and table browser are three different methods to search circBase. Simple search is used for search by sequence, gene explanation, or genomic location. List search helps users to find joint of a big number of query terms with circBase contents. The last method of search is table browser which can be used for conditional data recovery. After choosing the desired animal and experiment, users can additionally improve the results by some possibilities, like existence in a specific sample and quantity of supporting reads of the head-to-tail splice intersection [17].

CircInteractome is established by Dudekulay et al. and is openly available at http://circinteractome.nia.nih.gov. This database simplifies the investigation of circRNAs and their relations with more binding factors, principally miRNAs and RBPs [18]. CircInteractome offers users treasured features about circRNAs and their potential character in isolating miRNAs or RBPs and thus lessens their accessibility for mRNAs. CircInteractome similarly simplifies the primer design to assess circRNA by RT-Qpcr investigation. Furthermore, CircInteractome can be utilized to calculate RBP binding to sequences of the transcript, consequently possibly clarify the production of circRNAs. In addition, CircInteractome combines numerous features from other websites, such as StareBase 2.0, circBase, Primer3, and TargetScan 7.0. By combining these databases, CircInteractome allows researchers to realize the circRNA sequence and genomic site, circRNA-binding associates, primers, and siRNAs to analyze circRNA ranks, activity, and localization. Because all the information provided in CircInteractome are anticipated on the basis of sequence similarity and existence of circRNA structures, experimental confirmation is necessary to validate functional positions [19].

CircNet database is assembled by transcriptome sequencing datasets and is compiled by Liu et al., and the website is available at http://circnet.mbc.nctu.edu.tw/ [20]. In CircNet, human circRNAs are arranged with circRNA expression profiles through 464 human transcriptome samples. It provides circRNA-miRNA gene controlling networks and tissue-specific circRNA expression profiles. Moreover, it presents a combined controlling system that shows the arrangement among circRNAs, miRNAs, and genes. Generally, CircNet offers collective tools for researchers to simply access widespread data about genome location, expression analysis in different situations, and controlling complexes. In CircNet, researchers can select a desired miRNA or gene, and presented data containing the expression analysis, genomic location, and cohesive mRNA-miRNA-circRNA controlling system were gathered. In addition, CircNet would be an advantageous tool to investigate circRNA relationship to disease and also tissue specialized action [20].

Circ2Traits, a database of circRNAs hypothetically related to diseases in human, is developed by Ghosal et al. and is freely reachable at http://gyanxet-beta.com/circdb/ [21]. The current version of this database has classified 1951 human circRNAs possibly connected to 105 diverse diseases. circRNAs and their stored data in Circ2Traits are classified conferring to their possible relationship with diseases which was detected from the GWAS-associated SNPs. In addition, Circ2Traits stocks the whole putative miRNA-circRNA-mRNA-lncRNA interaction network for any disease. Users have several search options in this database. At first, the user can select a disease and observe a list of circRNAs related to the disease and moreover present the interaction table and the interaction network for each disease. The other search options are keyword search for circRNAs, miRNAs, lncRNAs, and protein-coding genes and search for GWAS traits connected to circRNAs [21].

CircR2Disease is a manually analyzed resource which is developed by Fan et al. and is openly reachable at http://bioinfo.snnu.edu.cn/CircR2Disease/ [22]. This tool delivers a broad database for circRNA dysregulation in different diseases, and collective indications have revealed that circRNAs have a key character in different levels of gene regulation including transcription, posttranscription, and translation levels. Based on previous studies, the irregular expression of circRNAs has been related with a set of diseases. Considering the enormous number of deregulated circRNAs in various diseases, it is necessary to make a superior resource to gather the circRNAs in diseases. The present version of CircR2Disease covers 725 relations among 100 diseases and 661 circRNAs by studying current articles. Every item in the CircR2Disease includes exhaustive data for the circRNA-disease association, comprising name of circRNA, name of disease, expression levels of circRNA, investigational methods, a concise explanation of the circRNA-disease connection, publication year, and the PubMed ID. CircR2Disease offers an easy-to-use platform to browse, explore, and transfer in addition to submit new disease-related circRNAs. CircR2Disease could be very helpful for users who study the process of disease-related circRNAs and discover the proper procedures for foreseeing novel relations [22].

TSCD (tissue-specific circRNA database) is compiled by Xia et al., and this resource is freely accessible at http://gb.whu.edu.cn/TSCD [23]. This database is the first overall observation of tissue specificity for circRNAs. At this point, TSCD accomplished the full investigation to distinguish the characteristic of human and mouse tissue-specific circRNAs. This database recognized altogether 302 853 tissue-specific circRNAs in the human and mouse genome and exhibited that the brain has the uppermost plethora of tissue-specific circRNAs. This resource additional established the presence of circRNAs by RT-PCR. TSCD similarly categorized the genomic position and preservation of these tissue-specific circRNAs and revealed that the mainstream of tissue-specific circRNAs is produced from exon areas. RNA binding protein and microRNAs which might bind to tissue-specific circRNA were recognized to more comprehend the possible activity of tissue-specific circRNAs. This procedure recommended their participation in progress and organ development [23].

CSCD (cancer-specific circRNA database) is developed by Xia et al. and is openly reachable at http://gb.whu.edu.cn/CSCD. CSCD is the first resource which fully explore the cancer-specific circRNAs [24]. This database recognized 272 152 cancer-specific circRNAs from 228 total RNA samples from cancer together normal cell lines. 170 909 circRNAs were recognized in normal and tumor samples which could be utilized as non-tumor background, and 950 962 circRNAs were known in normal samples only. CSCD foresees the RNA binding protein sites and microRNA response element sites for every circRNA to comprehend the practical properties of circRNAs. In addition, this database predicted possible ORFs (open reading frames) to recognize translatable circRNAs. Considering the properties of CSCD database could meaningfully provide to the investigation for the activity and control of cancer-related circRNAs [24].

Acknowledgments This research is supported by Mashhad University of Medical Sciences (No. 941245, 950909).

Competing Financial Interests The authors declare no competing financial interests.

References

1. Santosh B, Varshney A, Yadava PK (2015) Non-coding RNAs: biological functions and applications. Cell Biochem Funct 33(1):14–22
2. Han B, Chao J, Yao H (2018) Circular RNA and its mechanisms in disease: from the bench to the clinic. Pharmacol Ther. https://doi.org/10.1016/j.pharmthera.2018.01.010
3. Esteller M (2011) Non-coding RNAs in human disease. Nat Rev Genet 12(12):861
4. Rahimian A, Soleimani M, Kaviani S et al (2011) Bypassing the maturation arrest in myeloid cell line U937 by over-expression of microRNA-424. Hematology 16(5):298–302
5. Fallah P, Arefian E, Naderi M et al (2013) miR-146a and miR-150 promote the differentiation of CD133+ cells into T-lymphoid lineage. Mol Biol Rep 40(8):4713–4719
6. Attar M, Arefian E, Nabiuni M et al (2012) MicroRNA 17-92 expressed by a transposone-based vector changes expression level of cell-cycle-related genes. Cell Biol Int 36(11):1005–1012
7. Aghaee-Bakhtiari SH, Arefian E, Soleimani M et al (2016) Reproducible and reliable real-time PCR assay to measure mature form of miR-141. Appl Immunohistochem Mol Morphol 24(2):138–143
8. Aghaee-Bakhtiari SH, Arefian E, Naderi M et al (2015) MAPK and JAK/STAT pathways targeted by miR-23a and miR-23b in prostate cancer: computational and in vitro approaches. Tumour Biol 36(6):4203–4212
9. Arefian E, Kiani J, Soleimani M et al (2011) Analysis of microRNA signatures using size-coded ligation-mediated PCR. Nucleic Acids Res 39(12):e80
10. Chen L, Huang C, Wang X et al (2015) Circular RNAs in eukaryotic cells. Curr Genomics 16(5):312–318
11. Holdt LM, Kohlmaier A, Teupser D (2018) Molecular roles and function of circular RNAs in eukaryotic cells. Cell Mol Life Sci 75(6):1071–1098
12. Hansen TB, Jensen TI, Clausen BH et al (2013) Natural RNA circles function as efficient microRNA sponges. Nature 495(7441):384–388
13. Haque S, Harries LW (2017) Circular RNAs (circRNAs) in health and disease. Genes (Basel) 8(12):353
14. Bahn JH, Zhang Q, Li F et al (2015) The landscape of microRNA, Piwi-interacting RNA, and circular RNA in human saliva. Clin Chem 61(1):221–230
15. Memczak S, Papavasileiou P, Peters O et al (2015) Identification and characterization of circular RNAs as a new class of putative biomarkers in human blood. PLoS One 10(10):e0141214
16. Aghaee-Bakhtiari SH, Arefian E, Lau P (2017) miRandb: a resource of online services for miRNA research. Brief Bioinform. https://doi.org/10.1093/bib/bbw109
17. Glazar P, Papavasileiou P, Rajewsky N (2014) circBase: a database for circular RNAs. RNA 20(11):1666–1670
18. Dudekula DB, Panda AC, Grammatikakis I et al (2016) CircInteractome: a web tool for exploring circular RNAs and their interacting proteins and microRNAs. RNA Biol 13(1):34–42
19. Panda AC, Dudekula DB, Abdelmohsen K et al (2018) Analysis of circular RNAs using the web tool circInteractome. Methods Mol Biol 1724:43–56
20. Liu YC, Li JR, Sun CH et al (2016) CircNet: a database of circular RNAs derived from transcriptome sequencing data. Nucleic Acids Res 44(D1):D209–D215
21. Ghosal S, Das S, Sen R et al (2013) Circ2Traits: a comprehensive database for circular RNA potentially associated with disease and traits. Front Genet 4:283
22. Fan C, Lei X, Fang Z, et al (2018) CircR2Disease: a manually curated database for experimentally supported circular RNAs associated with various diseases. Database (Oxford) 2018.
23. Xia S, Feng J, Lei L et al (2017) Comprehensive characterization of tissue-specific circular RNAs in the human and mouse genomes. Brief Bioinform 18(6):984–992
24. Xia S, Feng J, Chen K et al (2018) CSCD: a database for cancer-specific circular RNAs. Nucleic Acids Res 46(D1):D925–D929

Part III

Biogenesis of Circular RNAs

Circular RNA Splicing

4

Nicole Eger, Laura Schoppe, Susanne Schuster,
Ulrich Laufs, and Jes-Niels Boeckel

Abstract

Circular RNAs (circRNAs) are covalently closed single-stranded RNA molecules derived from exons by alternative mRNA splicing. Circularization of single-stranded RNA molecules was already described in 1976 for viroids in plants. Since then several additional types of circular RNAs in many species have been described such as the circular single-stranded RNA genome of the hepatitis delta virus (HDV) or circular RNAs as products or intermediates of tRNA and rRNA maturation in archaea. CircRNAs are generally formed by covalent binding of the 5′ site of an upstream exon with the 3′ of the same or a downstream exon. Meanwhile, two different models of circRNA biogenesis have been described, the lariat or exon skipping model and the direct backsplicing model. In the lariat model, canonical splicing occurs before backsplicing, whereas in the direct backsplicing model, the circRNA is generated first. In this chapter, we will review the formation of circular RNAs and highlight the derivation of different types of circular RNAs.

Keywords

circRNA · RNA splicing · Circular RNA · Backsplicing

1 Introduction

Circular RNAs (circRNAs) are circular single-stranded RNA (ssRNA) molecules formed from exons of genes by alternative mRNA splicing (Figs. 4.1 and 4.3). Circular RNAs in eukaryotes were first detected by electron microscopy in human HeLa cells in 1979 [1]. However, at that time the origin of these circular RNA molecules was not known. The discovery of inverted orientated exons in ssRNA, referred to as "scrambled exons" in humans [2–4], gave first evidence that these molecules might be originated from protein-coding open reading frames [2, 5]. CircRNAs are generally formed by covalent linkage of the 5′ splice site of an upstream exon with the 3′ site of the same or a downstream exon [6]. However, also introns located in the open reading frame (ORF) of a protein-coding gene can circularize [7]. Spliced introns forming circular RNAs are referred to as circular intronic RNAs (ciRNAs) and are distinguished from those formed by exons from the open reading frame of a

Author contributed equally with all other contributors. Nicole Eger and Laura Schoppe

N. Eger · L. Schoppe
University of Heidelberg, Heidelberg, Germany

S. Schuster · U. Laufs · J.-N. Boeckel (✉)
Clinic and Polyclinic for Cardiology, University Hospital Leipzig, Leipzig, Germany
e-mail: Jes-Niels.Boeckel@medizin.uni-leipzig.de

J. Xiao (ed.), *Circular RNAs*, Advances in Experimental Medicine and Biology 1087,
https://doi.org/10.1007/978-981-13-1426-1_4

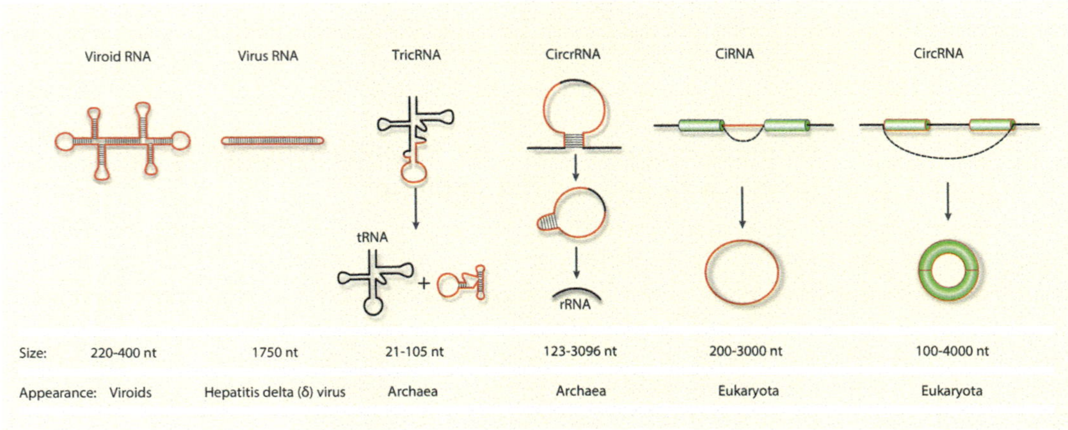

Fig. 4.1 Circular single-stranded RNA variants and their derivation
Various circular RNAs (here illustrated in red) have been reported. The single-stranded RNA circles are covalently closed, while sequence-dependent double-stranded base pairing can appear at multiple regions of the RNA circles. **The genome of viroids,** which does not code for proteins but has catalytic ribozyme properties, was shown to be formed of single-stranded RNAs with sizes of ~220–400 nucleotides length. Viroids replicate utilizing the host cell's machinery via a rolling-circle mechanism resulting in a concatemeric linear sequence which is then cleaved by endonuceases into multiple linear replicates and finally circularized by end-to-end ligation. **The Hepatitis delta (δ) virus (HDV)** was discovered as the first animal virus having a single-stranded circular RNA molecule as carrier of its genetic information with a length of 1750 nt. The host cell machinery synthesizes the circular viral RNA genome via a rolling-circle mechanism, while finally the host cell's ligases form 3′ to 5′ or 2′ to 5′ phosphodiester bonds between the respective ends of the HDV genome. tRNA introns occur within eukaryotes and prokaryotes alike, while **tRNA intronic circular (tric)RNAs** can appear in archaea as a product of tRNA splicing with lengths between 21 and 105 nt. The tRNA-splicing endonuclease (TSEN) complex in archaea, consisting of a specific endonuclease and ligase, has been shown to produce circularized excised tRNA introns. **Circular rRNA (circrRNA)** precursors were reported in archaea containing the premature rRNA sequence, which is later spliced out during the process of rRNA maturation. This splicing is autocatalytic and accompanied by a guanosine addition to the 5′ end. Mechanistically, rRNA splicing occurs in a similar fashion as splicing of circular tRNA introns. **Circular intronic RNAs (ciRNA)** can appear within eukaryotes as either products of spliceosomal splicing or intermediates of self-splicing introns. In canonic splicing intron sequences are removed by the spliceosome in lariat form, containing an internal 2′ to 5′ phosphodiester bond. Introns of group I and II are mobile genetic elements that catalyze their own splicing from DNA via their ribozymal properties. **Circular exonic RNAs (circRNA)** are produced via pre-mRNA processing in eukaryotic cell nuclei by an alternative splicing process referred to as backsplicing. A downstream exon's 3′ splice site performs a nucleophilic attack on its own or a different upstream exon's 5′ splice acceptor site, resulting in a 3′ to 5′ phosphodiester linked circular RNA molecule.

protein-coding gene, which are called circular RNAs (circRNAs) [8, 9] (Figs. 4.1 and 4.2b). Several circular RNA species have been reported in eukaryotes, prokaryotes, viruses, and subviral agents and were found to be synthesized by various ligation reactions. Circularization of single-stranded RNA (ssRNA) molecules in general was already described in 1976 for the genome of some subviral agents named "viroids" [10–12] (Fig. 4.1). In contrast, the genomes of viruses can consist of several different nucleotide structures, such as linear single-stranded RNA (ssRNA), linear or circular double-stranded DNA (dsDNA), and single-stranded DNA [13, 14]. The hepatitis delta virus (HDV) was discovered in 1986 as the first animal virus having a circular single-stranded RNA genome [15] (Fig. 4.1). Studies in archaea revealed the existence of circular RNAs as by-product and end product of tRNA maturation called tricRNAs [16–19] (Fig. 4.1). Archaea also give rise to circular single-stranded pre-rRNA intermediates of the 16S and 23S RNA [19, 20]. The predominant mechanism of circular RNA formation is 3′ to 5′ end ligation found in viroids, hepatitis delta virus, and the products or intermediates of tRNA and rRNA maturation in

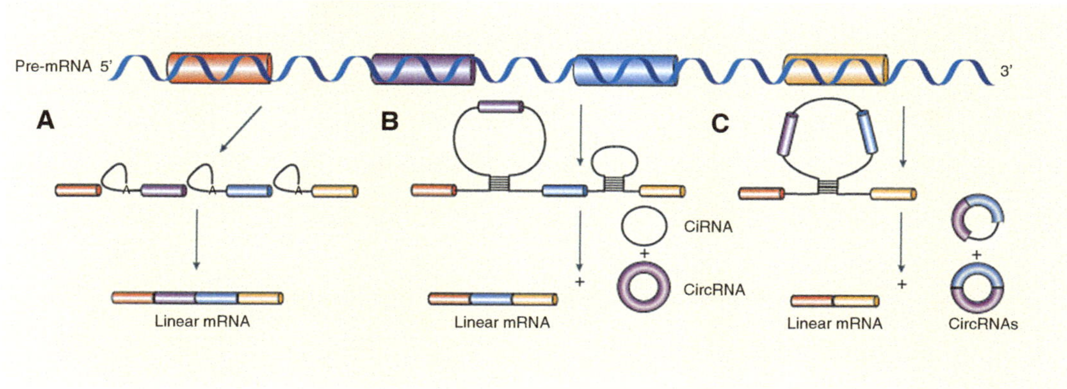

Fig. 4.2 Derivation of circular RNA molecules by alternative pre-mRNA splicing
Circular RNAs in eukaryotes can be derived from primary RNA transcripts via alternative splicing. (**a**) During canonical splicing by the spliceosome, intronic regions are removed in a two-step transesterification reaction with a lariat intermediate. The branch point adenosine's 2′ hydroxyl group performs a nucleophilic attack on an upstream exon's 3′ splice site within the same intron, forming an intronic 2′ to 5′ phosphodiester. In a second step, the upstream exon's 3′ hydroxyl group binds after deprotonation to the downstream exon's 5′ phosphate end, thereby displacing the intron in its lariat form from the mRNA. (**b**) Alternative splicing of pre-mRNA may also result in circular RNAs. Base pairing between intronic inverted repeat regions thereby connecting the two regions promotes circular RNA derivation. This internal secondary stem-loop structures within the pre-mRNA sterically facilitates backsplicing. Besides linear mRNA, circular intronic (ciRNA) and circular exonic RNA (circRNA) can be spliced from the same primary transcript. (**c**) CircRNAs can also be spliced in multiple different alternative variants. Up to six exons were reported to be spliced in inverted order within one circRNA. Interceding intronic regions may also be included in the final splice product.

archaea. However, also 2′ to 5′ end ligations involving a nucleophilic attack of the branch point were reported for ciRNAs in eukaryotes [8]. Two models were proposed for the biogenesis of circRNAs as a product of alternative exon splicing: the lariat model and the backsplicing model [21, 22] (Fig. 4.3). The essential difference between these two reactions is the order in which splicing events do occur.

In the following chapter, we will provide an overview about the different circular RNA classes arising from different species and their biogenesis by splicing and ligation.

2 The Origin and Splicing of Different Circular RNA Species

2.1 Viroid Circular RNA

The existence of viroids as proto-organisms containing circular RNAs as genomes has been known since the early twentieth century [11]. The first covalently closed RNA circles were discovered in the potato spindle tuber viroid in 1967 [23]. The genome of this viroid was first proposed to be a double-stranded RNA molecule but later on shown to be single-stranded [12]. These circular RNA containing viroids are in general small 220–400 nucleotides RNA-only plant pathogens which replicate autonomously within the nucleus or chloroplasts of their host's cells [10, 24, 25]. Their genome exist in circular conformation and further form self-complementary base-paired rod-like structures [10] (Fig. 4.1).

Viroid RNA does not code for proteins; however their circular single-stranded genomic RNA has catalytic ribozyme properties [12, 26]. The ribozyme is responsible for the infection of and proliferation within host cells [24]. After infection of the plant hosts, viroids also utilize the plant's enzymatic machinery in order to ensure their own functionality and the replication of their circular ssRNA genome [27]. Viroid replication ensues via a rolling-circle mechanism

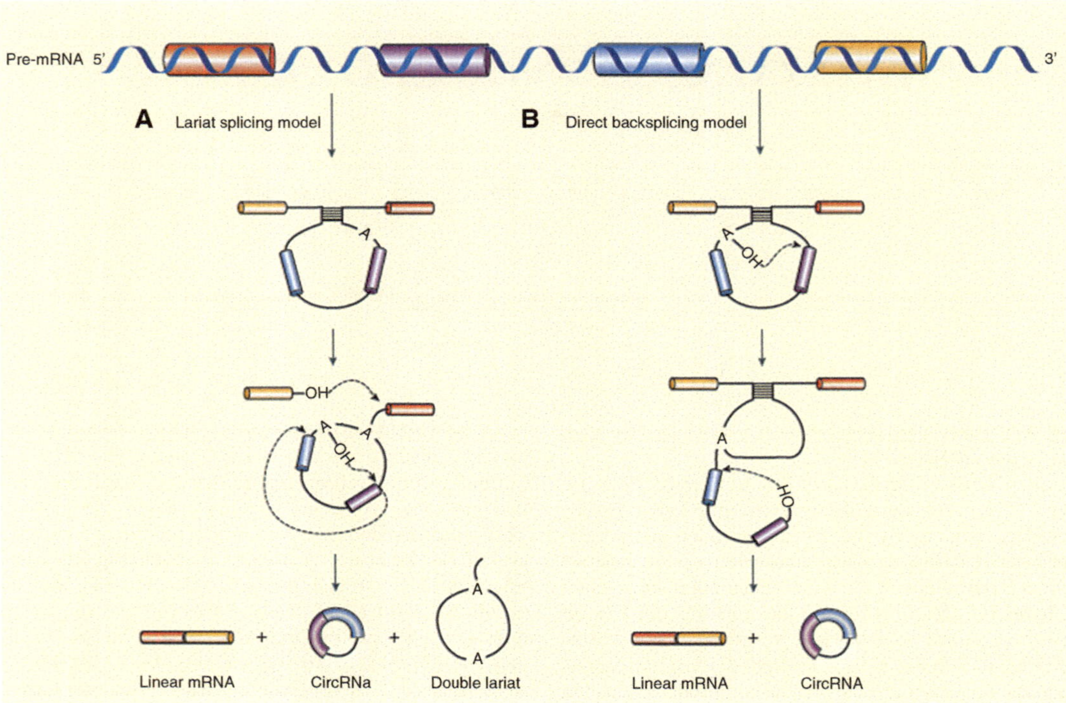

Fig. 4.3 Different backsplicing models
Two different mechanisms for derivation of circRNA have been described so far. In both models internal base pairing between inverted repeat regions facilitates the alternative splicing. The main difference is whether circularization of the circRNA is after canonical splicing or before. (**a**) **Lariat Splicing Model** According to the lariat splicing model, backsplicing occurs after canonical splicing. During canonical splicing alternative exons may be spliced out of the final mRNA product and end up contained within the excised lariat. The lariat then undergoes internal backsplicing. First an upstream branch point adenosine's 2′ hydroxyl group nucleophilically attacks the 3′ splice site of the downstream exon, forming a double lariat structure. The downstream exon's 3′ hydroxyl group is now free to attack the upstream 5′ splice acceptor of the exon, thereby circularizing it. (**b**) **Direct Backsplicing Model** In the direct backsplicing model, an upstream branch point adenosine 2′ hydroxyl group initiates the process by attacking a downstream exon's 3′ splice site, resulting in a Y-shaped intermediate. The free 3′ hydroxyl then attacks an upstream 5′ splice site resulting in circularization

enabled by their circular genomic structure [28, 29]. The circular genome is transcribed times in an iterative fashion forming one long, linear transcript of concatemers, which may later on be cleaved by endonucleases into multiple linear RNA replicates which are in turn circularized by end-to-end ligation [30, 31]. The common hypothesis explaining the genomic structure of viroids is that the circularity of their RNA allows for faster replication, since the rate-limiting initiation step must only be performed once, but creates multitudes of offspring.

2.2 Viral Circular RNA

Viruses are small nucleic acid-based intracellular pathogens. Outside a host they are covered by a protein coat encoded within their own genome which they shed during host infection. Viral genomes come in many different forms and conformations of both DNA and RNA molecules, while circular single-stranded RNA is one of them.

In 1986 the *hepatitis delta virus* (HDV) was discovered as the first animal virus having a circular RNA molecule as carrier of its genetic information [15] (Fig. 4.1). Comparable to viroids, HDV uses the host cell machinery for production of circular RNA molecules via the rolling-circle mechanism [30, 32]. Viruses utilize their host cell's ligases to form 3′ to 5′ or 2′ to 5′ phosphodiester bonds between the respective ends of their genomes [33]. This circularization process in turn protects the viral genome from digestion by intracellular exonucleases [34]. It further protects the viral RNA from other host immunity mechanisms, such as the MDA-5 and RIG-I receptors, which bind free nonhost cytoplasmic ssRNA in higher vertebrates [35, 36].

2.3 tRNA Intronic Circular (tric) RNAs

Transfer RNA (tRNA) introns occur within eukaryotes and prokaryotes alike. Archaeal tRNA primary transcripts can contain intronic regions that require splicing before acquiring functionality [37]. Also in yeast tRNA splicing occures but generates linear RNA intermediates [38]. Interestingly, tRNA intronic circular (tric)RNAs can appear in archaea as a product of tRNA splicing [39] (Fig. 4.1). The tRNA-splicing endonuclease (TSEN) complex in archaea [37], consisting of a specific endonuclease and ligase, has been shown to produce circularized excised tRNA introns [18, 19]. This complex recognizes the cleavage sites by a bulge-helix-bulge motif at exon-intron junctions [17, 18, 37]. One may hypothesize that circularization of functional noncoding RNAs in general may increase their stability and thereby prolong their regulatory effects within the cells. Like other circular RNAs, they are resistant to digestion by cytoplasmic exonucleases like RNAse R and therefore less likely to be degraded [40]. This effect has been used to the advantage of sequencing circular RNAs as well [21, 41]. Circular noncoding RNAs have been most prominently described in thermophilic archaea like *Haloferax volcanii* [18, 19]. Here, their circularity may protect from heat denaturation of the molecules since it does not offer a free terminal helix with steric flexibility at either its 3′ or 5′ end [42].

2.4 Circular rRNA (CircrRNA)

In 1981 Grabowski and colleagues discovered the existence of a circular RNA molecule in *Tetrahymena thermophila* as an intermediate of rRNA maturation [43]. Here the premature rRNA is part of a larger circular single-stranded ribosomal RNA precursor and later spliced out during proceeding rRNA maturation (Fig. 4.1). There have been several reports about ribosomal RNA maturation in archaea involving a procedure wherein rRNA precursors are first cleaved to derive linear pre-16S and pre-23S rRNAs [18, 20, 44, 45] (Fig. 4.1). The ends of these RNAs are then ligated, forming circular single-stranded RNA molecules, which are processed further until becoming mature, functional rRNAs [18]. This underlying splicing reaction is autocatalytic and accompanied by a guanosine addition to the 5′ end [46]. Mechanistically, this process is comparable to the previously described processing of the circular tRNA introns, where recognition occurs via a bulge-helix-bulge motif [20]. In accordance, the TSEN complex is also responsible for the final ligation of the circular molecules [37].

2.5 Circular Intronic RNAs (CiRNAs)

It was initially believed that spliced out introns were the main source of circular RNAs in higher eukaryotes [47–49]. Different forms and origin of intronic RNAs have been described so far. Circular intronic RNAs (ciRNAs) are defined as intronic sequences forming RNA circles [8], unlike circRNAs which can contain both exonic and intronic sequences. A regulatory function of ciRNA in relation to polymerase II transcription is suggested since knockdown of ciRNA showed altered expression at the corresponding gene locus [8].

2.5.1 Group I and Group II Self-Splicing-Derived Circular Intronic RNAs (CiRNAs)

The self-splicing group I introns are mainly located within genomic ribosomal RNA regions of eukaryotic microorganisms [50]. Unlike in spliceosomal or group II intron splicing, where the RNA hydroxyl groups act as internal nucleophiles, group I introns recruit an external guanosine as a nucleophile to initiate splicing [34]. During the process first a linear excised RNA is separated, which can then undergo 3′ to 5′ circularization [43] (Fig. 4.1).

Ribosomal introns of group II are mobile genetic elements that autocatalytically splice themselves from precursor RNAs by using a transesterification reaction similar to the classical spliceosomal splicing reaction [42, 51, 52]. Their structure is made up of catalytically active RNAs and an intron-encoded reverse transcriptase [51]. After sequence excision, circular lariat species with a 2′ to 5′ phosphodiester bond are formed [53]. Given the strong similarity in mechanism and characteristics of group II introns and eukaryotic spliceosomal introns, they are thought to be evolutionarily related [51].

2.5.2 Spliceosomal-Derived Circular Intronic RNAs (CiRNAs)

Spliceosomal splicing is the main, highly conserved, mechanism of mRNA processing within eukaryotic cells [54]. During intron excision a 2′ to 5′ transester is formed and released. Circular intronic RNAs (ciRNAs) are formed from such lariats, which have been additionally degraded from their 3′ end up to the branch point but have not been further degraded beyond this. Their sizes may vary from under 200 to over 3000 nucleotides [42]. The blockage of degradation beyond the branch point depends on a consensus motif of a 7 nucleotide GU-rich element near the 5′ splice site and an 11 nucleotide C-rich element near the branch point [8] (Fig. 4.1).

2.6 Circular RNA Spliced Exons

Circular exonic RNA molecules (circRNA) were first detected in 1991 in human cells and initially believed to be linear "scrambled exons" [2] (Fig. 4.1). Shortly afterward circRNA molecules were described as RNA products derived from the testis-determining gene *Sry* located to the cytoplasm of murine cells [55], as well as from the *Ets-1* gene in human cells [4, 5]. For many years they were described as rare events of transcription, being merely by-products of linear mRNA splicing [2, 4, 5, 55–58]. But in the last decade, evidences increase that circRNAs are derived from thousands of expressed protein-coding genes and moreover have functional roles in eukaryotic cells. Tissue- and cell-specific alternative mRNA splicing is a hallmark of cell specialization found in many eukaryotic organs and highly specialized cell types such as cardiomyocytes and neurons [59]. Alternative splicing was already reported in 1992 by the group of Bernard Bailleul for two circRNA products of the *Ets-1* gene but has been later shown to also appear in different tissues and cells [60] (Fig. 4.2). Interestingly, some circRNAs were reported to be even more abundantly expressed than their cognate linear mRNAs, while also the contribution of differences in molecular stability between the linear and circular products of a host gene cannot be excluded here [21, 41].

CircRNAs range from barely 100 to sizes over 4000 nucleotides [42]. They are mainly found within the cytoplasm of eukaryotes [61]. Circular transcripts usually encompass two to five "shuffled" or "scrambled" exons and bear the possibility to undergo alternative splicing [2]. Their genomic loci are usually flanked by repetitive complementary sequences which enhance the circularization efficiency [62].

CircRNAs are derived by a backsplicing mechanism, where a 3′ splice site of a downstream exon is effectively spliced to end up on an upstream exon or the same exon's 5′ splice site [21, 63]. The next chapter will give a more detailed insight into the different proposed mechanisms of backsplicing [61].

3 Splicing of circRNA from ORF Exons

3.1 Canonical Splicing of Pre-mRNA

The ~$3*10^9$ base pairs large human genome only consists of about 1% protein-coding sequences referred to as exons [64, 65]. The large rest of the genome, though not being translated into amino acid chains, fulfill many other functions, among them increasing overall genomic stability [66], controlling the chromatin accessibility via epigenetic modifications [67, 68], and regulation of transcription [69–71]. Protein-coding genes are transcribed in the nucleus by RNA polymerase II resulting in a single-stranded pre-mRNA containing introns and exons in genomic sequence [72]. During transcription there is no discrimination between exonic and intronic regions; both are equally transcribed after initiation of transcription at the promoter site; the RNA-polymerase complex is producing one continuous pre-mRNA transcript of the genomic DNA [73]. In the further process of mRNA maturation, a 5′ 7-methylguanosine cap and a 3′-poly-adenosine tail are added [72, 74–78].

In order to create functional coding mRNAs that can be translated by cytoplasmic ribosomes, the interceding intronic regions are removed from the primary transcript by RNA splicing. The multimeric enzyme complex made up of small nuclear uracil-rich ribonucleoprotein particles (U-snRNPs), referred to as the spliceosome [47], assembles in a stepwise manner on specific splice-signaling intronic guide sequences [79]. The mRNA splicing can be canonical or noncanonical, depending on the guiding sequence of the respective intron to be removed. Over 99% of all pre-mRNAs feature canonical splice sites 5′ GU and 3′ AG on the respective ends of the intronic regions [80]. These canonical splice sites function as recognition signals for U-snRNP binding, here the 5′ end of the U1-snRNP directly binds to the 5′ splice signal sequence on the pre-mRNA, thereby promoting the splicing reaction [81].

Mechanistically, the canonical pre-mRNA splicing is a two-step transesterification reaction. First, the 2′ branch point adenosine hydroxyl group, which is situated within the intron to be spliced, nucleophilically attacks the upstream 5′ end of the intron (Fig. 4.2a). This results in a lariat intermediate, which is covalently closed by a 2′ to 5′ phosphodiester bond. The upstream exon's 3′ hydroxyl group is now free to attack the downstream exon's 5′ phosphate, completely splicing out the intron in lariat form and leaving the exons linked together in a linear coding sequence manner.

3.2 CircRNAs Are a Product of an Alternative Splicing Process

The generation of circular RNAs (circRNA) from exons, is more similar to the formation of mRNAs as opposed to the generation of other circular RNA species (see, e.g., viral RNA genomes which are circularized by host ligases in the previous chapter), requires processing by the canonic spliceosome. CircRNAs were first identified as "scrambled exons" and thought to be splicing errors or waste products of the splicing process [2, 82, 83], since their order of exons is inverted compared to the exonic arrangement on the genomic open reading frame [2, 5, 7, 21]. Already in the year 1993, early after the first description of scrambled exons by the Vogelstein lab in 1991, circularity of exonic RNA was observed in eukaryotes by the Bailleul laboratory in RNA products of the *Ets-1* gene as well as by Capel and colleagues in RNA products of the *Sry* gene in adult mouse testis [4, 5, 55, 84]. Cocquerelle and colleagues moreover reported in their pioneering research work the generation of two circular RNAs from the *Ets-1* gene by incorporation of different exons into the different final RNA circles, thereby also reporting alternative splicing of circular exonic RNAs for the very first time [4] (Fig. 4.2c).

Alternative splicing allows for mRNAs with a variety of differently composed exons resulting from a single gene locus, which can code for multiple isoforms of a protein with different functional properties. Alternative splicing is further contributing to functional specification of proteins to certain tissue demands, as was predicted early on by Walter Gilbert in response to the discovery of the fact that genes consist of coding as well as noncoding sequences [85]. The ryanodine receptor gene, whose alternative mRNA splicing in cardiomyocytes allows for highly specialized contribution to intracellular calcium signaling by different resulting protein isoforms within the heart, is a good example in this regard [86–88]. The mechanism of alternative splicing, according to the combinatorial model, is controlled by regulatory factors that bind to pre-mRNA and induce certain splicing patterns, determining the inclusion or exclusion of a certain exon on the final mRNA product. It has been the proposed mechanism facilitating circRNAs formation in the first description of scrambled exons in 1991 [2, 89].

The number of exons in a single circRNA differs between one and five, with two or three exons being most frequently included in the circle. Interjacent intronic regions are mostly excised but are sometimes fully or in part included into the circularized molecule [62, 90]. Both splice sites of the canonical splice signal are required for successful exon circularization through the process of alternative splicing [84]. This so-called "trans-splicing" process which was already reported in 1985 is independent of the actual exon sequence but can be modulated by flanking intron structures and has at least been shown to be true for a subset of circRNAs [47, 91].

Repetitive inverted sequences upstream and downstream of the circularizing exon contribute to this circularization process. It has been proposed that the intronic base pairing of these repetetive inverted sequences facilitates the formation of a secondary RNA hairpin structures [92] (Figs. 4.2 and 4.3). These inverted repeat regions have been proven mandatory for circularization of the gene *Sry* in mice, where a 400 nucleotide sequence was minimally required for successful circle formation [9, 55]. In humans progressing next generation sequencing approaches revealed that these inverted repeats, especially *Alu* sequences, are two fold as likely to occur in regions bordering on circularizing exons compared to non-circularizing exons [9, 92]. These retro-transposed genomic elements have previously been demonstrated to contribute to alternative over canonical splicing [93, 94]. At least in higher eukaryotes, these regions seem to play a major role in circularization. For lower eukaryotes like *Saccharomyces*, these repetitive regions are less common, and therefore likely other mechanisms are at play [22, 95]. Studies in *Drosophila* however demonstrated that inverted repeats are not essential for circularization, at least in this organism [62]. Therefore, inverted sequences flanking the exon seem to promote circularization but not be mandatory for formation of all yet known circRNAs.

3.3 Backsplicing Models

The alternative splicing mechanism of backsplicing generates circRNAs by connection of the 3′ end of a containing exon to either their own or a different upstream exon's 5′ region [96]. This leads to an inverted or scrambled order of exons in comparison to the genomic sequence [2, 4, 5, 9] (Fig. 4.2).

Two models for the biogenesis of circRNA backsplicing have been proposed: the lariat model, sometimes also referred to as exon skipping model, and the direct backsplicing model [22, 58, 97] (Fig. 4.3a and b). The essential difference between these two mechanisms is the order in which splicing events do occur. In the lariat model, canonical splicing of the pre-mRNA occurs first, while backspliced circRNAs are products of further processing of lariat intermediates. In the direct backsplicing model on the other hand, circRNAs are generated first. Mechanistically, the two models can be distinguished by the different intermediates that occur during the splice processes. These can be analyzed and detected by their specific steric properties in order to identify which mechanism is responsible for the circularization of a specific circRNA [22, 97]. One method to analyze these properties would be two-dimensional denaturing polyacrylamide gel electrophoresis [95, 98].

According to the lariat model circRNAs are products of previous splicing, in which an alternative exon has been excised from the final mRNA product, ending up in an isolated lariat intermediate. This intermediate RNA molecule is not permanently stable, and spliced lariats are usually quickly degraded within the cell [99]. Debranching endonucleases specifically recognize the 2′ to 5′ phosphodiester bond that is characteristic for lariats and by digestion linearize them. Linear RNAs without protection of a polyadenosine tail or a 5′ 7-methylguanosine cap are easy targets for nucleases.

During generation of circRNAs via the lariat model splicing, a second double lariat intermediate occurs (Fig. 4.3a). First, the splicing of the linear pre-mRNA occurs as per usual, with the downstream branch point adenosine's 2′ hydroxyl group attacking an upstream splice acceptor site beyond an alternative exon. The result of this substitution reaction is a canonical lariat structure containing an exon and linked by a 2′ to 5′ phosphodiester bond. In a second step, this excised lariat undergoes internal backsplicing, and an upstream branch point attacks a downstream splice acceptor, resulting in a double lariat structure. The alternative exon's 3′ hydroxyl group is now able to perform a nucleophilic attack on its own upstream splice acceptor site, resulting in circularization by forming a 3′ to 5′ phosphodiester bond [22]. Circular RNA formation by the lariat model has been demonstrated in the yeast *Schizosaccharomyces pombe* gene Mrsp1. Yeast genomes rarely have repetitive sequences, which may be an explanation for the prevalence of lariat model circularization in these organisms [22, 95, 98]. Interestingly, RNA sequencing also revealed lariat species within human fibroblast RNA preparations [21].

Direct backsplicing presumes that circRNA biogenesis is a process independent from prior exon excision and lariat formation. This model proposes that inverted repeat regions on both sides of the prospectively circularized exon or exons, first results in a secondary stem-loop structure formation within the RNA transcript [21] (Fig. 4.3b). This brings an upstream branch point close to a downstream 5′ splice acceptor site and therefore sterically enables the 2′ hydroxyl group's nucleophilic attack, creating a Y-shaped intermediate. This frees the circularizing exon's 3′ hydroxyl group to in turn attack its 5′ phosphate. The final products of this reaction are a circularized RNA molecule containing the circularizing exon or exons, as well as the pre-mRNA transcript whose secondary hairpin structure now contains a 2′, 5′-phosphodiester linkage which can be removed by canonical splicing resulting in a linear mRNA [22].

The direct backsplicing model may explain the high expression of certain circRNAs, which for some genes even exceeds their linear counterparts, as has been shown for circular and linear RNAs species derived from the KIAA0182 and MAN1A2 genes in immune cells [41]. The majority of the RNAs expressed from these loci showed a scrambled order of exons. Interestingly, this finding was unrelated to alternative splicing, thereby giving evidence for a direct backsplicing mechanism without lariat intermediates. Contribution of intronic pairing of inverted repeat *Alu* sequences has been indicated to promote circularization even without exon skipping in human fibroblasts [21].

Acknowledgments This work was supported by the German Cardiac Society (Deutsche Gesellschaft für Kardiologie (DGK)) to Jes-Niels Boeckel.

Competing Financial Interests The authors declare no competing financial interests.

References

1. Hsu MT, Coca-Prados M (1979) Electron microscopic evidence for the circular form of RNA in the cytoplasm of eukaryotic cells. Nature 280:339–340
2. Nigro JM, Cho KR, Fearon ER et al (1991) Scrambled exons. Cell 64:607–613
3. Zaphiropoulos PG (1997) Exon skipping and circular RNA formation in transcripts of the human cytochrome P-450 2C18 gene in epidermis and of the rat androgen binding protein gene in testis. Mol Cell Biol 17:2985–2993

4. Cocquerelle C, Mascrez B, Hétuin D et al (1993) Mis-splicing yields circular RNA molecules. FASEB J 7:155–160
5. Cocquerelle C, Daubersies P, Majérus MA et al (1992) Splicing with inverted order of exons occurs proximal to large introns. EMBO J 11:1095–1098
6. Burd CE, Jeck WR, Liu Y et al (2010) Expression of linear and novel circular forms of an INK4/ARF-associated non-coding RNA correlates with atherosclerosis risk. PLoS Genet 6:e1001233
7. Aebi M, Weissman C (1987) Precision and orderliness in splicing. Trends Genet 3:102–107
8. Zhang Y, Zhang XO, Chen T et al (2013) Circular intronic long noncoding RNAs. Mol Cell 51:792–806
9. Dubin RA, Kazmi MA, Ostrer H (1995) Inverted repeats are necessary for circularization of the mouse testis Sry transcript. Gene 167:245–248
10. Sanger HL, Klotz G, Riesner D et al (1976) Viroids are single-stranded covalently closed circular RNA molecules existing as highly base-paired rod-like structures. Proc Natl Acad Sci U S A 73:3852–3856
11. Schultz ES, Folsom D (1923) A "spindling-tuber disease" of Irish potatoes. Science 57:149. https://doi.org/10.1126/science.57.1466.149
12. Diener TO (1971) Potato spindle tuber "virus". IV. A replicating, low molecular weight RNA. Virology 45:411–428
13. Stoler MH, Broker TR (1986) In situ hybridization detection of human papillomavirus DNAs and messenger RNAs in genital condylomas and a cervical carcinoma. Hum Pathol 17:1250–1258
14. Summers J, Jones SE, Anderson MJ (1983) Characterization of the genome of the agent of erythrocyte aplasia permits its classification as a human parvovirus. J Gen Virol 64(Pt 11):2527–2532
15. Kos A, Dijkema R, Arnberg AC et al (1986) The hepatitis delta (delta) virus possesses a circular RNA. Nature 323:558–560
16. Soma A, Onodera A, Sugahara J et al (2007) Permuted tRNA genes expressed via a circular RNA intermediate in Cyanidioschyzon merolae. Science 318:450–453
17. Lykke-Andersen J, Aagaard C, Semionenkov M et al (1997) Archaeal introns: splicing, intercellular mobility and evolution. Trends Biochem Sci 22:326–331
18. Salgia SR, Singh SK, Gurha P et al (2003) Two reactions of Haloferax volcanii RNA splicing enzymes: joining of exons and circularization of introns. RNA 9:319–330
19. Danan M, Schwartz S, Edelheit S et al (2012) Transcriptome-wide discovery of circular RNAs in Archaea. Nucleic Acids Res 40:3131–3142
20. Tang TH, Rozhdestvensky TS, d'Orval BC et al (2002) RNomics in Archaea reveals a further link between splicing of archaeal introns and rRNA processing. Nucleic Acids Res 30:921–930
21. Jeck WR, Sorrentino JA, Wang K et al (2013) Circular RNAs are abundant, conserved, and associated with ALU repeats. RNA 19:141–157
22. Barrett SP, Wang PL, Salzman J (2015) Circular RNA biogenesis can proceed through an exon-containing lariat precursor. Elife 4:e07540
23. Diener TO, Raymer WB (1967) Potato spindle tuber virus: a plant virus with properties of a free nucleic acid. Science 158:378–381
24. Flores R, Hernández C, Martínez de Alba AE et al (2005) Viroids and viroid-host interactions. Annu Rev Phytopathol 43:117–139
25. Bonfiglioli, McFadden, Symons RH (1994) In situ hybridization localizes avocado sunblotch viroid on chloroplast thylakoid membranes and coconut cadang cadang viroid in the nucleus. Plant J 6(1):99–103
26. Przybilski R, Gräf S, Lescoute A et al (2005) Functional hammerhead ribozymes naturally encoded in the genome of Arabidopsis thaliana. Plant Cell 17:1877–1885
27. Steger G, Perreault JP (2016) Structure and associated biological functions of viroids. Adv Virus Res 94:141–172
28. Flores R, Grubb D, Elleuch A et al (2011) Rolling-circle replication of viroids, viroid-like satellite RNAs and hepatitis delta virus: variations on a theme. RNA Biol 8(2):200–206
29. Flores R, Gas ME, Molina-Serrano D et al (2009) Viroid replication: rolling-circles, enzymes and ribozymes. Viruses 1:317–334
30. Branch AD, Robertson HD (1984) A replication cycle for viroids and other small infectious RNA's. Science 223:450–455
31. Flores (1999) Viroids with hammerhead ribozymes: some unique structural and functional aspects with respect to other members of the group. Biol Chem 380(7-8):849–854
32. Macnaughton TB, Shi ST, Modahl LE et al (2002) Rolling circle replication of hepatitis delta virus RNA is carried out by two different cellular RNA polymerases. J Virol 76:3920–3927
33. Reid CE, Lazinski DW (2000) A host-specific function is required for ligation of a wide variety of ribozyme-processed RNAs. Proc Natl Acad Sci U S A 97:424–429
34. Petkovic S, Müller S (2015) RNA circularization strategies in vivo and in vitro. Nucleic Acids Res 43:2454–2465
35. Barral PM, Sarkar D, Su ZZ et al (2009) Functions of the cytoplasmic RNA sensors RIG-I and MDA-5: key regulators of innate immunity. Pharmacol Ther 124:219–234
36. Yoneyama M, Onomoto K, Jogi M et al (2015) Viral RNA detection by RIG-I-like receptors. Curr Opin Immunol 32:48–53
37. Abelson J, Trotta CR, Li H (1998) tRNA splicing. J Biol Chem 273:12685–12688
38. Knapp G, Ogden RC, Peebles CL et al (1979) Splicing of yeast tRNA precursors: structure of the reaction intermediates. Cell 18:37–45
39. Lu Z, Filonov GS, Noto JJ, Schmidt CA, Hatkevich TL, Wen Y, Jaffrey SR, Matera AG (2015) Metazoan

tRNA introns generate stable circular RNAs in vivo. RNA 21:1554–1565
40. Suzuki H, Zuo Y, Wang J et al (2006) Characterization of RNase R-digested cellular RNA source that consists of lariat and circular RNAs from pre-mRNA splicing. Nucleic Acids Res 34:e63
41. Salzman J, Gawad C, Wang PL, Lacayo N, Brown PO (2012) Circular RNAs are the predominant transcript isoform from hundreds of human genes in diverse cell types. PLoS One 7:e30733
42. Lasda E, Parker R, Parker ROY (2014) Circular RNAs : diversity of form and function. RNA 20:1829–1842
43. Grabowski PJ, Zaug AJ, Cech TR (1981) The intervening sequence of the ribosomal RNA precursor is converted to a circular RNA in isolated nuclei of Tetrahymena. Cell 23:467–476
44. Durovic P, Dennis PP (1994) Separate pathways for excision and processing of 16S and 23S rRNA from the primary rRNA operon transcript from the hyperthermophilic archaebacterium Sulfolobus acidocaldarius: similarities to eukaryotic rRNA processing. Mol Microbiol 13:229–242
45. Dennis PP, Ziesche S, Mylvaganam S (1998) Transcription analysis of two disparate rRNA operons in the halophilic archaeon Haloarcula marismortui. J Bacteriol 180:4804–4813
46. Kruger K, Grabowski PJ, Zaug AJ et al (1982) Self-splicing RNA: autoexcision and autocyclization of the ribosomal RNA intervening sequence of Tetrahymena. Cell 31:147–157
47. Grabowski PJ, Seiler SR, Sharp PA (1985) A multicomponent complex is involved in the splicing of messenger RNA precursors. Cell 42:345–353
48. Qian L, Vu MN, Carter M, Wilkinson MF (1992) A spliced intron accumulates as a lariat in the nucleus of T cells. Nucleic Acids Res 20:5345–5350
49. Tabak HF, Van der Horst G, Smit J et al (1988) Discrimination between RNA circles, interlocked RNA circles and lariats using two-dimensional polyacrylamide gel electrophoresis. Nucleic Acids Res 16:6597–6605
50. Hedberg A, Johansen SD (2013) Nuclear group I introns in self-splicing and beyond. Mob DNA 4:1
51. Lambowitz AM, Zimmerly S (2011) Group II introns: mobile ribozymes that invade DNA. Cold Spring Harb Perspect Biol 3:1–19
52. Lehmann K, Schmidt U (2003) Group II introns: structure and catalytic versatility of large natural ribozymes. Crit Rev Biochem Mol Biol 38:249–303
53. Li-Pook-Than J, Bonen L (2006) Multiple physical forms of excised group II intron RNAs in wheat mitochondria. Nucleic Acids Res 34:2782–2790
54. Soesanto W, Lin HY, Hu E et al (2009) Mammalian target of rapamycin is a critical regulator of cardiac hypertrophy in spontaneously hypertensive rats. Hypertension 54:1321–1327
55. Capel B, Swain A, Nicolis S et al (1993) Circular transcripts of the testis-determining gene Sry in adult mouse testis. Cell 73:1019–1030
56. Pasman Z, Been MD, Garcia-Blanco MA (1996) Exon circularization in mammalian nuclear extracts. RNA 2:603–610
57. Caldas C, So CW, MacGregor A et al (1998) Exon scrambling of MLL transcripts occur commonly and mimic partial genomic duplication of the gene. Gene 208:167–176
58. Zaphiropoulos PG (1996) Circular RNAs from transcripts of the rat cytochrome P450 2C24 gene: correlation with exon skipping. Proc Natl Acad Sci U S A 93:6536–6541
59. Baralle FE, Giudice J (2017) Alternative splicing as a regulator of development and tissue identity. Nat Rev Mol Cell Biol 18:437–451
60. Salzman J, Chen RE, Olsen MN et al (2013) Cell-type specific features of circular RNA expression. PLoS Genet 9:e1003777
61. Zhang XO, Dong R, Zhang Y et al (2016) Diverse alternative back-splicing and alternative splicing landscape of circular RNAs. Genome Res 26:1277–1287
62. Zhang XO, Wang HB, Zhang Y et al (2014) Complementary sequence-mediated exon circularization. Cell 159:134–147
63. Chen LL (2016) The biogenesis and emerging roles of circular RNAs. Nat Rev Mol Cell Biol 17:205–211
64. 1000 Genomes Project Consortium, Auton A, Brooks LD et al (2015) A global reference for human genetic variation. Nature 526:68–74
65. Lynch M, Conery JS (2003) The origins of genome complexity. Science 302:1401–1404
66. Khanduja JS, Calvo IA, Joh RI, Hill IT, Motamedi M (2016) Nuclear noncoding RNAs and genome stability. Mol Cell 63:7–20
67. Böhmdorfer G, Wierzbicki AT (2015) Control of chromatin structure by long noncoding RNA. Trends Cell Biol 25:623–663
68. Magistri M, Faghihi MA, St Laurent G, Wahlestedt C (2012) Regulation of chromatin structure by long noncoding RNAs: focus on natural antisense transcripts. Trends Genet 28:389–396
69. Vance KW, Ponting CP (2014) Transcriptional regulatory functions of nuclear long noncoding RNAs. Trends Genet 30:348–355
70. Wilusz JE, Sunwoo H, Spector DL (2009) Long noncoding RNAs: functional surprises from the RNA world. Genes Dev 23:1494–1504
71. Bonasio R, Shiekhattar R (2014) Regulation of transcription by long noncoding RNAs. Annu Rev Genet 48:433–455
72. Lee TI, Young RA (2000) Transcription of eukaryotic protein-coding genes. Annu Rev Genet 34:77–137
73. Nogales E, Louder RK, He Y (2017) Structural insights into the eukaryotic transcription initiation machinery. Annu Rev Biophys 46:59–83
74. Jurado AR, Tan D, Jiao X et al (2014) Structure and function of pre-mRNA 5′-end capping quality control and 3′-end processing. Biochemistry 53:1882–1898
75. Moore MJ, Proudfoot NJ (2009) Pre-mRNA processing reaches back to transcription and ahead to translation. Cell 136:688–700

76. Furuichi Y, Shatkin AJ (2000) Viral and cellular mRNA capping: past and prospects. Adv Virus Res 55:135–184
77. Shandilya J, Roberts SGE (2012) The transcription cycle in eukaryotes: from productive initiation to RNA polymerase II recycling. Biochim Biophys Acta 1819:391–400
78. Wahle E, Rüegsegger U (1999) 3′-End processing of pre-mRNA in eukaryotes. FEMS Microbiol Rev 23:277–295
79. Frendewey D, Keller W (1985) Stepwise assembly of a pre-mRNA splicing complex requires U-snRNPs and specific intron sequences. Cell 42:355–367
80. Burset M, Seledtsov IA, Solovyev VV (2000) Analysis of canonical and non-canonical splice sites in mammalian genomes. Nucleic Acids Res 28:4364–4375
81. Hwang DY, Cohen JB (1996) Base pairing at the 5′ splice site with U1 small nuclear RNA promotes splicing of the upstream intron but may be dispensable for slicing of the downstream intron. Mol Cell Biol 16:3012–3022
82. Liu J, Liu T, Wang X, He A (2017) Circles reshaping the RNA world: from waste to treasure. Mol Cancer 16:58
83. Huang S, Yang B, Chen BJ et al (2017) The emerging role of circular RNAs in transcriptome regulation. Genomics 109:401–407
84. Starke S, Jost I, Rossbach O, Schneider T et al (2015) Exon circularization requires canonical splice signals. Cell Rep 10:103–111
85. Gilbert W (1978) Why genes in pieces? Nature 271:501
86. Valdivia HH (2007) One gene, many proteins: alternative splicing of the ryanodine receptor gene adds novel functions to an already complex channel protein. Circ Res 100:761–763
87. George CH, Rogers SA, Bertrand BMA et al (2007) Alternative splicing of ryanodine receptors modulates cardiomyocyte Ca2+ signaling and susceptibility to apoptosis. Circ Res 100:874–883
88. Lanner JT, Georgiou DK, Joshi AD, Hamilton SL (2010) Ryanodine receptors: structure, expression, molecular details, and function in calcium release. Cold Spring Harb Perspect Biol 2:a003996
89. Mabon SA, Misteli T (2005) Differential recruitment of pre-mRNA splicing factors to alternatively spliced transcripts in vivo. PLoS Biol 3:1893–1901
90. Boeckel J-N, Jaé N, Heumüller AW et al (2015) Identification and characterization of hypoxia-regulated endothelial circular RNA. Circ Res 117(10):884–890
91. Konarska MM, Padgett RA, Sharp PA (1985) Trans splicing of mRNA precursors in vitro. Cell 42:165–171
92. Liang D, Wilusz JE (2014) Short intronic repeat sequences facilitate circular RNA production. Genes Dev 28:2233–2247
93. Lev-Maor G, Ram O, Kim E et al (2008) Intronic Alus influence alternative splicing. PLoS Genet 4:1–12
94. Hu S, Wang X, Shan G (2016) Insertion of an Alu element in a lncRNA leads to primate-specific modulation of alternative splicing. Nat Struct Mol Biol 23:1011–1019
95. Barrett SP, Salzman J (2016) Circular RNAs: analysis, expression and potential functions. Development 143:1838–1847
96. Braun S, Domdey H, Wiebauer K (1996) Inverse splicing of a discontinuous pre-mRNA intron generates a circular exon in a HeLa cell nuclear extract. Nucleic Acids Res 24:4152–4157
97. Chen LL, Yang L (2015) Regulation of circRNA biogenesis. RNA Biol 12:381–388
98. Schindewolf C, Braun S, Domdey H (1996) In vitro generation of a circular exon from a linear pre-mRNA transcript. Nucleic Acids Res 24:1260–1266
99. Hesselberth JR (2013) Lives that introns lead after splicing. Wiley Interdiscip Rev RNA 4:677–691

Circular RNAs Biogenesis in Eukaryotes Through Self-Cleaving Hammerhead Ribozymes

5

Marcos de la Peña

Abstract

Circular DNAs are frequent genomic molecules, especially among the simplest life beings, whereas circular RNAs have been regarded as weird nucleic acids in biology. Now we know that eukaryotes are able to express circRNAs, mostly derived from back-splicing mechanisms, and playing different biological roles such as regulation of RNA splicing and transcription, among others. However, a second natural and highly efficient pathway for the expression in vivo of circRNAs has been recently reported, which allows the accumulation of abundant small (100–1000 nt) non-coding RNA circles through the participation of small self-cleaving RNAs or ribozymes called hammerhead ribozymes. These genome-encoded circRNAs with ribozymes seem to be a new family of small and nonautonomous retrotransposable elements of plants and animals (so-called retrozymes), which will offer functional clues to the biology and evolution of circular RNA molecules as well as new biotechnological tools in this emerging field.

Keywords

Circular RNA · Retrotransposons · Ribozyme

M. de la Peña (✉)
IBMCP (CSIC-UPV), Valencia, Spain
e-mail: rivero@ibmcp.upv.es

Abbreviations

circRNA	circular RNA
HHR	hammerhead ribozyme
LTR	long terminal repeat
PBS	primer binding site
PPT	polypurine tract
RT	retrotranscriptase
TSD	target site duplication

1 Introduction

Genomic circular DNAs are frequent macromolecules among simple organisms, from small prokaryotic plasmids to the larger genomes of many bacteriophages or viruses, bacteria, archaea and plastids. On the other hand, circular RNAs have been regarded as very rare nucleic acids in biology till very recently. Now we know that numerous life beings express stable circRNAs [1], and among them, it is noteworthy the recent discovery of a myriad of splicing-derived circRNAs in eukaryotes [2–4] with diverse functions in regulation of splicing [5] and transcription [6], small RNAs biology [7], RNA-mediated inheritance and epigenetics [8] and some others, as described in this book. However, it has been recently reported that eukaryotes have a second natural pathway that allows the expression in vivo of abundant circular RNAs [9, 10]. This alternative

J. Xiao (ed.), *Circular RNAs*, Advances in Experimental Medicine and Biology 1087,
https://doi.org/10.1007/978-981-13-1426-1_5

mechanism does not require any classical spliceosome reaction but the involvement of small self-cleaving RNAs or ribozymes called hammerhead ribozymes (HHRs) [11].

The finding of catalytic RNAs or ribozymes more than 30 years ago [12, 13] propelled the revolution in the RNA field and started with the uninterrupted discovery of the many different roles and capabilities of this macromolecule in biology. Moreover, the ground-breaking discovery of ribozymes strongly supported the hypothesis of the prebiotic RNA world [14], where RNAs carried out both informative (RNA genomes) and catalytic (ribozymes) roles. Somehow, it is thought that these primal RNA molecules would have evolved to present organisms based in DNA and proteins as the genetic material and catalytic machines, respectively [15–17]. Proofs supporting this hypothesis are the existence of RNA genomes among the simplest organisms (such as RNA viruses and viroids), as well as catalytic and regulatory ribofunctions among all living beings, where RNA itself is the final molecule in charge of the activity. A remarkable example of a catalytic RNA would be the central machine of life, the ribosome [18], which is the universal ribozyme that catalyses the peptide bond formation during protein synthesis in all known living entities. This fact allows to connect DNA and proteins through a catalytic RNA, which offers a solution to the chicken or the egg (or more precisely, the DNA or the protein) causality dilemma. Other key ribozymes and regulatory RNAs considered as ancient relics of the prebiotic RNA world would be the autocatalytic introns [12] and small ribozymes [19], the RNase P [13], the spliceosome [20], the riboswitches [21], most noncoding RNAs (such as those small RNA guides found in the CRISPR [22] and the RNAi [23] pathways) or even the circRNAs described in this book, which altogether confirm the extraordinary potential of any RNA molecule present in a living organism.

2 The Family of Small Self-Cleaving RNAs and the Singular Case of the Hammerhead Ribozyme

Among the simplest ribozymes so far described, it can be highlighted the enigmatic group of small (50–200 nt) self-cleaving RNAs, which all catalyse a sequence-specific intramolecular reaction of transesterification. This reaction starts by a nucleophilic attack of the 2′ oxygen to the adjacent 3′ phosphate, resulting in cleavage of the phosphodiester bond to form two RNA products with a 5′-hydroxyl and a 2′,3′-cyclic phosphate ends each (Fig. 5.1a). The family of small self-cleaving ribozymes is composed so far by nine different classes: hammerhead (HHR) [24, 25], hairpin (HPR) [26], human hepatitis-δ (HDV) [27], Varkud satellite (VS) [28], GlmS [29], twister [30], twister sister, hatchet and pistol [31] ribozymes. The HHR was the first discovered and one of the best known members of the family of small self-cleaving ribozymes. It is composed of a conserved catalytic core of 15 nucleotides surrounded by three double helixes (I to III), which adopt a γ-shaped fold where helix I interacts with helix II through tertiary interactions required for efficient in vivo activity (Fig. 5.1b) [32–34]. There are three possible circularly permuted topologies for the HHR, named type I, type II or type III, depending on the open-ended helix (Fig. 5.1c). The HHR were first found encoded in the small circRNA genomes of a group of infectious subviral agents of plants, such as viral RNA satellites and viroids [24, 25], where it catalyses a self-cleavage transesterification reaction required for the rolling-circle replication of these pathogens. Surprisingly, few other examples of HHR motifs were also found encoded in the genomes of some unrelated eukaryotes such as newts, trematodes or even some mammals, among others [35–39]. In 2010, different labs reported the widespread occurrence of HHR

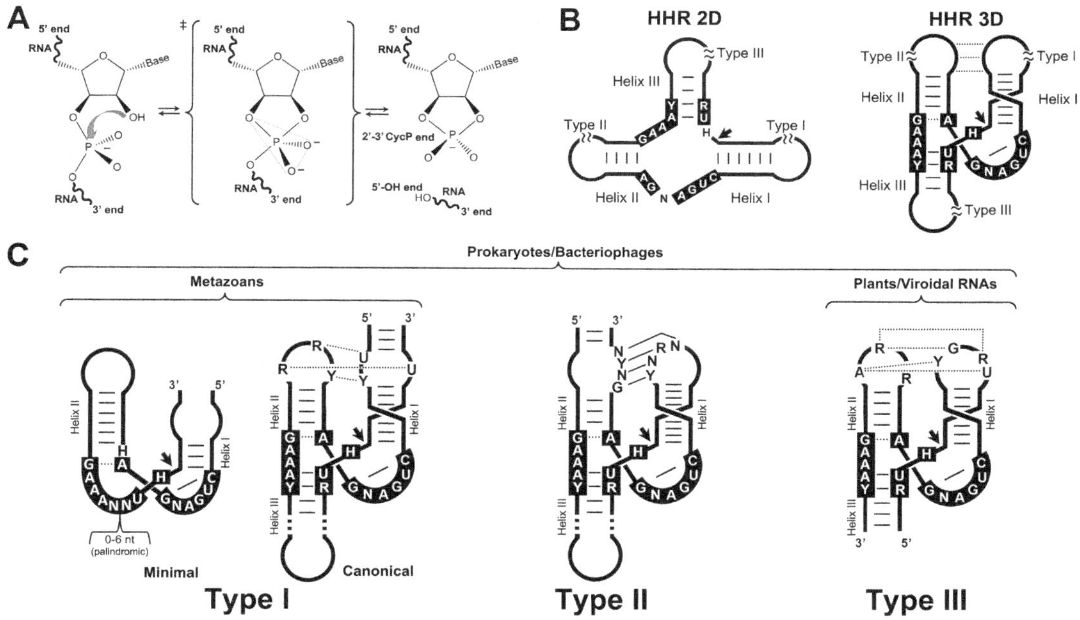

Fig. 5.1 RNA self-cleavage by the hammerhead ribozyme (**a**) Mechanism of internal transesterification in the RNA. The cleavage reaction starts with an attack of the 2′ hydroxyl to the 3′ phosphate, followed by a bipyramidal transition state. The cleavage products are a 2′,3′-cyclic phosphate at the 5′ RNA product and a 5′-hydroxyl at the 3′ RNA product. (**b**) Classic two (left)- and three (right)-dimensional diagrams of the hammerhead ribozyme motif. Black boxes indicate the highly conserved nucleotides (in white letters) at the catalytic core. (**c**) Representation of the three possible hammerhead ribozyme topologies (types I, II and III). Dotted and continuous lines refer to non-canonical and Watson-Crick base pairs, respectively. The three topologies have been reported in the genomes of bacteriophages and prokaryotes. Type-I hammerheads are mostly found in metazoan genomes, whereas typical type-III motifs are found in the plants and their infectious circRNAs (viroidal RNAs). N stands for any nucleotide, whereas R stands for purines (A or G), Y for pyrimidines (U or C) and H for either A, U or C

motifs in prokaryotic and eukaryotic genomes [40–43], including our own genome [44], which confirmed that the HHR was a ubiquitous catalytic RNA motif in all life kingdoms [45, 46]. Interestingly, the occurrence of genomic HHR motifs along the tree of life seems to follow a kind of structural or functional compartmentalization. This way, anyone of the three topologies of the HHR motif (types I, II and III, Fig. 5.1c) can be frequently detected in the genomes of prokaryotes and bacteriophages. However, metazoan genomes mostly show type-I HHR motifs, whereas plant genomes, as well as their subviral agents, almost exclusively show the presence of type-III HHR motifs.

Other small self-cleaving RNA motifs have been also found widespread in DNA genomes, such as HDV [47] or twister ribozymes [30], which confirms that small catalytic RNAs would be more frequent than previously thought. Although the precise biological roles of all these genomic self-cleaving ribozymes are still under study, a direct connection with retrotransposons and other mobile genetic elements has been reported for most of them [10, 48–51].

3 Hammerhead Ribozymes in Plant Genomes Promote circRNA Expression: The Retrozymes

Two examples of type-III HHR motifs were originally reported in the genome of A. thaliana [35]. Numerous copies of this ribozyme were also detected in the genomes of diverse flowering plants [40]. In many instances, these HHRs have been found as tandem repeats of several

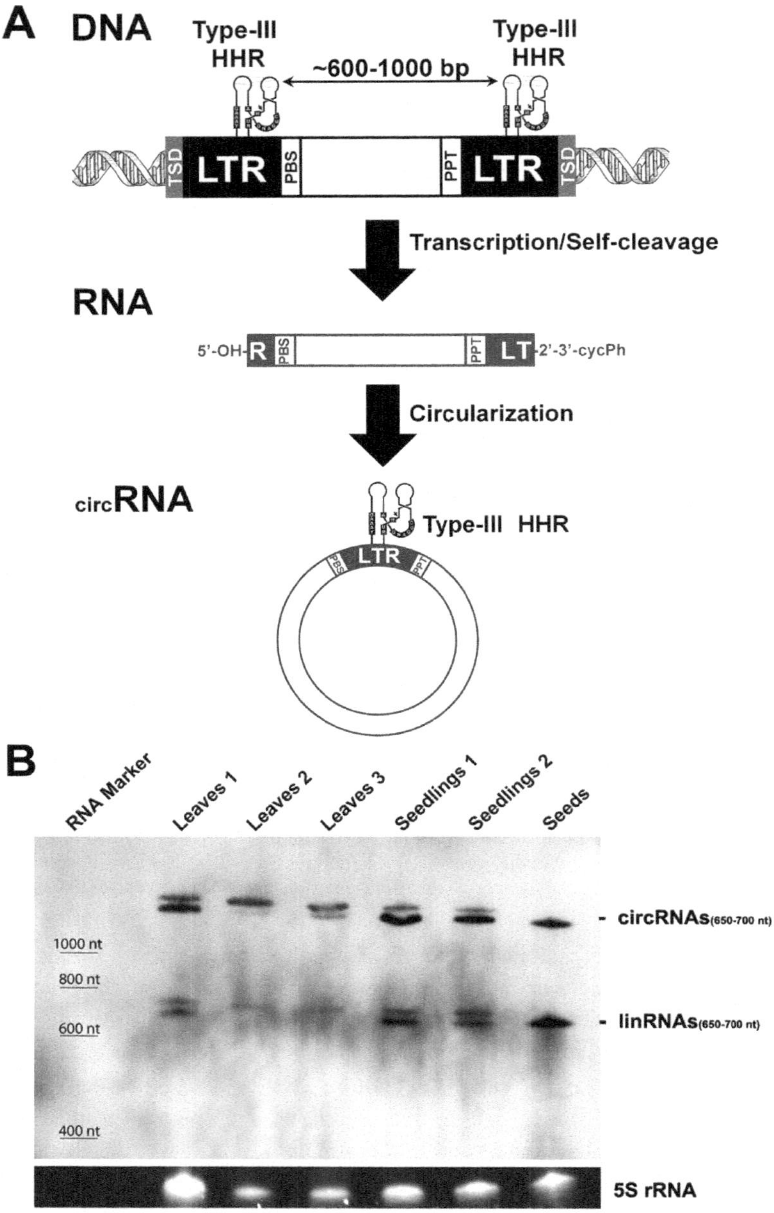

Fig. 5.2 Genomic plant retrozymes (**a**) Schematic representation (top) of a full genomic retrozyme element of plants. Target side duplications (TSDs) delimiting the retrozyme are shown in grey boxes. Long terminal repeats (LTRs) are shown in black boxes. The positions of the primer binding site (PBS), the polypurine tract (PPT), the hammerhead ribozymes (HHR) and the typical sizes encompassed by the ribozymes are indicated. The resulting self-cleaved retrozyme RNA after transcription (middle) and circularization (bottom) is indicated. (**b**) An example of a northern blot analysis of RNA extracts (~30 μg each) from physic nut (*Jatropha curcas*) leaves, young seedlings and seeds. Samples were run on a 5% denaturing PAGE and were detected using a digoxigenin-labelled *J. curcas* retrozyme fragment as a probe, which revealed the presence of both circular and linear RNA forms in each plant tissue as indicated at the right. Ethidium bromide staining of the 5S rRNA is shown at the bottom as a loading control

motifs (usually two or three) separated by a few hundred base pairs. These observations have been recently extended in our lab, and we have reported the occurrence of hundreds of type-III HHRs in more than 40 plant species [10]. Comparative genomics revealed that sequences flanked by tandem HHR motifs sized from 600 to 1000 bp with almost no identity. However, these genomic repetitive elements show a similar topology: they are delimited by 4 bp target site duplications (TSDs), whereas HHRs are embedded in direct long terminal repeats (LTRs) of ~350 bp. LTRs delimit a central region (~300–700 bp), which begins with the primer binding site (PBS, corresponding to the tRNAMet) and finishes with the polypurine tract (PPT) sequences characteristic of LTR retrotransposons [52] (Fig. 5.2a). Altogether, these elements

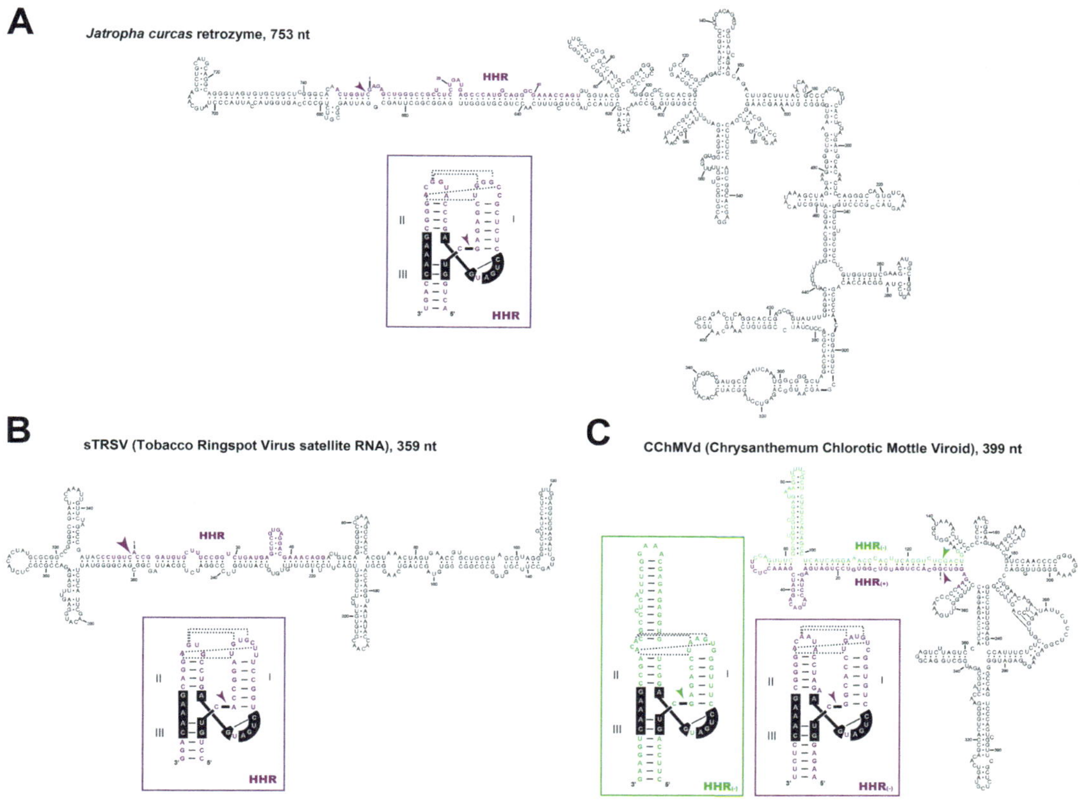

Fig. 5.3 Minimum free energy secondary structure predictions for (**a**) a retrozyme circRNA of *Jatropha curcas* (Entry KX273075.1), (**b**) the Nepovirus satellite RNA sTRSV (Entry M14879.1) and (**c**) the viroid CChMVd (Entry AJ878085.1). HHR sequences are shown in purple (positive polarity) and green (negative polarity). The corresponding structure of the HHRs motifs are shown under each circRNA structure, and dotted lines indicate predicted tertiary interactions between HHR loops based on previous models [60, 61]. Self-cleavage sites are indicated with arrowheads. Kissing-loop interactions described for CChMVd [62] are shown. Numbering for each circRNA starts at the self-cleavage site

were classified as a new family of nonautonomous retrotransposons with hammerhead ribozymes (so-called retrozymes) similar to other nonautonomous retroelements of plants like TRIMs [53] and SMARTs [54]. Most likely, autonomous retrotransposons of the Ty3-Gypsy family would mobilize the small retrozymes based on the sequence similarities (PBS and 5′ and 3′ LTR ends) between both types of retroelements.

Northern blot analysis and RT-PCR experiments of diverse somatic and reproductive tissues from several plant species, such as physic nut, strawberry, eucalyptus or citrus plants, revealed the presence of high levels of circular and linear RNAs (up to 1 ng per μg of total RNA) of the precise size encompassed by the HHR motifs (Fig. 5.2b), which strongly indicates an RNA self-processing activity by the ribozymes during in vivo transcription followed by RNA circularization. Although sequence identity between retrozymes from non-related plant species is very low, secondary structure predictions for these circRNAs show similar architecture and high stability (Fig. 5.3a). These structured circRNAs with type-III hammerhead ribozymes highly resemble those infectious circRNAs of plants, such as viral satellite RNAs and viroids (Fig. 5.3b and c), which indicates a clear evolutionary relationship between all of them [9].

4 Tandem Copies of Hammerhead Ribozymes in Metazoan Genomes

Previous bioinformatic searches in metazoan genomes have also revealed the widespread occurrence of the HHR motif in animals [10, 40, 45]. As observed in plants, these ribozymes are usually found in close tandem copies, suggesting that genomic retrozymes in animals may express similar circRNA molecules with HHRs, which would also accumulate in metazoan transcriptomes. However, several differences can be highlighted between plant and animal retrozymes. On the one hand, none of the characteristic sequences of plant retrozymes, such as LTRs, PBS or PPT, are present in their animal counterparts. Moreover, whereas plant retrozymes only show a few HHR motifs (usually, just two copies per retrozyme) of the type III, ribozymes in metazoan retrozymes occur as many copies (dozens to even hundreds) of type-I HHR motifs (Fig. 5.4). These type-I HHRs not only show a characteristic set of tertiary interactions but a very short or even no helix III at all, which indicates that these ribozymes may require in many instances the adoption of dimeric HHR conformations to self-cleave efficiently [55, 24]. The minimal type-I HHRs of metazoan retrozymes highly resemble those described in the pseudo-LTRs of the autonomous Penelope-like retroelements (PLEs) of metazoans and other eukaryotes [48], which somehow links these two families of retrotransposons. On the other hand, animal retrozymes are composed of smaller minimal repeats in tandem (150–300 bp), indicating that the expected animal circRNAs are also smaller than those described in plants. These repeats, however, are frequently, but not always, flanked by TSDs as well, although these are slightly larger (8–12 bp) than those found in plant retrozymes (4 bp) [9, 10].

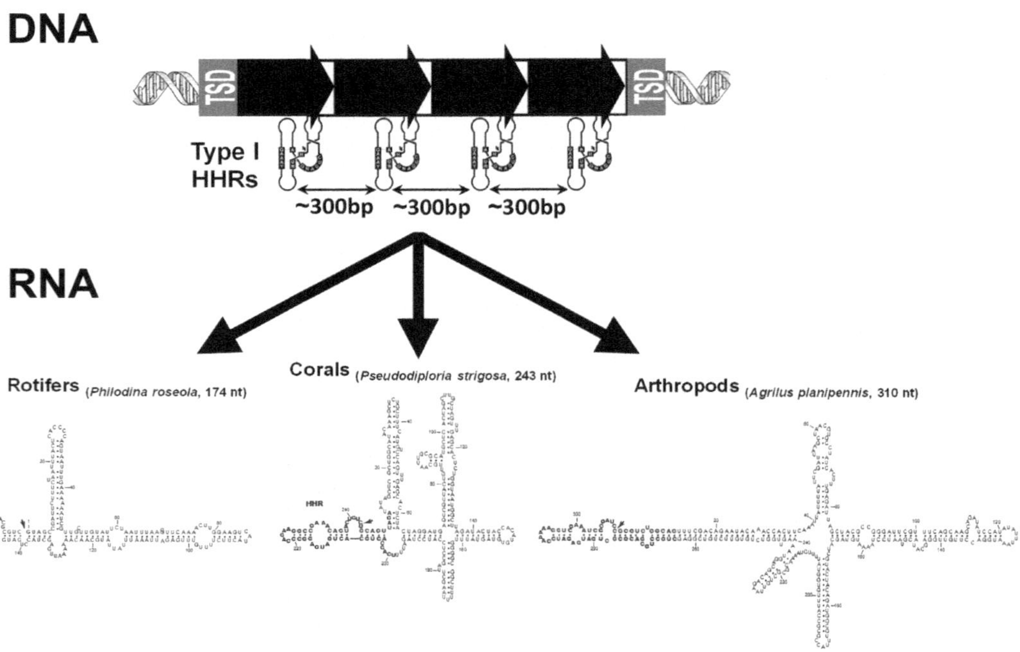

Fig. 5.4 Metazoan retrozymes
Schematic representation (top) of a typical genomic retrozyme element present in metazoan genomes.
Target side duplications (TSDs) delimiting the retrozyme are shown in grey boxes. Tandem repeats of around 300 bp are indicated with arrows. Typical type-I HHRs are shown. Minimum free energy secondary structure prediction of three examples of circRNAs derived from metazoan retrozymes (rotifers, corals and arthropods) are shown (bottom). HHR sequences in the circRNAs are shown in purple letters, and the self-cleavage sites are indicated with arrowheads. Numbering starts after the HHR self-cleavage site

Recent analysis done in our lab with diverse retrozyme-containing metazoans has confirmed that, as suspected, these organisms accumulate abundant circRNAs in most of the analysed tissues, in a similar way as described for plants (De la Peña and Cervera, to be published). Altogether, the resulting landscape offered by genomic HHRs (either type I or III) in eukaryotes indicates that close tandem copies of this ribozyme allows the expression of small circRNAs (100–1000 nt). Although the presence of other small self-cleaving ribozymes in eukaryotic genomes have been described, such as HDV and twister ribozymes [30, 47], the characteristic occurrence in close tandem repeats seems to be exclusively restricted to the case of the hammerhead ribozyme.

5 A Proposed Mechanism for the Expression and Spreading of Eukaryotic circRNAs with Hammerhead Ribozymes

Retrozymes are a new and atypical group of non-autonomous eukaryotic retroelements with self-cleaving hammerhead ribozymes. In plants, genomic retrozymes resemble other small nonautonomous LTR retrotransposons such as TRIMs [53] and SMARTs [54]. As nonautonomous retrotransposons, retrozymes do not show protein-coding regions but self-cleaving HHR motifs in their LTRs, which, most likely, are responsible of the accumulation in vivo of circular and linear RNAs of the precise size encompassed by the HHRs. Regarding the life cycle of retrozymes, the most plausible model would start with the transcription of the genomic retrozyme. Similar retroelements, such as TRIMs or autonomous LTR retrotransposons, are known to be generally transcribed by RNA Pol II, although examples of RNA Pol III-transcribed retrotransposons have been also reported [56]. Plant and metazoan retrozymes do not seem to contain any recognizable promoter, and in consequence, a feasible hypothesis could be that retrozymes may undergo Pol-driven (either I, II or III) read-through transcription depending on tissues and/or their genomic location. In any case, nascent RNA transcripts would follow co-transcriptional self-processing by tandem self-cleaving HHRs, producing linear RNAs with 5′-OH and 2′,3′-cyclic-phosphate ends. Whereas the step of self-cleavage is expected to occur with high efficiency for plant retrozymes carrying type-III HHRs, in the case of metazoan retrozymes, self-cleavage frequently requires the adoption of a dimeric conformation of minimal type-I HHRs, which is expected to be slightly less efficient than the monomeric version [55]. As summarized in Fig. 5.5, covalent circularization of the resulting self-cleaved RNAs through either the HHR itself or a host RNA ligase factor [57] would finish in stable circRNAs. As the most plausible model, these circRNAs are the final template for retrotranscription, whereas linear retrozyme RNAs would be intermediaries and/or by-products of the circRNAs. In the case of plants, circRNAs derived from LTR-like retrozymes could be primed by any cellular tRNAMet through their PBS motifs. Then, retrotranscriptases encoded by Ty3-Gypsy LTR retrotransposons would produce cDNAs of different lengths, thanks to the circular nature of the RNA template. In the case of metazoan retrozymes, retrotranscriptases encoded by non-LTR retrotransposon (such as PLEs or LINEs) would be responsible of carrying out this latter step of cDNA synthesis. Finally, the resulting cDNAs would be integrated in new genomic locations through the machinery of the autonomous retrotransposons (Fig. 5.5). A last question to be addressed is related to the very high levels of circRNAs with HHRs detected in most organisms analysed. Genomic retrozymes are frequently found as many copies (from dozens to thousands of repeats) within a given genome, which suggests that even low transcription activity would result in abundant levels of circRNAs. However, most of the obtained data indicates that only a few retrozyme copies would be transcriptionally active [10], which suggests that the higher stability of these structured circRNAs with ribozymes compared with linear RNAs would be the reason of their high levels of accumulation in vivo. Moreover, the presence of a

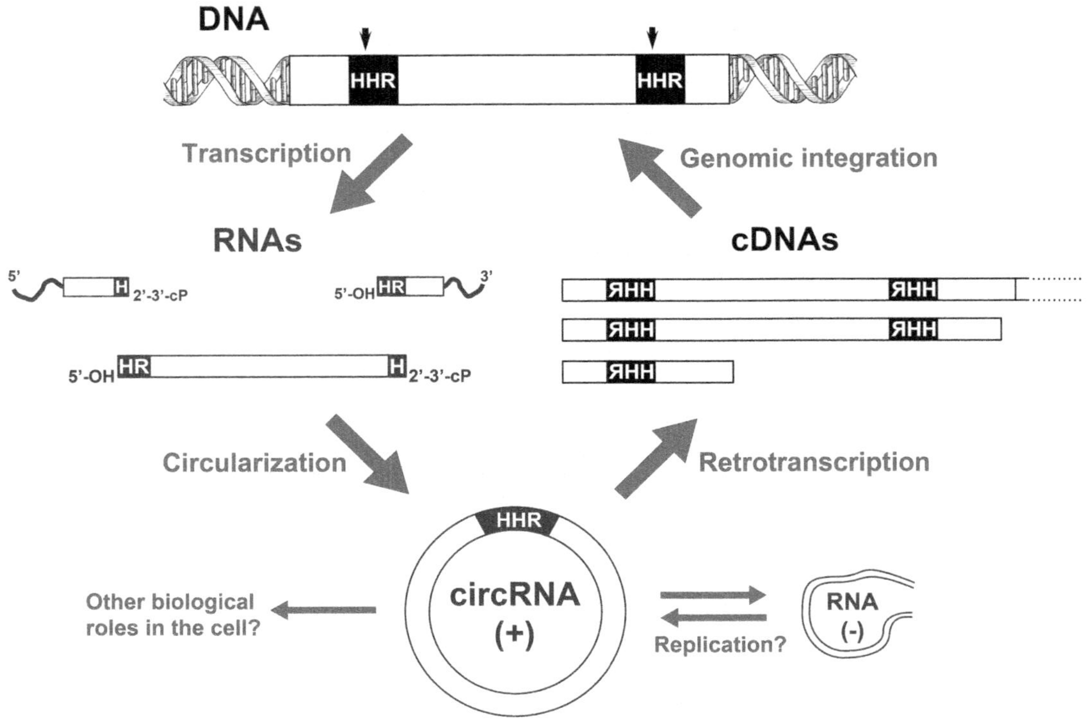

Fig. 5.5 Model for the life cycle of retrozymes
A full genomic retrozyme containing at least two HHRs in tandem is transcribed (top), and the resulting RNA would self-process through the HHRs to give a linear RNA with 5′-OH and 2′,3′-cyclic phosphate ends. The linear RNA would be circularized through an RNA ligase activity, and the resulting circRNA(+) could be recognized for either endogenous RNA polymerases (replication cycles), other cell factors (new biological roles), or retrotranscriptases encoded by autonomous retrotransposons. In the latter case, the resulting cDNAs from retrotranscription of a circular RNA template would have different lengths depending on the processivity of the retrotranscriptase. In a final step, the machinery of the retrotransposon would integrate the retrozyme DNAs at new genomic loci

high sequence heterogeneity observed for a population of circRNAs in a given organism, together with the presence of retrozyme RNAs of the negative polarity, also suggests the intriguing possibility of replication of the circRNAs through endogenous polymerases.

6 Functional and Biotechnological Applications of circRNAs with Ribozymes

Retrotransposons, and mobile genetic elements in general, constitute a major fraction of nuclear genomes of most eukaryotes. Historically, these genomic sequences have been regarded as junk DNA, but now we know that retroelements are major drivers of genome evolution with a role in shaping the genomes that they inhabit. In this regard, genomic retrozymes and their associated circRNAs would have similar evolutionary impact as any other retroelement. However, the atypically high accumulation levels of RNA circles encoded by genomic retrozymes in the transcriptomes of most eukaryotic tissues, either somatic or reproductive, suggest that other biological roles can be possible. In this regard, several genic circRNAs have been found to play a role as microRNA sponges [7], and a comparable role for retrozyme circRNAs would be feasible. Moreover, the highly structured circRNAs derived from genomic retrozymes are suitable to be recognized and processed by the RNAi machinery of the cell, and, consequently, these abundant circRNAs with ribozymes would be potential templates for the production of miRNA/siRNAs with specific regulatory roles in the biology of the organisms where they are expressed. At the same time, we already know many examples of co-option or domestication of the transposable elements by their hosts as adaptations to diverse problems [58]. Usually, these domestications are performed with transposon-derived proteins, but also small ribozymes such as 3′ UTR HHRs [39] and intronic HHRs [44] or HDV ribozymes [59] seem to be examples of retroelement domestication. Consequently, circRNAs with HHRs in some organisms could have been specifically co-opted to play precise functions, a possibility that should be studied in the future.

Regarding the biotechnological applications of tandem small ribozymes in the expression of circRNAs, it has to be pointed out that the mechanism of backsplicing described for the synthesis of most genic circRNAs seems to be a complex pathway, which still requires deeper study in order to fully understand and use for practical applications. In this regard, our current knowledge about small ribozymes allow us to design much easier approaches for the expression of circRNAs, which, moreover, could reach higher accumulation levels as observed for most eukaryotic retrozymes so far analysed. Moreover, in vitro synthesis of circRNAs through self-cleaving ribozymes may also offer a straightforward approach for the production and study of specific genic circRNAs from eukaryotes.

7 Conclusions and Future Prospects

Genomic retrozymes are a new family of eukaryotic retrotransposons, which spread through circRNAs with hammerhead ribozymes among plant and animal genomes [10]. In plants, retrozymes seem to be mostly restricted to eudicots, although the presence of putative retrozymes with tandem HHR copies were also detected in some monocots, primitive land plants (like the spikemoss *Selaginella moellendorffii*) and algae (such as *Chlamydomonas reinhardtii*) [10, 40]. However, most of these HHRs in the genomes of primitive plants are type-I motifs, which are more related to those found in metazoan than in angiosperm genomes. This observation would indicate that retrozymes in flowering plants may have a different origin, either due to a de novo origin in angiosperms by chance or through horizontal transfer from other organisms containing type-III HHRs such as bacteria [40–42]. In any case, all these data suggest that genome-encoded circRNAs with HHRs would be more frequent molecules in eukaryotic transcriptomes than previously thought. Moreover, circRNAs with type-III HHRs in plants allow to propose an evolutionary path for the origin of the small infectious circRNAs with HHRs of plants (viroids and viral satellite RNAs), which may come by chance from the abundant reservoirs of circRNAs present in plant transcriptomes [9]. In contrast, metazoan retrozymes with type-I HHRs seem to indicate that this HHR topology would be more efficient for circRNA expression in animals. Future in vivo experiments will help us to better understand this new tool in the knowledge of the biology and biotechnology of eukaryotic circRNAs.FundingFunding for this work was provided by the Ministerio de Economía y Competitividad of Spain and FEDER funds (BFU2014-56094-P and BFU2017-87370-P).

References

1. Lasda E, Parker R (2014) Circular RNAs: diversity of form and function. RNA 20(12):1829–1842
2. Salzman J, Gawad C, Wang PL et al (2012) Circular RNAs are the predominant transcript isoform from hundreds of human genes in diverse cell types. PloS One 7(2):e30733
3. Jeck WR, Sorrentino JA, Wang K et al (2013) Circular RNAs are abundant, conserved, and associated with ALU repeats. RNA 19(2):141–157
4. Wang PL, Bao Y, Yee MC et al (2014) Circular RNA is expressed across the eukaryotic tree of life. PloS One 9(6):e90859
5. Ashwal-Fluss R, Meyer M, Pamudurti NR et al (2014) circRNA biogenesis competes with pre-mRNA splicing. Mol Cell 56(1):55–66
6. Memczak S, Jens M, Elefsinioti A et al (2013) Circular RNAs are a large class of animal RNAs with regulatory potency. Nature 495(7441):333–338

7. Hansen TB, Jensen TI, Clausen BH et al (2013) Natural RNA circles function as efficient microRNA sponges. Nature 495(7441):384–388
8. Talhouarne GJ, Gall JG (2014) Lariat intronic RNAs in the cytoplasm of Xenopus tropicalis oocytes. RNA 20(9):1476–1487
9. De la Peña M, Cervera A (2017) Circular RNAs with hammerhead ribozymes encoded in eukaryotic genomes: the enemy at home. RNA Biol 14(8):985–991
10. Cervera A, Urbina D, de la Peña M (2016) Retrozymes are a unique family of non-autonomous retrotransposons with hammerhead ribozymes that propagate in plants through circular RNAs. Genome Biol 17(1):135
11. De la Peña M, Garcia-Robles I, Cervera A (2017) The hammerhead ribozyme: a long history for a short RNA. Molecules 22(1):78–89
12. Kruger K, Grabowski PJ, Zaug AJ et al (1982) Self-splicing RNA: autoexcision and autocyclization of the ribosomal RNA intervening sequence of Tetrahymena. Cell 31(1):147–157
13. Guerrier-Takada C, Gardiner K, Marsh T et al (1983) The RNA moiety of ribonuclease P is the catalytic subunit of the enzyme. Cell 35(3 Pt 2):849–857
14. Gilbert W (1986) The RNA world. Nature 319:618
15. Crick FH (1968) The origin of the genetic code. J Mol Biol 38(3):367–379
16. Orgel LE (1968) Evolution of the genetic apparatus. J Mol Biol 38(3):381–393
17. Woese CR (1968) The fundamental nature of the genetic code: prebiotic interactions between polynucleotides and polyamino acids or their derivatives. Proc Natl Acad Sci U S A 59(1):110–117
18. Steitz TA, Moore PB (2003) RNA, the first macromolecular catalyst: the ribosome is a ribozyme. Trends Biochem Sci 28(8):411–418
19. Ferre-D'Amare AR, Scott WG (2010) Small self-cleaving ribozymes. Cold Spring Harb Perspect Biol 2(10):a003574
20. Valadkhan S, Manley JL (2001) Splicing-related catalysis by protein-free snRNAs. Nature 413(6857):701–707
21. Winkler W, Nahvi A, Breaker RR (2002) Thiamine derivatives bind messenger RNAs directly to regulate bacterial gene expression. Nature 419(6910):952–956
22. Mojica FJ, Diez-Villasenor C, Garcia-Martinez J et al (2005) Intervening sequences of regularly spaced prokaryotic repeats derive from foreign genetic elements. J Mol Evol 60(2):174–182
23. Fire A, Xu S, Montgomery MK et al (1998) Potent and specific genetic interference by double-stranded RNA in Caenorhabditis elegans. Nature 391(6669):806–811
24. Prody GA, Bakos JT, Buzayan JM et al (1986) Autolytic processing of dimeric plant virus satellite RNA. Science 231(4745):1577–1580
25. Hutchins CJ, Rathjen PD, Forster AC et al (1986) Self-cleavage of plus and minus RNA transcripts of avocado sunblotch viroid. Nucleic Acids Res 14(9):3627–3640
26. Buzayan JM, Gerlach WL, Bruening G (1986) Non-enzymatic cleavage and ligation of RNAs complementary to a plant virus satellite RNA. Nature 323:349–353
27. Kuo MY, Sharmeen L, Dinter-Gottlieb G et al (1988) Characterization of self-cleaving RNA sequences on the genome and antigenome of human hepatitis delta virus. J Virol 62(12):4439–4444
28. Saville BJ, Collins RA (1990) A site-specific self-cleavage reaction performed by a novel RNA in Neurospora mitochondria. Cell 61(4):685–696
29. Winkler WC, Nahvi A, Roth A et al (2004) Control of gene expression by a natural metabolite-responsive ribozyme. Nature 428(6980):281–286
30. Roth A, Weinberg Z, Chen AG et al (2014) A widespread self-cleaving ribozyme class is revealed by bioinformatics. Nat Chem Biol 10(1):56–60
31. Weinberg Z, Kim PB, Chen TH et al (2015) New classes of self-cleaving ribozymes revealed by comparative genomics analysis. Nat Chem Biol 11(8):606–610
32. De la Peña M, Gago S, Flores R (2003) Peripheral regions of natural hammerhead ribozymes greatly increase their self-cleavage activity. EMBO J 22(20):5561–5570
33. Khvorova A, Lescoute A, Westhof E et al (2003) Sequence elements outside the hammerhead ribozyme catalytic core enable intracellular activity. Nat Struct Biol 10(9):708–712
34. Martick M, Scott WG (2006) Tertiary contacts distant from the active site prime a ribozyme for catalysis. Cell 126(2):309–320
35. Przybilski R, Graf S, Lescoute A et al (2005) Functional hammerhead ribozymes naturally encoded in the genome of Arabidopsis thaliana. Plant Cell 17(7):1877–1885
36. Ferbeyre G, Smith JM, Cedergren R (1998) Schistosome satellite DNA encodes active hammerhead ribozymes. Mol Cell Biol 18(7):3880–3888
37. Rojas AA, Vazquez-Tello A, Ferbeyre G et al (2000) Hammerhead-mediated processing of satellite pDo500 family transcripts from Dolichopoda cave crickets. Nucleic Acids Res 28(20):4037–4043
38. Epstein LM, Gall JG (1987) Self-cleaving transcripts of satellite DNA from the newt. Cell 48(3):535–543
39. Martick M, Horan LH, Noller HF et al (2008) A discontinuous hammerhead ribozyme embedded in a mammalian messenger RNA. Nature 454(7206):899–902
40. De la Peña M, Garcia-Robles I (2010) Ubiquitous presence of the hammerhead ribozyme motif along the tree of life. RNA 16(10):1943–1950
41. Jimenez RM, Delwart E, Luptak A (2011) Structure-based search reveals hammerhead ribozymes in the human microbiome. J Biol Chem 286(10):7737–7743
42. Perreault J, Weinberg Z, Roth A et al (2011) Identification of hammerhead ribozymes in all domains of life reveals novel structural variations. PLoS Comput Biol 7(5):e1002031

43. Seehafer C, Kalweit A, Steger G et al (2011) From alpaca to zebrafish: hammerhead ribozymes wherever you look. RNA 17(1):21–26
44. De la Peña M, Garcia-Robles I (2010) Intronic hammerhead ribozymes are ultraconserved in the human genome. EMBO Rep 11(9):711–716
45. Hammann C, Luptak A, Perreault J et al (2012) The ubiquitous hammerhead ribozyme. RNA 18(5):871–885
46. Garcia-Robles I, Sanchez-Navarro J, De la Peña M (2012) Intronic hammerhead ribozymes in mRNA biogenesis. Biol Chem 393(11):1317–1326
47. Webb CH, Riccitelli NJ, Ruminski DJ et al (2009) Widespread occurrence of self-cleaving ribozymes. Science 326(5955):953
48. Cervera A, De la Peña M (2014) Eukaryotic penelope-like retroelements encode hammerhead ribozyme motifs. Mol Biol Evol 31(11):2941–2947
49. Eickbush DG, Eickbush TH (2010) R2 retrotransposons encode a self-cleaving ribozyme for processing from an rRNA cotranscript. Mol Cell Biol 30(13):3142–3150
50. Ruminski DJ, Webb CH, Riccitelli NJ et al (2011) Processing and translation initiation of non-long terminal repeat retrotransposons by hepatitis delta virus (HDV)-like self-cleaving ribozymes. J Biol Chem 286(48):41286–41295
51. Kennell JC, Saville BJ, Mohr S et al (1995) The VS catalytic RNA replicates by reverse transcription as a satellite of a retroplasmid. Genes Dev 9(3):294–303
52. Gorinsek B, Gubensek F, Kordis D (2004) Evolutionary genomics of chromoviruses in eukaryotes. Mol Biol Evol 21(5):781–798
53. Witte CP, Le QH, Bureau T et al (2001) Terminal-repeat retrotransposons in miniature (TRIM) are involved in restructuring plant genomes. Proc Natl Acad Sci U S A 98(24):13778–13783
54. Gao D, Chen J, Chen M et al (2012) A highly conserved, small LTR retrotransposon that preferentially targets genes in grass genomes. PloS one 7(2):e32010
55. Forster AC, Davies C, Sheldon CC et al (1988) Self-cleaving viroid and newt RNAs may only be active as dimers. Nature 334(6179):265–267
56. Kalendar R, Tanskanen J, Chang W et al (2008) Cassandra retrotransposons carry independently transcribed 5S RNA. Proc Natl Acad Sci U S A 105(15):5833–5838
57. Flores R, Grubb D, Elleuch A et al (2011) Rolling-circle replication of viroids, viroid-like satellite RNAs and hepatitis delta virus: variations on a theme. RNA Biol 8(2):200–206
58. Jangam D, Feschotte C, Betran E (2017) Transposable element domestication as an adaptation to evolutionary conflicts. Trends Genet 33(11):817–831
59. Salehi-Ashtiani K, Luptak A, Litovchick A et al (2006) A genomewide search for ribozymes reveals an HDV-like sequence in the human CPEB3 gene. Science 313(5794):1788–1792
60. Chi YI, Martick M, Lares M et al (2008) Capturing hammerhead ribozyme structures in action by modulating general base catalysis. PLoS Biol 6(9):e234
61. Dufour D, de la Peña M, Gago S et al (2009) Structure-function analysis of the ribozymes of chrysanthemum chlorotic mottle viroid: a loop-loop interaction motif conserved in most natural hammerheads. Nucleic Acids Res 37(2):368–381
62. Gago S, De la Peña M, Flores R (2005) A kissing-loop interaction in a hammerhead viroid RNA critical for its in vitro folding and in vivo viability. RNA 11(7):1073–1083

Part IV

Molecular Mechanisms and Gene Regulation of Circular RNAs

Circular RNAs Act as miRNA Sponges

6

Amaresh Chandra Panda

Abstract

Majority of RNAs expressed in animal cells lack protein-coding ability. Unlike other cellular RNAs, circular (circ)RNAs include a large family of noncoding (nc)RNAs that lack the 5′ or 3′ ends. The improvements in high-throughput RNA sequencing and novel bioinformatics tools have led to the identification of thousands of circRNAs in various organisms. CircRNAs can regulate gene expression by influencing the transcription, the mRNA turnover, and translation by sponging RNA-binding proteins and microRNAs. Given the broad impact of circRNA on miRNA activity, there is huge interest in understanding the impact of miRNA sponging by circRNA on gene regulation. In this review, we summarize our current knowledge of the miRNA-circRNA interaction and mechanisms that influence gene expression.

Keywords

mRNA · miRNA · circRNA · Competing endogenous RNA · Translation · miRNA sponge

A. C. Panda (✉)
Institute of Life Sciences,
Bhubaneswar, Odisha, India

1 Introduction

RNA molecules were conventionally believed to transfer the genetic information coded in the genomic DNA into specific proteins [1]. However, the protein-coding mRNAs represent only ~5% of the human transcriptome, while the rest of the transcriptome is noncoding (nc)RNAs [2]. The vast majority are ribosomal (r)RNA and transfer (t)RNA, both involved in translation [1, 3]. The other categories of ncRNAs include microRNAs (miRNAs), pseudogenes, long (l)ncRNAs, and circular (circ)RNAs [4–6]. In 1976, electron microscopy of plant viroid discovered covalently closed single-stranded RNA molecules for the first time [7]. Later, the hepatitis delta virus (HDV) was found to have a circRNA genome [8]. Another report suggested the expression of scrambled exon RNA from tumor suppressor gene DCC in human cells [9]. Due to lack of RNA-sequencing technologies and inability to map the circRNAs to the genome, circRNAs were completely neglected for last two decades. Traditional molecular biology techniques used for RNA analysis cannot differentiate circRNAs from linear RNAs [10, 11]. Interestingly, the innovations in next-generation sequencing associated with new computational pipelines to map circRNAs to the genome have moved circRNA to the forefront of RNA research [12–15]. Most of the circRNAs are found to be abundant, conserved across species, and often show tissue-specific

J. Xiao (ed.), *Circular RNAs*, Advances in Experimental Medicine and Biology 1087,
https://doi.org/10.1007/978-981-13-1426-1_6

expression pattern [13, 16]. Recent studies established that covalently closed circRNAs are generated from the canonical splicing machinery by a process called backsplicing [17]. CircRNAs are categorized as exonic (E), intronic (I), and exon-intron (EI) circRNAs based on the primary transcript sequence they are generated from [15, 17–19]. circRNAs are very stable due to lack of the 5′-3′ ends which makes them resistant to exonucleases [20, 21]. Recent studies reported that some may act as sponges for miRNAs, sponges for RBPs, compete with linear splicing, and translated into peptides [4, 18, 22]. Growing evidence indicated that circRNAs involved in various cellular events including proliferation, differentiation, apoptosis, and metastasis [23]. This review briefly discusses the impact of circRNAs on key cellular processes by acting as a competing endogenous RNA (ceRNA) for target miRNAs.

2 MiRNA Sponging by circRNA

MicroRNAs are small, evolutionarily conserved ncRNA molecules predicted to regulate ~30% of the protein-coding genes [24, 25]. There are ~1800 miRNAs which have been identified in humans [26]. The miRNA genes are transcribed into primary miRNA (pri-miRNA) by RNA polymerase II followed by processing with Drosha and DGCR8 to generate pre-miRNA [27]. The pre-miRNA is exported to the cytoplasm and processed by Dicer to produce ~19–22 nt mature miRNA. The functional miRNA is incorporated in the effector RNA-induced silencing complex (RISC) and activates the RISC complex to bind the target mRNA that has sequence complementarity with the loaded miRNA. The miRNAs usually target the 3′ untranslated regions (UTRs) of specific mRNA targets and regulate their stability and/or translation [28]. The level of complementarity between the mRNA and miRNA determines the mechanism of miRNA action on target mRNA. As each miRNAs can have complementarity with many mRNAs, they have the potential to regulate multiple genes [29]. miRNAs are shown to be involved in posttranscriptional regulation of gene expression in nearly all cellular events including cell proliferation, migration, differentiation, and apoptosis [29–33]. Given their importance in gene expression regulation, there is enormous interest in understanding the regulatory mechanisms that can regulate miRNA function. Accumulating evidence indicates that circRNAs play a crucial role in gene expression regulation partly by inhibiting miRNA activity (Fig. 6.1). In this review, several circRNA-miRNA interactions and their function are listed as follows (Table 6.1).

CDR1as The CDR1as is generated from the antisense transcript of CDR1 gene (also termed as ciRS-7). CDR1as was the first miRNA sponge reported to negatively regulate miR-7 and found to be expressed in brain tissue, neuroblastoma, astrocytoma, HeLa cells, and lung carcinoma [14, 34–36]. The CDR1as has more than 60 binding sites for miR-7, and its resistance to miRNA-mediated RNA degradation makes this a perfect ceRNA for miR-7 [12, 14]. Expression of CDR1as inhibits miR-7 activity that leads to increase in expression of miR-7 targets. Coexpression of CDR1as and miR-7 is necessary for sponging activity of circRNA. The sponging of miR-7 by ciRS-7 is reported to affect the expression of ubiquitin protein ligase A (UBE2A) in Alzheimer's disease [37]; Myrip and Pax6 in insulin secretion and synthesis, respectively [38]; and epidermal growth factor receptor (EGFR) in cancer [36, 14, 39]. Upregulation of CDR1as in gastric cancer suppresses miR-7 activity which leads to more aggressive oncogenic phenotype mediated by PTEN/PI3K/AKT pathway [40]. Another study reported the upregulation of CDR1as and downregulation of miR-7 in hepatocellular carcinoma (HCC) tissue compared with the adjacent non-tumor tissues [34]. The overexpression of miR-7 or silencing of CDR1as inhibited the HCC cell proliferation and invasion by inhibiting the expression of target genes CCNE1 and PIK3CD. Together, CDR1as act as an oncogene in HCC through sponging miR-7 [34].

cir-SRY The sex-determining region Y gene produces a circRNA known as cir-SRY. The cir-SRY is highly expressed in adult mouse testis

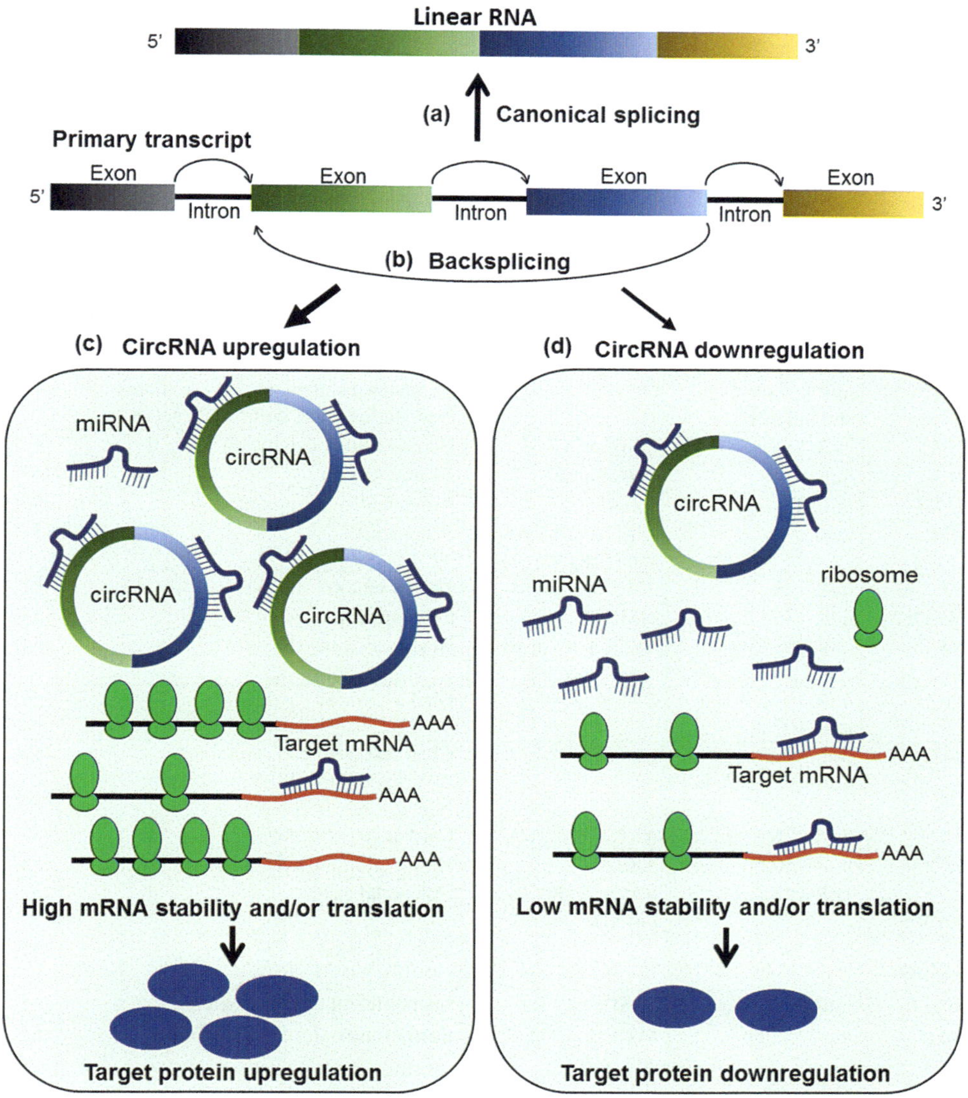

Fig. 6.1 Schematic representation of circRNA biogenesis and their impact on gene expression by sponging miRNA
(**a**) The canonical splicing machinery generates mature linear RNA from the primary transcript by removing intervening introns. (**b**) The circRNA is generated by the "head-to-tail" backsplicing of the circularizing exons. (**c**) Most circRNA sponges are enriched in miRNA-binding sites. Overexpression of circRNA leads to inactivation of miRNAs, thereby upregulating miRNA target gene expression. (**d**) The decrease in circRNA expression allows higher levels of functional miRNAs to suppress the expression of mRNAs containing miRNA-binding sites

[41]. A recent study reported that there are 16 putative miR-138-binding sites present in cir-SRY which can act as ceRNA and thus potentially regulate expression of miR-138 target genes [14, 42].

cir-ITCH Cir-ITCH (itchy E3 ubiquitin protein ligase) is generated from the ICTH gene. Cir-ICTH was reported to enrich in miRNA regulatory elements (MREs) for miR-7, miR-17, and miR-214 and act as a sponge for these miRNAs,

Table 6.1 Potential circRNA-miRNA-mRNA regulatory networks

CircRNA name	Sponged miRNA	miRNA target gene	Diseases/tissue	References
CDR1as/ ciRS-7	miR-7	UBE2A	Alzheimer's disease	[37]
CDR1as/ ciRS-7	miR-7	Myrip and Pax6	Diabetes	[38]
CDR1as/ ciRS-7	miR-7	CCNE1 and PIK3CD	Hepatocellular carcinoma	[34]
CDR1as/ ciRS-7	miR-7	EGFR	Glioblastoma	[14, 39]
CDR1as/ ciRS-7	miR-7	PI3K	Gastric cancer	[40]
cir-SRY	miR-138	TWIST2	Colorectal cancer	[14, 42]
cir-ITCH	miR-7, miR-17, and miR-214	ITCH	Esophageal squamous cell carcinoma, bladder, lung, and colorectal cancer	[43–46]
circHIPK3	miR-124	Aquaporin 3	Hepatocellular carcinoma	[48]
circHIPK3	miR-558	Heparanase	Bladder cancer	[49]
circHIPK3	miR-379	IGF-1	Non-small cell lung cancer	[50]
circPVT1	let-7	IGF2BP1, KRAS, and HMGA2	Senescence	[51]
circPVT1	miR-125	E2F2	Gastric cancer	[52]
circRNA-CER	miR-136	MMP13	Osteoarthritis	[53]
circRNA-MYLK	miR-29a	VEGFA	Bladder carcer	[54]
circTCF25	miR-103a-3p and miR-107	CDK6	Bladder cancer	[55]
circHIAT1	for miR-195-5p/29a-3p/29c-3p	CDC42	Clear cell renal cell carcinoma	[56]
HRCR	miR-223	ARC	Cardiac hypertrophy	[57]
cir-ZNF609	miR-150-5p	AKT3	Hirschsprung disease	[58]
hsa_ circ_001569	miR-145	BAG4, E2F5, and FMNL2	Colorectal cancer	[59]
hsa_ circ_001564	miR-29c-3p		Osteosarcoma	[60]
circVMA21	miR-200c	XIAP	Intervertebral disc degeneration	[61]
circRNA_ Atp9b	miR-138-5p	MMP13, COX-2, and IL-6	Osteoarthritis	[62]
circDOCK1	miR-196a-5p	BIRC3	Oral squamous cell carcinoma	[63]
circMTO1	miR-9	P21	Hepatocellular carcinoma	[64]
hsa_ circ_0005986	miR-129-5p	Notch1	Hepatocellular carcinoma	[65]
hsa_ circ_000984	miR-106b	CDK6	Colorectal cancer	[66]
hsa_ circ_0020397	miR-138	TERT and PD-L1	Colorectal cancer	[67]
hsa_ circ_0009910	miR-449a	IL6R	Osteosarcoma	[68]
circGFRA1	miR-34a	GFRA1	Triple negative breast cancer	[69]
hsa-circ-0016347	miR-24	caspase-1	Osteosarcoma	[70]
circWDR77	miR-124	FGF2	Vascular smooth muscle cells	[71]

(continued)

Table 6.1 (continued)

CircRNA name	Sponged miRNA	miRNA target gene	Diseases/tissue	References
circACTA2	miR-548f-5p	α-SMA	Vascular smooth muscle cells	[72]
hsa_circ_0012673	miR-22	ErbB3	Lung adenocarcinoma	[73]
circRNA_LARP4	miR-424	LATS1	Gastric cancer	[74]
circ-ABCB10	miR-1271		Breast cancer	[75]

leading to upregulation of miRNA target gene ITCH. Cir-ITCH was reported to have antitumor activity by suppressing the Wnt/β-catenin signaling by regulating ICTH expression in various cancers including esophageal squamous cell carcinoma and bladder, lung, and colorectal cancer [43–46].

circHIPK3 The second exon of homeodomain-interacting protein kinase 3 (*HIPK3*) gene generates a circRNA called circHIPK3 that can act as a sponge for nine miRNAs including miR-124. Silencing of CircHIPK3 reduced cell growth through suppressing miR-124 activity which directly interacts with circHIPK3 [47]. CircHIPK3 was upregulated in HCC tissues [48]. CircHIPK3 acted as a sponge for miR-124 and suppressed miR-124 activity in HCC, leading to upregulation of miR-124 target gene aquaporin 3 (AQP3). Further, increase in AQP3 expression promoted cell proliferation and migration in HCC cells. Together, these results suggested that circHIPK3 regulated HCC growth through the miR-124-AQP3 axis [48]. CircHIPK3 was also found to be downregulated in bladder cancer tissues compared with normal bladder tissues, and the level of circHIPK3 negatively correlates with bladder cancer grade [49]. Increase in circHIPK3 level led to inhibition of migration, invasion, and angiogenesis of bladder cancer cells in vitro. CircHIPK3 acted as a ceRNA for miR-558 and inhibit miR-558 activity, thereby regulating the expression of heparanase (HPSE) in bladder cancer cells [49]. Another study reported the expression pattern of circHIPK3 in six non-small cell lung cancer (NSCLC) cell lines [50]. The NSCLC cell lines NCI-H2170 and NCI-H1299 had the highest and lowest expression level, respectively. The circHIPK3 overexpression in NCI-H1299 promoted cell proliferation and circHIPK3 silencing in NCI-H2170-inhibited cell proliferation. CircHIPK3 was found to sequester miR-379 and increase the expression levels of miR-379 target IGF1, leading to increase in cell proliferation [50].

circPVT1 Circular RNA expression pattern in proliferating (early-passage) and senescent (late-passage) human diploid WI-38 fibroblasts were analyzed and identified hundreds of differentially expressed senescence-associated circRNAs (SAC-RNAs) [51]. One of the SAC-RNA called circPVT1 was significantly downregulated in senescent fibroblasts. Further, circPVT1 silencing in proliferating fibroblasts promoted cellular senescence. CircPVT1 selectively sponged let-7 and promoted the expression of let-7 target genes including IGF2BP1, KRAS, and HMGA2. Together, these data suggested that the SAC-RNA circPVT1, elevated in the proliferating cells, inhibits endogenous let-7 activity to enable a proliferative phenotype [51]. Another report suggested that the circPVT1 is often upregulated in gastric cancer (GC) tissues compared with matched normal tissues [52]. The circPVT1 acted as a sponge for the members of the miR-125 family and promoted cell proliferation. Further, circPVT1 expression level was correlated with the survival of patients with gastric cancer. In sum, circPVT1 acts as a proliferative factor and prognostic marker in gastric cancer [52].

circRNA-CER A recent study explored the circRNA expression pattern and function of chondrocyte extracellular matrix (ECM)-related circRNAs (circRNA-CER) in cartilage. Several circRNAs were differentially expressed in osteoarthritis samples compared to normal cartilage. The circRNA-CER expression was upregulated in chondrocytes upon interleukin-1 (IL-1) and tumor necrosis factor α (TNFα) treatment [53]. The circRNA-CER was found to have five putative MREs for miR-636, miR-665, miR-217, miR-646, and miR-136. The MMP13 gene was regulated by miR-136 which is sponged by circRNA-CER. The silencing of circRNA-CER increased chondrocyte extracellular matrix formation by inhibiting MMP13 expression [53].

circRNA-MYLK A recent study found that the circRNA-MYLK and VEGFA were significantly upregulated in bladder carcinoma. The circRNA-MYLK directly binds to miR-29a that leads to increase in expression of miR-29a target VEGFA. Overexpression of circRNA-MYLK activated VEGFA/VEGFR2 and downstream Ras/ERK signaling pathway by acting as a ceRNA for miR-29a [54].

circTCF25 The circRNA circTCF25 was predicted to sponge miR-103a-3p and miR-107. The overexpression of circTCF25 sequestered miR-103a-3p and miR-107, leading to upregulation of their target CDK6 which in turn enhanced the proliferation and migration of bladder cancer cells [55].

circHIAT1 The expression of circHIAT1 was downregulated in clear cell renal cell carcinoma (ccRCC) compared to the adjacent normal tissues. Androgen receptor (AR) downregulated the expression of circHIAT1 by suppressing transcription of its host gene, hippocampus abundant transcript 1 (HIAT1). The circHIAT1 may act as a sponge for miR-195-5p/29a-3p/29c-3p to modulate CDC42 expression which is linked to ccRCC cell migration and invasion [56].

HRCR A recent work suggested that a heart-related circRNA (HRCR) could act as a sponge for miR-223 to suppress cardiac hypertrophy. ARC was identified to be the downstream target to mediate the function of miR-223 in cardiac hypertrophy. The circRNA HRCR was found to sequester and inhibit the function miR-223, leading to upregulation of ARC expression which is involved in cardiac hypertrophy and heart failure [57].

cir-ZNF609 In Hirschsprung disease (HSCR) the expression of cir-ZNF609 was lower as compared with normal bowel tissues. The silencing of cir-ZNF609 led to suppression of proliferation and migration of cells. Further, cir-ZNF609 was found to regulate the expression of AKT3 by acting as a decoy for miR-150-5p. These findings suggest that the cir-ZNF609-miR-150-5p-AKT3 axis plays a critical role in the onset of HSCR [58].

has_circ_001569 The expression of hsa_circ_001569 was upregulated in colorectal cancer tissues and predicted to sponge miR-145. The miR-145 can modulate the expression of BAG4, E2F5, and FMNL2 transcripts in colorectal cancer cells. Expression of circ_001569 upregulated the expression of BAG4, E2F5, and FMNL2 by sponging miR-145, leading to colorectal cancer cell proliferation and invasion [59].

hsa_circ_001564 The circRNA hsa_circ_0001564 is derived from the gene called CANX and significantly upregulated in osteosarcoma cell lines. Silencing of hsa_circ_0001564 reduced the proliferation by inducing cell cycle arrest and apoptosis in HOS and MG-63 cells. Further, hsa_circ_0001564 was found to be a sponge for miR-29c-3p which could reverse the tumorigenic effect of circ_0001564. These data suggested that hsa_circ_0001564 plays a critical role in osteosarcoma by acting as a ceRNA for miR-29c-3p [60].

circVMA21 The role of circVMA21 was explored in nucleus pulposus (NP) cells and degenerative NP tissues from intervertebral disc degeneration (IVDD) patients. The circVMA21 was found to directly interact with miR-200c which inhibits the expression of the target gene X-linked inhibitor-of-apoptosis protein (XIAP). The NP cell function and viability was regulated by miR-200c through suppression of XIAP. Together, circVMA21 could inhibit NP cell apoptosis through miR-200c-XIAP axis [61].

circRNA_Atp9b In osteoarthritis, circRNA_Atp9b was overexpressed in mouse chondrocytes upon interleukin-1 beta (IL-1β) treatment. Further, circRNA_Atp9b silencing upregulated type II collagen expression and suppressed the expression of MMP13, COX-2, and IL-6. miR-138-5p was found to be sponged by circRNA_Atp9b, and their expression levels are negatively correlated. Moreover, the effects of circRNA_Atp9b on extracellular matrix catabolism and inflammation were partly reversed by inhibition of miR-138-5p. In sum, these data suggested that the extracellular matrix in chondrocytes is regulated by circRNA_Atp9b through sponging miR-138-5p [62].

circDOCK1 A TNF-α-induced apoptotic model was developed for oral squamous cell carcinoma (OSCC) to study the impact of circDOCK1 on apoptosis. The silencing of circDOCK1 increased the apoptosis in OSCC cells. Bioinformatics analysis predicted the interaction of circDOCK1 with miR-196a-5p which targets BIRC3. Interestingly, both overexpression of miR-196a-5p or knockdown of circDOCK1 led to suppression of BIRC3 which is a negative regulator of apoptosis. Together, these results suggested that the apoptosis of OSCC cells is regulated by circDOCK1 through sponging miR-196a-5p [63].

circMTO1 A recent study reported the circRNA expression profile in HCC and identified circMTO1, generated from the gene mitochondrial translation optimization 1 (MTO1). The level of circMTO1 was found to be downregulated in HCC tissues and positively correlated with survival of HCC patients. Biochemical assays revealed the interaction of miR-9 with circMTO1in HCC cells. The circMTO1 silencing led to the suppression of p21 which is a target of miR-9, leading to increase in proliferation and invasion of HCC cells. Taken together, these data suggest that circMTO1 inhibits HCC progression by acting as ceRNA for miR-9 to upregulate the expression of p21 [64].

hsa_circ_0005986 The circRNA hsa_circ_0005986 was found to be downregulated in HCC tissue samples compared with adjacent non-tumorous tissues [65]. Furthermore, hsa_circ_0005986 expressions were significantly downregulated in HCC cell lines, HepG2, SMMC7721, and HCCLM3 compared to normal hepatic cell line L02. Interestingly, the level of hsa_circ_0005986 downregulation was found to be correlated with Barcelona clinic liver cancer (BCLC) stage, chronic hepatitis B family history, and tumor diameters. miR-129-5p was one of the miRNA that could be sponged by hsa_circ_0005986 and regulated the target gene Notch1. Silencing of hsa_circ_0005986 increased miR-129-5p activity and downregulated Notch1 expression, leading to increase in cell proliferation and tumorigenesis in HCC [65].

hsa_circ_000984 The circular RNA hsa_circ_000984 generated from the CDK6 gene was significantly overexpressed in colorectal cancer (CRC) tissues from patients as well as in the CRC cell lines [66]. Further, the expression level of hsa_circ_000984 was positively associated with CRC advancement. The knockdown of hsa_circ_000984 expression led to inhibition of cell proliferation, migration, and invasion in CRC cell lines. Hsa_circ_000984 could act as a sponge for miR-106b, leading to upregulation of miR-106b target CDK6. Together, these data suggest that the hsa_circ_000984 could upregulate CDK6 by inhibiting miR-106b activity, thereby promoting colon cancer growth and metastasis [66].

hsa_circ_0020397 Another study reported that the circRNA hsa_circ_0020397 was upregulated, while its target miR-138 was downregulated in CRC cells. Furthermore, overexpression of hsa_circ_0020397 could inhibit the miR-138 activity indicating that hsa_circ_0020397 act as a ceRNA for miR-138. The hsa_circ_0020397 promoted the expression of miR-138 targets telomerase reverse transcriptase (TERT) and programmed death-ligand 1 (PD-L1) by sponging miR-138, thereby promoting cell viability and invasion of CRC cells. Together, these data suggest that hsa_circ_0020397 plays a crucial role in CRC pathogenesis by acting as a sponge for miR-138 [67].

hsa_circ_0009910 In osteosarcoma, the hsa_circ_0009910 expression was found to be upregulated and silencing of circ_0009910 promoted cell cycle arrest and apoptosis. The miR-449a expression was found to be suppressed in osteosarcoma cells and predicted to be sponged by circ_0009910. Further, miR-449a could target and downregulate the expression of IL6R in osteosarcoma cells, thereby promoting inhibition of cell proliferation, cell cycle arrest, and apoptosis. The transcript level of IL6R was also found to be negatively correlated with the level of miR-449a in osteosarcoma cells. Taken together, these data suggested that the carcinogenesis of osteosarcoma cells was induced by the circ_0009910/miR-449a/IL6R axis [68].

circGFRA1 A little is known about the role of circRNA in triple negative breast cancer (TNBC). A recent study reported that the circGFRA1 was upregulated in TNBC cell lines and tissues. Kaplan-Meier survival analysis suggested that the level of circGFRA1 was negatively correlated with survival. CircGFRA1 silencing led to the suppression of proliferation and induced apoptosis in TNBC cells. Further, biochemical assays found that circGFRA1 can directly bind and inhibit miR-34a activity, leading to increase in miR-34a target gene GFRA1. In sum, circGFRA1 act as a sponge for miR-34a to regulate GFRA1 expression in TNBC [69].

hsa-circ-0016347 Hao J et al. reported that the circ-0016347 acted as a ceRNA for miR-214 in osteosarcoma cell leading to upregulation of miR-24 target caspase-1. Moreover, circ-0016347 was found to induce proliferation and invasion of osteosarcoma cells. These data suggested that the circ-0016347 plays a crucial role in osteosarcoma progression by acting as a sponge for miR-124, which could be used as a potential target for development of therapy for osteosarcoma [70].

circWDR77 The circRNA expression profiling in glucose-induced vascular smooth muscle cells (VSMCs) discovered hundreds of differentially expressed circRNAs. CircWDR77 is one of the upregulated circRNAs whose silencing led to inhibition of proliferation and migration of VSMCs. Computational prediction suggested the interaction of circWDR77 with miR-124. Furthermore, circWDR77 inhibited miR-124 activity and upregulated the expression of miR-124 target fibroblast growth factor 2 (FGF2) in VSMCs. Together, these results indicated that the proliferation and migration of VSMCs were regulated through circWDR77/miR-124/FGF2 axis [71].

circACTA2 Neuregulin-1 (NRG-1) was found to be upregulated and cleaved in response to transforming growth factor-β1 in VSMCs. NRG-1 was also known to promote the expression of an extracellular epidermal growth factor-like domain and intracellular domain (NRG-1-ICD) which induced circular ACTA2 (alpha-actin-2; circACTA2) expression. Further, circACTA2 acted as a sponge for miR-548f-5p which upregulated α-SMA expression, leading to increase in stress fiber formation and cell contraction in VSMCs. Together, these data indicated that circACTA2 fine-tunes the α-SMA expression and VSMC contraction through the NRG-1-ICD/circACTA2/miR-548f-5p axis [72].

hsa_circ_0012673 The hsa_circ_0012673 expression was upregulated in lung adenocarcinoma (LAC) tissues compared with adjacent non-tumor tissues. Further, the expression level of circ_0012673 was positively correlated with tumor size. Biochemical assays revealed that hsa_circ_0012673 promoted LAC proliferation by acting as a sponge for miR-22, which inhibits erb-b2 receptor tyrosine kinase 3 (ErbB3) [73].

circRNA_LARP4 The large tumor suppressor kinase 1 (LATS1) acts as a tumor suppressor in gastric cancer by regulating the Hippo signaling pathway. The circRNA_ LARP4 was downregulated in gastric cancer and predicted to sponge miR-424 computationally. The direct interaction between miR-424 and LATS1 or circLARP4 was verified by various biochemical assays. Overexpression of miR-424 suppressed the expression of miR-424 target gene LATS1 which led to increase in proliferation and invasion of gastric cancer cells. In sum, circLARP4 act as a novel tumor suppressive factor by sponging miR-424-5p which modulate the expression of LATS1 in gastric cancer cells [74].

circ-ABCB10 The circ-ABCB10 was significantly overexpressed in breast cancer tissue. Further experiments suggested that silencing of circ-ABCB10 inhibited the proliferation and promoted apoptosis of breast cancer cells. Bioinformatics and biochemical analysis found that miR-1271 can be sponged by circ-ABCB10. Furthermore, the effect of circ-ABCB10 on breast cancer cells was rescued by miR-1271. These data indicated that circ-ABCB10 promotes breast cancer pathogenesis via sponging miR-1271 [75].

3 Web Tools for Analysis of miRNA-circRNA Interaction

Besides in-depth studies on functional circRNAs, several circRNA databases have been developed to explore the interaction of circRNAs with miRNAs. The databases such as CircInteractome, Circ2Traits, CircNet, and StarBase v2.0 provide excellent platforms to predict the miRNA-circRNA interactions (Table 6.2).

circInteractome The CircInteractome (http://circinteractome.nia.nih.gov) database is the first web tool developed to predict the miRNAs-circRNA interactions for circRNAs listed in CircBase [76–78]. Furthermore, this is the only web tool available to date for designing divergent primer and siRNA against circRNAs. Together, the CircInteractome web tool provides bioinformatic analyses of miRNA-binding sites on circRNAs and predicts potential miRNA sponge circRNAs that are crucial for posttranscriptional gene regulation [76].

circ2Trait The circ2Traits (http://gyanxet-beta.com/circdb/) is a database of 1951 human circRNAs and their association with 105 human diseases [79]. The circRNAs were categorized based on number of disease-associated SNPs, AGO interaction sites, and interaction with disease-associated miRNA. Circ2Trait also checks the enrichment of sets of genes in the miRNA-circRNA interactome that is associated with particular diseases. Circ2Trait also provides the complete information of miRNA-circRNA-mRNA-lncRNA interaction networks for the human diseases.

Table 6.2 Web tools for prediction of circRNA-miRNA interactions

Database name	URL	References
CircInteractome	https://circinteractome.nia.nih.gov/	[76, 77]
Circ2Traits	http://gyanxet-beta.com/circdb/	[79]
CircNet	http://circnet.mbc.nctu.edu.tw/	[80]
StarBase v2.0	http://starbase.sysu.edu.cn/	[81]

CircNet The CircNet (http://circnet.mbc.nctu.edu.tw/) is a database to explore circRNA expression in specific tissues, circRNAs isoforms, circRNA sequences, and circRNA-miRNA interactions. The CircNet database was the first database to report the expression of tissue-specific circRNAs and circRNA-miRNA-gene interaction network. In sum, the CircNet is an interactive web interface for visualization and analysis of the regulatory network of circRNA, miRNA, and genes [80].

starBase v2.0 The starBase v2.0 (http://starbase.sysu.edu.cn/) is the first database to systematically identify the regulatory RNA-RNA and protein-RNA interaction networks using the experimentally supported CLIP-Seq data. This study identified ~9000 miRNA-circRNA regulatory interactions. Moreover, starBase v2.0 provides CLIP-supported miRNA target sites for interacting circRNAs. This web server predicts the functional interaction of miRNA-circRNA and their coordinated regulatory networks [81].

4 Conclusion and Future Directions

With the advancement in circRNA enrichment and sequencing technology, a huge number of circRNAs have been identified in various organisms, and the number will most likely increase. The circRNAs are currently one of the focus areas in biological research, and the field is still in its early stages. There is huge interest in how circRNAs are generated and what are their biological functions. Although thousands of circRNAs have been identified, only a handful of circRNAs has been reported to have a biological function. Almost all studies on circRNAs largely focused on their interactions with miRNAs and RBPs. The expression of vast number and types of circular RNAs increases the difficulty level for understanding their regulatory mechanisms. Further intensive studies are required to understand the biogenesis of circRNAs, functional interaction of circRNAs with genome and mRNA transcripts, and their translatability into proteins. With ever-increasing evidence of functional circRNA and discovery of novel molecular mechanisms of action, there is no doubt that circRNAs will be used for disease diagnosis and treatment in near future.

Acknowledgments This work was supported by the Science and Engineering Research Board, a statutory body of the Department of Science and Technology (DST), Government of India (SERB/F/6890/2017-18).

Conflicts of Interest The authors have no conflicts of interest to declare.

References

1. Crick F (1970) Central dogma of molecular biology. Nature 227(5258):561–563
2. Derrien T, Johnson R, Bussotti G et al (2012) The GENCODE v7 catalog of human long noncoding RNAs: analysis of their gene structure, evolution, and expression. Genome Res 22(9):1775–1789
3. Noller HF (1991) Ribosomal RNA and translation. Annu Rev Biochem 60:191–227
4. Hsiao KY, Sun HS, Tsai SJ (2017) Circular RNA – new member of noncoding RNA with novel functions. Exp Biol Med (Maywood) 242(11):1136–1141
5. Wery M, Kwapisz M, Morillon A (2011) Noncoding RNAs in gene regulation. Wiley Interdiscip Rev Syst Biol Med 3(6):728–738
6. Yoon JH, Abdelmohsen K, Gorospe M (2013) Posttranscriptional gene regulation by long noncoding RNA. J Mol Biol 425(19):3723–3730
7. Sanger HL, Klotz G, Riesner D et al (1976) Viroids are single-stranded covalently closed circular RNA molecules existing as highly base-paired rod-like structures. Proc Natl Acad Sci U S A 73(11):3852–3856
8. Kos A, Dijkema R, Arnberg AC et al (1986) The hepatitis delta (delta) virus possesses a circular RNA. Nature 323(6088):558–560
9. Nigro JM, Cho KR, Fearon ER et al (1991) Scrambled exons. Cell 64(3):607–613
10. Jeck WR, Sharpless NE (2014) Detecting and characterizing circular RNAs. Nat Biotechnol 32(5):453–461
11. Panda AC, Abdelmohsen K, Gorospe M (2017) RT-qPCR detection of senescence-associated circular RNAs. Methods Mol Biol 1534:79–87
12. Memczak S, Jens M, Elefsinioti A et al (2013) Circular RNAs are a large class of animal RNAs with regulatory potency. Nature 495(7441):333–338
13. Jeck WR, Sorrentino JA, Wang K et al (2013) Circular RNAs are abundant, conserved, and associated with ALU repeats. RNA 19(2):141–157

14. Hansen TB, Jensen TI, Clausen BH et al (2013) Natural RNA circles function as efficient microRNA sponges. Nature 495(7441):384–388
15. Panda AC, De S, Grammatikakis I et al (2017) High-purity circular RNA isolation method (RPAD) reveals vast collection of intronic circRNAs. Nucleic Acids Res 45(12):e116
16. Xia S, Feng J, Lei L et al (2017) Comprehensive characterization of tissue-specific circular RNAs in the human and mouse genomes. Brief Bioinform 18(6):984–992
17. Chen LL, Yang L (2015) Regulation of circRNA biogenesis. RNA Biol 12(4):381–388
18. Panda AC, Grammatikakis I, Munk R et al (2017) Emerging roles and context of circular RNAs. Wiley Interdiscip Rev RNA 8(2):e1413
19. Li Z, Huang C, Bao C et al (2015) Exon-intron circular RNAs regulate transcription in the nucleus. Nat Struct Mol Biol 22(3):256–264
20. Chen LL (2016) The biogenesis and emerging roles of circular RNAs. Nat Rev Mol Cell Biol 17(4):205–211
21. Zhang Y, Xue W, Li X et al (2016) The biogenesis of nascent circular RNAs. Cell Rep 15(3):611–624
22. Abdelmohsen K, Panda AC, Munk R et al (2017) Identification of HuR target circular RNAs uncovers suppression of PABPN1 translation by CircPABPN1. RNA Biol 14(3):361–369
23. Haque S, Harries LW (2017) Circular RNAs (circRNAs) in health and disease. Genes (Basel) 8(12):353
24. Berezikov E, Guryev V, van de Belt J et al (2005) Phylogenetic shadowing and computational identification of human microRNA genes. Cell 120(1):21–24
25. Rajewsky N (2006) microRNA target predictions in animals. Nat Genet 38(Suppl):S8–S13
26. Kozomara A, Griffiths-Jones S (2011) miRBase: integrating microRNA annotation and deep-sequencing data. Nucleic Acids Res 39(Database issue):D152–D157
27. Siomi H, Siomi MC (2010) Posttranscriptional regulation of microRNA biogenesis in animals. Mol Cell 38(3):323–332
28. Pillai RS (2005) MicroRNA function: multiple mechanisms for a tiny RNA? RNA 11(12):1753–1761
29. Zhou X, Yang PC (2012) MicroRNA: a small molecule with a big biological impact. Microrna 1(1):1
30. Lee KP, Shin YJ, Panda AC et al (2015) miR-431 promotes differentiation and regeneration of old skeletal muscle by targeting Smad4. Genes Dev 29(15):1605–1617
31. Panda AC, Abdelmohsen K, Gorospe M (2017) SASP regulation by noncoding RNA. Mech Ageing Dev 168:37–43
32. Panda AC, Sahu I, Kulkarni SD et al (2014) miR-196b-mediated translation regulation of mouse insulin2 via the 5′UTR. PLoS One 9(7):e101084
33. Munk R, Panda AC, Grammatikakis I et al (2017) Senescence-associated microRNAs. Int Rev Cell Mol Biol 334:177–205
34. Yu L, Gong X, Sun L et al (2016) The circular RNA Cdr1as act as an oncogene in hepatocellular carcinoma through targeting miR-7 expression. PLoS One 11(7):e0158347
35. Tang W, Ji M, He G et al (2017) Silencing CDR1as inhibits colorectal cancer progression through regulating microRNA-7. Onco Targets Ther 10:2045–2056
36. Peng L, Yuan XQ, Li GC (2015) The emerging landscape of circular RNA ciRS-7 in cancer (review). Oncol Rep 33(6):2669–2674
37. Lukiw WJ (2013) Circular RNA (circRNA) in Alzheimer's disease (AD). Front Genet 4:307
38. Xu H, Guo S, Li W et al (2015) The circular RNA Cdr1as, via miR-7 and its targets, regulates insulin transcription and secretion in islet cells. Sci Rep 5:12453
39. Kefas B, Godlewski J, Comeau L et al (2008) microRNA-7 inhibits the epidermal growth factor receptor and the Akt pathway and is down-regulated in glioblastoma. Cancer Res 68(10):3566–3572
40. Pan H, Li T, Jiang Y et al (2018) Overexpression of circular RNA ciRS-7 abrogates the tumor suppressive effect of miR-7 on gastric cancer via PTEN/PI3K/AKT signaling pathway. J Cell Biochem 119(1):440–446
41. Capel B, Swain A, Nicolis S et al (1993) Circular transcripts of the testis-determining gene Sry in adult mouse testis. Cell 73(5):1019–1030
42. Long L, Huang G, Zhu H et al (2013) Down-regulation of miR-138 promotes colorectal cancer metastasis via directly targeting TWIST2. J Transl Med 11:275
43. Huang G, Zhu H, Shi Y et al (2015) cir-ITCH plays an inhibitory role in colorectal cancer by regulating the Wnt/beta-catenin pathway. PLoS One 10(6):e0131225
44. Yang C, Yuan W, Yang X et al (2018) Circular RNA circ-ITCH inhibits bladder cancer progression by sponging miR-17/miR-224 and regulating p21, PTEN expression. Mol Cancer 17(1):19
45. Li F, Zhang L, Li W et al (2015) Circular RNA ITCH has inhibitory effect on ESCC by suppressing the Wnt/beta-catenin pathway. Oncotarget 6(8):6001–6013
46. Wan L, Zhang L, Fan K et al (2016) Circular RNA-ITCH suppresses lung cancer proliferation via inhibiting the Wnt/beta-catenin pathway. Biomed Res Int 2016:1579490
47. Zheng Q, Bao C, Guo W et al (2016) Circular RNA profiling reveals an abundant circHIPK3 that regulates cell growth by sponging multiple miRNAs. Nat Commun 7:11215
48. Chen G, Shi Y, Liu M et al (2018) circHIPK3 regulates cell proliferation and migration by sponging miR-124 and regulating AQP3 expression in hepatocellular carcinoma. Cell Death Dis 9(2):175

49. Li Y, Zheng F, Xiao X et al (2017) Circ HIPK3 sponges mi R-558 to suppress heparanase expression in bladder cancer cells. EMBO Rep 18(9):1646–1659
50. Tian F, Wang Y, Xiao Z et al (2017) Circular RNA circHIPK3 promotes NCI-H1299 and NCI-H2170 cell proliferation through miR-379 and its target IGF1. Zhongguo Fei Ai Za Zhi 20(7):459–467
51. Panda AC, Grammatikakis I, Kim KM et al (2017) Identification of senescence-associated circular RNAs (SAC-RNAs) reveals senescence suppressor CircPVT1. Nucleic Acids Res 45(7):4021–4035
52. Chen J, Li Y, Zheng Q et al (2017) Circular RNA profile identifies circPVT1 as a proliferative factor and prognostic marker in gastric cancer. Cancer Lett 388:208–219
53. Liu Q, Zhang X, Hu X et al (2016) Circular RNA related to the chondrocyte ECM regulates MMP13 expression by functioning as a MiR-136 'Sponge' in human cartilage degradation. Sci Rep 6:22572
54. Zhong Z, Huang M, Lv M et al (2017) Circular RNA MYLK as a competing endogenous RNA promotes bladder cancer progression through modulating VEGFA/VEGFR2 signaling pathway. Cancer Lett 403:305–317
55. Zhong Z, Lv M, Chen J (2016) Screening differential circular RNA expression profiles reveals the regulatory role of circTCF25-miR-103a-3p/miR-107-CDK6 pathway in bladder carcinoma. Sci Rep 6:30919
56. Wang K, Sun Y, Tao W et al (2017) Androgen receptor (AR) promotes clear cell renal cell carcinoma (ccRCC) migration and invasion via altering the circHIAT1/miR-195-5p/29a-3p/29c-3p/CDC42 signals. Cancer Lett 394:1–12
57. Wang K, Long B, Liu F et al (2016) A circular RNA protects the heart from pathological hypertrophy and heart failure by targeting miR-223. Eur Heart J 37(33):2602–2611
58. Peng L, Chen G, Zhu Z et al (2017) Circular RNA ZNF609 functions as a competitive endogenous RNA to regulate AKT3 expression by sponging miR-150-5p in Hirschsprung's disease. Oncotarget 8(1):808–818
59. Xie H, Ren X, Xin S et al (2016) Emerging roles of circRNA_001569 targeting miR-145 in the proliferation and invasion of colorectal cancer. Oncotarget 7(18):26680–26691
60. Song YZ, Li JF (2018) Circular RNA hsa_circ_0001564 regulates osteosarcoma proliferation and apoptosis by acting miRNA sponge. Biochem Biophys Res Commun 495(3):2369–2375
61. Cheng X, Zhang L, Zhang K et al (2018) Circular RNA VMA21 protects against intervertebral disc degeneration through targeting miR-200c and X linked inhibitor-of-apoptosis protein. Ann Rheum Dis 77(5):770–779
62. Zhou ZB, Du D, Huang GX et al (2018) Circular RNA Atp9b, a competing endogenous RNA, regulates the progression of osteoarthritis by targeting miR-138-5p. Gene 646:203–209
63. Wang L, Wei Y, Yan Y et al (2018) CircDOCK1 suppresses cell apoptosis via inhibition of miR196a5p by targeting BIRC3 in OSCC. Oncol Rep 39(3):951–966
64. Han D, Li J, Wang H et al (2017) Circular RNA circMTO1 acts as the sponge of microRNA-9 to suppress hepatocellular carcinoma progression. Hepatology 66(4):1151–1164
65. Fu L, Chen Q, Yao T et al (2017) Hsa_circ_0005986 inhibits carcinogenesis by acting as a miR-129-5p sponge and is used as a novel biomarker for hepatocellular carcinoma. Oncotarget 8(27):43878–43888
66. Xu XW, Zheng BA, Hu ZM et al (2017) Circular RNA hsa_circ_000984 promotes colon cancer growth and metastasis by sponging miR-106b. Oncotarget 8(53):91674–91683
67. Zhang XL, Xu LL, Wang F (2017) Hsa_circ_0020397 regulates colorectal cancer cell viability, apoptosis and invasion by promoting the expression of the miR-138 targets TERT and PD-L1. Cell Biol Int 41(9):1056–1064
68. Deng N, Li L, Gao J et al (2018) Hsa_circ_0009910 promotes carcinogenesis by promoting the expression of miR-449a target IL6R in osteosarcoma. Biochem Biophys Res Commun 495(1):189–196
69. He R, Liu P, Xie X et al (2017) circGFRA1 and GFRA1 act as ceRNAs in triple negative breast cancer by regulating miR-34a. J Exp Clin Cancer Res 36(1):145
70. Jin H, Jin X, Zhang H et al (2017) Circular RNA hsa-circ-0016347 promotes proliferation, invasion and metastasis of osteosarcoma cells. Oncotarget 8(15):25571–25581
71. Chen J, Cui L, Yuan J et al (2017) Circular RNA WDR77 target FGF-2 to regulate vascular smooth muscle cells proliferation and migration by sponging miR-124. Biochem Biophys Res Commun 494(1–2):126–132
72. Sun Y, Yang Z, Zheng B et al (2017) A novel regulatory mechanism of smooth muscle alpha-actin expression by NRG-1/circACTA2/miR-548f-5p axis. Circ Res 121(6):628–635
73. Wang X, Zhu X, Zhang H et al (2018) Increased circular RNA hsa_circ_0012673 acts as a sponge of miR-22 to promote lung adenocarcinoma proliferation. Biochem Biophys Res Commun 496(4):1069–1075
74. Zhang J, Liu H, Hou L et al (2017) Circular RNA_LARP4 inhibits cell proliferation and invasion of gastric cancer by sponging miR-424-5p and regulating LATS1 expression. Mol Cancer 16(1):151
75. Liang HF, Zhang XZ, Liu BG et al (2017) Circular RNA circ-ABCB10 promotes breast cancer proliferation and progression through sponging miR-1271. Am J Cancer Res 7(7):1566–1576
76. Dudekula DB, Panda AC, Grammatikakis I et al (2016) CircInteractome: a web tool for exploring circular RNAs and their interacting proteins and microRNAs. RNA Biol 13(1):34–42

77. Panda AC, Dudekula DB, Abdelmohsen K et al (2018) Analysis of circular RNAs using the web tool CircInteractome. Methods Mol Biol 1724:43–56
78. Glazar P, Papavasileiou P, Rajewsky N (2014) circBase: a database for circular RNAs. RNA 20(11):1666–1670
79. Ghosal S, Das S, Sen R et al (2013) Circ2Traits: a comprehensive database for circular RNA potentially associated with disease and traits. Front Genet 4:283
80. Liu YC, Li JR, Sun CH et al (2016) CircNet: a database of circular RNAs derived from transcriptome sequencing data. Nucleic Acids Res 44(D1):D209–D215
81. Li JH, Liu S, Zhou H et al (2014) starBase v2.0: decoding miRNA-ceRNA, miRNA-ncRNA and protein-RNA interaction networks from large-scale CLIP-Seq data. Nucleic Acids Res 42(Database issue):D92–D97

Regulation of Transcription by Circular RNAs

7

Rumela Bose and Rupasri Ain

Abstract

Circular RNAs (circRNAs) are a class of non-coding RNA that are present in wide variety of cells in various tissue types across species. They are non-polyadenylated, single-stranded, covalently closed RNAs. CircRNAs are more stable than other RNAs due to lack of 5′ or 3′ end leading to resistance to exonuclease digestion. The length of circRNAs varies from 1 to 5 exons with retention of introns in mature circRNAs with ~25% frequency. They are primarily found in the cytosol within the cell although the mechanism of their nuclear export remains elusive. However, there is a subpopulation of circRNAs that remain in the nucleus and regulate RNA-Pol-II-mediated transcription. Bioinformatic approaches mining RNA sequencing data enabled genome-wide identification of circRNAs. In mammalian genome over 20% of the expressed genes in cells and tissues can produce these transcripts. Owing to their abundance, stability, and diverse expression profile, circRNAs likely play a pivotal role in regulatory pathways controlling lineage determination, cell differentiation, and function of various cell types. Yet, the impact of circRNA-mediated regulation on various cell transcriptome remains largely unknown. In this chapter, we will review the regulatory effects of circRNAs in the transcription of their own or other genes. Also, we will discuss the association of circRNAs with miRNAs and RNA-binding proteins (RBPs), with special reference to *Drosophila* circMbl and their role as an "mRNA trap," which might play a role in its regulatory potential transcriptionally or posttranscriptionally.

Keywords

circRNAs · Noncoding RNA · Transcription regulation · Splicing · miRNA sponge

1 Introduction

With the advent of our understanding of molecular regulation of gene expression in last three decades, the layers of complexity have increased in every possible step in gene regulation. Circular RNAs have emerged as a new player of gene regulation intervening in transcriptional as well as posttranscriptional processes. Although transport of circRNAs is not well understood, their cellular compartmentalization dictates their role in gene regulation.

Circular RNAs exist in diverse isoforms in wide variety of organisms ranging from

R. Bose · R. Ain (✉)
Division of Cell Biology and Physiology, CSIR-Indian Institute of Chemical Biology,
Kolkata, West Bengal, India
e-mail: rupasri@iicb.res.in

J. Xiao (ed.), *Circular RNAs*, Advances in Experimental Medicine and Biology 1087,
https://doi.org/10.1007/978-981-13-1426-1_7

self-replicating viroids to humans. The simplest type of circular RNAs with 250–400 nucleotides exist as single-stranded RNA genome of viroids, which are plant pathogens. Yet another example of viral genome circular RNA is that of hepatitis delta virus, which is a human pathogen, and length of the circle is approximately 1.7 kb. Second types of circRNAs primarily comprised of excised intronic circles. Based on the type of introns, they can be of four different types: (a,b) circRNAs containing either excised group I or group II introns formed by ribozymes in some eukaryotes, bacteria, and viruses as RNA processing by-product; (c) circular intronic RNAs (ciRNAs) formed in eukaryotes by spliceosome mediated 2′-5′ branchpoint attack followed by degradation of downstream intronic sequence; and (d) circRNAs containing excised tRNA introns found in some archaea as tRNA processing by-product. The third type of circRNAs arise either as by-product of rRNA processing or as intermediate during tRNA processing in some archaea and algae [1]. The first discovery of an intron-derived circular RNA was made from an intervening sequence in the pre-rRNA transcript of the ciliate *Tetrahymena* [2]. Transcriptome-wide study of the archaeon *Sulfolobus solfataricus* P2 revealed several novel circular transcripts derived from rRNA and tRNA intronic sequences [3]. The other types of circRNAs are primarily found in higher eukaryotic cells. Most of the circular RNAs in eukaryotes are produced by a process termed as backsplicing, wherein a downstream splice donor site is joined to an upstream splice acceptor site to form circular molecules (Fig. 7.1).

This gives rise to three broadly defined isoforms: circular intronic RNAs (ciRNA), exonic circular RNAs (ecircRNA), and exon-intron circular RNAs (EIciRNA) which have different modes of biogenesis (Fig. 7.1). Intronic circRNAs are derived from intron lariats produced during pre-mRNA splicing [4], exonic circular RNAs are produced co-transcriptionally compromising the linear mRNA splicing [5], while the exon-intron circRNAs form circles with the intervening intron retained between them.

In addition, eukaryotic circRNAs may also arise either by trans-splicing or exon scrambling. In trans-splicing event, exons from two separate mRNAs are joined together, followed by backsplicing-mediated rearrangements of the exons in a circular orientation [6]. Scrambled exons may arise due to genomic rearrangement, tandem duplication, trans-splicing, or backsplicing [1]. With the help of RT-PCR with divergent

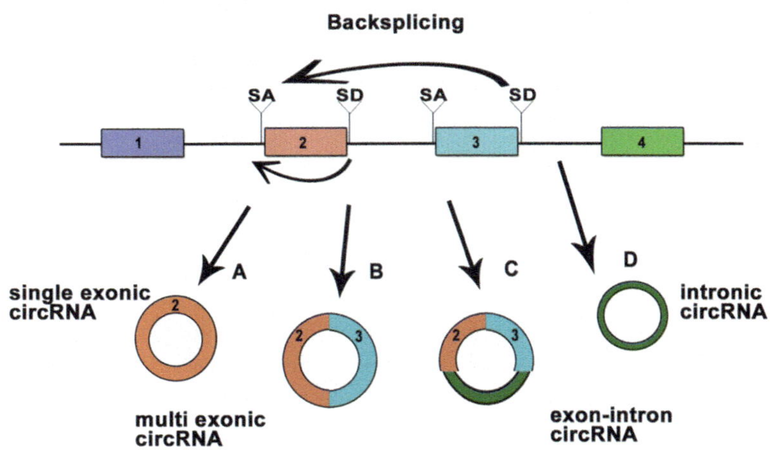

Fig. 7.1 Biogenesis and diversity of circular RNAs Circular RNAs are produced by backsplicing in which a downstream splice donor site is joined to an upstream splice acceptor site. These circular RNAs show a wide range of diverse isoforms. They may consist of a single (**a**) or multiple exons (**b**) or retain an unspliced intron between two exons (**c**). Sometimes they can be entirely composed of an intron (**d**). Colored bars, exons; black lines, introns; *SD* splice donor, *SA* splice acceptor

primers, as well as computational algorithms, the order of some exons in matured eukaryotic mRNAs were found to be inconsistent with respect to their gene sequences. Such exons are called scrambled exons. Patrick Brown and his team statistically evaluated the origin of circular RNAs from scrambled exons. About 32% of the scrambled exon junctions indicated toward the occurrence of a hypothetical circular RNA associated with them. They tested for six genes including nine scramble exon-containing isoforms. Consistent with the statistical findings, each of these transcripts were resistant to RNaseR further confirming their circular nature [7].

Bioinformatic analysis by Jeck et al. showed that sequences flanking exons that form circular RNAs are more likely to contain complementary Alu elements [8]. More than 12,000 and 6000 circRNAs have been found in *Arabidopsis thaliana* and *Oryza sativa*, respectively, which are derived from intronic, exonic, as well as other sequences [9]. CircRNAs accumulate during heat stress in Arabidopsis, implying its potential role in heat stress [10]. Evolutionary significance of circRNAs lies in the fact that several of them are conserved in sequences as well as function within same or different species. For example, circular RNAs produced from the HIPK family are conserved between humans and rodents [8]; muscleblind circular RNA regulates alternative splicing across *Drosophila* species in a conserved manner [5]. Various databases have been designed to identify and for quick access to unified data about different aspects of circRNAs. Some of the commonly used databases are circ2Traits [11], circBase [12], circNet [13], circRNADb [14], etc.

Once produced, circRNAs are either exported to the cytoplasm or retained inside the nucleus. Molecular mechanisms as well as factors governing circRNAs' nuclear export are not well understood. As the potential function of a circRNA is much dependent on the cellular milieu to which it is exposed, it is important to determine its export mechanism. Emerging evidences suggest that they may surpass the nucleus during mitosis [6] or are exported by mRNA transporters recruited at the exon-exon junction during splicing [15]. Additionally, intron-containing EIciRNAs may be retained in the nucleus in a manner similar to incompletely spliced mRNAs [16]. This is a very interesting yet unexplored area of circular RNA biology.

Nuclear localized circRNAs (ciRNAs, EIciRNAs) primarily function at the transcription level as demonstrated by their interaction with Pol II and U1 snRNP complex. On the other hand, certain circRNAs can alternatively splice their own pre-mRNA to produce circular transcripts. This circularization of the exons inhibits the production of the canonical protein from their locus thus regulating the expression of these genes [17]. Majority of the circRNAs are exported to the cytoplasm, where they interact with RBPs and miRNAs sequestering them into specific cellular locations [18–21]. These circRNAs have several conserved binding sites for RBPs and miRNAs, allowing them to posttranscriptionally regulate expression of the target genes. The list of circRNAs acting as RBP and miRNA sponge is given in the Table 7.1 and 7.2.

Although circRNAs are "noncoding" RNAs, there are instances where they may code protein. One such naturally occurring protein encoding circular RNA is found in the hepatitis delta virus, where a circRNA codes for a 122 amino acid protein in infected mammalian cells [22]. Insertion of internal ribosomal entry site (IRES) in circRNAs leads to translation by endogenous ribosome [23]. In their experiment, the ribosomes traversed the circles repeatedly and gave rise to high molecular weight multimeric polypeptides. By introducing thrombin cleavage site upstream of the IRES sequence, these high molecular weight products could be cleaved to produce single polypeptides of 40kD size (as was predicted to be encoded by the circular RNA) [23].

2 The Transcription Machinery in Eukaryotes

2.1 Eukaryotic RNA Polymerases

Eukaryotic nuclei contain three RNA polymerases which have different locations within the nucleus and also different responses to salt and

Table 7.1 CircRNAs that act as RBP sponge

Name	Protein it binds	Mode of action	Reference
circMbl	MBL	Regulates its own biosynthesis	[5]
circPABPN1	HuR	Prevents binding of HuR to PABPN1 mRNA and lowers its translation	[49, 50]
circFoxo3	Id-1, E2F1, HIF-α, and FAK	Reduced nuclear and mitochondrial translocation of these factors during cardiac stress causing cardiac senescence	[50, 51]
	P21, CDK2	Facilitates inhibition of CDK2 by p21, inhibiting cell cycle progression	[52]
	MDM2, p53	Facilitates MDM2-mediated ubiquitination of p53	[18]
circANRIL	PES1	Competes with rRNA binding of PES1 impairing ribosome biogenesis	[50, 53]
circAmotl1	c-myc, STAT3, PDK1, AKT1	Translocates them to the nucleus affecting transcription	[17, 19–21]
ciRS-7/Cdr1as	AGO	Inhibits miRNA-mediated degradation of target mRNAs	[54]
circ CDYL, circNFATC3, circANKRD17	IMP-3	–	[55]

Table 7.2 CircRNAs that act as miRNA sponge

Name	miRNA it binds	Mode of action	Reference
ciRS-7/ Csr1as	miR-7	Prevents downregulation of target genes	[54]
circZNF609	miR-150-5p	Modulates AKT expression in Hirschsprung's disease	[56]
circSry	miR-138	Regulates hypoxia-induced apoptosis in cardiac myocytes	[17, 57]
circZNF91	miR-23, miR-181, miR-199	Prevents downregulation of *ZNF225, ZNF486, and ZNF85* (miR-23) tumor suppressor genes RB1 and RBAK (miR-181) *ZNF20 and ZNF791 (miR-199)*	[5, 58]
circHRCR	miR-223	Removes the translation inhibition of apoptosis inhibitor with CARD domain (ARC)	[17, 59]
circMFACR	miR-652-3p	Regulates mitochondrial dynamics, apoptosis of cardiac myocytes, and myocardial infarction	[17, 60]
circWDR77	miR-124	Vascular smooth muscle cell proliferation	[17, 61]
hsa-circ-000595	miR-19a	Decreased apoptosis in human aortic smooth muscle cells	[62]
hsa-circ-001569	miR-145	Positively regulates cell proliferation	[63]

divalent cations, and RNA polymerase I, which transcribes rRNA genes, is located in the nucleolus, while RNA polymerase II and III, which mainly transcribe mRNA and tRNAs, respectively, are located in the nucleoplasm. These three enzymes are differently sensitive to the poison alpha-amanitin [24]. The subunit structures of these three polymerases from different eukaryotes have been well studied. All these structures contain multiple subunits, some of which are common to all three polymerases. Twelve subunits of yeast, *Saccharomyces cerevisiae*, RNA Pol II, have been discovered till date [25, 26]. Richard Young categorized the originally identified subunits into three broad classifications: the core subunits which are indispensible for polymerase structure and function, the common subunits which are found in all three polymerases, and the nonessential subunits, which are conditionally dispensable for enzyme activity. The three polypeptides Rpb1, Rpb2, and Rpb3 are absolutely required for enzyme activity. The Rpb1 subunit, identified as the functional homologue of the bacterial polymerase β' subunit, is involved in DNA binding. Under physiological conditions, the Rpb1 subunit exists in two isoforms IIo and IIa based upon the state of phosphorylation. The amino acid sequence of IIa subunit shows a repeat string of seven amino acids with the following consensus sequence: Tyr-Ser-Pro-Thr-Ser-Pro-Ser. Because this sequence is found at the carboxy terminus of the IIa subunit, it is named as the carboxy terminal domain (CTD). This CTD is likely to get phosphorylated at the Ser, Thr residues, transforming IIa to the IIo subunit. The existence of two forms of Rpb1 subunit in the cells implies that they serve different purpose in transcription. It is indeed the case wherein Pol II with Rpb1 IIa (referred to as Pol IIA) can bind to promoter and thus initiate transcription, and Pol II with Rpb1 IIo (referred to as Pol IIo) is the species that carries out elongation [27, 28]. The Rpb2 subunit is involved in nucleotide binding at the active site of the enzyme in all the three polymerases. There is one 20-amino acid region of Rpb3 that shares great similarity to E. coli α-subunit. Also same kinds of polymerase assembly defects are seen in RPB3 mutant yeasts as in E. coli α-subunit mutants. Thus it has been predicted that Rpb3 is required for the appropriate assembly of the RNA polymerase holoenzyme. Five subunits – Rpb5, Rpb6, Rpb8, Rpb10, and Rpb12 – are common to all three polymerases and serve general purpose of transcription like processivity or fidelity. The two nonessential subunits Rpb4 and 9 are not absolutely required for polymerase activity under normal conditions, but RPB4 and 9 mutants are inviable at high temperatures.

2.2 Eukaryotic Promoters

The three polymerases have different structures and transcribe different genes and therefore recognize different promoters. The promoters recognized by Pol II are termed as class II promoters which have two parts: the core promoter and an upstream proximal element. The core promoters generally constitute of the TATA box, an initiator site located approximately 25–30 bp upstream of the transcription start site. However, TATA-less promoters are frequently found in the cells, as in the case of house-keeping genes and developmentally regulated homoeotic genes. Promoter proximal elements include CCAAT box and GC box (GGGCGG) located at ~100 bp and ~200 bp upstream of the transcription start site. The core promoter drives basal level of transcription and is the binding site for TATA-binding protein (TBP) and associated factors (TAFs), whereas the promoter proximal elements regulate true level of transcription and are binding sites for gene-specific transcription factors. Yet another cis-acting element that regulates transcription is enhancer element located upstream or downstream of transcription start site. Activators or repressors of transcription bind to these elements and regulate rate of transcription.

The class I promoters, recognized by Pol I, are not well conserved across species. It consists of a core element surrounding the transcription start site and an upstream control element about 100 bp further upstream. The spacing between the two elements is very important as insertion or deletion of bases between them greatly reduces promoter strength [29]. The classical genes are transcribed by Pol III promoters that are located within the genes. The internal promoter of 5S rRNA gene is split into three regions: box A, a short intermediate element, and box C. One class III gene, 7SL, contains a weak internal promoter and a sequence at the 5′-flanking region of the gene that is required for high level transcription. Other class III genes (e.g., 7SK and U6 RNA) completely lack internal promoters and contain class II-like promoters that lie in the 5′-flanking region and contains TATA box [30].

2.3 Transcription Factors

General transcription factors drive basal level of transcription by binding to core promoter element and the RNA polymerase to form a pre-initiation complex (PIC). The minimal PIC includes RNA pol-II and six general transcription factors that are TFIIA, TTFIIB, TFIID, TFIIE,

TFIIF, and TFIIH. Binding and sequential recruitment of TFs have been demonstrated by various scientists using gel mobility shift and various other assays [31, 32]. The largest general TF, TFIID, contains various subunits that include TBP and 16 TAFs. The saddle-shaped TBP binds to the promoter in the DNA minor groove creating a bend in the DNA followed by recruitment of TFIIA and TFIIB, respectively. TBP mutants are not only deficient in class II genes but also in class I and III genes suggesting that TBP is a universal transcription factor required by all three RNA Pols. RNA Pol-II binds to TFIIF to form the Pol-II complex. TFIIB recruits the PolII complex to the promoter and helps the complex bind correctly. This is followed by binding of TFIIE and TFIIH binding to the complex resulting in formation of basal PIC. TFIIH subunits possess ATPase and helicase activity that create negative super helical tension resulting in unwinding of one turn of DNA to form the transcription bubble. TFIIA, B, E, F, and H leave once RNA elongation begins, but TFIID stays till the end of elongation [33, 34].

In addition to general transcription factors that drive basal level of transcription, there are about 2600 transcription factors coded by human genome. These transcription factors possess DNA binding domain and transcription activation domain and interact with other proteins and increase the level of transcription as much as 100-fold. They drive tissue-specific and cell type-specific expression or repression of various genes.

3 Circular RNAs Act as Potent Regulators of Transcription

CircRNAs regulate transcription at the initiation as well as the elongation step. In addition they also regulate gene expression posttranscriptionally. An elaborate interplay among diverse protein coding and noncoding RNA species during transcription has been described [35–40]. The various roles played by different circular RNAs at different stages of transcription have been described below.

3.1 Regulation of Transcription at Initiation Step: Role of Exon–Intron Circular RNAs

Eukaryotic transcription can be primarily tuned at the initiation step, which involves formation of the pre-initiation complex at the promoter. Almost all known transcription factors assemble at this point further stabilizing the complex, stimulating the rate of transcription. To test whether noncoding RNAs can regulate transcription, Li et al. [41] performed cross-linking followed by immunoprecipitation (CLIP) using RNA Pol II-specific antibody. RNA sequencing of Pol II CLIP samples and further bioinformatics analyses revealed as many as 111 circRNAs to be associated with Pol II. Out of these 111 circRNAs, 15 were EIciRNAs. Fluorescence in situ hybridization (FISH) revealed that 2 of these 15 EIciRNAs, circEIF3J and circPAIP2, were exclusively localized in the nucleus. Knocking down of these two EIciRNAs, using either short interfering RNA (siRNA) or RNase H-based antisense oligonucleotides (ASO), resulted in decrease in the parent transcripts (eif3j and paip2) in both HeLa and HEK293 cells, without any effect on the neighboring genes' transcripts (ctdspl2 and matr3). To understand whether this decrease was due to decrease in the transcription of the respective mRNAs, nuclear run-on experiments were performed with nuclei extracted from circEIF3J and circPAIP2 knockdown cells. It was found that knockdown of circEIF3J and circPAIP2 indeed resulted in lower EIF3J and PAIP2 transcription, whereas knockdown of EIF3J and PAIP2 with siRNA had no effect on their transcription. Not only that circEIF3J and circPAIP2 were found to co-localize with the genomic loci of their parental genes as revealed by RNA-DNA double FISH. These data collectively indicates that circEIF3J and circPAIP2 may regulate the expression of their parental genes in cis. However, these EIciRNAs were not confined to their parental gene loci only, thus leaving a possibility of trans effects of these circRNAs on other loci as well.

The obvious question that would arise is whether these EIciRNAs directly associate with Pol II or other factors of the pre-initiation complex to exert their effects on transcription. Pull down experiments with specific oligos corresponding to different regions of either circEIF3J or circPAIP2 led to coprecipitation of U1A and U1C snRNPs, U1 snRNA, along with Pol II suggesting these interactions to be specific. Interestingly, sites within the promoter and also regions of the first exon of parent genes coprecipitated in these experiments. U1 snRNA is a core-splicing component that associates specifically with TFIIH which is a general transcription initiation factor [42]. The role of U1 snRNA in transcription initiation is to stimulate the formation of the first phosphodiester bond by Pol II [42]. Each of these EIciRNAs has one U1 snRNA-binding site in their retained intron. Sterically blocking this site not only decreased interaction of U1 snRNA with EIciRNA but also with Pol II and EIciRNA and Pol II with promoters of the parental genes of circEIF3J and circPAIP2. As a result, the transcription of the parental genes of the corresponding circRNAs also decreased. Conversely, the binding of Pol II with specific U1 snRNPs (U1A and U1C) to their parent gene promoters also requires the presence of the EIciRNAs [41]. Pull down with U1A- and U1C-specific antibodies, but not with any other snRNP (like Lsm10), or auxiliary factors (U2AF65 and U2AF35), led to coprecipitation of substantial amount of EIciRNAs. Furthermore, chromatin immunoprecipitation (ChIP) experiments revealed that only U1A and U1C and not U2AF65 and U2AF35 interacted with the promoter regions of some genes like EIF3J and PAIP2, but not their neighboring genes. In line with these findings, U1 snRNA was found to be co-localized with majority of the circEIF3J or circPAIP2 within the nucleus using dual RNA FISH.

Available experimental evidence therefore suggest that specific RNA-RNA interaction between U1 snRNA and the EIciRNAs followed by interaction of the EIciRNA-snRNP complex with the Pol II at the promoter site leads to upregulation of transcription of their parent genes (Fig. 7.2).

3.2 Regulation of Transcription During Elongation: Interaction of Intronic circRNAs with Elongating Pol II

Although in most cases transcription is regulated at the initiation stage, transcriptional regulation can also take place during elongation. Circular RNAs, intronic circRNAs in particular, provide such example where they control transcription at the elongation step. It is generally believed introns are unused part of the mRNA which are unstable and rapidly degraded. One way by which intronic RNAs can prevent their degradation and accumulate in the cells is by circularization, known as intronic circular RNAS (ciRNAs). These ciRNAs do not have their own promoters but are derived from the spliced introns of their parent transcripts, sometimes enhancing the production of the latter. One such ciRNA identified is the ci-ankrd52 [4]. The ci-ankrd52 is derived from the second intron of the ankrd52 gene, which codes for a protein with unknown function, containing a large ankyrin repeat domain. Synthetic antisense oligodeoxynucleotides (ASO), which target the intron-derived ci-ankrd52, successfully downregulated the expression of these circular RNAs leading to a decrease in the parent mRNA level as well. ASO specific to ci-ankrd52, being complementary to the ankrd52 pre-mRNA intron, may bind to the pre-mRNA and subsequently leads to its degradation resulting in decreased mRNA levels. Inability of ASOs against introns, adjacent to the ci-ankrd52 to reduce ankrd52 mRNA, excludes this possibility. In addition, co-expression of ASOs and corresponding intronic RNAs except ci-ankrd52 failed to reduce ankrd52 mRNA levels confirming the specificity of ci-ankrd52 in regulating its parental mRNA expression. Similar results were obtained with ASO-mediated knockdown of two other ci-RNAs, ci-mcm5, and ci-sirt7.

In the quest to find out the mechanism of ciRNA-mediated reduction of parental mRNAs, authors tested three hypotheses.

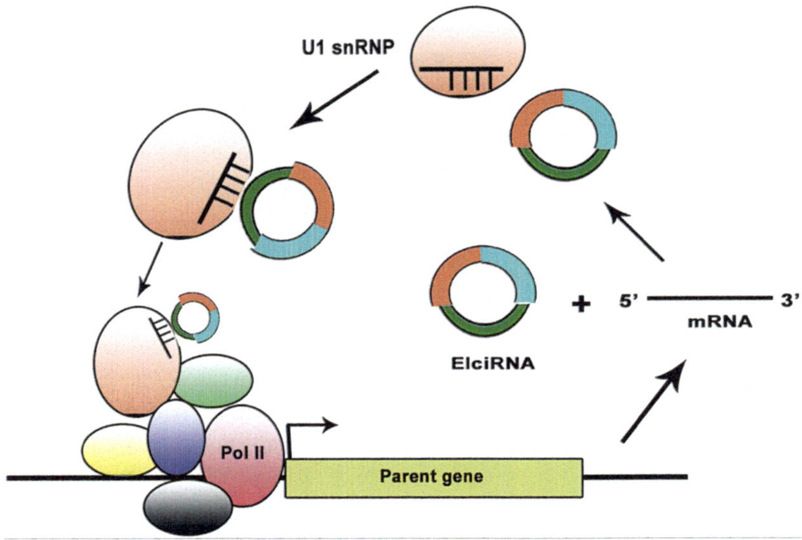

Fig. 7.2 Regulation of transcription initiation by ElciRNAs
Exon-intron circular RNAs are composed of exons and unspliced introns retained in between the exons. They are produced as by-product of gene transcription. They interact with U1-snRNA through specific RNA-RNA interactions. This ElciRNA-U1snRNP complex further interacts with the Pol II transcription initiation complex at the promoter of parent genes and promote their transcription. Orange and light blue bars, exons; green bar, intron

A. CiRNAs acting as miRNA sponge: Presence of only a few miRNA-binding sites on these ciRNAs and their exclusive nuclear localization exclude the possibility of these circRNAs acting as miRNA sponges.
B. CiRNAs required for proper mRNA processing: Analysis of relative abundance of splicing intermediates revealed that the ciRNA and its downstream introns are processed at a similar rate. However, knockdown of ci-ankrd52 gave rise to new isoforms of the mRNA with retained introns containing stop codons that lead to nonsense-mediated decay (NMD) of the parent mRNAs. These results indicate that ciRNAs regulate mRNA processing.
C. Transcriptional regulation by ciRNAs: DNA-RNA dual FISH revealed presence of ciRNAs in the elongating transcript of the parent gene.

Furthermore, biotinylated ci-ankrd52 interacted with phosphorylated RNA Pol II in in vitro pull down assay. This was further substantiated by co-immunoprecipitation of phosphorylated RNA Pol II with ci-ankrd52 in PA1 cell line. Pol II interaction with ci-mcm5 and ci-sirt7 was demonstrated using similar assays.

Phosphorylation of Pol II is pivotal to transcription elongation process [27]. Association of ciRNAs with phosphorylated Pol II therefore confirms their regulatory role in transcription elongation (Fig. 7.3).

3.3 Circular RNAs Posttranscriptionally Regulate Gene Expression by Acting as miRNA Sponges

CircRNAs that are exported from the nucleus and are located in the cytoplasm have several binding sites for miRNAs and compete with the target mRNAs for miRNA binding in the cytoplasm thus regulate gene expression at the posttranscriptional level. The best characterized circRNA that acts as miRNA sponge is the vertebrate ciRS-7, also known as Cdr1as that acts as sponge for miR-7. Produced from the vertebrate cerebellar degeneration-related 1 (CDR1) antisense transcript, Cdr1as is preferentially expressed in

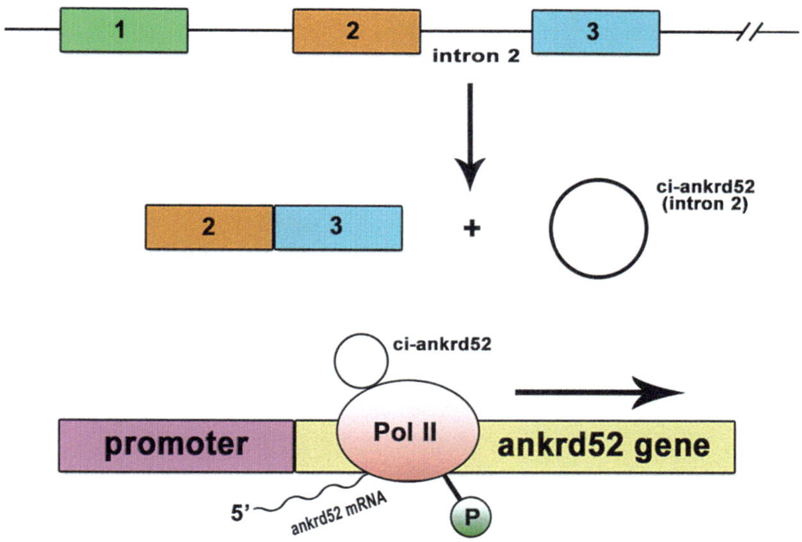

Fig. 7.3 Regulation of transcription elongation by intronic circRNA Intronic ci-ankrd52 is produced from the second intron of the ankrd52 gene. After its synthesis it accumulates at its site of transcription where it interacts with phosphorylated RNA Pol II and enhances transcription elongation of its parent gene. Colored bars, exons; black lines, introns

human and mouse brains [14]. It has over 60 binding sites for miR-7 as analyzed by PAR-CLIP experiments with human AGO [43]. Cdr1as and miR-7 are co-expressed in neuronal tissues, pancreas, and pituitary gland and also in murine pancreatic tissue-derived MIN6 cell line. Specifically, high level of co-expression is found in the developing midbrain of D13.5 mouse embryos. As expected, downregulation of Cdr1as led to downregulation of miR-7 targets along with house-keeping genes in HEK293 cells, indicating miR-7-mediated repression of targets in absence of Cdr1as.

Interestingly, Cdr1as has been shown to regulate insulin transcription and secretion in mouse islets [44]. Stimulation of islet cells with either forskolin or PMA led to increased expression of Cdr1as, but miR-7 was reduced under the same condition. Overexpression of miR-7 and Cdr1as separately, in MIN6 and isolated islet cells, led to decrease and increase in insulin secretion, respectively, as evaluated by glucose-stimulated insulin secretion assay (GSIS). These results indicate that miR-7 directly regulate levels of insulin transcript in the cell, whereas Cdr1as binds and to miR-7 and acting as a sponge reverses the effect of miR-7 on insulin transcript levels. As expected, in miR-7 overexpressing MIN6 cells, there was ~25% decrease in insulin content as compared to a ~70% increase in Cdr1as overexpressing cells. In mouse islet cells, the percentage was as high as ~90%. Thus, Cdr1as affects the insulin secretion by upregulating its biosynthesis. Interestingly, Myrip and Pax6, which are involved in insulin biosynthesis and secretion, are also potential targets of miR-7. Thus by acting as miR-7 sponge, Cdr1as upregulates levels of Myrip and Pax6 in the cell. In line with this argument, Cdr1as overexpression led to significant upregulation of Myrip and Pax6 mRNA in MIN6 cells by up to 70% and 50%, respectively. An even better result was observed in mouse islets. As expected, ectopic overexpression of miR-7 led to decrease in Myrip and Pax6 mRNA levels by 40–50%. Therefore, overexpressed Cdr1as could bind and sequester miR-7 in the cytoplasm and hence abolish its inhibitory effects on the Myrip and Pax6 mRNAs, which in turn elevates insulin biosynthesis and secretion (Fig. 7.4).

3.4 CircularMbl RNA Posttranscriptionally Regulates Its Own Expression by Acting as an RBP Sponge

CircRNAs in some cases regulate its own expression by tuning the posttranscriptional events.

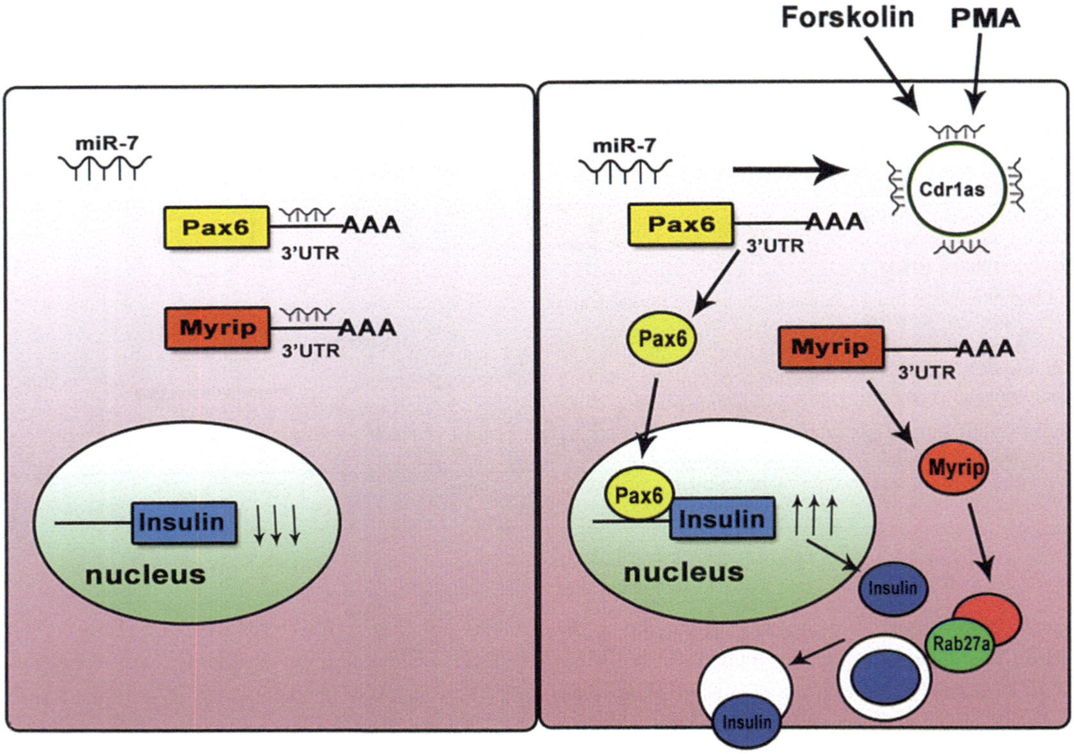

Fig. 7.4 Schematic diagram showing Cdr1as as a potent regulator of insulin transcription in mouse pancreatic β-cells and its secretion

In absence of any secretagogue, miR-7 binds to the 3′UTR of its targets Pax6 and Myrip mRNA, thereby decreasing the expression and secretion of insulin in mouse β-cells (left). In the presence of secretagogues-like forskolin or PMA, increased expression of Cdr1as leads to sequestration and inhibition of miR-7. Thus Pax6 mRNA is translated to produce a transcriptional activator, which translocates to the nucleus and promotes the transcription of insulin gene. On the other hand, the product of Myrip gene is involved in translocation and secretion of insulin, which is also increased due to Cdr1as overexpression (right)

Drosophila muscleblind (Mbl) circular RNA is one such transcript which drives its own expression through alternative splicing of its precursor RNA. The MUSCLEBIND protein (MBL) in *Drosophila* is required for the development of muscle and photoreceptor cells in the fly eye. It is expressed in the cells of embryonic muscle, and Mbl deficiency is embryonic lethal [45]. MBL protein promotes the splicing of the second exon of its own pre-mRNA into a circular transcript, circMbl, It thus competes with mbl mRNA production thereby decreasing the levels of MBL protein [46]. At low levels of MBL protein, the Mbl mRNA is spliced to produce the linear transcript which is translated to give rise to MBL protein. As the level of the MBL protein builds up, it binds to the mbl pre-mRNA and causes it to backsplice into circMbl. As a consequence the level of linear mbl transcript is diminished and so is the MBL protein. Furthermore, the circMbl itself contains several MBL protein-binding sites and thus can act as an RBP sponge. The circMbl binds to and sequesters MBL protein, lowering its free cellular concentration so that it can no longer produce circMbl transcripts thereby lowering its own level (Fig. 7.5). Thus circMbl regulates its own expression via MBL sequestration in a negative feedback mechanism.

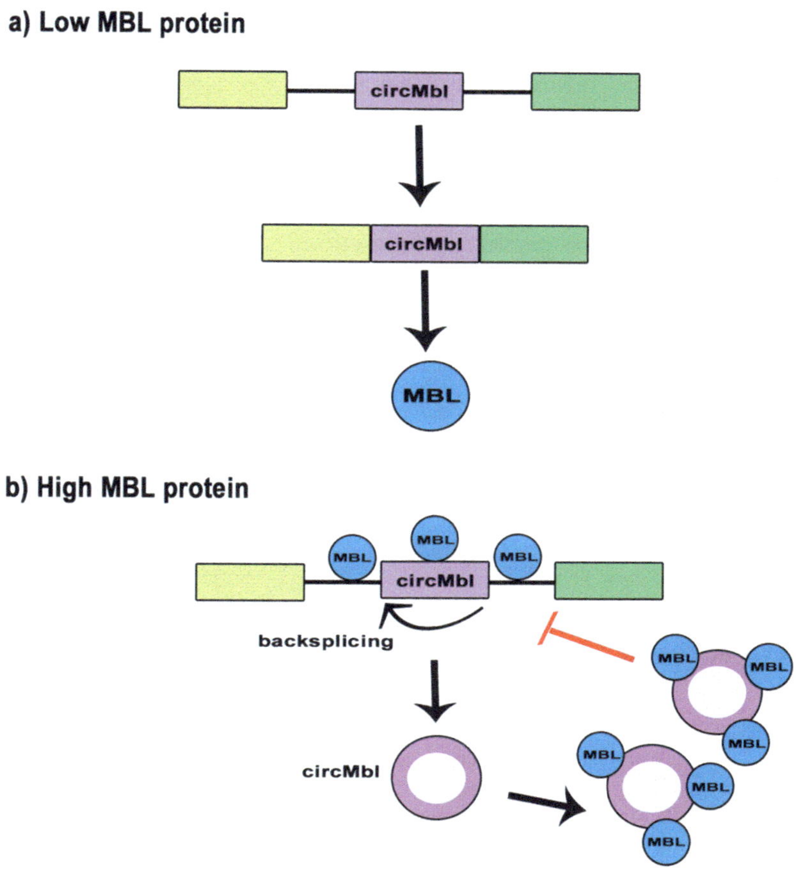

Fig. 7.5 *Drosophila* circMbl negatively regulates the expression of its own gene (**a**) When MBL protein is low, the mbl transcript is spliced to produce a linear mRNA which is translated to produce MBL protein. (**b**) As the amount of MBL protein rises, it binds to its pre-mRNA and causes the second exon to circularize to form circMbl thereby competing with splicing and synthesis of native MBL protein. Furthermore, circMbl binds to and sequesters MBL protein by its several MBL-binding sites. This gradually lowers its free cellular concentration and consequently decreasing its own expression in a negative feedback loop

3.5 Circular RNAs Act as "mRNA Trap": A Novel Mechanism in Regulating Gene Expression

The previous section describes how circRNA regulates the expression of a functional protein thereby regulating its own expression. In this section we will discuss about how some circRNAs sequester the translation start site on their linear transcripts to monitor their protein expression level. The Formin (fmn) gene, which is responsible for development of limbs and kidney in mouse, was reported to produce circular exonic RNAs (ecircRNAs) comprising exons 4 and 5 [47]. Fmn mutant mice with deletion of exon 4 and 5 did not produce the circfmn transcripts. Phenotypically, they had normal limb development but had renal agenesis with incomplete penetrance. The authors put forth the model of "mRNA trap" from these observations. In this model, the circular Fmn RNAs sequesters the linear fmn transcripts from being translated into functional FORMIN protein.

The "mRNA trap" phenomenon is also observed in patients with dystrophinopathy. The dystrophin (DMD) gene produces several circular transcripts. In the skeletal muscle of patients, scrambled RNAs in the form of circular dystrophin RNAs are produced, at the expense of linear in-frame transcripts reducing the levels of functional proteins [6, 48].

In mouse and humans, HIPK2 and HIPK3 loci generate exonic circular RNAs from the exon that contains ATG start codon, yet they are not translated to yield any protein product. For HIPK3 locus in particular, the circular isoform is more abundant than its linear isoform, which does not translate into any protein. Thus this mechanism can be viewed as yet another example of "mRNA-trapping" event by circular RNAs in order to modulate gene expression at posttranscriptional level [6, 8].

Acknowledgment Supported by CSIR-Indian Institute of Chemical Biology internal support grant, Rumela Bose is a recipient of Shyama Prasad Mukherjee predoctoral fellowship from the Council of Scientific and Industrial Research, India.

Conflict of Interest The authors declare that there is no conflict of interest.

References

1. Lasda E, Parker R (2014) Circular RNAs: diversity of form and function. RNA 20(12):1829–1842
2. Grabowski PJ, Zuag AJ, Cech TR (1981) The intervening sequence of the ribosomal RNA precursor is converted to a circular RNA in isolated nuclei of Tetrahymena. Cell 23(2):467–476
3. Danan M, Schwartz S, Edelheit S et al (2012) Transcriptome-wide discovery of circular RNAs in Archaea. Nucleic Acids Res 40(7):3131–3142
4. Zhang Y, Zhang XO, Chen T et al (2013) Circular intronic long noncoding RNAs. Mol Cell 51(6):792–806
5. Ashwal-Fluss R, Meyer M, Pamudurti NR et al (2014) circRNA biogenesis competes with pre-mRNA splicing. Mol Cell 56(1):55–66
6. Jeck WR, Sharpless NE (2014) Detecting and characterizing circular RNAs. Nat Biotechnol 32(5):453–461
7. Salzman J, Gawad C, Wang PL et al (2012) Circular RNAs are the predominant transcript isoform from hundreds of human genes in diverse cell types. PLoS One 7(2):e30733
8. Jeck WR, Sorrentino JA, Wang K et al (2013) Circular RNAs are abundant, conserved, and associated with ALU repeats. RNA 19(2):141–157
9. Ye CY, Chen L, Liu C et al (2015) Widespread noncoding circular RNAs in plants. New Phytol 208(1):88–95
10. Pan T, Sun X, Liu Y et al (2017) Heat stress alters genome-wide profiles of circular RNA in Arabidopsis. Plant Mol Biol 96(3):217–229
11. Ghoshal S, Das S, Sen R et al (2013) Circ2Traits: a comprehensive database for circular RNA potentially associated with disease and traits. Front Genet 4(283):283
12. Glazar P, Papavasileiou P, Rajewsky N (2014) circBase: a database for circular RNAs. RNA 20(11):1666–1670
13. Liu YC, Li JR, Sun CH et al (2016) CircNet: a database of circular RNAs derived from transcriptome sequencing data. Nucleic Acids Res 44(D1):D209–D215
14. Chen X, Han P, Zhou T et al (2016) circRNADb: a comprehensive database for human circular RNAs with protein-coding annotations. Sci Rep 6:34985
15. Le Hir H, Gatfield D, Izaurralde E et al (2001) The exon-exon junction complex provides a binding platform for factors involved in mRNA export and nonsense-mediated mRNA decay. EMBO J 20(17):4987–4997
16. Ebbesen KK, Kjems J, Hansen TB (2016) Circular RNAs: identification, biogenesis and function. Biochim Biophys Acta 1859(1):163–168
17. Han B, Chao J, Yao H (2018) Circular RNA and its mechanisms in disease: from the bench to the clinic. Pharmacol Ther 187:31. https://doi.org/10.1016/j.pharmthera.2018.01.010
18. Du WW, Fang L, Yang W et al (2017) Induction of tumor apoptosis through a circular RNA enhancing Foxo3 activity. Cell Death Differ 24(2):357–370
19. Yang ZG, Awan FM, Du WW et al (2017) The circular RNA interacts with STAT3, increasing its nuclear translocation and wound repair by modulating Dnmt3a and miR-17 function. Mol Ther 25(9):2062–2074
20. Yang Q, Du WW, Wu N et al (2017) A circular RNA promotes tumorigenesis by inducing c-myc nuclear translocation. Cell Death Differ 24(9):1609–1620
21. Zeng Y, Du WW, Wu Y et al (2017) A circular RNA binds to and activates AKT phosphorylation and nuclear localization reducing apoptosis and enhancing cardiac repair. Theranotics 7(16):3842–3855
22. Kos A, Dijkema R, Arnberg AC et al (1986) The hepatitis delta (δ) virus possesses a circular RNA. Nature 323(6088):558–560
23. Chen CY, Sarnow P (1995) Initiation of protein synthesis by the eukaryotic translational apparatus on circular RNAs. Science 268(5209):415–417
24. Weinmann R, Roeder RG (1974) Role of DNA-dependent RNA polymerase III in the transcription of the tRNA and 5S RNA genes. Proc Nalt Acad Sci USA 71(5):1790–1794
25. Woychik NA, Liao SM, Kolodziej PA et al (1990) Subunits shared by eukaryotic nuclear RNA polymerases. Genes Dev 4(3):313–323
26. Kolodziej PA, Woychik N, Liao SM et al (1990) RNA polymerase II subunit composition, stoichiometry, and phosphorylation. Mol Cell Biol 10(5):1915–1920
27. Payne JM, Laybourn PJ, Dahmus ME (1989) The transition of RNA polymerase II from initiation to

elongation is associated with phosphorylation of the carboxy-terminal domain of subunit IIa. J Biol Chem 264(33):19621–19629
28. Corden JL (1990) Tails of RNA polymerase II. Trends Biochem Sci 15:383–387
29. Haltiner MM, Smale ST, Tjian R (1986) Two distinct promoter elements in the human rRNA gene identified by linker scanning mutagenesis. Mol Cell Biol 6(1):227–235
30. Murphy S, Di Liegro C, Melli M (1987) The in vitro transcription of the 7SK RNA gene by RNA polymerase III is dependent on the presence of an upstream promoter. Cell 51(1):81–87
31. Buratowski S, Hahn S, Guarente L et al (1989) Five intermediate complexes in transcription initiation by RNA polymerase II. Cell 56(4):549–561
32. Flores O, Lu H, Killeen M et al (1991) The small subunit of transcription factor IIF recruits RNA polymerase II into the preinitiation complex. Proc Natl Acad Sci USA 88(22):9999–10003
33. Goodrich JA, Tjian R (1994) Transcription factors IIE and IIH and ATP hydrolysis direct promoter clearance by RNA polymerase II. Cell 77(1):145–156
34. Allison LA (2012) Fundamental molecular biology, 2nd edn. Wiley, Hoboken
35. Chen H, Zhang J, Yuan G et al (2014) Complex interplay among DNA modification, noncoding RNA expression and protein-coding RNA expression in Salvia miltiorrhiza chloroplast genome. PLoS One 9(6):e99314
36. Salmena L, Poliseno L, Tay Y et al (2011) A ceRNA hypothesis: the Rosetta stone of a hidden RNA language? Cell 146(3):353–358
37. Franco-Zorrilla JM, Valli A, Todesco M et al (2007) Target mimicry provides a new mechanism for regulation of microRNA activity. Nat Genet 39(8):1033–1037
38. Flynn RA, Martin L, Spitale RC et al (2015) Dissecting noncoding and pathogen RNA-protein interactomes. RNA 21(1):135–143
39. Hamilton MJ, Young MD, Sauer S et al (2015) The interplay of long non-coding RNAs and MYC in cancer. AIMS Biophys 2(4):794–809
40. Tay Y, Rinn J, Pandolfi PP (2014) The multilayered complexity of ceRNA crosstalk and competition. Nature 505(7483):344–352
41. Li Z, Huang C, Bao C et al (2015) Exon-intron circular RNAs regulate transcription in the nucleus. Nat Struct Mol Biol 22(3):256–264
42. Kwek KY, Murphy S, Furger A et al (2002) U1 snRNA associates with TFIIH and regulates transcriptional initiation. Nat Struct Biol 9(11):800–805
43. Memczak S, Jens M, Elefsinioti A et al (2013) Circular RNAs are a large class of animal RNAs with regulatory potency. Nature 495(7441):333–338
44. Xu H, Guo S, Li W et al (2015) The circular RNA Cdr1as, via miR-7 and its targets, regulates insulin transcription and secretion in islet cells. Sci Rep 5:12453
45. Begemann G, Paricio N, Artero R et al (1997) Muscleblind, a gene required for photoreceptor differentiation in Drosophila, encodes novel nuclear Cys3His-type zinc-finger-containing proteins. Development 124(21):4321–4331
46. Barrette SP, Salzman J (2016) Circular RNAs: analysis, expression and potential functions. Development 143(11):1838–1847
47. Chao CW, Chan DC, Kuo A et al (1998) The mouse Formin (Fmn) gene: abundant circular RNA transcripts and gene-targeted deletion analysis. Mol Med 4(9):614–628
48. Gualandi F, Trabanelli C, Rimessi P et al (2003) Multiple exon skipping and RNA circularization contribute to the severe phenotypic expression of exon 5 dystrophin deletion. J Med Genet 40(8):e100
49. Abdelmohsen K, Panda AC, Munk R et al (2017) Identification of HuR target circular RNAs uncovers suppression of PABPN1 translation by CircPABPN1. RNA Biol 14(3):361–369
50. Du WW, Zhang C, Yang W et al (2017) Identifying and characterizing circRNA-protein interaction. Theranostics 7(17):4183–4191
51. Du WW, Yang W, Chen Y et al (2017) Foxo3 circular RNA promotes cardiac senescence by modulating multiple factors associated with stress and senescence responses. Eur Heart J 38(18):1402–1412
52. Du WW, Yang W, Liu E et al (2016) Foxo3 circular RNA retards cell cycle progression via forming ternary complexes with p21 and CDK2. Nucleic Acids Res 44(6):2846–2858
53. Holdt LM, Stahringer A, Sass K et al (2016) Circular non-coding RNA ANRIL modulates ribosomal RNA maturation and atherosclerosis in humans. Nat Commun 7:12429
54. Hansen TB, Jensen TI, Clausen BH et al (2013) Natural RNA circles function as efficient microRNA sponges. Nature 495(7441):384–388
55. Schneider T, Hung LH, Schreiner S et al (2016) CircRNA-protein complexes: IMP3 protein component defines subfamily of circRNPs. Sci Rep 6:3131
56. Peng L, Chen G, Zhu Z et al (2017) Circular RNA ZNF609 functions as a competitive endogenous RNA to regulate AKT3 expression by sponging miR-150-5p in Hirschsprung's disease. Oncotarget 8(1):808–818
57. He S, Liu P, Jian Z et al (2013) miR-138 protects cardiomyocytes from hypoxia-induced apoptosis via MLK3/JNK/c-Jun pathway. Biochem Biophys Res Commun 441(4):763–769
58. Schnall-Levin M, Rissland OS, Johnston WK et al (2011) Unusually effective microRNA targeting within repeat-rich coding regions of mammalian mRNAs. Gnome Res 21(9):1395–1403
59. Wang K, Long B, Liu F et al (2016) A circular RNA protects the heart from pathological hypertrophy

and heart failure by targeting miR-223. Eur Heart J 37(33):2602–2611
60. Wang K, Gan TY, Li N et al (2017) Circular RNA mediates cardiomyocyte death via miRNA-dependent upregulation of MTP18 expression. Cell Death Differ 24(6):1111–1120
61. Chen J, Cui L, Yuang J et al (2017) Circular RNA WDR77 target FGF-2 to regulate vascular smooth muscle cells proliferation and migration by sponging miR-124. Biochem Biophys Res Commun 494(1–2):126–132
62. Zheng C, Niu H, Li M et al (2015) Cyclic RNA hsa-circ--000595 regulates apoptosis of aortic smooth muscle cells. Mol Med Rep 12(5):6656–6662
63. Xie H, Ren X, Xin S et al (2016) Emerging roles of circRNA_001569 targeting miR-145 in the proliferation and invasion of colorectal cancer. Oncotarget 7(18):26680–26691

Functional Analysis of Circular RNAs

8

Shanmugapriya, Hisham Alkatib Huda, Soundararajan Vijayarathna, Chern Ein Oon, Yeng Chen, Jagat R. Kanwar, Mei Li Ng, and Sreenivasan Sasidharan

Abstract

Circular RNAs characterize a class of widespread and diverse endogenous RNAs which are non-coding RNAs that are made by back-splicing events and have covalently closed loops with no polyadenylated tails. Various indications specify that circular RNAs (circRNAs) are plentiful in the human transcriptome. However, their participation in biological processes remains mostly undescribed. To date thousands of circRNAs have been revealed in organisms ranging from *Drosophila melanogaster* to *Homo sapiens*. Functional studies specify that these transcripts control expression of protein-coding linear transcripts and thus encompass a key component of gene expression regulation. This chapter provide a comprehensive overview on functional validation of circRNAs. Furthermore, we discuss the recent modern methodologies for the functional validation of circRNAs such as RNA interference (RNAi) gene silencing assay, luciferase reporter assays, circRNA gain-of-function investigation via overexpression of circular transcript assay, RT-q-PCR quantification, and other latest applicable assays. The methods described in this chapter are demonstrated on the cellular model.

Keywords

CircRNAs · Functional validation · Cellular model

1 Introduction

Circular RNAs (circRNAs) are closed RNA transcripts made by back-splicing of a single pre-mRNA that is found in all higher eukaryotes including mammals. The first circRNA was discovered in the early 1990s [1] as an obviously befalling family of non-coding RNAs that is

Shanmugapriya · H. A. Huda · S. Vijayarathna
C. E. Oon · S. Sasidharan (✉)
Institute for Research in Molecular Medicine (INFORMM), Universiti Sains Malaysia, Pulau Pinang, Malaysia

Y. Chen
Faculty of Dentistry, Dental Research & Training Unit, and Oral Cancer Research and Coordinating Centre (OCRCC), University of Malaya, Kuala Lumpur, Malaysia

J. R. Kanwar
Faculty of Health, Nanomedicine-Laboratory of Immunology and Molecular Biomedical Research (LIMBR), School of Medicine (SoM), Deakin University, Geelong, VIC, Australia

M. L. Ng
Integrative Medicine Cluster, Advanced Medical and Dental Institute (AMDI), Universiti Sains Malaysia, Pulau Pinang, Malaysia

J. Xiao (ed.), *Circular RNAs*, Advances in Experimental Medicine and Biology 1087,
https://doi.org/10.1007/978-981-13-1426-1_8

vastly denoted in the eukaryotic transcriptome [2, 3]. Previously these circRNAs had generally been considered to be errors or by-products of RNA splicing [4] with little biological function. However, circRNAs have now been accepted as another type of endogenous non-coding RNA species with the development of next-generation sequencing (NGS) which are plentiful and maintained in various biological systems. CircRNAs are extremely stable in vivo compared with their linear counterpart RNAs due to the absence of a 2′ to 5′ carbon linkage and free 3′ or 5′ ends, respectively [2], and are mostly found in the cytoplasm and exosomes [5]. Interestingly, a huge number of circRNAs have been effectively identified in recent time in numerous cell lines and across diverse species [6, 7].

Various properties of circRNAs have been efficaciously characterized from the time of its first discovery in the early 1990s, but a comprehensive understanding of their biological function remains unclear. Various evidences from latest research findings also suggest a possible role of circRNAs in diverse human diseases [8]. In spite of these findings in support of circRNAs' significant purposeful roles, their influence on biological processes is still mostly unknown. Various lab-based functional studies specify that these circRNA transcripts control expression of protein-coding linear RNA transcripts and therefore encompass a significant constituent of gene expression regulation. Several circRNAs with changed expression patterns are recognized in numerous types of diseases. CircRNAs thus function as healing targets for treatment because of their action on various target genes and proteins. Validation of these circRNAs permits further study and is within the scope of this chapter which focuses on cellular model. To study the role of circRNA variation in cells, the most upregulated or downregulated circRNAs that were the mostly affected biological function of the cells can be selected for further validation. Furthermore, the circRNAs important target proteins that could assist as potential targets for disease prevention also can be identified. Assessing the potential of circRNAs to modulate disease could assist in the detection and development of new healing policies against various diseases in human. Hence, in this chapter we discuss the latest methodologies for the validation of circRNA transcripts which will eventually lead to the utilization of circRNA as molecular markers of complex diseases in human and gene therapy agent. Various experimental validation assays for functional analysis of circular RNAs such as RNA interference (RNAi) gene silencing assay, luciferase reporter assays, CircRNA gain-of-function analysis through over expression of circular transcripts assay, RT-q-PCR quantification and other applicable assay were discussed in this chapter with appropriate examples in cellular model. Figure 8.1 depicts the proposed model to study the role of circRNA in a cellular model by using latest methodologies for the validation of circRNA transcripts.

2 Principle of Functional Analysis of Circular RNAs

2.1 CircRNA Quantification

Quantification of circRNA abundance required application of bioinformatics to precisely analyse circRNA datasets generated by deep sequencing [9]. Currently, RNA-seq is regarded as a high-throughput sequencing technique for quantification and functional analysis of the transcriptome [9, 10]. RNA-seq analyses structural features of circRNA based on RNA-seq-derived datasets including find_circ, MapSplice [11], CIRCexplorer [12], circRNAFinder and CIRI [13]. A range of computational methods have been utilized to recognize scrambled sequences in RNA-seq datasets. Such approach consuming paired-end RNA-seq reads that line up to a custom database of all probable exon-exon pair junctions [14]. Utilizing computational method to map read ends from a genomic anchor site to a "breakpoint" edged by GU/AG sequence [15], or MapSplice algorithm that segments reads to identify the "back-splice" events. In addition, novel algorithms designed to detect novel splicing or structurally mutated transcriptome are developed. Recent library cloning uses oligo(dT) primer in cDNA synthesis to allow detection of both

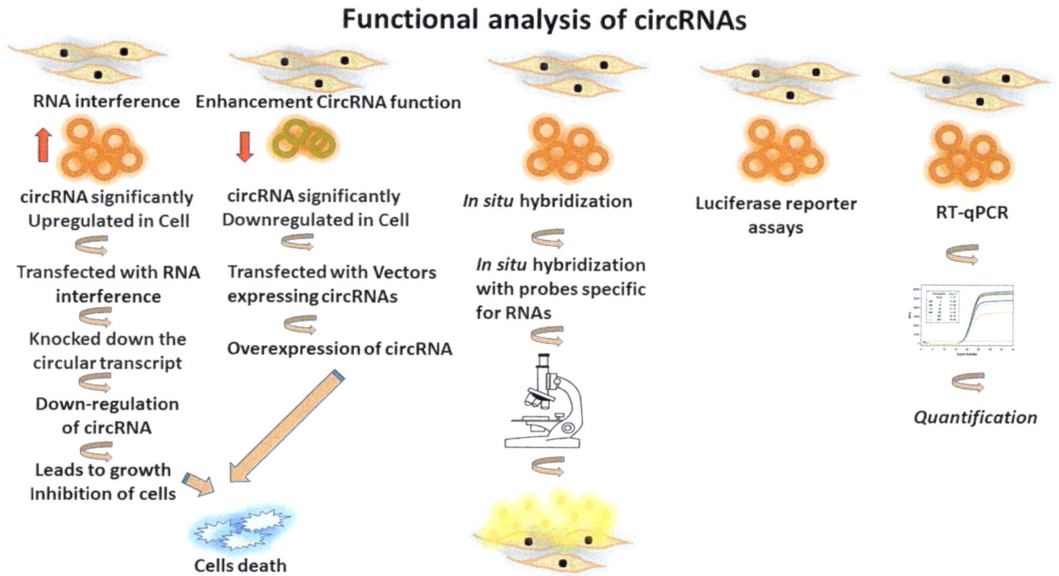

Fig. 8.1 Schematically summarizes various modern experimental validation methods for circRNAs

unpolyadenylated and polyadenylated fractions of long circRNAs [1, 16]. Quantification of circRNA uses random priming because circRNA generally has lower abundance in the cells. The easiest quantification of circRNA expression is to count split-back spice read and include reads that do not align directly to exons [17–19]. The use of algorithms to annotate and quantify circRNA from RNA-seq data can yield different results. Less than 1% of split-back-spliced reads are generated from a total RNA-seq experiment, whereas 99% of the reads are aligned to express transcriptome [20, 21]. Generally, circRNAs are structurally different from linear RNA because circRNAs are uncapped and unpolyadenylated. Consequently RT-PCR reactions or RNA-seq of cap-enriched, poly(A) selected or oligo(dT) reverse-transcription-primed samples favourably identify linear scrambled exon RNAs. Additionally, scrambled exon RNAs resulting from DNA rearrangements or tandem duplications would be detectable in genomic DNA PCR reactions. These methods, used in combination with RNase R, a 3′ to 5′ exoribonuclease specifically degrades linear RNAs, are a useful tool to validate and enrich circRNAs in a total RNA sequencing library [20, 22]. However, RNase R may cause artifactual enrichment of circRNA levels due to endogenous nicking of the RNA or contaminating nucleases, and some linear RNAs are resistant to RNase R degradation [23], and different circRNAs can show drastically different levels of RNase R resistance [24], or potentially due to interference with reverse transcription reactions [20]. For this reason, it is imperative to verify its efficacy by qRT-PCR. Hence, quantitative reverse transcription PCR (qRT-PCR) is a powerful tool to quantitatively assess the relative abundance of circRNAs. In this method, the circRNA is reversely transcripted to a cDNA molecule that contains an exon-exon junctional sequence which can be specifically targeted and amplified by primers. These primers are recognized as "inverse" or "outward-facing" primers since it stops the amplification of RNA species that do not comprise the exon-exon junction when aligned to the genome. However, it is imperative to verify the amplified sequence because amplification errors might occur during RT step such as template switching [25], splicing between two separate pre-mRNA molecules [26], unexpected genomic duplication and rolling circle RT. Compared to PCR, Northern blots allow precise monitoring of species mobility [27]. This technique utilizes probes which are intended to specifically target the circularized RNA exonic sequence in separate blots. When RNA is run on a polyacrylamide gel matrix, differential mobility forms two bands if the RNA is circular or three bands if the RNA is linear [27]. In addition, two-dimensional denaturing polyacrylamide gel electrophoresis can be also utilized to identify the circRNAs. Total RNA is run on a 2D gel comprising various percentages of polyacrylamide in each dimension; circRNA runs in an arc, which can be sequenced and enriched in next-generation sequencing (NGS) [28]. Otherwise, the 2D gel can be probed through Northern blotting to quantity specific circRNAs. Furthermore, ribosomal RNA (rRNA) depletion and polyA-depletion are general approaches utilized to enrich for circRNAs in sequencing libraries. However, neither promises that the enriched sequences are absolutely circular since numerous forms of non-coding RNA will also survive in these selections. "TRAP electrophoresis" is another mean of verifying circRNAs. In this method, circRNAs are separated based on characteristic variances in movement on a gel paralleled with their corresponding linear molecules [15, 29]. Additionally, one- and two-dimensional polyacrylamide gel electrophoresis consuming diverse percentage gels can also distinguish characteristic movement patterns of circular RNAs as single-hit nicking or targeted RNase H cleavage must transform the circle RNAs to linear species with predictable electrophoretic mobility. On the contrary, split of a linear RNA will produce two products on a gel electrophoresis [30–32].

2.2 Interrogate the Biological Function of CircRNAs

Ectopic circRNA expression plasmid is a convenient tool to interrogate the biological function of circRNA [15, 33]. CircRNAs are frequently overexpressed on plasmid by utilizing gene fragments under the control of a strong promoter. Circularization is prompted by inverted repeats (IR) flanking the circularized exon, which seemingly brings the splice acceptor (SA) and splice donor (SD) into closeness for back-splicing. Furthermore, mammalian genes can be studied by overexpressing mammalian vectors comprise the circularized exon(s) along with flanking splicing signals and intronic sequences, which harbour inverted repeats to assist their linking into a circle [10, 34, 35]. However, it's difficult to determine whether the given phenotypes are exclusively driven by circRNA because the vector also overexpresses linear RNA. In addition, a common artefact may arise from rolling circle transcription of the plasmid. Hypothetically, the expression plasmid harbours circularized exon of a gene with flanking introns. If the transcription termination signals in the vector are circumvented, the RNA polymerase will remain to transcribe around the entire plasmid, generating a concatemer of the RNA sequence contained in the plasmid. This transcript piece can lead to off-target effects on the cell and spurious circRNA quantification. Therefore, attention has been given in vector design to lessen the amplification of inaccurate products [10]. For additional uses, such as establishing translation or analysing function of a circRNAs, these artefacts should be entirely removed. Furthermore, suppressing circRNAs function can be achieved by siRNA knockdown. In this method, siRNA is designed to specifically disrupt circRNA expression without affecting linear protein-coding RNAs.

2.3 Validation Through Biological Function of CircRNAs

Expression of circRNAs and its isoforms is often specific to cell type, tissue and developmental stage. While the abundance of the circRNAs can be assessed by biochemical methods, many biological functions of circRNAs are unidentified. Multiple lines of evidence have shown that circRNAs function as "microRNA sponge". Notably, the circRNAs ciRS-7/CDR1as contains many highly conserved target sites for microRNA miR-7 that leads to decreased miRNA activity [15]. Likewise, Sry is another abundantly expressed circRNA in mouse testis that has been previously clearly demonstrated to suppress the miRNA-138. This points to the role of circRNA in regulatory framework of post-transcriptional gene expression, supporting the function of circRNAs as miRNA decoys. Additional evidence for functional circRNAs sponge has been demonstrated by the downregulation of circRNA, named HRCR (mm9-circ-012559) in the mice expressing miR-223 transgene [35], suggesting that role of HRCR in preventing heart failure. Other circRNAs have been identified to promote cell proliferation in cancer. A circRNA from the HIPK3 gene (circHIPK3) was found to sponge miR-124 to enhance cell growth [36]. Other studies have linked circRNA circRNA-CER which sponges miR-136 [37] and circRNA_001569 which sponges miR-145 to promote cell survival [38]. Another circRNA, circZNF292 was found to enhance proliferation. However, it's unknown as whether circRNA could be a general function of miRNA decoys. This is probably attributed by a highly conserved sequence in the codon of circulating exons [15, 26] or decreased single-nucleotide polymorphisms in microRNA target sites of circularized exons [39].This is too a matter of competing endogenous RNA (ceRNAs) which also binds and destabilizes miRNAs. Regardless of this endogenous competition, circRNAs remain as more effective "miRNA sponge" than ceRNAs because of the circular structure of circRNAs that protect from exonucleases degradation [40]. Furthermore, the circular structure also confers intrinsic resistance against miRNA-mediated destabilization. This shows that circRNAs function as important regulator for miRNA expression [41, 42]. A novel subclass of circRNAs, named EIciRNAs, has been recently identified as positive regulator for gene transcription, through an interaction with U1 small nuclear ribonucleoprotein (snRNP) and RNA polymerase II in the promoter region of the host gene [33]. In this regard, circRNAs are able to induce gene expression in cis or trans to regulate other genome loci [43]. Other functions of circRNAs include acting as protein decoys [42, 44]. CircRNAs are mostly localized in the cytoplasm that can sequester protein to prevent entry into nucleus. In this regard, Circ-Foxo3 was found to reduce nuclear concentration of stress-related proteins FAK and HIF-1α [45, 46]. Circ-Foxo3 has been shown to interact with cell-cycle proteins CDK2 and P21 to reduce cancer cells growth [46]. In addition, circRNA coined circMbl harbours binding sites for MBL protein itself that reduces mbl mRNA and protein production [47]. Other potential function for circRNAs has been described in subcellular transportation and as stable molecule scaffold for assembly of complexes [15]. In short, the identification of circRNAs contributing to the post-transcriptional regulation of gene expression and RBP sponge highlight a good capacity of circRNA associated functionalities, as demonstrated by the conserved nature of circRNA expression in tissue-specific abundance. Unrevealing these molecule functionalities should be the main focus in the future directions for circRNA research field.

3 Modern Experimental Validation Assays and Functional Analysis of Circular RNAs

3.1 Validation of CircRNA

Validation of circRNA can be performed through RTqPCR, in situ hybridization and Northern blotting with the incorporation of circRNA-specific primers and probes.

3.1.1 RTqPCR

CircRNA identified through RNA-seq and bioinformatics can be validated by performing the reverse transcription quantitative polymerase chain reaction (RTqPCR). Enrichment of circRNA through RNase R treatment from the total RNA is confirmed as real circle and nonlinear RNA products by designing the divergent primers of the circRNA [29] (Fig. 8.2). Nevertheless, RNase R digestion is not an essential step in qPCR since the divergent primers designed is not expected to amplify the linear RNAs [51]. After that, reverse transcription (RT) of the RNA will be conducted to synthesize complementary DNA (cDNA) with the utilization of random hexamers for priming instead of

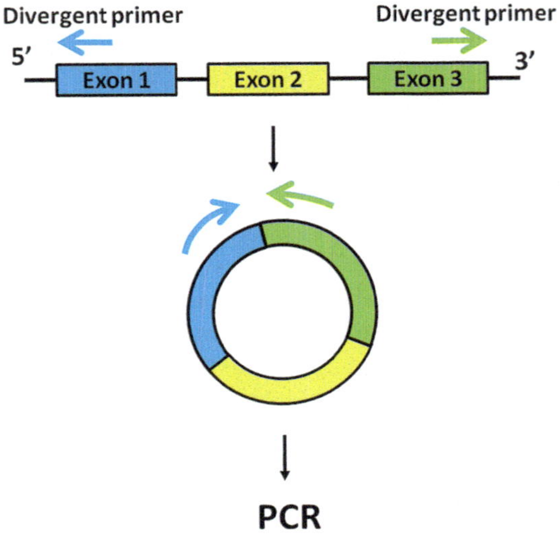

Fig. 8.2 Schematic depiction of amplification process of circular RNA from back-splicing with the aid of divergent primers

oligo(dT) priming on the ground that poly(A) tail is in paucity in circRNA [48]. There are two popular methods to perform qPCR, namely, SYBR Green or TaqMan. Although SYBR Green is considered to be cost-effective and straightforward [49] as compared to TaqMan, SYBR Green is non-specific and highly believed to produce false positive as SYBR Green dye binds to all double-stranded DNA (dsDNA) including primer-dimers [50, 51]. However, specific outward-facing primers and TaqMan probes complementary to the back-splice junction of the circRNA will be designed to validate the specific circRNA via TaqMan RTqPCR [52]. The quantification and validation of circRNA can be analysed by the amplification plot representing the fluorescent intensity versus number of cycle. Ct value known as the threshold cycle is where all the quantification data begins as the first distinguishable fluorescent escalates [53].

3.1.2 In Situ Hybridization

Localization, quantification and validation of specific circRNA can be analysed through in situ hybridization (ISH). Specific probe complementary to the back-splicing junction of circRNA will be designed with fluorescent label or radioactive label (Fig. 8.3). The labelled probes will then be introduced to be hybridized after the denaturation process. Labelled probes will be hybridized to the complementary back-splicing of circRNA. Then, the cells will be washed, and the probes can be detected with the aid of cross-linking agent. Cells will be counter strained before acquiring the images [29, 40].

3.2 Functional Analysis of CircRNA

Functional analysis of circRNA can be performed by using loss-of-function, gain-of-function investigation and luciferase reporter assay.

3.2.1 RNA Interference

Gene silencing method is one of the commonly used techniques to determine the function of specific circRNA. Knockdown of circRNA can be performed with the utilization of small interfering RNAs (siRNAs). The siRNA will be designed so as to be complementary with the back-splice sequence of the circRNA. There are several ways to transfect the siRNA into the cells which include chemical transfection (Lipofectamine), mechanical transfection (electroporation) and viral-mediated delivery (expression vectors). siRNA directed circRNA knockdown will eventually lead to gene silencing, followed by inhibition of protein translation via upregulation of miRNA [54] (Fig. 8.4). This is because circRNAs play an important role as microRNA sponges [15].

3.2.2 Enhancement CircRNA Function

Enhancement CircRNA function can be performed by increasing the expression of circRNA in the cells. Recapitulating the circRNA will be accomplished with the incorporation of expression vectors in which unabridged intron responsible for circRNA production is infused together with its natural splice sites and exons [55]. The expression vectors containing the circRNA expressing sequences will then be transiently transfected to the cells [34]. The expression of circRNA from the plasmid can be distinguished by performing Northern blot analy-

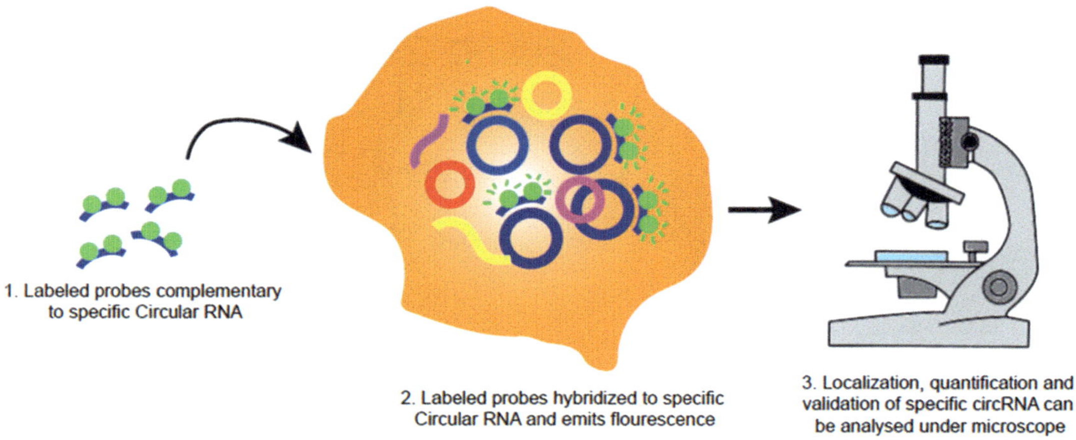

Fig. 8.3 Schematic diagram of in situ hybridization

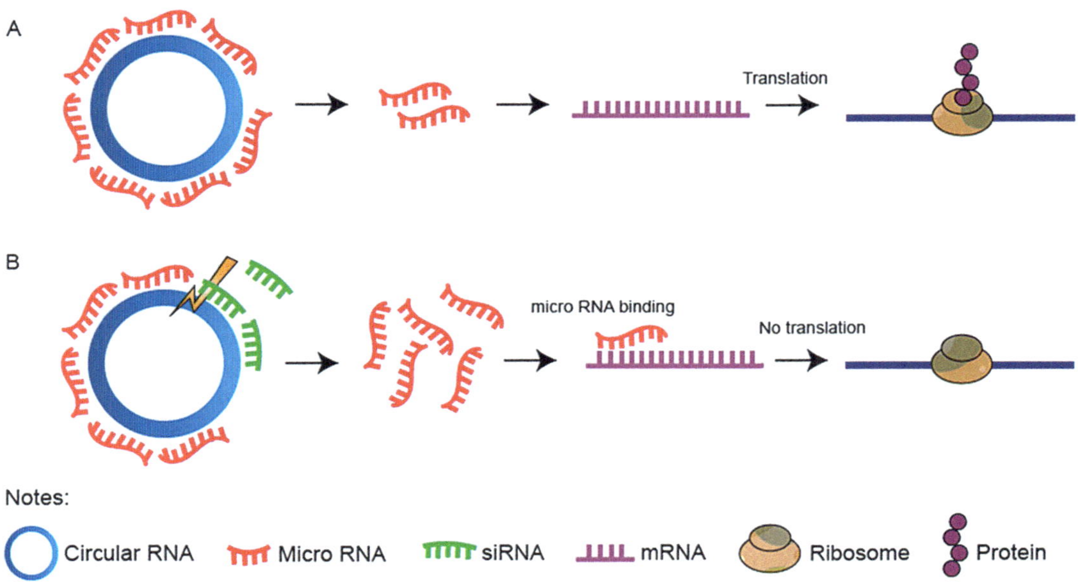

Fig. 8.4 Schematic image illustrates (**a**) the role of CircRNA as microRNA sponges which eventually downregulates or inhibits the microRNA. This consequently causes an efficient transcription of mRNA and successful translation of proteins. (**b**) However, the introduction of small interfering RNA (siRNA) causes knockdown of the circRNA and releases the microRNA which will bind to its mRNA target and silences the gene. As a result, protein translation is inhibited

sis, and the production of bona fide circRNA can be validated through RNase R treatment as the back-splicing reaction of circular RNA transcripts makes it RNase R resistant with increased stability, unlike the linear RNA [29, 56, 57]. Therefore, downstream investigations of circRNA overexpression can be conducted.

3.2.3 Luciferase Reporter Assays

Luciferase reporter assays can be utilized to investigate the regulation miRNA complementary to specific circRNA. Briefly, the luciferase vector will be constructed by inserting the specific circRNA sequence in the 3' UTR into the promoter-driven luciferase reported gene

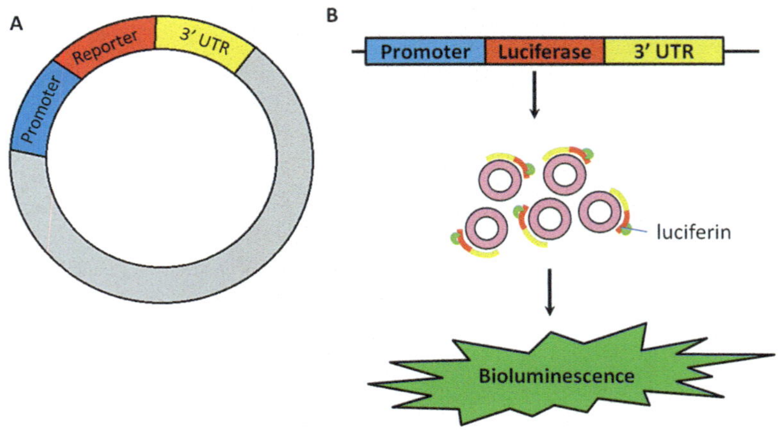

Fig. 8.5 Schematic diagram depicts the (**a**) construction of luciferase reporter vector with the insertion of specific circRNA sequence in the 3′ UTR. (**b**) After transfection of vectors into the cells, the circRNA complementary sequences binds to the 3′ UTR together with the reporter gene, luciferase which emits bioluminescence in the presence of its substrate, luciferin

(Fig. 8.5). Cells will then be cotransfected with the luciferase reporter vector and the miRNA mimics in order to establish the characteristic of circRNA as miRNA sponges by comparing the luciferase activity with the negative controls [55]. In addition, the relationship between circRNA and miRNA can also be analysed through luciferase reporter analysis [58].

3.3 Bioinformatic Analysis of Circular RNA

To detect circRNAs from RNA-seq data, there are around 11 software available. This software's package could be commonly separated into two groups in line with the main approaches to detect circRNA. For example, KNIFE, NCLscan and PTESFinder require the circRNA sequences with the gene annotation info in order to identify the circRNA. This approach is named "pseudo-reference based" or "candidate-based" strategy. NCLScan and PTESFinder construct the assumed circRNA sequences achieving the mapping info of the segmented anchors found after alignment to the genome or transcriptome. While, KNIFE instantly builds all the possible out-of-order exon-exon junction from gene annotation information before alignment. However, the other software category specifically circRNA_finder, CIRCexplorer, DCC, MapSplice and segemehl could be assigned to a subcategory because they invent spliced alignment algorithms to identify and investigate the back-splicing events under the approach of "fragmented-based" or "segmented read approach" which recognized back-splicing junctions from the mapping info of a multiple-split read's alignment to the genome. However, find_circ and UROBORUS could be categorized together because both develop back-splicing events from the mapping information of these anchors after collecting the unmapped reads. Lastly, CIRI is exclusive; it can identify the paired chiastic clipping (PCC) signals from the mapping information of reads by local alignment with BWA-MEM combined through orderly filtering steps to get rid of possible false positives [59]. Figure 8.6 explains the downstream bioinformatics analysis which can be performed to further investigate the functional annotation of the circRNA.

4 Conclusion

Functional validation of circRNAs is an important step for circRNA-based research in order to ascertain their role in various diseases. In order to

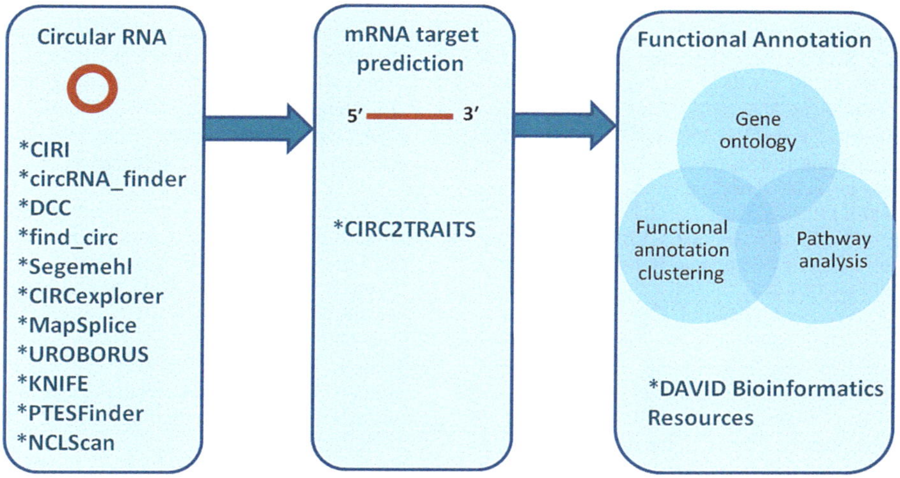

Fig. 8.6 Bioinformatics analysis flowchart

validate the function of circRNAs, various methods were discussed in this chapter. To ensure the possible vital role of circRNAs in diverse human diseases, functional validation of circRNAs is the first step towards establishing its identity in various ailments and development of novel therapeutic strategies against various diseases in human. In conclusion, this chapter discussed various reliable and modern methods to validate the function of circRNAs which will result in further acceleration of research on this unique circRNAs.

Acknowledgements Shanmugapriya was supported by the Graduate Research Assistance Scheme from Universiti Sains Malaysia, Malaysia.

Competing Financial Interests The authors declare no competing financial interests.

References

1. Nigro JM, Cho KR, Fearon ER et al (1991) Scrambled exons. Cell 64(3):607–613
2. Jeck WR, Sharpless NE (2014) Detecting and characterizing circular RNAs. Nat Biotechnol 32(5):453–461
3. Lasda E, Parker R (2014) Circular RNAs: diversity of form and function. RNA 20(12):1829–1842
4. Cocquerelle C, Mascrez B, Hetuin D et al (1993) Mis-splicing yields circular RNA molecules. FASEB J 7(1):155–160
5. Li Y, Zheng QP, Bao CY et al (2015) Circular RNA is enriched and stable in exosomes: a promising biomarker for cancer diagnosis. Cell Res 25(8):981–984
6. Glazar P, Papavasileiou P, Rajewsky N (2014) circBase: a database for circular RNAs. RNA 20(11):1666–1670
7. Rybak-Wolf A, Stottmeister C, Glazar P et al (2015) Circular RNAs in the mammalian brain are highly abundant, conserved, and dynamically expressed. Mol Cell 58(5):870–885
8. Westholm JO, Miura P, Olson S et al (2014) Genome-wide analysis of drosophila circular RNAs reveals their structural and sequence properties and age-dependent neural accumulation. Cell Rep 9(5):1966–1980
9. Conn SJ, Pillman KA, Toubia J et al (2015) The RNA binding protein quaking regulates formation of circRNAs. Cell 160(6):1125–1134
10. Kramer MC, Liang D, Tatomer DC et al (2015) Combinatorial control of Drosophila circular RNA expression by intronic repeats, hnRNPs, and SR proteins. Genes Dev 29(20):2168–2182
11. Koopman P, Munsterberg A, Capel B et al (1990) Expression of a candidate sex-determining gene during mouse testis differentiation. Nature 348(6300):450–452
12. Cocquerelle C, Daubersies P, Majerus MA et al (1992) Splicing with inverted order of exons occurs proximal to large introns. EMBO J 11(3):1095–1098
13. Guo J (2014) Transcription: the epicenter of gene expression. J Zhejiang Univ Sci B 15(5):409–411
14. Salzman J, Gawad C, Wang PL et al (2012) Circular RNAs are the predominant transcript isoform from

hundreds of human genes in diverse cell types. PLoS One 7(2):e30733
15. Memczak S, Jens M, Elefsinioti A et al (2013) Circular RNAs are a large class of animal RNAs with regulatory potency. Nature 495(7441):333–338
16. Grabherr MG, Haas BJ, Yassour M et al (2011) Full-length transcriptome assembly from RNA-Seq data without a reference genome. Nat Biotechnol 29(7):644–U130
17. Lu T, Cui L, Zhou Y et al (2015) Transcriptome-wide investigation of circular RNAs in rice. RNA 21(12):2076–2087
18. Saad FA, Vitiello L, Merlini L et al (1992) A 3′ consensus splice mutation in the human dystrophin gene detected by a screening for intra-exonic deletions. Hum Mol Genet 1(5):345–346
19. Bailleul B (1996) During in vivo maturation of eukaryotic nuclear mRNA, splicing yields excised exon circles. Nucleic Acids Res 24(6):1015–1019
20. Hansen TB, Wiklund ED, Bramsen JB et al (2011) miRNA-dependent gene silencing involving Ago2-mediated cleavage of a circular antisense RNA. EMBO J 30(21):4414–4422
21. Zaphiropoulos PG (1997) Exon skipping and circular RNA formation in transcripts of the human cytochrome P-450 2C18 gene in epidermis and of the rat androgen binding protein gene in testis. Mol Cell Biol 17(6):2985–2993
22. Houseley JM, Garcia-Casado Z, Pascual M et al (2006) Noncanonical RNAs from transcripts of the Drosophila muscleblind gene. J Hered 97(3):253–260
23. Li XF, Lytton J (1999) A circularized sodium-calcium exchanger exon 2 transcript. J Biol Chem 274(12):8153–8160
24. Hooper JE (2014) A survey of software for genome-wide discovery of differential splicing in RNA-Seq data. Hum Genomics 8:3
25. Kulpa D, Topping R, Telesnitsky A (1997) Determination of the site of first strand transfer during Moloney murine leukemia virus reverse transcription and identification of strand transfer-associated reverse transcriptase errors. EMBO J 16(4):856–865
26. Agabian N (1990) Trans splicing of nuclear pre-mRNAs. Cell 61(7):1157–1160
27. Capel B, Swain A, Nicolis S et al (1993) Circular transcripts of the testis-determining gene Sry in adult mouse testis. Cell 73(5):1019–1030
28. Awan AR, Manfredo A, Pleiss JA (2013) Lariat sequencing in a unicellular yeast identifies regulated alternative splicing of exons that are evolutionarily conserved with humans. Proc Natl Acad Sci U S A 110(31):12762–12767
29. Jeck WR, Sorrentino JA, Wang K et al (2013) Circular RNAs are abundant, conserved, and associated with ALU repeats. RNA 19(2):141–157
30. Sanger HL, Klotz G, Riesner D et al (1976) Viroids are single-stranded covalently closed circular RNA molecules existing as highly base-paired rod-like structures. Proc Natl Acad Sci U S A 73(11):3852–3856
31. Ruskin B, Krainer AR, Maniatis T et al (1984) Excision of an intact intron as a novel lariat structure during pre-mRNA splicing in vitro. Cell 38(1):317–331
32. Tabak HF, Van der Horst G, Smit J et al (1988) Discrimination between RNA circles, interlocked RNA circles and lariats using two-dimensional polyacrylamide gel electrophoresis. Nucleic Acids Res 16(14A):6597–6605
33. Li Z, Huang C, Bao C et al (2015) Exon-intron circular RNAs regulate transcription in the nucleus. Nat Struct Mol Biol 22(3):256–264
34. Liang D, Wilusz JE (2014) Short intronic repeat sequences facilitate circular RNA production. Genes Dev 28(20):2233–2247
35. Wang K, Long B, Liu F et al (2016) A circular RNA protects the heart from pathological hypertrophy and heart failure by targeting miR-223. Eur Heart J 37(33):2602–2611
36. Zheng QP, Bao CY, Guo WJ et al (2016) Circular RNA profiling reveals an abundant circHIPK3 that regulates cell growth by sponging multiple miRNAs. Nat Commun 7:11215
37. Liu Q, Zhang X, Hu XQ et al (2016) Circular RNA related to the chondrocyte ECM regulates MMP13 expression by functioning as a MiR-136 'Sponge' in human cartilage degradation. Sci Rep 6:22572
38. Xie H, Ren X, Xin S et al (2016) Emerging roles of circRNA_001569 targeting miR-145 in the proliferation and invasion of colorectal cancer. Oncotarget 7(18):26680–26691
39. Thomas LF, Saetrom P (2014) Circular RNAs are depleted of polymorphisms at microRNA binding sites. Bioinformatics 30(16):2243–2246
40. Zhang Y, Xue W, Li X et al (2016) The biogenesis of nascent circular RNAs. Cell Rep 15(3):611–624
41. Zhang XO, Wang HB, Zhang Y et al (2014) Complementary sequence-mediated exon circularization. Cell 159(1):134–147
42. Ashwal-Fluss R, Meyer M, Pamudurti NR et al (2014) circRNA biogenesis competes with pre-mRNA splicing. Mol Cell 56(1):55–66
43. Holt CE, Schuman EM (2013) The central dogma decentralized: new perspectives on RNA function and local translation in neurons. Neuron 80(3):648–657
44. Denzler R, Agarwal V, Stefano J et al (2014) Assessing the ceRNA hypothesis with quantitative measurements of miRNA and target abundance. Mol Cell 54(5):766–776
45. Du WW, Yang W, Chen Y et al (2017) Foxo3 circular RNA promotes cardiac senescence by modulating multiple factors associated with stress and senescence responses. Eur Heart J 38(18):1402–1412
46. Du WW, Yang W, Liu E et al (2016) Foxo3 circular RNA retards cell cycle progression via forming ternary complexes with p21 and CDK2. Nucleic Acids Res 44(6):2846–2858
47. Chen CY, Sarnow P (1995) Initiation of protein synthesis by the eukaryotic translational apparatus on circular RNAs. Science 268(5209):415–417

48. Panda AC, Abdelmohsen K, Gorospe M (2017) RT-qPCR detection of senescence-associated circular RNAs. Methods Mol Biol 1534:79–87
49. Orlando C, Pinzani P, Pazzagli M (1998) Developments in quantitative PCR. Clin Chem Lab Med 36(5):255–269
50. Valasek MA, Repa JJ (2005) The power of real-time PCR. Adv Physiol Educ 29(3):151–159
51. Wong M, Medrano J (2005) Real-time PCR for mRNA quantitation. BioTechniques 39(1):75–85
52. Bachmayr-Heyda A, Reiner AT, Auer K et al (2015) Correlation of circular RNA abundance with proliferation–exemplified with colorectal and ovarian cancer, idiopathic lung fibrosis, and normal human tissues. Sci Rep 5:8057
53. Tajadini M, Panjehpour M, Javanmard SH (2014) Comparison of SYBR green and TaqMan methods in quantitative real-time polymerase chain reaction analysis of four adenosine receptor subtypes. Adv Biomed Res 3:85
54. Hansen TB, Kjems J, Damgaard CK (2013) Circular RNA and miR-7 in Cancer. Cancer Res 73(18):5609–5612
55. Zhang Y, Zhang XO, Chen T et al (2013) Circular intronic long noncoding RNAs. Mol Cell 51(6):792–806
56. Suzuki H, Tsukahara T (2014) A view of pre-mRNA splicing from RNase R resistant RNAs. Int J Mol Sci 15(6):9331–9342
57. Suzuki H, Zuo YH, Wang JH et al (2006) Characterization of RNase R-digested cellular RNA source that consists of lariat and circular RNAs from pre-mRNA splicing. Nucleic Acids Res 34(8):e63
58. Li Y, Chen B, Huang S (2018) Identification of circRNAs for miRNA targets by Argonaute2 RNA immunoprecipitation and luciferase screening assays. Methods Mol Biol 1724:209–218
59. Zeng X, Lin W, Guo M et al (2017) A comprehensive overview and evaluation of circular RNA detection tools. PLoS Comput Biol 13(6):e1005420

Part V

Circular RNAs as Potential Disease Biomarkers

Circular RNA in Exosomes

9

Daniele Fanale, Simona Taverna, Antonio Russo, and Viviana Bazan

Abstract

Circular RNAs (circRNAs) are a novel family of non-coding endogenous RNAs discovered in all eukaryotic cells and generated through a particular mechanism of alternative splicing called "back-splicing". These molecules show multiple functions, by acting as modulators of gene and miRNA expression, and may have a role in several biological processes, such as cell proliferation and invasion with, tumour development and progression, and in several mechanisms underlying other diseases. Their presence has been shown to be abundant in several body fluids such as blood and saliva. Based on their biogenesis mechanism, circRNAs may be categorized into five classes: exonic circRNAs, intronic circRNAs, antisense circRNAs, sense overlapping circRNAs and intergenic circRNAs. Recently, the presence of circRNAs, in addition to that of miRNAs and long non-coding RNAs, has been detected also in small extracellular vesicles called exosomes. Investigating the presence and expression levels of serum exosomal circRNAs could allow us, in future, to discriminate cancer patients from healthy individuals, identifying new potential exosome-based cancer biomarkers.

In this chapter, we briefly will describe the major features and functions of exosomal circRNAs, discussing their potential role as molecular biomarkers for diagnosis, prognosis and monitoring of complex diseases, including cancer.

Keywords

Biomarkers · CDR1as · Circular RNAs (circRNAs) · Exosomes · Non-coding RNAs

1 Introduction

In addition to non-coding RNAs such as small RNAs (microRNAs) and long non-coding RNAs (LncRNAs) [1–10], circular RNAs (circRNAs) represent a novel and large family of non-coding endogenous RNAs recently discovered in all eukaryotic cells and arising from a particular alternative splicing mechanism of precursor mRNAs (pre-mRNAs) [11–14]. However, some few circRNAs have been shown to have the ability to be translated into proteins via insertion of an internal ribosomal entry site [15, 16]. Differently from linear RNAs, circRNAs are covalently

Daniele Fanale and Simona Taverna contributed equally to this work

D. Fanale · S. Taverna · A. Russo (✉) · V. Bazan
Section of Medical Oncology, Department of Surgical, Oncological and Oral Sciences, University of Palermo, Palermo, Italy

J. Xiao (ed.), *Circular RNAs*, Advances in Experimental Medicine and Biology 1087,
https://doi.org/10.1007/978-981-13-1426-1_9

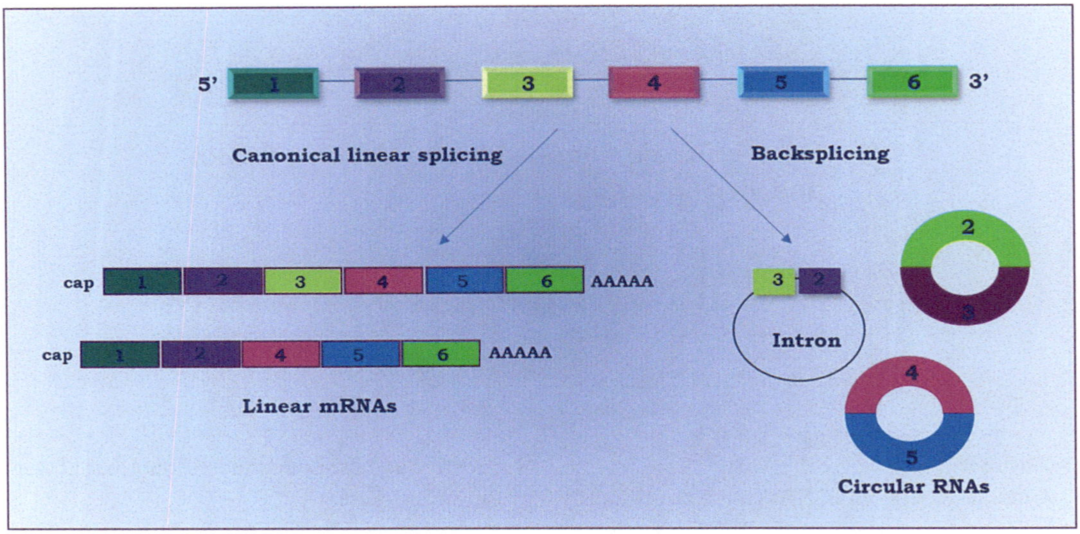

Fig. 9.1 Biogenesis of circular RNAs
Canonical splicing (left panel) leads to linear mRNAs production, while back-splicing (right panel) leads to circRNAs lacking free 5′ and 3′ ends

closed single-stranded transcripts produced from exonic, intronic or intergenic regions and lacking of the typical terminal structures (5′cap and 3′ polyadenylated tails). The lack of these structures makes them more stable and resistant to exonuclease R than linear RNAs [17]. Three different biogenesis mechanisms to explain the origin of circRNAs have suggested that these RNAs are cyclized through a back-splicing process, in which an upstream splice acceptor is linked to a downstream splice donor through direct splice or splice skipping [18–20] (Fig. 9.1). To date, circRNAs are categorized into five classes: exonic circRNA, intronic circRNA, antisense circRNA, sense overlapping circRNA, and intergenic circRNA [21, 22]. Most of circRNAs are amply conserved and stable across different species and show specific features according to the tissue/cell type and developmental stage [23, 24]. Many circRNAs have been found in numerous body fluids such as blood and saliva. Since they have been shown to be tissue-specific and have hallmark properties, this feature could make them potential and useful biomarkers for diagnosis, prognosis and monitoring of several diseases, such as cancer, osteoarthritis, diabetes and neurodegenerative pathologies [25–27]. Although the discovery of circRNAs dates back some decades ago, they were initially considered only as non-functional artefacts of aberrant RNA splicing [23, 28, 29]. Only thanks to the introduction of recent bioinformatics and RNA deep sequencing technologies, several circRNAs have been detected, acquiring the deserved importance which they hold today [18, 19]. CircRNAs play several and crucial functions, as they can work as microRNA (miRNA) sponges, negatively modulating miRNA expression, as regulators of splicing and transcriptional and posttranscriptional events, and as modifiers of parental gene expression [30] (Fig. 9.2). For example, CDR1as (also called ciRS-7) functions as a miR-7 sponge, regulating, via miR-7 targets, the insulin transcription and secretion in pancreatic islet cells and thus opposing to the development of diabetes induced by miR-7 overexpression [31]. The human/mouse ciRS-7/CDR1as and mouse Sry are the two most representative circRNAs acting as miRNA sponges [32]. The function of miRNA sponge enables circRNAs to control their activity and indirectly regulate the target mRNA stability [33]. In addition, circRNAs may have a role in the regulation of cell growth and invasion processes in several tumours, including gastric, colon and oesophageal cancers and

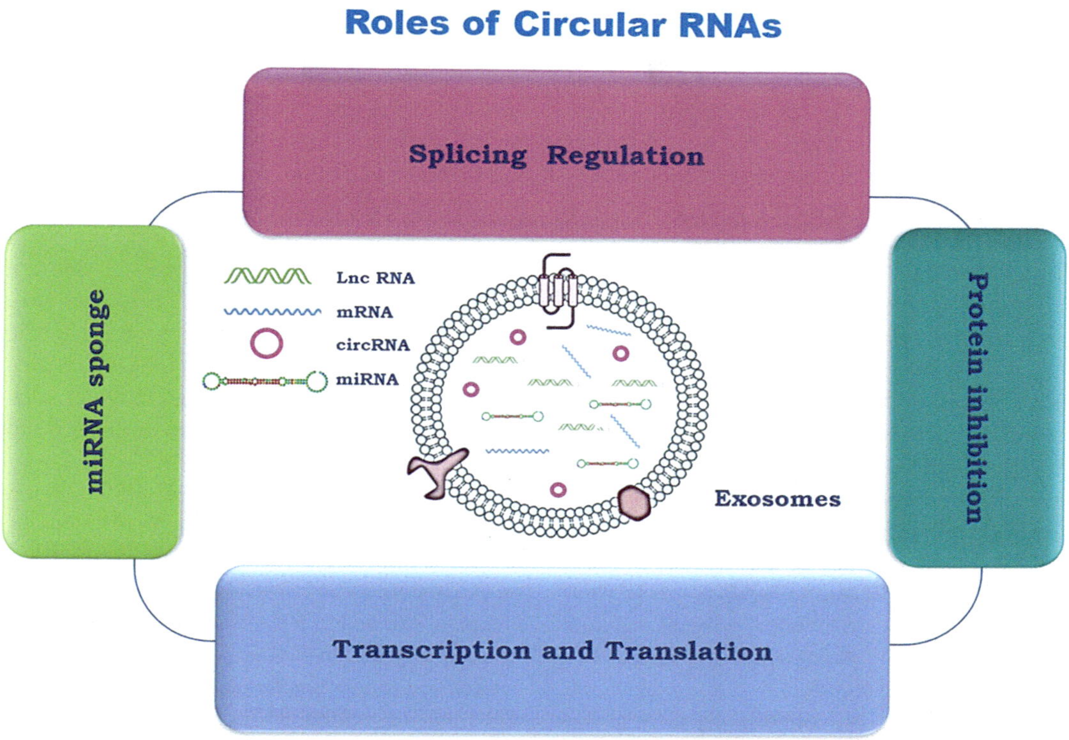

Fig. 9.2 Schematic representation of circular RNA biological functions
CircRNAs contained in exosomes can have several functions such as transcription and translation regulation, splicing regulation, miRNA sponge and protein inhibition

may allow to develop new approaches for cancer detection and therapy [34]. However, the biological functions of most circRNAs remain yet not totally understood.

Most of cell types secretes nanosize-extracellular vesicles (EVs) of endolysosomal origin called exosomes, containing a specific load of mRNAs, microRNAs, and proteins able to affect the cell behaviour and potentially useful for diagnosis of several human diseases [35]. Recently, RNA-seq analyses proved, for the first time, the presence of several circRNAs with potential biological function in exosomes. In particular, human serum exosomes have been shown to contain more than 1000 circRNAs, probably arising from the entry into the bloodstream of circRNAs present in tumour [36]. Investigating the presence and expression levels of serum exosomal circRNAs could allow us to differentiate cancer patients from healthy individuals, identifying new potential exosome-based cancer biomarkers [37].

In this chapter, we will focus on the major progress in the field of circRNA biology, reporting the current knowledge about their presence and biological role in exosomes and discussing their potential significance as molecular biomarkers for diagnosis, prognosis, and monitoring of complex diseases, including cancer.

2 Exosomes

One of the most attractive methods of cell-to-cell communication is mediated by EVs considered as an alternative to the paracrine and endocrine cellular system [38, 39]. The two better characterized classes of EVs are exosomes and microvesicles. Exosomes are the most deeply studied subpopulation of EVs [40]. Although exosomes were initially considered as the “garbage bins” of cells [41, 42], in the last decades, the attention of the researchers on the functions of these vesicles is growing exponentially. Exosomes are nanoscale EVs with lipid bilayer and are released into extracellular space after fusion of multivesicular bodies (MVBs) with plasma membrane [43]. Exosome formation is a mechanism consisting of four stages: initiation, endocytosis, MVBs formation, and exosome secretion [44]. In the first stage, early endosomes mature into late endosomes or MVBs; during this process the endosomal membrane invaginates to produce intraluminal vesicles (ILVs) in the lumen of MVBs [45]. In this mechanism is involved ESCRT machinery that consists of four protein complexes: ESCRT-0, ESCRT-I, ESCRT-II, ESCRT-III and its associated proteins, such as TSG101 and Alix that are used as markers of exosomal population [46].

Exosomes may be collected by several biological fluids such as blood, urine, saliva, breast milk, synovial fluid, amniotic fluid, bronchoalveolar lavage fluid, malignant ascites and semen [47, 48]. Exosomes can be internalized by target cells in the closeness or cells at significant distance from parental cells. These vesicles can have different fates: they may interact with cells, by acting as messenger shuttles, in order to transfer information that can modulate the phenotype of target cells. Several mechanisms mediate exosomal uptake, including exosome fusion with the plasma membrane of target cells, leading to the release of exosomal contents into the cytoplasm, endocytosis by phagocytosis and juxtacrine signalling through receptor-ligand interactions. It was demonstrated that cancer cells released about 10 folds more exosomes than normal cell to mediate tumour progression [49].

Exosomes carry bioactive cargos, including common and donor cell-specific proteins, lipids and RNA and DNA molecules that reflected cells and tissue of origin and provided a snapshot of cells at the time of release [50]. Exosomal composition can be different from parental cells thanks to the selective sorting of the cargos. ExoCarta database [51] lists 9769 proteins, 3408 mRNAs, and 2838 miRNAs contained in exosomes collected from 286 published studies (www.exocarta.org). This data reflects the number of targets that can be modulated by exosomes and highlights the importance of studying them. Proteomic studies have demonstrated that exosomes contained cytosolic, cytoskeletal and membrane proteins, integrins, enzymes, adhesion and signalling molecules. Among exosomal proteins, tetraspanins and heat shock proteins are the most conserved molecules [52, 53]. Exosomal proteins maintain their biological activities such as antigen presentation, protein cleavage and pathway activation.

Moreover, RNA populations in exosomes were identified using high-throughput RNA-seq, including messenger RNAs (mRNAs) and many types of non-coding RNAs, such as circRNAs, miRNAs, transfer RNAs (tRNAs), ribosomal RNAs (rRNAs), lncRNAs, small nuclear RNAs (snRNAs), small nucleolar RNAs (snoRNAs) and piwi-interacting RNAs (piRNAs) [54, 55]. These RNAs can be shuttled from parental to target cells, where they modulate target genes or are the templates for protein synthesis. It was reported that exosomes act as nano-shuttles for miRNAs with a dual role in cancer progression. In this context, they can have oncogenic and tumour-suppressor functions [56]. Recently, circRNAs were found enriched and stable in cancer exosomes. The transport of nucleic acids by exosomes ensures the protection against degradation and dilution in the extracellular space, allowing long-distance distribution through the bloodstream or interstitial fluid [57].

Several papers reported a pleiotropic role of tumour-derived exosomes, as they are involved in tumour growth, angiogenesis, metastasis, modulation of the microenvironment, pre-metastatic niche formation, immunomodulation and drug resistance [58, 59]. Since exosomes shuttle their typical cargo through the bloodstream and mediate the horizontal transfer of genetic material from parental to target cells [56, 60], the idea of “liquid biopsy” encouraged studies on exosomes-based biomarkers especially in cancer as potential cancer biomarkers and theranostic devices [61].

Recently, a position paper by the International Society for Extracellular Vesicles (ISEV) summarized the recent application and current findings on the EVs-based therapies [62]. The translation of these vesicles in clinical practice requires a classification of EVs-based therapies in agreement with supervisory outlines [62]. Substantial improvement in exosomes studies has directed to upgraded and standardized protocols for purification and storage, as well as methods and standards for quality analyses of exosome-based cures [63]. Clinical trials proposing EVs as theranostic nanoparticles have been described in the early 2000s; exosome power on clinical research is established by numerous current clinical trials (https://clinicaltrials.gov/). Nowadays, 20 clinical trials investigate on EVs as biomarkers for diagnosis, prognosis or devices for drug delivery. Exosomes are also used as a new tool for clinical evaluation and screening system in liquid biopsy approaches [64, 65]. Only one clinical trial with circular RNAs is ongoing. The aim of this study is to develop a slightly invasive analysis to identify pancreatic cancer at initial stage of neoplasia and check the response to treatment, but there are no clinical trials that investigate exosomal circRNAs.

2.1 Circular RNA in Exosomes

In 2015, Li et al. [62] reported, for the first time, that exosomes contain abundant circRNAs. Genome-wide RNA-seq analyses discovered that circRNAs were enriched in exosomes compared to parental cells. It was reported that circRNA sorting to exosomes may be controlled by modulation of associated miRNA levels in parental cells and may transfer biological activity to target cells. Considering that circRNAs sponge miRNAs, the correlation between circRNAs and miRNAs, about circRNA shuttled with exosomes, was investigated. The circRNA, CDR1as, is known to work as a miR-7 sponge, because miR-7 mimics were introduced into HEK293T and MCF-7 cell lines and the level of CDR1as in exosomes and parental cells was determined. It was described that CDR1as level was deeply decreased in exosomes and improved in cells, upon ectopic expression of miR-7 in both HEK293T and MCF-7 cells. Exosomal CDR1as maintained biological activity also in exosomes abrogating miR-7-induced growth suppression in target cells [66, 67].

Since circRNAs are abundant in exosomes, they can be collected by human blood. In order to test if exo-circRNA enters into the circulation and is quantifiable for cancer diagnosis, Li and colleagues [68] used a xenograft mouse model of human MHCC-LM3 cancer cells. These cells were inoculated in mice, and 7 weeks later, serum from mice was harvested, and exosomal circRNAs were isolated and quantified by qRT-PCR analysis. The human CDYL circRNA was detected in serum from tumour-bearing mice, and the amount of this circRNA in xenografted mice was correlated with tumour mass [68]. To confirm the idea that circRNAs enriched in exosomes may represent biomarkers for cancer diagnosis and prognosis, the expression profile of serum exosomal circRNAs was explored in cancer patients and healthy donors. The expression of circRNAs in serum from 11 colorectal cancer patients, tested by RNA-seq analysis, was significantly different from healthy donors; in cancer patients, 67 circRNAs were lost, and 257 new circRNA types were found compared to healthy individuals [31].

In human serum more than 1000 exosomal circRNAs useful to discriminate patients with tumour from healthy controls were identified. These data suggest that circRNAs derived from human cancer can enter in the bloodstream and be easily quantified in serum. CircRNA expression profiles have been also performed in both cells and exosomes from three isogenic colorectal cancer cell lines that vary in *KRAS* mutational status. Although circRNAs have a tendency to be enriched in exosomes, circRNA concentration decreased at global level in mutant-*KRAS* cell lines, indicating a modulation of circRNAs during colorectal cancer progression and a possible contribution of circRNAs in oncogenesis [69].

Furthermore, RNA-seq technique allows to profile circRNA expression in EVs isolated from serum of patients with endometrial cancer and healthy controls. It was found that the number of upregulated circRNAs was higher than that of downregulated circRNAs in EVs from patients compared to healthy subjects. Xu et al. [25] reported that circRNAs may act as competing endogenous RNAs in receipt cells after internalization of EVs from cancer cells. They identified 209 upregulated and 66 downregulated circRNAs in EVs from serum of patients with endometrial cancer compared to those from healthy controls. The roles of differently expressed circRNAs by using KEGG pathway enrichment analysis were investigated. The expression of two circRNAs, hsa circ 0109046 and hsa circ 0002577, was confirmed by RT-qPCR, and the circRNA/miRNA interactions for these two circRNAs were also predicted. Overall, these data indicate that exosomal circRNAs can influence target cells contributing to the identification of new mechanisms of cancer development [25]. Recently, the studies on circRNAs have increasing value in the field of genomic research and their hallmark properties convinced the researchers to explore the powers of these molecules as biomarkers for complex diseases such as cancer [17, 70].

3 Conclusions and Future Perspectives

Exosomes, considered "diamonds in the rough", in combination with circRNAs for their peculiarity and high specificity can increase the potential use of both exosomes and circRNAs as markers of diagnosis and prognosis for cancer patients. The biomarkers are biological molecules contained in blood, tissues and body fluids that can be objectively evaluated and measured as indicators for normal and pathological conditions [71]. The use of biomarkers is important for early detection and diagnosis of different diseases as well as for monitoring the responses to treatments [72]. The typical features of the biomarkers such as stability, sensitivity and specificity allow their use in clinical practice. The goal of precision medicine, in particular the liquid biopsies, is the discovery of cancer biomarkers with high powerful detection and monitoring strategies for cancer risk indication, useful for patients to receive the most appropriate therapy and for clinicians to monitor the disease progression, regression and recurrence. Recent studies indicate exosomes as potential biomarkers for diagnosis, prognosis and prediction in cancer. The goal of exosomes used as biomarker is the substantial reduction of sample complexity, when compared to whole body fluids, and the low invasiveness in a liquid biopsy scenario [65, 73]. Recently, exosomal circRNAs have been suggested as potential biomarkers in cancer for their stability and high specificity. These new findings could be translated in clinical practice in order to discriminate patients with cancer from healthy individuals with high accuracy.

Competing Financial Interests The authors declare no competing financial interests.

References

1. Caruso S, Bazan V, Rolfo C, Insalaco L, Fanale D, Bronte G et al (2012) MicroRNAs in colorectal cancer stem cells: new regulators of cancer stemness? Oncogene 1(11):e32. https://doi.org/10.1038/oncsis.2012.33
2. Corsini LR, Bronte G, Terrasi M, Amodeo V, Fanale D, Fiorentino E et al (2012) The role of microRNAs in cancer: diagnostic and prognostic biomarkers and targets of therapies. Expert Opin Ther Targets 16(sup2):S103–S1S9. https://doi.org/10.1517/14728222.2011.650632
3. Rolfo C, Fanale D, Hong D, Tsimberidou A, Piha-Paul S, Pauwels P et al (2014) Impact of microRNAs in resistance to chemotherapy and novel targeted agents in non-small cell lung cancer. Curr Pharm Biotechnol 15(5):475–485. https://doi.org/10.2174/1389201015666140519123219
4. Amodeo V, Bazan V, Fanale D, Insalaco L, Caruso S, Cicero G et al (2013) Effects of anti-miR-182 on TSP-1 expression in human colon cancer cells: there is a sense in antisense? Expert Opin Ther Targets 17(11):1249–1261. https://doi.org/10.1517/14728222.2013.832206
5. Bronte F, Bronte G, Fanale D, Caruso S, Bronte E, Bavetta MG et al (2016) HepatomiRNoma: the proposal of a new network of targets for diagnosis, prognosis and therapy in hepatocellular carcinoma. Crit Rev Oncol Hematol 97:312–321. https://doi.org/10.1016/j.critrevonc.2015.09.007.

6. Fanale D, Amodeo V, Bazan V, Insalaco L, Incorvaia L, Barraco N et al (2016) Can the microRNA expression profile help to identify novel targets for zoledronic acid in breast cancer? Oncotarget 7(20). https://doi.org/10.18632/oncotarget.8722
7. Cabibi D, Caruso S, Bazan V, Castiglia M, Bronte G, Ingrao S et al (2016) Analysis of tissue and circulating microRNA expression during metaplastic transformation of the esophagus. Oncotarget 7(30). https://doi.org/10.18632/oncotarget.10291
8. Fanale D, Barraco N, Listì A, Bazan V, Russo A (2016) Non-coding RNAs functioning in colorectal cancer stem cells. Adv Exp Med Biol 937:93–108. https://doi.org/10.1007/978-3-319-42059-2_5
9. Fanale D, Castiglia M, Bazan V, Russo A (2016) Involvement of non-coding RNAs in chemo- and radioresistance of. Colorectal Cancer 937:207–228. https://doi.org/10.1007/978-3-319-42059-2_11
10. Rizzo S, Cangemi A, Galvano A, Fanale D, Buscemi S, Ciaccio M et al (2017) Analysis of miRNA expression profile induced by short term starvation in breast cancer cells treated with doxorubicin. Oncotarget 8(42). https://doi.org/10.18632/oncotarget.18028
11. Zhong Z, Lv M, Chen J (2016) Screening differential circular RNA expression profiles reveals the regulatory role of circTCF25-miR-103a-3p/miR-107-CDK6 pathway in bladder carcinoma. Sci Rep 6:1–12. https://doi.org/10.1038/srep30919
12. Chen L, Huang C, Wang X, Shan G (2015) Circular RNAs in eukaryotic cells. Curr Genomics 16(5):312–318. https://doi.org/10.2174/138920291666615070716 1554
13. Barrett SP, Salzman J (2016) Circular RNAs: analysis, expression and potential functions. Development 143(11):1838–1847. https://doi.org/10.1242/dev.128074
14. Gao Y, Wang J, Zheng Y, Zhang J, Chen S, Zhao F (2016) Comprehensive identification of internal structure and alternative splicing events in circular RNAs. Nat Commun 7:12060. https://doi.org/10.1038/ncomms12060
15. Wang Y, Wang Z (2015) Efficient backsplicing produces translatable circular mRNAs. RNA 21(2):172–179. https://doi.org/10.1261/rna.048272.114
16. Chen C, Sarnow P (1995) Initiation of protein synthesis by the eukaryotic translational apparatus on circular RNAs. Science 268(5209):415–417. https://doi.org/10.1126/science.7536344
17. Huang S, Yang B, Chen BJ, Bliim N, Ueberham U, Arendt T et al (2017) The emerging role of circular RNAs in transcriptome regulation. Genomics 109(5–6):401–407. https://doi.org/10.1016/j.ygeno.2017.06.005
18. Jeck WR, Sorrentino JA, Wang K, Slevin MK, Burd CE, Liu J et al (2012) Circular RNAs are abundant, conserved, and associated with ALU repeats. RNA 19(2):141–157. https://doi.org/10.1261/rna.035667.112
19. Ashwal-Fluss R, Meyer M, Pamudurti Nagarjuna R, Ivanov A, Bartok O, Hanan M et al (2014) circRNA Biogenesis competes with Pre-mRNA splicing. Mol Cell 56(1):55–66. https://doi.org/10.1016/j.molcel.2014.08.019
20. Conn Simon J, Pillman Katherine A, Toubia J, Conn Vanessa M, Salmanidis M, Phillips Caroline A et al (2015) The RNA binding protein quaking regulates formation of circRNAs. Cell 160(6):1125–1134. https://doi.org/10.1016/j.cell.2015.02.014
21. Memczak S, Jens M, Elefsinioti A, Torti F, Krueger J, Rybak A et al (2013) Circular RNAs are a large class of animal RNAs with regulatory potency. Nature 495(7441):333–338. https://doi.org/10.1038/nature11928
22. Zheng Q, Bao C, Guo W, Li S, Chen J, Chen B et al (2016) Circular RNA profiling reveals an abundant circHIPK3 that regulates cell growth by sponging multiple miRNAs. Nat Commun 7:11215. https://doi.org/10.1038/ncomms11215
23. Hou L-D, Zhang J (2017) Circular RNAs: an emerging type of RNA in cancer. Int J Immunopathol Pharmacol 30(1):1–6. https://doi.org/10.1177/0394632016686985
24. Moran JV, Salzman J, Chen RE, Olsen MN, Wang PL, Brown PO (2013) Cell-type specific features of circular RNA expression. PLoS Genet 9(9):e1003777. https://doi.org/10.1371/journal.pgen.1003777
25. Xu Z, Yan Y, Zeng S, Dai S, Chen X, Wei J et al (2017) Circular RNAs: clinical relevance in cancer. Oncotarget 9(1). https://doi.org/10.18632/oncotarget.22846
26. Kristensen LS, Hansen TB, Venø MT, Kjems J (2017) Circular RNAs in cancer: opportunities and challenges in the field. Oncogene 37(5):555–565. https://doi.org/10.1038/onc.2017.361
27. Greene J, Baird A-M, Brady L, Lim M, Gray SG, McDermott R et al (2017) Circular RNAs: biogenesis, function and role in human diseases. Front Mol Biosci 4. https://doi.org/10.3389/fmolb.2017.00038
28. Arnberg AC, Van Ommen GJB, Grivell LA, Van Bruggen EFJ, Borst P (1980) Some yeast mitochondrial RNAs are circular. Cell 19(2):313–319. https://doi.org/10.1016/0092-8674(80)90505-x
29. Capel B, Swain A, Nicolis S, Hacker A, Walter M, Koopman P et al (1993) Circular transcripts of the testis-determining gene Sry in adult mouse testis. Cell 73(5):1019–1030. https://doi.org/10.1016/0092-8674(93)90279-y
30. Haque S, Circular LWH (2017) RNAs (circRNAs) in health and disease. Genes 8(12):353. https://doi.org/10.3390/genes8120353
31. Xu H, Guo S, Li W, Yu P (2015) The circular RNA Cdr1as, via miR-7 and its targets, regulates insulin transcription and secretion in islet cells. Sci Rep 5(1). https://doi.org/10.1038/srep12453
32. Chen I, Chen C-Y, Chuang T-J (2015) Biogenesis, identification, and function of exonic circular RNAs. Wiley Interdiscip Rev RNA 6(5):563–579. https://doi.org/10.1002/wrna.1294
33. Kulcheski FR, Christoff AP, Margis R (2016) Circular RNAs are miRNA sponges and can be used as a new

class of biomarker. J Biotechnol 238:42–51. https://doi.org/10.1016/j.jbiotec.2016.09.011

34. Zhu L-P, He Y-J, Hou J-C, Chen X, Zhou S-Y, Yang S-J et al (2017) The role of circRNAs in cancers. Biosci Rep 37(5):BSR20170750. https://doi.org/10.1042/bsr20170750
35. Pant S, Hilton H, Burczynski ME (2012) The multifaceted exosome: biogenesis, role in normal and aberrant cellular function, and frontiers for pharmacological and biomarker opportunities. Biochem Pharmacol 83(11):1484–1494. https://doi.org/10.1016/j.bcp.2011.12.037
36. Li S, Li Y, Chen B, Zhao J, Yu S, Tang Y et al (2018) exoRBase: a database of circRNA, lncRNA and mRNA in human blood exosomes. Nucleic Acids Res 46(D1):D106–DD12. https://doi.org/10.1093/nar/gkx891
37. Siegel RL, Sahar L, Portier KM, Ward EM, Jemal A (2015) Cancer death rates in US congressional districts. CA Cancer J Clin 65(5):339–344
38. Maia J, Caja S, Strano Moraes MC, Couto N, Costa-Silva B (2018) Exosome-based cell-cell communication in the tumor microenvironment. Front Cell Dev Biol 20(6):18. https://doi.org/10.3389/fcell.2018.00018
39. Taverna S, Pucci M, Alessandro R (2017) Extracellular vesicles: small bricks for tissue repair/regeneration. Ann Transl Med 5(4):83. https://doi.org/10.21037/atm.2017.01.53
40. Tkach M, Théry C (2016) Communication by extracellular vesicles: where we are and where we need to go. Cell 164(6):1226–1232. https://doi.org/10.1016/j.cell.2016.01.043
41. Wolf P (1967) The nature and significance of platelet products in human plasma. Br J Haematol 13(3):269–288. https://doi.org/10.1111/j.1365-2141.1967.tb08741.x
42. Pan B-T, Johnstone RM (1983) Fate of the transferrin receptor during maturation of sheep reticulocytes in vitro: selective externalization of the receptor. Cell 33(3):967–978. https://doi.org/10.1016/0092-8674(83)90040-5
43. Colombo M, Raposo G, Théry C (2014) Biogenesis, secretion, and intercellular interactions of exosomes and other extracellular vesicles. Annu Rev Cell Dev Biol 30(1):255–289. https://doi.org/10.1146/annurev-cellbio-101512-122326
44. Kowal J, Tkach M, Théry C (2014) Biogenesis and secretion of exosomes. Curr Opin Cell Biol 29:116–125. https://doi.org/10.1016/j.ceb.2014.05.004
45. Huotari J, Helenius A (2011) Endosome maturation. EMBO J 30(17):3481–3500. https://doi.org/10.1038/emboj.2011.286
46. Henne WM, Stenmark H, Emr SD (2013) Molecular mechanisms of the membrane sculpting ESCRT pathway. Cold Spring Harb Perspect Biol 5(9):a016766. https://doi.org/10.1101/cshperspect.a016766
47. Boukouris S, Mathivanan S (2015) Exosomes in bodily fluids are a highly stable resource of disease biomarkers. PROTEOMICS Clin Appl 9(3–4):358–367. https://doi.org/10.1002/prca.201400114
48. Rodriguez-dorantes M, Romero-Cordoba S, Peralta-Zaragoza O, Salido-Guadarrama I, Hidalgo-Miranda A (2014) MicroRNAs transported by exosomes in body fluids as mediators of intercellular communication in cancer. OncoTargets Ther 7:1327. https://doi.org/10.2147/ott.s61562
49. Li W, Li C, Zhou T, Liu X, Liu X, Li X et al (2017) Role of exosomal proteins in cancer diagnosis. Mol Cancer 16(1). https://doi.org/10.1186/s12943-017-0706-8
50. Hessvik NP, Llorente A (2017) Current knowledge on exosome biogenesis and release. Cell Mol Life Sci 75(2):193–208. https://doi.org/10.1007/s00018-017-2595-9
51. Keerthikumar S, Chisanga D, Ariyaratne D, Al Saffar H, Anand S, Zhao K et al (2016) ExoCarta: a web-based compendium of exosomal cargo. J Mol Biol 428(4):688–692. https://doi.org/10.1016/j.jmb.2015.09.019
52. Haraszti RA, Didiot M-C, Sapp E, Leszyk J, Shaffer SA, Rockwell HE et al (2016) High-resolution proteomic and lipidomic analysis of exosomes and microvesicles from different cell sources. J Extracell Vesicles 5(1):32570. https://doi.org/10.3402/jev.v5.32570
53. Simona F, Laura S, Simona T, Riccardo A (2013) Contribution of proteomics to understanding the role of tumor-derived exosomes in cancer progression: state of the art and new perspectives. Proteomics 13(10–11):1581–1594. https://doi.org/10.1002/pmic.201200398
54. Nolte-'t Hoen ENM, Buermans HPJ, Waasdorp M, Stoorvogel W, Wauben MHM, 't Hoen PAC. Deep sequencing of RNA from immune cell-derived vesicles uncovers the selective incorporation of small non-coding RNA biotypes with potential regulatory functions. Nucleic Acids Res 2012;40(18):9272–9285. doi:https://doi.org/10.1093/nar/gks658.
55. Kim KM, Abdelmohsen K, Mustapic M, Kapogiannis D (2017) Gorospe M. RNA in extracellular vesicles. Wiley Interdisciplinary Reviews RNA 8(4):e1413. https://doi.org/10.1002/wrna.1413
56. Pucci M, Reclusa Asiáin P, Duréndez Sáez E, Jantus-Lewintre E, Malarani M, Khan S et al (2018) Extracellular vesicles as miRNA nano-shuttles: dual role in tumor progression. Target Oncol. https://doi.org/10.1007/s11523-018-0551-8
57. Morton DJ, Kuiper EG, Jones SK, Leung SW, Corbett AH, Fasken MB (2018) The RNA exosome and RNA exosome-linked disease. RNA 24(2):127–142. https://doi.org/10.1261/rna.064626.117
58. Ruivo CF, Adem B, Silva M, Melo SA (2017) The biology of cancer exosomes: insights and new perspectives. Cancer Res 77(23):6480–6488. https://doi.org/10.1158/0008-5472.can-17-0994
59. Steinbichler TB, Dudás J, Riechelmann H, Skvortsova I-I (2017) The role of exosomes in cancer metastasis. Semin Cancer Biol 44:170–181. https://doi.org/10.1016/j.semcancer.2017.02.006

60. Valadi H, Ekström K, Bossios A, Sjöstrand M, Lee JJ, Lötvall JO (2007) Exosome-mediated transfer of mRNAs and microRNAs is a novel mechanism of genetic exchange between cells. Nat Cell Biol 9(6):654–659. https://doi.org/10.1038/ncb1596
61. He C, Zheng S, Luo Y, Wang B (2018) Exosome theranostics: biology and translational medicine. Theranostics 8(1):237–255. https://doi.org/10.7150/thno.21945
62. Lener T, Gimona M, Aigner L, Börger V, Buzas E, Camussi G et al (2015) Applying extracellular vesicles based therapeutics in clinical trials – an ISEV position paper. J Extracell Vesicles 4(1):30087. https://doi.org/10.3402/jev.v4.30087
63. Coumans FAW, Brisson AR, Buzas EI, Dignat-George F, Drees EEE, El-Andaloussi S et al (2017) Methodological Guidelines to Study Extracellular Vesicles. Circ Res 120(10):1632–1648. https://doi.org/10.1161/circresaha.117.309417
64. Reiner AT, Witwer KW, van Balkom BWM, de Beer J, Brodie C, Corteling RL et al (2017) Concise review: developing best-practice models for the therapeutic use of extracellular vesicles. Stem Cells Transl Med 6(8):1730–1739. https://doi.org/10.1002/sctm.17-0055
65. Taverna S, Giallombardo M, Gil-Bazo I, Carreca AP, Castiglia M, Chacártegui J et al (2016) Exosomes isolation and characterization in serum is feasible in non-small cell lung cancer patients: critical analysis of evidence and potential role in clinical practice. Oncotarget 7(19):28748–28760. https://doi.org/10.18632/oncotarget.7638
66. Hansen TB, Kjems J, Damgaard CK (2013) Circular RNA and miR-7 in cancer. Cancer Res 73(18):5609–5612. https://doi.org/10.1158/0008-5472.can-13-1568
67. Peng LI, Yuan XQ, Li GC (2015) The emerging landscape of circular RNA ciRS-7 in cancer (Review). Oncol Rep 33(6):2669–2674. https://doi.org/10.3892/or.2015.3904
68. Li Y, Zheng Q, Bao C, Li S, Guo W, Zhao J et al (2015) Circular RNA is enriched and stable in exosomes: a promising biomarker for cancer diagnosis. Cell Res 25(8):981–984. https://doi.org/10.1038/cr.2015.82
69. Dou Y, Cha DJ, Franklin JL, Higginbotham JN, Jeppesen DK, Weaver AM et al (2016) Circular RNAs are down-regulated in KRAS mutant colon cancer cells and can be transferred to exosomes. Sci Rep 6(1):37982. https://doi.org/10.1038/srep37982
70. Qu S, Zhong Y, Shang R, Zhang X, Song W, Kjems J et al (2016) The emerging landscape of circular RNA in life processes. RNA Biol 14(8):992–999. https://doi.org/10.1080/15476286.2016.1220473
71. Henry NL, Hayes DF (2012) Cancer biomarkers. Mol Oncol 6(2):140–146. https://doi.org/10.1016/j.molonc.2012.01.010
72. Mayeux R (2004) Biomarkers potential uses and limitations. NeuroRx 1(2):182–188. https://doi.org/10.1602/neurorx.1.2.182
73. Rolfo C, Castiglia M, Hong D, Alessandro R, Mertens I, Baggerman G et al (2014) Liquid biopsies in lung cancer: the new ambrosia of researchers. Biochim Biophys Acta 1846(2):539–546. https://doi.org/10.1016/j.bbcan.2014.10.001

Circular RNAs in Blood

10

Angela Vea, Vicenta Llorente-Cortes,
and David de Gonzalo-Calvo

Abstract

Recent advances in RNA sequencing and bioinformatic analysis have allowed the development of a new research field: circular RNAs (circRNAs). These members of the non-coding transcriptome are generated by backsplicing, which results in a covalently closed, single-stranded RNA molecule. To date, thousands of circRNAs have been identified in different human cell types. CircRNAs are evolutionarily conserved, highly stable, cell-/developmental stage-specific and have longer half-lives compared with linear RNAs. Interestingly, different studies have demonstrated that circRNAs are abundantly expressed in the bloodstream. In this chapter, we review the current knowledge of circRNA biology in blood cells and the cell-free compartment, including extracellular vesicles. The potential clinical application of blood circRNAs in the biomarker and therapy fields is also discussed. Finally, perspectives for future studies are proposed.

Keywords

Circular RNA · Blood · Serum · Plasma · Extracellular vesicles

A. Vea
Biomedical Research Institute Sant Pau (IIB Sant Pau), Barcelona, Spain

V. Llorente-Cortes · D. de Gonzalo-Calvo (✉)
Biomedical Research Institute Sant Pau (IIB Sant Pau), Barcelona, Spain

Institute of Biomedical Research of Barcelona (IIBB) – Spanish National Research Council (CSIC), Barcelona, Spain

CIBERCV, Institute of Health Carlos III, Madrid, Spain

1 Introduction

Non-coding RNAs (ncRNAs) are a heterogeneous group of RNA molecules that do not encode proteins but have key regulatory and structural functions. During the last years, most studies have been focused on members of this family, such as microRNAs (miRNAs) and long non-coding RNAs (lncRNAs). Recent advances in high-throughput RNA sequencing (RNA-Seq) and computational analysis have drawn attention to a new class of ncRNAs, circular RNAs (circRNAs), as a natural feature of the cell expression programme.

CircRNAs are single-stranded and covalently closed RNA molecules that lack of free caps or poly(A) tails [1]. CircRNAs are generated by a process called backsplicing in which a splice donor site is joined to a splice acceptor site upstream in the primary transcript. These ncRNAs are mainly formed by exons, but they can also be derived from intronic, non-coding, antisense, untranslated or intergenic genomic

J. Xiao (ed.), *Circular RNAs*, Advances in Experimental Medicine and Biology 1087,
https://doi.org/10.1007/978-981-13-1426-1_10

regions [2]. Their size ranges from hundreds to thousands of nucleotides. The biogenesis of mammalian circRNAs is not fully understood [1]. CircRNAs are derived from pre-mRNAs, which are transcribed by RNA polymerase II. The process of backsplicing requires the canonical spliceosomal machinery and seems to depend on sequencing motifs within flanking introns and RNA-binding proteins (RBPs) such as quaking [3, 4]. Different circular isoforms with different expression patterns can be produced from a given gene [5]. Interestingly, circRNA production is regulated independently of the underlying linear RNA gene [6]. In some cases, the expression level of circRNAs is significantly higher than that of their corresponding linear RNA isoforms from the same gene [7]. The identification of circRNAs supports the concept that genes are complex transcriptional units that contain multiple and overlapping information [8, 9].

CircRNAs were initially discovered in plant viroids as early as the 1970s [10]. This class of ncRNAs were originally considered as splicing by-products, background noise or specific to a few pathogens such as viruses [11, 12]. Recently, transcriptome-wide circRNA analysis has identified and characterized thousands of circRNAs in diverse human cells [2, 7, 13], suggesting the relevance of circRNAs in the ncRNA family. Studies of circRNAs in different species have shown that the majority of circRNAs are evolutionarily conserved at the sequence level [7], pointing to a key role in relevant biological processes. CircRNAs are predominantly cytoplasmic [2], although their presence has also been described in the nucleus [14], and tend to accumulate in cells with a low proliferation rate, such as neurons [15]. Circularity confers specific properties to circRNAs. In comparison to linear RNAs, circRNAs are highly stable and less susceptible to degradation by ribonuclease R (RNase R) and have longer half-lives in cells [7]. Mounting data have demonstrated that circRNAs are expressed in a cell-, developmental stage- and disease-specific manner [13, 16, 17].

Evidence of potential functions in the regulation of the transcriptome and proteome is continuously emerging. Cytoplasmic circRNAs can inhibit miRNA function, acting as miRNA sponges by complementary base pairing [18], and participating in RNA networks, acting as competing endogenous RNAs (ceRNAs) [19]. By contrast, circRNA could also serve as a miRNA reservoir by stabilizing miRNAs [20]. CircRNAs regulate RNA transcription by binding to RNA polymerase II [14] or DNA [21] and can interact with RBPs participating in their storage, localization and function [22]. Additionally, recent studies revealed that several circRNAs could function as coding transcripts [23].

Despite the great advances achieved during the last years, current knowledge of the biological functions and potential clinical implications of circRNAs remain limited. The modulation of circRNAs on gene expression plays a significant role in a great variety of pathological conditions, including cancer and cardiovascular disease [24, 25]. Thus, circRNAs are promising therapeutic targets for future drugs. Interestingly, circRNAs have been consistently identified in the bloodstream and therefore are potential minimally invasive biomarkers. In this chapter, we provide an overview of the presence of circRNAs in human blood and their biological role (Fig. 10.1). We also discuss their future clinical application as biomarkers and therapeutics targets.

2 Circular RNAs in Blood

The RNA profile of the bloodstream can reflect the transcriptomic changes in blood cells. Furthermore, the circulating cell-free RNA can be informative of the alterations in the gene expression of different non-haematopoietic and haematopoietic cells. Therefore, the circulating transcriptomic biosignature is a promising tool that may adequately reflect the molecular fingerprint of the subject phenotype. The results from a number of publications point to circRNAs as novel regulatory elements of blood cell biology and biomarkers with potential clinical application.

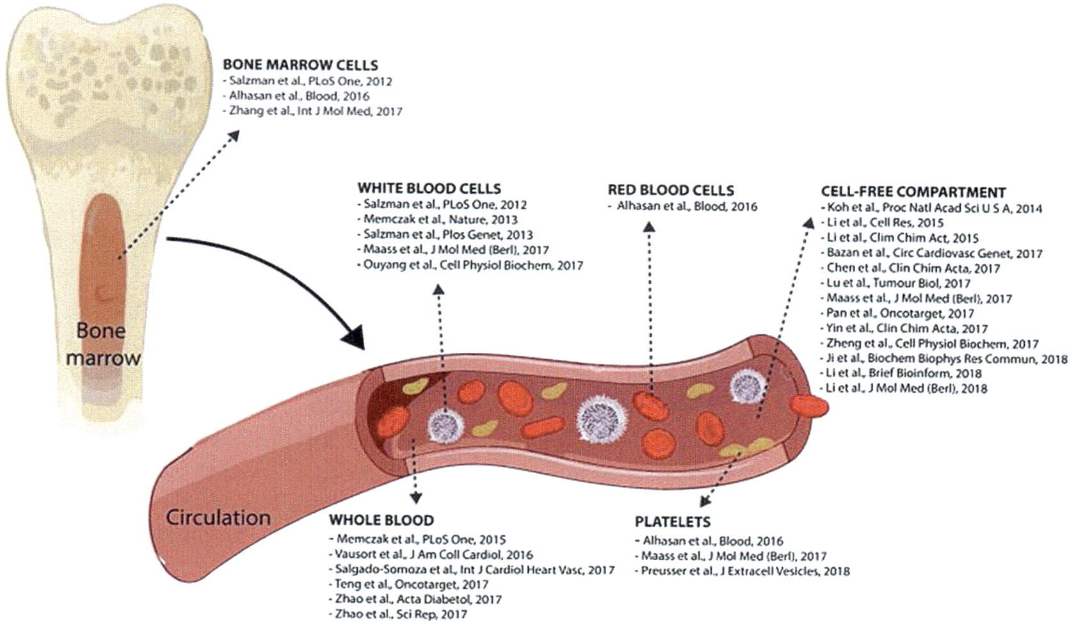

Fig. 10.1 Summary of publications demonstrating the presence of circular RNAs in blood
The presence of circular RNAs has been described in blood cells (red blood cells, white blood cells and platelets) and in cell-free compartments, including extracellular vesicles. Circular RNAs have also been detected in bone marrow cells

2.1 Circular RNA in Whole Blood

Previous investigations proposed that the whole blood is enriched in circRNAs. RNA-Seq analysis of two independent human whole blood samples, which were processed following standard procedures, identified 4550 and 4105 unique circRNA candidates in each sample, with approximately 2400 circRNAs reproducibly detected [26]. Most blood circRNAs were derived from protein-coding exonic regions or 5' UTR sequences. CircRNA expression levels were comparable to the circRNA-rich tissue cerebellum and > 15-fold higher compared to the liver. The predicted spliced length of the blood circRNAs (median = 343 nt) was similar to that in the liver or cerebellum (median = 394 nt and 448 nt, respectively). However, the number of circRNAs per gene was higher in whole blood. Blood circRNAs partially overlapped circRNAs expressed in the cerebellum and liver at approximately 30% and 10%, respectively, but also contain a considerable number of specific circRNAs. The top expressed blood circRNAs and the same number of top linear RNAs showed significant enrichment of different biological function annotations, which suggests that circRNA expression levels are independent of the linear RNA isoform abundance. Most functions of blood circRNAs were related to transcription regulation. Seven candidates were further evaluated using alternative methodology, including PCR. These candidates were expressed from loci that were not related to specific blood-related functions, which generated new questions about the function of circRNAs in the bloodstream. Importantly, hundreds of circRNAs were much more highly expressed—at least 30-fold—than the cognate linear isoforms. In contrast with the liver and cerebellum, blood circular RNA isoforms were detectable even while the corresponding linear gene products showed low abundances. Therefore, authors proposed that circRNA levels in human blood could be informative of the coding gene activity that could not be evaluated using classical RNA analysis. Different studies have subsequently detected a number of circRNAs in the peripheral blood [27–31].

The presence of exogenous circRNAs in the bloodstream should also be taken into account. Broadbent et al. [32] reported the expression of 1381 circRNAs during the blood stage development of *Plasmodium falciparum*. Interestingly, their experimentally validated circRNA candidates contained predicted human miRNA binding sites, which indicate a potential parasite-host communication mechanism in malaria.

Whole blood is composed of a plethora of different cells. In addition, the cell-free compartment can contain a number of circRNAs from cells of diverse non-haematopoietic origin. Despite their great potential as a source of biomarkers, a more detailed analysis of the circRNA biology in the blood components is necessary.

2.2 Circular RNA in Platelets

Although platelets are anucleated, they contain RNAs in the form of non-coding transcripts. Previous investigations have demonstrated that the platelet transcriptome is significantly enriched for circRNAs. Using 3 publicly available RNA-Seq datasets, Alhasan et al. [33] identified 33,829 structures consistent with circRNAs. Authors showed that circRNAs were 17- to 188-fold enriched in human platelets compared to nucleated tissues and identified 3162 genes significantly enriched for circRNAs. Approximately 27% of circRNAs were platelet-specific when they compared their findings with previous RNA-Seq datasets. The mean number of circRNAs per gene was higher in platelets (5 circRNAs per gene) compared to different nucleated tissues and cell lines (1–2 circRNAs per gene). The expression levels of ten selected circRNAs were higher than their corresponding linear structures, with circRNA isoforms from SMARCA5, UBXN7 and PNN ranging from 50- to 1000-fold more abundance. In contrast, all evaluated circRNA isoforms were expressed at an equivalent or lower level than their linear counterparts in nucleated cells. Experimental evidence also established that circRNAs are not enriched in cultured megakaryocytes. Thus, circRNAs seem to be produced in platelets rather than being inherited from their precursor cells. Authors proposed that the enrichment of circRNAs in platelets was associated with the degradation/decay of linear RNAs during the lifetime of the platelets. The analysis of platelet circRNAs in circulation may thus provide insights into megakaryocyte function in the bone marrow.

Maass et al. [34] generated a circRNA resource catalogue by sequencing ribosomal RNA-depleted total RNA in 20 human tissues highly relevant to disease-related research, including platelets isolated from whole blood from a single subject. According to their results, the platelets expressed a total of 3324 circRNAs with 2339 unique circRNAs. Supporting previous evidence, the number of circRNAs observed in platelets was more abundant than in any other evaluated tissue, including the cortex, atrium, fat, or muscle, among others. Furthermore, low overlap in the circRNA pattern was observed with other blood cells (neutrophils) or in the cell-free compartment (serum and plasma), which again suggested the tissue-specific expression of circRNAs. In general, circular-to-linear RNA ratios were high in the tissues with abundant circRNA expression. For example, the platelet circRNA in the SMARCA5 gene showed a circular-to-linear ratio of 151:1 in platelets. Furthermore, the authors reported that approximately 100 genes hosted more than five different circRNA isoforms. In some cases, such as PTPN12 or TTN genes, they detected 18 circRNAs isoforms in the platelets, atrium and vena cava. Since platelets are translationally competent [35], the authors hypothesized that platelet circRNAs could serve as templates for translation. However, their experimental results using mass spectrometry were inconclusive.

These results are also consistent with recent findings. Characterization of circRNAs using an RNA-Seq approach in human platelets revealed that, compared to other haematopoietic cell types, including monocytes, macrophages, T cells and megakaryocytes, circRNAs are abundant in platelets [36]. A large set of circular isoforms are predominantly expressed in platelets (55–70%), and 95% of these abundant circRNAs were identical in resting platelets and platelets activated by thrombin receptor activator peptide-6 (TRAP-6). Supporting the cell specificity of circRNAs, the most abundant circRNA in platelets, Plt-circR4, was exclusively expressed in platelets when compared to ten different cell lines.

Different genes implicated in blood vessel relaxation and platelet aggregation express platelet circRNAs [34]. Unfortunately, the function of this class of ncRNAs in platelets remains unclear. Since the deregulation of platelet activation is associated with a number of relevant diseases, including myocardial infarction and stroke, and ncRNAs seems to play a key role in platelet biology, further investigations should evaluate whether circRNAs mediate relevant biological effects in platelets.

2.3 Circular RNA in White Blood Cells

The presence of circRNAs in circulating leukocyte populations has been described in leukocytes isolated from the blood [37] and bone marrow [38]. In a seminal study, Salzmann et al. [13] performed RNA-Seq on ribosomal RNA-depleted total RNA from the diagnostic bone marrow of five children between the ages of 2 and 6 with hyperdiploid B-precursor acute lymphoblastic leukaemia. They identified a hundred genes with a permutated exon order (scrambled exons) that were predicted to be circRNAs. More than 700 isoforms with scrambled exons were estimated to comprise more than 10% of all transcript isoforms produced from a comparable number of genes. It should be noted that, due to their experimental design, an underestimation of the prevalence of circular RNA isoforms was expected. Using RT-qPCR, they confirmed the results of the most abundant circRNAs: ESYT2, FBXW4, CAMSAP1, KIAA0368, CLNS1A, FAM120A, MAP3K1, ZKSCAN1, MANBA, ZBTB46, NUP54, RARS and MGA. CircRNAs were not a specific feature of leukaemic cells, since PCR results verified the presence of scrambled exons in HeLa cells and normal primary human cells, including peripheral blood collected from the same patients in remission, and H9 ES cells. In addition, the authors reported evidence for scrambled transcripts comprising at least 10% of the transcripts from more than 800 genes in specific cell populations isolated from the bone marrow of a single individual: naive B cells (CD19+), haematopoietic stem cells (CD34+) and neutrophils. The presence of circRNAs has also been recently corroborated in neutrophils isolated from peripheral whole blood from a single donor, with a total of 274 circRNAs, including 58 unique circRNAs [34]. Differences in the relative abundance of circRNAs have been observed between different leukocyte types. For example, the most abundant circRNAs in CD19+, CD34+ and neutrophils samples were KIAA0182, MAN1A2 and CCDC126, respectively [13]. CircRNAs represented more than half of all transcripts produced by these genes. Interestingly, authors reported the expression of scrambled isoforms from ncRNAs. These results are supported by later findings from the same group that demonstrated the circular/linear RNA ratio and the pattern of circRNA isoforms from each gene, in addition to the repertoire of genes expressing circRNAs, which were cell-type specific by analysing 15 different cancer and non-cancer cell lines, including the leukaemia cell line K562 [6], and the results from Memczak et al. [2] who suggested that the expression of circRNAs were in part cell- and developmental stage-specific. Indeed, these authors reported specific circRNA patterns with 939 exclusively expressed in CD19+ cells, 333 in CD34+ and 194 in neutrophils [2]. For example, the hsa-circRNA 2149 was detected in CD19+ leukocytes but not in CD34+ leukocytes or neutrophils [2].

Recent evidence suggested that circRNAs may play a relevant role in leukocyte biology. Using publicly available RNA-Seq data from mouse macrophages, Ng et al. [39] identified an LPS-inducible circRNA, mcircRasGEF1B, that regulates the expression and stability of ICAM-1 mRNA. Several TLR pathways regulate the expression of mcircRasGEF1B, including TLR4, TLR9, TLR3 and TLR2/TLR1, in RAW264.7

cells but not in MEF cells. Interestingly, this circRNA has a human homologue with similar properties. Authors proposed that circRNAs may participate in the fine-tuning immune responses and protection against microbial infection. Different expression circRNA profiles were observed in CD28(+)CD8(+) T cells and CD28(-) CD8(+) T cells isolated from healthy elderly or adult control subjects [40]. *In silico* prediction results suggested that the circRNA 100783 may play a role in phosphoprotein-associated functions during CD28-related CD8(+) T-cell ageing. circRNA 100783 may therefore constitute a biomarker for the longitudinal tracking of T-cell ageing and global immunosenescence. Zhang et al. [38] compared the circRNA expression profiles of bone marrow-derived macrophages under distinct polarizing conditions. Authors showed that 189 circRNAs were differentially expressed between M1 and M2 macrophages and proposed that circRNAs may be implicated in macrophage differentiation and polarization.

2.4 Circular RNA in Red Blood Cells

Despite red blood cells (RBCs) being the most abundant cell type in the blood, the knowledge about the presence of circRNAs in this cell type is limited. Nonetheless, the biology of circRNAs in this cell type seems to be similar to that observed in other anucleated cells such as platelets. Indeed, circRNAs are also highly enriched in mature RBCs relative to nucleated cells [33]. Again, the expression levels of circRNAs are higher than linear RNAs [33]. Since RBCs are not able to synthesise proteins, the circRNA profile may reflect the biological processes of the erythropoietic progenitor cells from the bone marrow.

2.5 Circular RNA in the Blood Cell-Free Compartment

Although the exact number of circRNAs that can be detected in the plasma remains unknown, different studies have provided exhaustive evidence about the presence of cell-free circRNAs in the circulation. Koh et al. [41] detected 19 circRNAs in plasma samples from pregnant women using an approach based on RNA-Seq and microarrays. Maass et al. [34] also demonstrated the expression of 57 circRNAs in plasma and 39 circRNAs in serum using RNA-seq. Notably, these authors reported 51 and 37 unique circRNAs in plasma and serum, respectively, compared to other clinically relevant tissues. Using a circRNA microarray, a recent study proposed the presence of a higher number of circRNAs, more than 10,000, in each of the 21 plasma samples obtained from patients with cervical cancer [42]. The presence of circRNA in circulating extracellular vesicles has also been described (Fig. 10.2). Using RNA-Seq analysis, Li et al. [43] identified 1215 circRNAs in human exosomes isolated from a pool of serum obtained from three healthy donors. Most circRNAs (90%) were derived from protein-coding exons but also consisted of introns, lncRNAs, unannotated regions and antisense regions. Similar to previous findings, the median length was 350 nt. Three candidates selected for further analyses with RT-qPCR, circ-N4BP2L2, circ-GSE1 and circ-SMARCA5 were detected in serum-derived exosomes but not in exosome-depleted serum, which suggested that circRNAs may be transported by specific mechanisms in circulation. Supporting the high stability of circRNAs, the incubation of serum at room temperature for up to 24 h had minimal effects on exosomal circRNA levels. Interestingly, circRNAs originated from human MHCC-LM3 cancer cells in a xenograft mouse model, such as human circRNA CDYL, could be detected in the mouse serum and correlated with tumour weight, which provided a relevant clue about the release of circRNA from tissues to the circulation and the potential of circRNAs as biomarkers. Indirect evidence has been provided by independent studies that suggests a change in plasma levels of circRNAs in postoperative gastric cancer patients compared to preoperative patients [44, 45]. The expression of approximately 2700 plasma circRNAs was also significantly changed after surgical removal of cervical tumours [42]. The

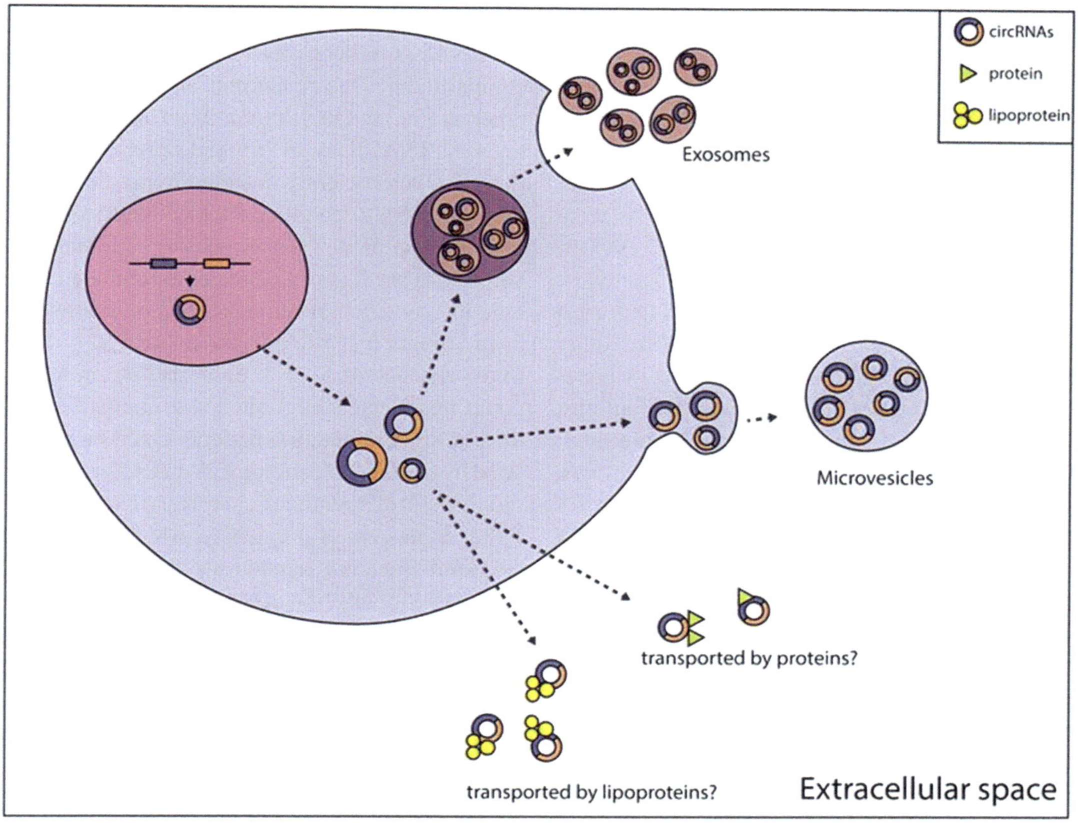

Fig. 10.2 Mechanisms of circular RNA release to the extracellular space
Similar to other non-coding RNAs, circular RNAs are released in the extracellular space into exosomes or microvesicles. Future studies should evaluate whether circular RNAs are transported by proteins and/or lipoproteins

presence of circRNAs in the serum and plasma has been corroborated by a considerable number of biomarker-based studies using different methodologies: RNA-Seq, microarray, RT-qPCR or RT-ddPCR [19, 44–57].

Results observed in the cell-free compartment suggest the circRNA secretion to the extracellular space/circulation, as it was shown for other ncRNAs such as miRNAs [58]. These hypotheses have been validated by different studies. CircRNAs were detected in cell-derived exosomes released by MHCC-LM3 liver cancer cells [43]. The expression level was enriched in exosomes compared to cells (at least twofold). Additionally, circular-to-linear RNA ratios in exosomes were approximately sixfold higher than those in cells, suggesting that circRNAs were incorporated into exosomes more than linear RNAs. The level of circRNAs in exosomes was only moderately correlated with that of cellular circRNAs. Importantly, circRNAs contained in exosomes retained biological activity. The exosomes containing the circRNA CDR1 abrogate the miR-7-induced *in vitro* inhibition of cell proliferation in receipt SMCC-7721 cancer cells. These results indicate the participation of ncRNAs in cell-to-cell communication [59]. Nonetheless, further investigations should corroborate these findings. Authors suggested that the sorting of circRNAs into exosomes may be regulated, at least in part, by changes in associated intracellular miRNA levels. The selective packaging and release of circRNAs within extracellular vesicles (microvesicles and exosomes) have also been reported in platelets [36]. Since whole blood has been defined as the main contributor (~40%) towards the cell-free RNA transcriptome [41], and platelets have been reported as a major source of ncRNAs in the circulation [35], these results are especially relevant. Interestingly, a group of selected circRNAs were preferentially released in exosomes (FAM13B, DYRK1A, AMD1 and TMEM30) and microvesicles (AMD1 and DYRK1A) compared with their corresponding linear RNA. Nonetheless, other circRNAs were preferentially retained in platelets (ASAP1). Based on the size distributions, released circRNAs were smaller than preferentially retained circRNAs (mean of 283 nt vs. 459 nt for microvesicles; 286 nt vs. 435 nt for exosomes). These results suggest that circRNA size may be an additional determinant for selective vesicle export. Further investigations should evaluate other factors that may affect sorting, such as sequence motifs, as demonstrated for miRNAs [60]. An alternative hypothesis to explain circRNA secretion has been proposed by Lasda et al. [61]. Authors hypothesized that, due to the long half-live of circRNAs in cells, the release of circRNAs in extracellular vesicles may constitute one possible mechanism to clear cellular circRNAs. Indeed, the higher circular/linear RNA ratio in extracellular vesicles compared to the producer cell may provide evidence in this sense. Overall, the results suggested that the secretion of circRNAs in extracellular vesicles seems to be a common property of many cell types [61]. Supporting this hypothesis, circRNAs have been detected in exosomes released from three different colon cancer lines [62] and microparticles secreted by vascular smooth muscle cells [46].

3 Clinical Application

The development of blood-based biomarkers is of great interest for clinical practice due to the relatively simple blood withdrawal procedure, compared to the more invasive tissue biopsy, and the fast and cost-effective analysis [63]. In this context, circRNAs may constitute a new entity of biomarkers. First, the presence of thousands of circRNAs has been described in the cell-free compartment. This constitutes an advantage over other ncRNAs. Only hundreds of miRNAs, the main ncRNA class in biomarker-based studies [64], can be efficiently detected in plasma/serum samples [65]. Second, circularity confers excellent biochemical properties as biomarkers. CircRNAs are free of exonuclease-mediated degradation, are cell-specific, are more stable and have a longer half-life than most linear RNAs due to the absence of 5′ or 3′ ends. Third, circRNAs could be detected in clinical specimens. Fourth, the circRNA circulation patterns could be modulated by different physiological states. Circulating circRNAs are specifically expressed during different trimesters of pregnancy, which suggests a temporal dynamic regulation of these ncRNAs [41]. Additionally, the deregulation of circulating circRNA levels in pathological conditions are supported by a number of publications that have evaluated the expression of circRNAs in whole blood, blood cells and circulating cell-free compartments in a wide array of diseases, including coronary artery disease [19, 30], type 2 diabetes mellitus [31], diabetes retinopathy [54], rheumatoid arthritis [37], colorectal cancer [43], gastric cancer [44], breast cancer [53], acute myeloid leukaemia [66], systemic lupus erythaematosus [67], intracranial aneurysm [28], pulmonary tuberculosis [68] and primary biliary cholangitis [56]. Indeed, the results from different studies point to the potential clinical application of circRNAs as biomarkers. In a recent investigation with a large sample size (N = 769), circRNA_025016 was upregulated in patients with new-onset AF after isolated off-pump coronary artery bypass grafting with high diagnostic accuracy (AUC = 0.802) [57]. Vasourt et al. [29] reported that the blood levels of MICRA (myocardial infarction-associated circular RNA) were a strong predictor of left ventricle dysfunction 3–4 months after myocardial infarction, even after adjusting by potential confounding factors, in peripheral blood samples from two independent cohorts totalling 642 patients. MICRA showed an incremental predictive value on top of established clinical parameters and biomarkers.

The same group has recently validated these findings using an alternative stratification criteria [27]. In breast cancer, compared with commonly used biomarkers for diagnosis of carcinoembryonic antigen (CEA, AUC = 0.562) and carbohydrate antigen 15-3 (CA15-3, AUC = 0.629), peripheral blood circ_0001785 had higher diagnostic accuracy (AUC = 0.784) [53]. Furthermore, circRNAs seem to not only be biomarkers themselves but can also be combined with other ncRNAs. The serum ratio of circRNA-284 to miR-221, a miRNA for which circRNA-284 has a binding site, was increased in patients presenting with an acute carotid-related ischaemic event and showed great performance in terms of discrimination (AUC = 0.820) as diagnostic biomarkers for carotid-related cerebrovascular ischaemia [46]. These results amplify the potential of circRNAs as biomarkers of disease. Given the inverse putative functional relationship between miR-221 and circR-284, these results may also provide valuable information about the pathological mechanism linked to the disease. Supporting their role as biomarkers, unique fusion-circRNAs (f-circRNAs) derived from the exons of genes affected by cancer-associated chromosomal translocation have been detected in particular pathological conditions such as leukaemia [69]. The evaluation of this f-circRNA in blood cells and the cell-free compartment represents an interesting diagnostic tool.

Despite the progress in circRNAs, whether circRNAs can be potential therapeutic targets for blood conditions remains elusive [8]. As circRNAs may be involved in a wide range of biological processes, deregulation of circRNA expression may affect a number of pathological mechanisms and therefore may play a causative role in a number of diseases [20, 70]. Previous evidence suggests the potential of circRNAs in novel therapeutic approaches. F-circRNAs contribute to tumour progression by increasing cell proliferation and clonogenicity and protect leukaemia cells from the cytotoxic effects of cytarabine, a drug used for the treatment of leukaemia [69]. Therefore, interventions aimed to block f-circRNAs could provide novel therapeutic strategies. Furthermore, since extracellular ncRNAs could modulate the phenotype and gene expression of recipient cells [71], circRNAs emerged as tools with great potential application in therapeutics.

4 Limitations and Perspectives

Given the emerging role for blood circRNAs as biomarkers to aid in the management of patients, the development of independent and multicentre studies with large population sizes to explore the real clinical application of circRNAs in diagnosis and prognosis seems mandatory. The effect of potential sources of variation on circRNA levels, including age, sex, comorbidities, genetic background and disease stage, among others, deserves particular attention. Concerning extracellular circRNAs, the identification of cellular sources and cellular targets, in addition to their function in health and disease, is an interesting research field that should be addressed. Indeed, it is not known whether extracellular circRNAs are casually involved in the pathophysiology of the underlying disease. Thus, it is necessary to evaluate whether circRNAs are mediators in cell-to-cell communication or their secretion in extracellular vesicles is merely a circRNA discard pathway. Since circRNAs interact with RNA-binding proteins [72], future work should also investigate the transport of circRNAs complexed with proteins, or even lipoproteins, similar to that observed for miRNAs (Fig. 10.2) [73, 74]. It should be noted that circRNAs are relatively well-conserved in a broad range of species, which facilitates the investigation of their biological function in different cellular and animal models. Overall, more functional studies are needed to elucidate the functions of circRNAs and their possible role in blood disease. In addition, although it has been proposed that the half-lives of cellular circRNAs can be longer than 48 h [7], it is not clear which is the real half-life in extracellular fluids [75]. The evidence presented here suggested a long stability in the cell-free compartment.

Implementation of ncRNAs is currently not feasible in clinical laboratories due to technological limitations and the high variability in the preanalytical phases. The analysis of the RNA family in plasma/serum has strong methodological limitations, mainly due to the low concentration. The evaluation of ncRNAs in blood cells has become recognized as an interesting alternative [76]. Nonetheless, due to the presence of circRNAs in blood cells, particularly RBCs, the possible cross-contamination in haemolytic samples should be taken into account when analysing the cell-free compartment. There are some controversies related to the methodology used for genome-wide profiling of circRNAs. RNA-Seq is a common method, but its detection efficiency is limited [77]. This may explain the differences observed in the number of circRNAs detected from the same type of samples, especially in the cell-free compartment. Recently, it has been proposed that circRNA microarrays could be an interesting alternative [42]. However, contrary to RNA-Seq, microarrays can only detect known circRNAs. More investigations are needed to identify the most sensitive and accurate technology. The development of standard operating procedures and guidelines for best practices in addition to automated and standardized assays is needed. A detailed description of the methods used for circRNA analysis is fundamental to ensure the reproducibility of the results. A special effort should be performed to clarify the circRNA nomenclature. In the current chapter, we have used the name given by each publication. Nonetheless, there is strong diversity among different publications.

5 Conclusions

Recent advances in transcriptomics have placed the focus on a new species of ncRNAs called circRNAs. Due to their biological properties, circRNAs have emerged as a new source of blood-based biomarkers and therapeutic approaches. However, more research is needed to elucidate the biological function of circRNAs in the blood as well as their real potential as clinical indicators.

Acknowledgements DdG-C was a recipient of Juan de la Cierva-Incorporación grants from the Ministerio de Economía y Competitividad (IJCI-2016-29393). CIBER Cardiovascular (CB16/11/00403 to DdG-C and VL-C) is a project of the Instituto de Salud Carlos III.

VLl-C and DdG-C are members of the Cardiolinc™ network.

Competing Financial Interests The authors declare no competing financial interests.

References

1. Chen LL (2016) The biogenesis and emerging roles of circular RNAs. Nat Rev Mol Cell Biol 17(4):205–211
2. Memczak S, Jens M, Elefsinioti A et al (2013) Circular RNAs are a large class of animal RNAs with regulatory potency. Nature 495(7441):333–338
3. Conn SJ, Pillman KA, Toubia J et al (2015) The RNA binding protein quaking regulates formation of circRNAs. Cell 160(6):1125–1134
4. Zhang XO, Wang HB, Zhang Y et al (2014) Complementary sequence-mediated exon circularization. Cell 159(1):134–147
5. Khan MA, Reckman YJ, Aufiero S et al (2016) RBM20 regulates circular RNA production from the Titin gene. Circ Res 119(9):996–1003
6. Salzman J, Chen RE, Olsen MN et al (2013) Cell-type specific features of circular RNA expression. PLoS Genet 9(9):e1003777
7. Jeck WR, Sorrentino JA, Wang K et al (2013) Circular RNAs are abundant, conserved, and associated with ALU repeats. RNA 19(2):141–157
8. Bonizzato A, Gaffo E, Te Kronnie G et al (2016) CircRNAs in hematopoiesis and hematological malignancies. Blood Cancer J 6(10):e483
9. Tuck AC, Tollervey D (2011) RNA in pieces. Trends Genet 27(10):422–432
10. Sanger HL, Klotz G, Riesner D et al (1976) Viroids are single-stranded covalently closed circular RNA molecules existing as highly base-paired rod-like structures. Proc Natl Acad Sci U S A 73(11):3852–3856
11. Cocquerelle C, Mascrez B, Hetuin D et al (1993) Mis-splicing yields circular RNA molecules. FASEB J 7(1):155–160
12. Kos A, Dijkema R, Arnberg AC et al (1986) The hepatitis delta (delta) virus possesses a circular RNA. Nature 323(6088):558–560
13. Salzman J, Gawad C, Wang PL et al (2012) Circular RNAs are the predominant transcript isoform from hundreds of human genes in diverse cell types. PLoS One 7(2):e30733
14. Li Z, Huang C, Bao C et al (2015) Exon-intron circular RNAs regulate transcription in the nucleus. Nat Struct Mol Biol 22(3):256–264

15. Rybak-Wolf A, Stottmeister C, Glazar P et al (2015) Circular RNAs in the mammalian brain are highly abundant, conserved, and dynamically expressed. Mol Cell 58(5):870–885
16. Cao S, Wei D, Li X et al (2017) Novel circular RNA expression profiles reflect progression of patients with hypopharyngeal squamous cell carcinoma. Oncotarget 8(28):45367–45379
17. Xu T, Wu J, Han P et al (2017) Circular RNA expression profiles and features in human tissues: a study using RNA-seq data. BMC Genomics 18(Suppl 6):680
18. Hansen TB, Jensen TI, Clausen BH et al (2013) Natural RNA circles function as efficient microRNA sponges. Nature 495(7441):384–388
19. Pan RY, Liu P, Zhou HT et al (2017) Circular RNAs promote TRPM3 expression by inhibiting hsa-miR-130a-3p in coronary artery disease patients. Oncotarget 8(36):60280–60290
20. Wang K, Sun Y, Tao W et al (2017) Androgen receptor (AR) promotes clear cell renal cell carcinoma (ccRCC) migration and invasion via altering the circHIAT1/miR-195-5p/29a-3p/29c-3p/CDC42 signals. Cancer Lett 394:1–12
21. Conn VM, Hugouvieux V, Nayak A et al (2017) A circRNA from SEPALLATA3 regulates splicing of its cognate mRNA through R-loop formation. Nat Plants 3:17053
22. Du WW, Yang W, Liu E et al (2016) Foxo3 circular RNA retards cell cycle progression via forming ternary complexes with p21 and CDK2. Nucleic Acids Res 44(6):2846–2858
23. Yang Y, Fan X, Mao M et al (2017) Extensive translation of circular RNAs driven by N(6)-methyladenosine. Cell Res 27(5):626–641
24. Devaux Y, Creemers EE, Boon RA et al (2017) Circular RNAs in heart failure. Eur J Heart Fail 19(6):701–709
25. Xu Z, Yan Y, Zeng S et al (2018) Circular RNAs: clinical relevance in cancer. Oncotarget 9(1):1444–1460
26. Memczak S, Papavasileiou P, Peters O et al (2015) Identification and characterization of circular RNAs as a new class of putative biomarkers in human blood. PLoS One 10(10):e0141214
27. Salgado-Somoza A, Zhang L, Vausort M et al (2017) The circular RNA MICRA for risk stratification after myocardial infarction. Int J Cardiol Heart Vasc 17:33–36
28. Teng L, Chen Y, Chen H et al (2017) Circular RNA hsa_circ_0021001 in peripheral blood: a potential novel biomarker in the screening of intracranial aneurysm. Oncotarget 8(63):107125–107133
29. Vausort M, Salgado-Somoza A, Zhang L et al (2016) Myocardial infarction-associated circular RNA predicting left ventricular dysfunction. J Am Coll Cardiol 68(11):1247–1248
30. Zhao Z, Li X, Gao C et al (2017) Peripheral blood circular RNA hsa_circ_0124644 can be used as a diagnostic biomarker of coronary artery disease. Sci Rep 7:39918
31. Zhao Z, Li X, Jian D et al (2017) Hsa_circ_0054633 in peripheral blood can be used as a diagnostic biomarker of pre-diabetes and type 2 diabetes mellitus. Acta Diabetol 54(3):237–245
32. Broadbent KM, Park D, Wolf AR et al (2011) A global transcriptional analysis of Plasmodium falciparum malaria reveals a novel family of telomere-associated lncRNAs. Genome Biol 12(6):R56
33. Alhasan AA, Izuogu OG, Al-Balool HH et al (2016) Circular RNA enrichment in platelets is a signature of transcriptome degradation. Blood 127(9):e1–e11
34. Maass PG, Glazar P, Memczak S et al (2017) A map of human circular RNAs in clinically relevant tissues. J Mol Med (Berl) 95(11):1179–1189
35. Sunderland N, Skroblin P, Barwari T et al (2017) MicroRNA biomarkers and platelet reactivity: the clot thickens. Circ Res 120(2):418–435
36. Preusser C, Hung LH, Schneider T et al (2018) Selective release of circRNAs in platelet-derived extracellular vesicles. J Extracell Vesicles 7(1):1424473
37. Ouyang Q, Wu J, Jiang Z et al (2017) Microarray expression profile of circular RNAs in peripheral blood mononuclear cells from rheumatoid arthritis patients. Cell Physiol Biochem 42(2):651–659
38. Zhang Y, Zhang Y, Li X et al (2017) Microarray analysis of circular RNA expression patterns in polarized macrophages. Int J Mol Med 39(2):373–379
39. Ng WL, Marinov GK, Liau ES et al (2016) Inducible RasGEF1B circular RNA is a positive regulator of ICAM-1 in the TLR4/LPS pathway. RNA Biol 13(9):861–871
40. Wang YH, Yu XH, Luo SS et al (2015) Comprehensive circular RNA profiling reveals that circular RNA100783 is involved in chronic CD28-associated CD8(+)T cell ageing. Immun Ageing 12:17
41. Koh W, Pan W, Gawad C et al (2014) Noninvasive in vivo monitoring of tissue-specific global gene expression in humans. Proc Natl Acad Sci U S A 111(20):7361–7366
42. Li S, Teng S, Xu J et al (2018) Microarray is an efficient tool for circRNA profiling. Brief Bioinform. https://doi.org/10.1093/bib/bby006
43. Li Y, Zheng Q, Bao C et al (2015) Circular RNA is enriched and stable in exosomes: a promising biomarker for cancer diagnosis. Cell Res 25(8):981–984
44. Li P, Chen S, Chen H et al (2015) Using circular RNA as a novel type of biomarker in the screening of gastric cancer. Clin Chim Acta 444:132–136
45. Li T, Shao Y, Fu L et al (2018) Plasma circular RNA profiling of patients with gastric cancer and their droplet digital RT-PCR detection. J Mol Med (Berl) 96(1):85–96
46. Bazan HA, Hatfield SA, Brug A et al (2017) Carotid plaque rupture is accompanied by an increase in the ratio of serum circR-284 to miR-221 levels. Circ Cardiovasc Genet 10(4):e001720
47. Chen S, Li T, Zhao Q et al (2017) Using circular RNA hsa_circ_0000190 as a new biomarker in the diagnosis of gastric cancer. Clin Chim Acta 466:167–171

48. Huang M, He YR, Liang LC et al (2017) Circular RNA hsa_circ_0000745 may serve as a diagnostic marker for gastric cancer. World J Gastroenterol 23(34):6330–6338
49. Ji W, Qiu C, Wang M et al (2018) Hsa_circ_0001649: a circular RNA and potential novel biomarker for colorectal cancer. Biochem Biophys Res Commun 497(1):122–126
50. Li WH, Song YC, Zhang H et al (2017) Decreased expression of Hsa_circ_00001649 in gastric cancer and its clinical significance. Dis Markers 2017:4587698
51. Lu R, Shao Y, Ye G et al (2017) Low expression of hsa_circ_0006633 in human gastric cancer and its clinical significances. Tumour Biol 39(6):1010428317704175
52. Sun H, Tang W, Rong D et al (2018) Hsa_circ_0000520, a potential new circular RNA biomarker, is involved in gastric carcinoma. Cancer Biomark 21(2):299–306
53. Yin WB, Yan MG, Fang X et al (2017) Circulating circular RNA hsa_circ_0001785 acts as a diagnostic biomarker for breast cancer detection. Clin Chim Acta S0009–8981(17):30407–30402
54. Zhang SJ, Chen X, Li CP et al (2017) Identification and characterization of circular RNAs as a new class of putative biomarkers in diabetes retinopathy. Invest Ophthalmol Vis Sci 58(14):6500–6509
55. Zhao Q, Chen S, Li T et al (2018) Clinical values of circular RNA 0000181 in the screening of gastric cancer. J Clin Lab Anal 32(4): e22333
56. Zheng J, Li Z, Wang T et al (2017) Microarray expression profile of circular RNAs in plasma from primary biliary cholangitis patients. Cell Physiol Biochem 44(4):1271–1281
57. Zhang J, Xu Y, Xu S et al (2018) Plasma circular RNAs, Hsa_circRNA_025016, predict postoperative atrial fibrillation after isolated off-pump coronary artery bypass grafting. J Am Heart Assoc 7:e006642
58. de Gonzalo-Calvo D, Cenarro A, Garlaschelli K et al (2017) Translating the microRNA signature of microvesicles derived from human coronary artery smooth muscle cells in patients with familial hypercholesterolemia and coronary artery disease. J Mol Cell Cardiol 106:55–67
59. Hergenreider E, Heydt S, Treguer K et al (2012) Atheroprotective communication between endothelial cells and smooth muscle cells through miRNAs. Nat Cell Biol 14(3):249–256
60. Villarroya-Beltri C, Gutierrez-Vazquez C, Sanchez-Cabo F et al (2013) Sumoylated hnRNPA2B1 controls the sorting of miRNAs into exosomes through binding to specific motifs. Nat Commun 4:2980
61. Lasda E, Parker R (2016) Circular RNAs co-precipitate with extracellular vesicles: a possible mechanism for circRNA clearance. PLoS One 11(2):e0148407
62. Dou Y, Cha DJ, Franklin JL et al (2016) Circular RNAs are down-regulated in KRAS mutant colon cancer cells and can be transferred to exosomes. Sci Rep 6:37982
63. de Gonzalo-Calvo D, Quezada M, Campuzano O et al (2017) Familial dilated cardiomyopathy: a multidisciplinary entity, from basic screening to novel circulating biomarkers. Int J Cardiol 228:870–880
64. de Gonzalo-Calvo D, Iglesias-Gutierrez E, Llorente-Cortes V (2017) Epigenetic biomarkers and cardiovascular disease: circulating MicroRNAs. Rev Esp Cardiol (Engl Ed) 70(9):763–769
65. Umu SU, Langseth H, Bucher-Johannessen C et al (2018) A comprehensive profile of circulating RNAs in human serum. RNA Biol 15(2):242–250
66. Li W, Zhong C, Jiao J et al (2017) Characterization of hsa_circ_0004277 as a new biomarker for acute myeloid leukemia via circular RNA profile and bioinformatics analysis. Int J Mol Sci 18(3):597
67. Li H, Li K, Lai W et al (2018) Comprehensive circular RNA profiles in plasma reveals that circular RNAs can be used as novel biomarkers for systemic lupus erythematosus. Clin Chim Acta 480:17–25
68. Zhuang ZG, Zhang JA, Luo HL et al (2017) The circular RNA of peripheral blood mononuclear cells: Hsa_circ_0005836 as a new diagnostic biomarker and therapeutic target of active pulmonary tuberculosis. Mol Immunol 90:264–272
69. Guarnerio J, Bezzi M, Jeong JC et al (2016) Oncogenic role of fusion-circRNAs derived from cancer-associated chromosomal translocations. Cell 165(2):289–302
70. Gupta SK, Garg A, Bar C et al (2018) Quaking inhibits doxorubicin-mediated cardiotoxicity through regulation of cardiac circular RNA expression. Circ Res 122(2):246–254
71. Bang C, Batkai S, Dangwal S et al (2014) Cardiac fibroblast-derived microRNA passenger strand-enriched exosomes mediate cardiomyocyte hypertrophy. J Clin Invest 124(5):2136–2146
72. Schneider T, Hung LH, Schreiner S et al (2016) CircRNA-protein complexes: IMP3 protein component defines subfamily of circRNPs. Sci Rep 6:31313
73. Arroyo JD, Chevillet JR, Kroh EM et al (2011) Argonaute2 complexes carry a population of circulating microRNAs independent of vesicles in human plasma. Proc Natl Acad Sci U S A 108(12):5003–5008
74. Vickers KC, Palmisano BT, Shoucri BM et al (2011) MicroRNAs are transported in plasma and delivered to recipient cells by high-density lipoproteins. Nat Cell Biol 13(4):423–433
75. Jeck WR, Sharpless NE (2014) Detecting and characterizing circular RNAs. Nat Biotechnol 32(5):453–461
76. Devaux Y (2017) Transcriptome of blood cells as a reservoir of cardiovascular biomarkers. Biochim Biophys Acta 1864(1):209–216
77. Szabo L, Salzman J (2016) Detecting circular RNAs: bioinformatic and experimental challenges. Nat Rev Genet 17(11):679–692

11 Circular RNA in Saliva

Farinaz Jafari Ghods

Abstract

Although the type and amount of salivary components are influenced by many factors, due to easy, quick, cheap, and noninvasive sampling method alongside with the existence of the vast majority of the substances found in peripheral blood and urine in it, in recent years saliva has been considered as an ideal biofluid for disease research. Salivary circular RNA (circRNA), as an endogenous RNA molecule with a great variety of regulatory potency, is becoming a novel focus for detecting wide range of local or systemic diseases. Expectantly, with characterization of many more circRNAs in saliva, their motifs, and target sites, they can be used routinely in personalized medicine.

Keywords

Circular RNA · Saliva · Noninvasive sampling · Biofluid

F. Jafari Ghods (✉)
Department of Molecular Biology and Genetics, Faculty of Science, Istanbul University, Istanbul, Turkey

1 Introduction

In the middle of the eighteenth century, Langley argued that according to Nuck's belief, the effect of the brain through the nerves on salivary glands causes the flow of saliva [1]. Saliva has wide range of functions such as lubrication, speech facilitation, preliminary food digestion, controlling of dental/oral infections by balancing demineralization/remineralization, oral tissue repair, and antimicrobial peptides [2–5]. Saliva has been considered as a research material in recent years due to its potential to detect bacterial, viral, and systemic diseases.

While 99% of the total volume of saliva is water, the remaining 1% consists of organic and inorganic compounds. The major salivary glands secrete 93% of saliva, and 7% is salivated by the minor salivary glands [6, 7]. Being an acidic (pH = 6–7) multi-constituent body fluid, minerals, electrolytes, buffers, enzymes, enzyme inhibitors, growth factors, cytokines, IgM, IgG, sIgA, and a group of glycoproteins all make key components of the human saliva [8–10]. Most of these components are added to saliva after filtering, processing and secreting from the vasculature that nourish salivary glands [11–13]. Saliva is initially sterile, but as soon as it releases into the oral cavity, it would be exposed to oral microorganisms (bacteria, viruses, and fungi), microbial products, leukocytes, erythrocytes, desquamated oral epithelial cells and cellular

J. Xiao (ed.), *Circular RNAs*, Advances in Experimental Medicine and Biology 1087,
https://doi.org/10.1007/978-981-13-1426-1_11

products, nucleic acids, food debris, upper respiratory tract secretions, oral mucous, and gingival crevicular fluid forming whole saliva (WS) [6, 14–17]. The type and amount of salivary components, regardless of whether they are transcripts, proteins, metabolites, or oral microbes, are influenced by factors such as age, gender, salivary gland development, microbial colonization pattern, nutritional status, and tooth development status [18–21].

Cell-free and exosomal DNA fragments and different kinds of RNAs (coding and noncoding) also exist in saliva [22–24]. Applying high-throughput RNA-Seq, the first global characterization of human saliva transcriptome was done in 2012, and it was shown that saliva encompasses more than 4000 RNAs belonging to variety of RNA species [25]. In saliva an intricate composition of extracellular RNA has been transpired, including mostly mRNAs, long ncRNAs (≥200 nucleotide-long), and small ncRNAs (<200-nucleotide-long) such as miRNAs (19–23-nucleotide-long), piRNAs (24–30-nucleotide-long), snoRNAs (60–300-nucleotide-long), circRNAs (less than 100- to over 4000-nucleotide-long), etc. [26–28]. They may be originated from the apoptosis or necrosis. Interestingly, their degradation occurs much more slowly than exogenous species.

Special characteristics of saliva such as easy, quick, cheap, and noninvasive sampling, straightforward storage and transportation, lack of clotting, high security for both the patient and the health personnel, convenient analysis, and, most importantly, existence of the vast majority of the substances found in peripheral blood and urine in it have made saliva as an ideal biofluid for performing investigations. In the same vein, in various studies conducted and ongoing, changes in the expression levels of certain RNA molecules, coding and noncoding (microRNAs, snoRNAs, piRNAs, circular RNAs), have been associated with the susceptibility or the development of certain diseases [29–34]. One of the major impetuses in ongoing researches is to use salivary ncRNAs as diagnostic biomarkers either for local or systemic diseases.

2 Extracellular RNAs in Human Saliva

2.1 Origin of Salivary RNAs

RNA molecules found in saliva may originate from a variety of sources, for example, from the cell lysis that occurs in the salivary ducts, gingival pockets, or desquamated epithelial cells of the oral cavity. Gingival crevice fluid (GCF) harbors various cell types such as blood leukocytes and erythrocytes and their cell contents [35]. Products derived from the lysis of these cells are considered as important sources of salivary RNAs. RNAs that are actively secreted can also be considered as another source of RNAs in saliva. RNAs that have been produced in secretary cells or anywhere else in the body enter the circulatory system and are secreted through the filtering and processing into the saliva [12, 36, 37]. Since the mouth of any healthy person contains approximately 500 million bacterial cells belonging to 700 different colony species, it can be concluded that the genome and RNA contents of the oral microorganisms, including bacteria, viruses, and fungi, can be a major source for variety of salivary DNA and RNA molecules [38–40]. When the purpose of the study is to evaluate and measure salivary RNAs for the diagnosis of oral or systemic diseases, due to the presence of a high fraction of microbial RNAs, a significant difference in the amount of RNA composition will be detected in whole saliva (WS) compared with the cell-free saliva (CFS) [25]. This in turn reduces the sensitivity of the study performed in whole saliva (WS). To overcome this problem, adding subsequent steps such as low-speed centrifugation would be helpful to subtract microbial RNAs and cell debris.

2.2 Salivary RNAs Stability

In saliva, the most RNA molecules are degraded, and only a percent of RNAs remain intact. In each person, the percentage of RNAs that have been broken is associated with the types of microorganisms present in their oral cavity and to the amount and type of the endonucleases and exonucleases contents [41]. A more in-depth study of fragmentated RNAs suggests that endonuclease enzymes overcome exonucleases in fragmentation process. Because the exonucleases are much more progressive, in most cases, the RNA molecule is completely disintegrated [42, 43]. On the other hand, it has been seen that degradation of salivary RNAs occurs much more slowly than exogenous species, and this suggests mechanisms to protect salivary RNA [44]. One of these mechanisms is the association of these salivary RNAs with macromolecules. For example, salivary mucus contains oligomeric structures of MUC5B and MUC7proteins [45]. The association of salivary RNAs with glycosylated oligomer is a protective mechanism for the preservation of salivary RNAs [44]. In a study by Turchinovich et al., it was shown that large part of the miRNAs in the extracellular environment was associated with Ago2 proteins [46]. The stability of the salivary RNAs has also been attributed to placement of them into the extracellular vesicles (EVs), also known as exosomes, which are small membrane vesicles (30–100 nm in diameter) and have potential to carry diverse biomolecules. In 2007, Valadi et al. showed that mRNA and miRNA molecules can be entrapped into the EVs, transferred between cells, and be functional in that new environment [47]. This finding was later confirmed by other investigations [48, 49]. In another study comparing EVs fraction to EV-depleted salivary supernatant in order to investigate whether body fluid miRNAs are circulating freely or via exosomes, predominant existence of miRNAs in EVs was proven [50]. Since miRNAs, piRNAs, and snoRNAs had been detected only in WS, it was conceived that small RNAs are not associated with exosomes. Using next-generation sequencing (NGS) of small RNAs in salivary exosomes expression of known and novel miRNAs alongside with piRNAs, snoRNAs, and other small RNAs in exosomes was defined for the first time [51]. Formerly conducted studies found that GW182 protein, a component of the RNA-induced silencing complex (RISC) binds to and stabilize the miRNAs in exosomes [46, 52].

3 Importance of Emerging Technologies in Salivary Diagnostics

The rapid and growing development of knowledge and information on the "omics" constituents of saliva increased the hopes for the development of biomarkers and personalized medicine, and, for the first time in 2008, the term "salivaomics" was introduced, which represents the study of the five main salivary diagnostic components as genome/epigenome, transcriptome, proteome, metabolome, and microbiome [53].

With the development of the aforementioned branches and the application of new technologies such as automated extraction, purification, the whole-genome sequencing, the profiling of DNA and various types of RNAs using microarray analysis, quantitative PCR (qPCR), 2-D gel electrophoresis, mass spectrometry, blotting methods, data and bioinformatics (ranging from rapid short read aligners to detailed examination of RNA expression patterns), etc., personalized medicine was raised more strongly [25, 54–61]. As in this new approach, instead of considering the term "one size fits all," the health status of individuals is determined based on inherited differences, environmental conditions, and lifestyle. Discovery of novel biomarkers is of increasing importance for personalized medicine in which the ultimate goals are to match the right molecular marker to the underlying processes involved in the disease pathology and to design experimental assays that provide valuable information about diagnosis, prognosis, and response to the therapy on drug discovery.

Saliva was called "the mirror of the body" because it contains most of the compounds in the blood and urine and has potential to reflect the current physiological state of an individual [62, 63]. However, the presence of some of these substances at generally lower concentrations in saliva makes their detection problematic [64]. But by advances in highly sensitive technologies, detection of minute quantities of these substances in saliva has become possible [3].

4 CircRNAs Serve as a Fingerprint in Various Human Diseases

After discovering of salivary exRNAs in approximately 10 years ago, myriad of studies have been pursued to provide insights on potential use of these molecules to detect wide range of diseases such as oral cancer, ovarian cancer, breast cancer, Sjögren syndrome, etc. [29, 65–68]. Recently the circular RNA (circRNA), as an endogenous RNA molecule with a great variety of regulatory potency, is becoming a novel focus in this field [69].

Although circular RNAs were first detected about 30 years ago, due to their closed loop structure, their direct mapping on the genome was not possible through traditional methods of analyzing RNAs. This, in turn, has led to a delay in the discovery of such relatively new RNA species and limited information about them [70]. In fact, these species of RNA molecules originally deciphered in ribosomal-depleted RNA-Seq data, and circRNAs tracing was possible in excised exons or introns using deep RNA sequencing and after rRNA depletion, size selection, unique mapping, and classification of them according to their relative abundance using BPKM (bases per kilobase of gene model per million mapped bases) [71]. After the development of these technologies, an increasing number of circRNAs from different tissues, such as serum and plasma, semen, saliva, exosomes, cancer tumors, etc., were detected [72]. By determining the ratio of these circRNAs to the linear counterparts, it was found that circular types are expressed at higher levels. On the other hand, studies have shown that the half-life of circRNAs varies considerably and is, on average, about 2.5-fold longer than the median half-life of their linear host transcripts and even can be up to 50 h [73]. CircRNAs, titer miRNAs, interfere with splicing and regulate transcription through varying mechanisms such as miRNA sponges, RNA-binding proteins sponges, or scaffolding molecules, thus expanding the complexity of downstream gene expression [65]. In a study by Hansen et al. (2013), two circRNAs were identified acting as miRNA sponges and play role on miRNAs targeting [74]. Besides, although these molecules are classified in the noncoding RNAs group, a study in 2015 showed that some circRNAs are likely code for proteins [75]. All of these findings along with their cell-type-specific and tissue-specific expression led to the logical assumption that presence of altered expression patterns of circRNAs in different body fluids and tissues might be the reason or the consequence of various human disease. At present, several studies are under initial discovery of circRNAs to clarify their role in various human diseases and to introduce them as diagnostic, predictive, or therapeutic targets biomarkers (Table 11.1).

Since sampling of saliva is painless and stress-free and negates the need for trained medical staff, application of saliva as diagnostic medium is most practical for the population of pediatric patients. However, as the diversities in biomarker levels in individuals are affected by factors such as age and diet, as well as the absence of determining thresholds that differentiate health and disease status, there will be limitations in applicability and translatability of assays in pediatric population [21, 89]. So, investigators should be aware of these biological changes when defining normative values and designing salivary assays. Assays' interpretation also should reflect these findings appropriately.

5 Salivary CircRNAs, Pros, Cons, and Potential Solutions

In 2015, Bahn et al., for the first time, conducted and validated the existence of circRNAs using high-throughput RNA sequencing and an in-depth bioinformatics analysis following construction of circRNA libraries in cell-free fraction of saliva [90]. Since desquamated epithelial cells

Table 11.1 A selection of recent studies in which circRNAs were reported to be related with human diseases

Disease	circRNA(s)	Tissue Type	Potential value	References
Colorectal Cancer	hsa_circ_001569	Human CRC cell lines, CRC tissue samples	Biomarker	[76]
	circ-BANP	CRC tissue samples	Prognostic and therapeutic biomarker	[77]
Pre-eclampsia	circ_101222	Blood corpuscles	Prognostic biomarker	[78]
Ischemic heart disease	hsa_circ_0124644	Peripheral blood samples	Diagnostic biomarker	[79]
Alzheimer's disease	CircPVT1	WI-38 human fibroblasts	Biomarker	[80]
Pancreatic ductal adenocarcinoma	209 circRNAs up- or downregulated	PDAC tissue samples	circRNAs expression profiling	[81]
Gastric cancer	hsa_circ_002059	Gastric cancer patients tissue, plasma samples	Diagnostic biomarker	[82]
Cardiomyopathy	circFOX-3	Heart samples of aged patients and mice	Therapeutic, target	[83]
MI with or without left ventricular dysfunction	MIRC	Peripheral blood samples	Biomarker	[84]
Hypopharyngeal squamous cell carcinoma	2392 circRNAs up- or downregulated	HSCC tumor tissues	circRNAs expression profiling	[85]
Rheumatoid arthritis	circRNA_104871, circRNA_003524, circRNA_101873, circRNA_103047	Peripheral blood mononuclear cells	circRNAs expression profiling, diagnostic biomarkers	[86]
Systemic lupus erythematosus	207 circRNAs up- or downregulated	Plasma samples	circRNAs expression profiling	[87]
ChronicCD28-associated CD8(+) T-cell aging	circRNA100783	CD28(−)CD8(+)T cells and CD28(+) CD8(+)T cell	Biomarker	[88]

and the rest of the cell types mentioned earlier may release their RNA content into the extracellular space interrupting exRNAs profiling, in order to enriching for physiological extracellular RNAs from environmentally originated RNAs, cell-free saliva was used. To gain better results circRNAs enrichment procedure was done on extracted total RNA samples and total RNA treated with RNase H for 20 min at 37 ° C. Thereinafter RNA extraction was done by acid phenol/chloroform (pH = 4.5) [91, 92]. In this study, according to the presence of rich oral microorganisms in the saliva, despite the assessing of cell-free saliva, 58.5% of the reads belonged to the microbial RNA sequences. However, 422 putative circRNAs were identified from which 327 circRNAs were noncanonical ones. To validate the existence of circRNAs, RT-PCR and TOPO-cloned PCR [93] were done on a selective number of them. Low degree of overlap of these 422 circRNAs was observed among the different individuals, which suggested their high individual specificity. Finally, with purpose of exploring the functional relation of the circRNAs, gene ontology analysis of genes overlapping putative circRNAs was conducted. According to the results, it was suggested that these salivary circRNAs probably contribute to intercellular signaling pathways and inflammatory responses [90].

Application of saliva-based circular RNAs in personalized medicine has myriad benefits including (1) they can be traced in saliva due to their stability, longer half-live, and relatively abundance in saliva and exosomes; (2) the evolution and utilization of saliva-based circRNA biomarkers offer easy, inexpensive, reliable, noninvasive, and safer approach for patients, physicians, and medical staff, making them much more favorable; and (3) they exhibit tremendous potential to discriminate between patients and healthy individuals according to different levels of their expression. As a result, they can be harnessed as sensitive and specific biomarkers for identification of high-risk populations, cancer susceptibility evaluations, prognosis or diagnosis of diseases, developing targeted therapies, assessing responses to treatment, detecting of treatment resistance, and preliminary tracing of disease recurrence [94, 95]. Along with the benefits mentioned above, this rapidly expanding field of research has challenges and ambiguities that need to be addressed including:

1. Still little is known about the genetic networks and pathways regulated by salivary circRNAs, so that deeper understanding of them may allow investigators to knock out circRNAs by genome-editing tools and diminish their destructive activities or by applying functional knockout rescue experiments and design and produce ectopic circRNA expression plasmids in order to induce their overexpression [96].
2. It has shown that circular species of RNA molecules form as by-products through missplicing events; circularization of RNAs and artifacts results from use of reverse transcriptase enzyme in sequencing technique [97].

In a recent study five common algorithms compared and it was indicated that up to 40% of predicted circRNAs were only flagged by one algorithm while less than 20% of the circRNAs included into this study were identified by all five [98]. Thus, the necessity of establishment of statistical methods in order to overcome these technological barriers to estimate false detection rates and discriminate true circ-RNAs and by-products is evident. Hopefully, by solving aforementioned problems alongside with characterization of many more circRNAs in saliva, their motifs and target sites, they can be used routinely in personalized medicine.

Acknowledgments The authors thank Prof. Dr. Ahad Jafari Ghods, MD., FACP, (Iran University of Medical Sciences) for critical reading and comments on the manuscript.

Competing Financial Interests The author declares no competing financial interests.

References

1. Garrett JR (1975) Changing attitudes on salivary secretion—a short history on spit. Proc R Soc Med 68:553–560
2. Schipper RG, Silletti E, Vingerhoeds MH (2007) Saliva as research material: biochemical, physicochemical and practical aspects. Arch Oral Biol 52:1114–1135
3. Farnaud SJ, Kosti O, Getting SJ et al (2010) Saliva: physiology and diagnostic potential in health and disease. Sci World J 10(16):434–456
4. Malathi N, Mythili S, Vasanthi HR (2014) Salivary diagnostics: a brief review. ISRN Dent 2014:158786
5. Khurshid Z, Naseem M, Sheikh Z et al (2016) Oral antimicrobial peptides: types and role in the oral cavity. Saudi Pharm J 24(5):515–524
6. Hofman LF (2001) Human saliva as a diagnostic specimen. J Nutr 131:1621S–1625S
7. Llena-Puy C (2006) The rôle of saliva in maintaining oral health and as an aid to diagnosis. Med Oral Patol Oral Cir Bucal 11(5):E449–E455
8. Lawrence HP (2002) Salivary markers of systemic disease: noninvasive diagnosis of disease and monitoring of general health. J Can Dent Assoc 68(3):170–174
9. Almstahl A, Wikstrom M (2003) Electrolytes in stimulated whole saliva in individuals with hyposalivation of different origins. Arch Oral Biol 48:337–344
10. Aps JK, Martens LC (2005) Review: the physiology of saliva and transfer of drugs into saliva. Forensic Sci Int 150:119–131
11. Haeckel R, Hanecke P (1996) Application of saliva for drug monitoring: an in vivo model for transmembrane transport. Eur J Clin Chem Clin Biochem 34:171–191
12. Forde M, Koka S, Ecker S et al (2006) Systemic assessments utilizing saliva, part I: general considerations and current assessments. Int J Prosthodont 19:43–52
13. Lee YH, Wong DT (2009) Saliva: an emerging biofluid for early detection of diseases. Am J Dent 22(4):241–248
14. Humphrey SP, Williamson RT (2001) A review of saliva: normal composition, flow, and function. J Prosthet Dent 85:162–169

15. Aps JK, Van Den Maagdenberg K, Delanghe JR et al (2002) Flow cytometry as a new method to quantify the cellular content of human saliva and its relation to gingivitis. Clin Chim Acta 321:35–41
16. Falcão DP, da Mota LM, Pires AL et al (2013) Sialometry: aspects of clinical interest. Rev Bras Reumatol 53(6):525–531
17. Carpenter GH (2013) The secretion, components, and properties of saliva. Annu Rev Food Sci Technol 4:267–276
18. Ben-Aryeh H, Fisher M, Szargel R et al (1990) Composition of whole unstimulated saliva of healthy children: changes with age. Arch Oral Biol 35:929–931
19. Salvolini E, Mazzanti L, Martarelli D et al (1999) Changes in the composition of human unstimulated whole saliva with age. Aging 11:119–122
20. Nuriel-Ohayon M, Neuman H, Koren O (2016) Microbial changes during pregnancy, birth, and infancy. Front Microbiol 7:1031
21. Hassaneen M, Maron JL (2017) Salivary diagnostics in pediatrics: applicability, translatability, and limitations. Front Public Health 20(5):83
22. Vandewoestyne M, Van Hoofstat D, Franssen A et al (2013) Presence and potential of cell free DNA in different types of forensic samples. Forensic Sci Int Genet 7:316–320
23. Tu M, Wei F, Yang J et al (2015) Detection of exosomal biomarker by electric field-induced release and measurement (ERIFM). J Vis Exp 95:52439
24. Yao W, Mei C, Nan X et al (2016) Evaluation and comparison of in vitro degradation kinetics of DNA in serum, urine and saliva: a qualitative study. Gene 15:142–148
25. Spielmann N, Ilsley D, Gu J et al (2012) The human salivary RNA transcriptome revealed by massively parallel sequencing. Clin Chem 58:1314–1321
26. Brosnan CA, Voinnet O (2009) The long and the short of noncoding RNAs. Curr Opin Cell Biol 21:416–425
27. Esteller M (2011) Non-coding RNAs in human disease. Nat Rev Genet 12:861–874
28. Lasda E, Parker R (2014) Circular RNAs: diversity of form and function. RNA 20:1829–1842
29. Li Y, St John MA, Zhou X et al (2004) Salivary transcriptome diagnostics for oral cancer detection. Clin Cancer Res 10:8442–8450
30. Dalmay T (2008) MicroRNAs and cancer. J Intern Med 263:366–375
31. Mercer TR, Dinger ME, Mattick JS (2009) Long non-coding RNAs: insights into functions. Nat Rev Genet 10:155–159
32. Reis EM, Verjovski-Almeida S (2012) Perspectives of long non-coding RNAs in cancer diagnostics. Front Genet 3:32
33. Tang H, Wu Z, Zhang J et al (2013) Salivary lncRNA as a potential marker for oral squamous cell carcinoma diagnosis. Mol Med Rep 7:761–766
34. Jafari Ghods F, Topal Sarikaya A, Arda N et al (2017) MiRNA and mRNA profiling in systemic lupus reveals a novel set of cytokine – related miRNAs and their target genes in cases with and without renal involvement. Kidney Blood Press Res 42(6):1322–1337
35. Delima AJ, Van Dyke TE (2003) Origin and function of the cellular components in gingival crevice fluid. Periodontol 2000 31:55–76
36. Kaufman E, Lamster IB (2002) The diagnostic applications of saliva—a review. Crit Rev Oral Biol Med 13:197–212
37. Streckfus CF, Bigler L, Dellinger T et al (2004) The expression of the c-erbB-2 receptor protein in glandular salivary secretions. J Oral Pathol Med 33:595–600
38. Fabian TK, Fejerdy P, Csermely P (2008) Salivary genomics, transcriptomics, and proteomics: the emerging concept of the oral ecosystem and their use in the early diagnosis of cancer and other diseases. Curr Genomics 9:11–21
39. Parahitiyawa NB, Scully C, Leung WK et al (2010) Exploring the oral bacterial flora: current status and future directions. Oral Dis 16:136–145
40. Lif Holgerson P, Öhman C, Rönnlund A et al (2015) Maturation of oral microbiota in children with or without dental caries. PLoS One 10:e0128534
41. Li Y, Zhou X, St John MA et al (2004) RNA profiling of cell-free saliva using microarray technology. J Dent Res 83:199–203
42. Meyer S, Temme C, Wahle E (2004) Messenger RNA turnover in eukaryotes: pathways and enzymes. Crit Rev Biochem Mol Biol 39:197–216
43. Wong BC, Chiu RW, Tsui NB et al (2005) Circulating placental RNA in maternal plasma is associated with a preponderance of 5mRNA fragments: implications for noninvasive prenatal diagnosis and monitoring. Clin Chem 51:1786–1795
44. Park NJ, Li Y, Yu T et al (2006) Characterization of RNA in saliva. Clin Chem 52(6):988–994
45. Lamblin G, Degroote S, Perini JM et al (2001) Human airway mucin glycosylation: a combinatory of carbohydrate determinants which vary in cystic fibrosis. Glycoconj J 18:661–684
46. Turchinovich A, Weiz L, Langheinz A et al (2011) Characterization of extracellular circulating microRNA. Nucleic Acids Res 39:7223–7233
47. Valadi H, Ekström K, Bossios A et al (2007) Exosome-mediated transfer of mRNAs and microRNAs is a novel mechanism of genetic exchange between cells. Nat Cell Biol 9:654–659
48. Michael A, Bajracharya SD, Yuen PS et al (2010) Exosomes from human saliva as a source of microRNA biomarkers. Oral Dis 16:34–38
49. Koga Y, Yasunaga M, Moriya Y et al (2011) Exosome can prevent RNase from degrading microRNA in feces. J Gastrointest Oncol 2:215–222
50. Gallo A, Tandon M, Alevizos I et al (2012) The majority of microRNAs detectable in serum and saliva is concentrated in exosomes. PLoS One 7:e30679
51. Ogawa Y, Taketomi Y, Murakami M et al (2013) Small RNA transcriptomes of two types of exosomes in human whole saliva determined by next generation sequencing. Biol Pharm Bull 36:66–75

52. Ogawa Y, Miura Y, Harazono A et al (2011) Proteomic analysis of two types of exosomes in human whole saliva. Biol Pharm Bull 34:13–23
53. Wong DT (2012) Salivaomics. J Am Dent Assoc 143:19S–24S
54. Ramachandran P, Boontheung P, Xie Y et al (2006) Identification of N-linked glycoproteins in human saliva by glycoprotein capture and mass spectrometry. J. Proteome Res 5:1493–1503
55. Park NJ, Zhou X, Yu T et al (2007) Characterization of salivary RNA by cDNA library analysis. Arch Oral Biol 52:30–35
56. Whitelegge JP, Zabrouskov V, Halgand F et al (2007) Protein-sequence polymorphisms and post-translational modifications in proteins from human saliva using top–down fourier-transform ion cyclotron resonance mass spectrometry. Int J Mass Spectrom 268:190–197
57. Hu Z, Zimmermann BG, Zhou H et al (2008) Exon-level expression profiling: a comprehensive transcriptome analysis of oral fluids. Clin Chem 54:824–832
58. Wei F, Wang J, Liao W et al (2008) Electrochemical detection of low-copy number salivary RNA based on specific signal amplification with a hairpin probe. Nucleic Acids Res 36:e65
59. Sondej M, Denny PA, Xie Y et al (2009) Glycoprofiling of the human salivary proteome. Clin Proteomics 5:52–68
60. Hu S, Jiang J, Wong DT (2010) Proteomic analysis of saliva: 2D gel electrophoresis, LC–MS/MS, and western blotting. Methods Mol Biol 666:31–41
61. Oshlack A, Robinson MD, Young MD (2010) From RNA-seq reads to differential expression results. Genome Biol 11:220
62. Mandel ID (1993) Salivary diagnosis: promises, promises. Ann N Y Acad Sci 694:1–10
63. Kaczor-Urbanowicz KE, Martin Carreras-Presas C, Aro K et al (2017) Saliva diagnostics – current views and directions. Exp Biol Med (Maywood) 242(5):459–472
64. Miller SM (1994) Saliva testing--a non-traditional diagnostic tool. Clin Lab Sci 7(1):39–44
65. Majem B, Rigau M, Reventós J, et al (2015) Non-coding RNAs in saliva: emerging biomarkers for molecular diagnostics. Farah CS, ed. Int J Mol Sci 16(4):8676–8698
66. Lee YH, Kim JH, Zhou H et al (2012) Salivary transcriptomic biomarkers for detection of ovarian cancer: for serous papillary adenocarcinoma. J Mol Med (Berl) 90:427–434
67. Zhang L, Xiao H, Karlan S et al (2010) Discovery and preclinical validation of salivary transcriptomic and proteomic biomarkers for the noninvasive detection of breast cancer. PLoS One 5:e15573
68. Hu S, Wang J, Meijer J et al (2007) Salivary proteomic and genomic biomarkers for primary Sjogren's syndrome. Arthritis Rheum 56:3588–3600
69. Li Y, Zheng Q, Bao C et al (2015) Circular RNA is enriched and stable in exosomes: a promising biomarker for cancer diagnosis. Cell Res 25(8):981–984
70. Jeck WR, Sharpless NE (2014) Detecting and characterizing circular RNAs. Nat Biotechnol 32(5):453–461
71. Yang L, Duff MO, Graveley BR et al (2011) Genomewide characterization of non-polyadenylated RNAs. Genome Biol 12(2):R16
72. Memczak S, Papavasileiou P, Peters O et al (2015) Identification and characterization of circular RNAs as a new class of putative biomarkers in human blood. PLoS One 10(10):e0141214
73. Enuka Y, Lauriola M, Feldman ME et al (2016) Circular RNAs are long-lived and display only minimal early alterations in response to a growth factor. Nucleic Acids Res 44(3):1370–1383
74. Hansen TB, Jensen TI, Clausen BH et al (2013) Natural RNA circles function as efficient microRNA sponges. Nature 495:384–388
75. Wang Y, Wang Z (2015) Efficient back splicing produces translatable circular mRNAs. RNA 21:172–179
76. Xie H, Ren X, Xin S et al (2016) Emerging roles of circRNA_001569 targeting miR-145 in the proliferation and invasion of colorectal cancer. Oncotarget 7(18):26680–26691
77. Zhu M, Xu Y, Chen Y et al (2017) Circular BANP, an up-regulated circular RNA that modulates cell proliferation in colorectal cancer. Biomed Pharmacother 88:138–144
78. Zhang YG, Yang HL, Long Y, Li WL (2016) Circular RNA in blood corpuscles combined with plasma protein factor for early prediction of pre-eclampsia. BJOG 123:2113. https://doi.org/10.1111/1471-0528.13897
79. Zhao Z, Li X, Gao C et al (2017) Peripheral blood circular RNA hsa_circ_0124644 can be used as a diagnostic biomarker of coronary artery disease. Sci Rep 7:39918
80. Panda AC, Grammatikakis I, Kim KM et al (2017) Identification of senescence-associated circular RNAs (SAC-RNAs) reveals senescence suppressor CircPVT1. Nucleic Acids Res 45(7):4021–4035
81. Li H, Hao X, Wang H et al (2016) Circular RNA expression profile of pancreatic ductal adenocarcinoma revealed by microarray. Cell Physiol Biochem 40(6):1334–1344
82. Li P, Chen S, Chen H et al (2015) Using circular RNA as a novel type of biomarker in the screening of gastric cancer. Clin Chim Acta 444:132–136
83. Du WW, Yang W, Chen Y et al (2017) Foxo3 circular RNA promotes cardiac senescence by modulating multiple factors associated with stress and senescence responses. Eur Heart J 38:1402–1412
84. Vausort M, Salgado-Somoza A, Zhang L et al (2016) Myocardial infarction-associated circular RNA predicting left ventricular dysfunction. J Am Coll Cardiol 68:1247–1248
85. Cao S, Wei D, Li X et al (2017) Novel circular RNA expression profiles reflect progression of patients with hypopharyngeal squamous cell carcinoma. Oncotarget 8(28):45367–45379
86. Ouyang Q, Wu J, Jiang Z et al (2017) Microarray expression profile of circular RNAs in peripheral blood mononuclear cells from rheumatoid arthritis patients. Cell Physiol Biochem 42(2):651–659

87. Li H, Li K, Lai W et al (2018) Comprehensive circular RNA profiles in plasma reveals that circular RNAs can be used as novel biomarkers for systemic lupus erythematosus. Clin Chim Acta 480:17–25
88. Wang YH, Yu XH, Luo SH et al (2015) Comprehensive circular RNA profiling reveals that circularRNA100783 is involved in chronic CD28-associated CD8(+) T cell ageing. Immun Ageing 12:17
89. Turner-Cobb JM, Rixon L, Jessop DS (2008) A prospective study of diurnal cortisol responses to the social experience of school transition in four-year-old children: anticipation, exposure, and adaptation. Dev Psychobiol 50:377–389
90. Bahn JH, Zhang Q, Li F et al (2015) The landscape of microRNA, Piwi-interacting RNA, and circular RNA in human saliva. Clin Chem 61(1):221–230
91. Suzuki H, Zuo Y, Wang J et al (2006) Characterization of RNase R-digested cellular RNA source that consists of lariat and circular RNAs from pre-mRNA splicing. Nucleic Acids Res 34:e63
92. Jeck WR, Sorrentino JA, Wang K et al (2013) Circular RNAs are abundant, conserved, and associated with ALU repeats. RNA 19(2):141–157
93. Cortés-López M, Gruner MR, Cooper DA et al (2018) Global accumulation of circRNAs during aging in Caenorhabditis elegans. BMC Genomics 19(1):8
94. Abu N, Jamal R (2016) Circular RNAs as promising biomarkers: a mini-review. Front Physiol 7:355
95. Greene J, Baird AM, Brady L et al (2017) Circular RNAs: biogenesis, function and role in human diseases. Front Mol Biosci 4:38
96. Barrett SP, Salzman J (2016) Circular RNAs: analysis, expression and potential functions. Development 143(11):1838–1847
97. Chen I, Chen CY, Chuang TJ (2015) Biogenesis, identification, and function of exonic circular RNAs. WIREs RNA 6:563–579
98. Hansen TB, Venø MT, Damgaard CK et al (2016) Comparison of circular RNA prediction tools. Nucleic Acids Res 44(6):e58

12 Emerging Role of Circular RNAs as Potential Biomarkers for the Diagnosis of Human Diseases

Rupal Ojha, Raj Nandani, Nina Chatterjee, and Vijay Kumar Prajapati

Abstract

In the eukaryotic transcriptome, the evolutionary conserved circular RNAs naturally occur from the family of noncoding RNAs. Circular RNAs possess a unique feature to interact with nucleic acids and ribonucleoproteins and are establishing themselves as an obligatory composition for the regulatory messages which are encoded by the genome. The back-splicing mechanism leads to the formation of circularized RNA, and because of this they become resistant to exonuclease-mediated degradation. The differential and aberrant expression of circular RNAs can be detected with the help of various profiling methods by using serum, plasma, and tissue samples. In this chapter, we have highlighted the role of circular RNAs as putative biomarker for the detection of various human diseases along with its profiling methods. Here we have discussed the differentially expressed circular RNAs in neurological disorders and infectious diseases along with cancer diseases. For instance, in case of pulmonary tuberculosis, hsa_circRNA_001937 was upregulated, while hsa_circRNA_102101 got downregulated; Hsa_circ_000178 was depicted to get upregulated in breast cancer which is associated with disease progression. Furthermore, it has been observed that circRNAs are abundantly present within the mammalian brain tissues. In epileptic condition, Circ-EFCAB2 was observed to get notably upregulated within patients. Taking the above conditions into consideration, circular RNAs have proven themselves as promising noninvasive biomarker for the detection of human diseases.

Keywords

Circular RNA · Biomarker · Diagnosis · Human diseases

1 Introduction

The noncoding single-stranded circular RNAs are stable, bountiful, and evolutionary conserved. Like other forms of RNA, the circular RNAs also originated in the nucleus and are then exported to cytoplasm with the help of nuclear pore complex. The presence of circular RNA in cytoplasm is ten times greater in comparison to the linear RNAs. The role of circular RNA is still unclear, but some circular RNA play a major character in gene regulation as well as in pathophysiological processes by conducting itself as miRNA sponge, exporter of RNA, and binding protein molecules. Apart

R. Ojha · R. Nandani · N. Chatterjee · V. K. Prajapati (✉)
Department of Biochemistry, School of Life Sciences, Central University of Rajasthan, Bandarsindri, Ajmer, Rajasthan, India
e-mail: vkprajapati@curaj.ac.in

J. Xiao (ed.), *Circular RNAs*, Advances in Experimental Medicine and Biology 1087,
https://doi.org/10.1007/978-981-13-1426-1_12

from gene regulation, they also play a key role in the normal homeostasis, for example, in regulation and control of cell cycle, development of embryo, regulation of metabolic activities, and stress condition of cell [1]. Maintenance of conventional homeostasis is very important for normal functioning of living organisms. The circumstances under which circular RNAs are formed are quite different from those of the linear ones because circular RNAs form a covalent closed-loop for circularization, and it is possible because they lack 5′ (cap) and 3′ (polyadenylation) end. Due to the circularization process, they possess resistance to exonuclease-mediated degradation. The unique feature makes the circular RNA highly stable and allows its interaction to many molecules including miRNAs (miRNA sponges) and spliceosomes complex (RNA-protein complex) which further helps in the transcription process [2]. The splicing mechanism of circular RNA is mediated by alternative and back-splicing of primary mRNA which leads to the formation of circular RNA in which the exon part gets shuffled and forms different protein products. In back-splicing, the donor splice site is joined to acceptor splice site which exists at the upstream region and forms circRNAs as product, whereas in case of normal splicing, the donor splice site is usually joined to downstream acceptor splice site [3]. Most of the circRNAs can be procured from exonic or intronic forms, as well as from 5′-3′ untranslated regions. The difference between the exonic and intronic circRNAs is that the former one connects the exons and forms truncated but functional protein products, whereas the intronic circRNAs formed by the joining of introns is meagre in eukaryotes [2]. During the intronic splicing, the lariat structure formed is different from exonic splicing because of the formation of 2′-5′ carbon linkage at the splicing junction. Later on, it was observed that the lariat structure formed from intronic RNAs is very stable and possesses the properties to degrade the 3′ appendages and leave the remains behind [4]. This devised lariat products are known as circular intronic RNAs, while the RNAs which exhibit 3′-5′ junctions are known as circular exonic RNAs [5] (Fig. 12.1).

In 1979, circRNAs were primarily observed in mammalian cells with the help of electron microscopy [6]. Further, the presence of circRNA was reported in yeast and viroid viruses, but till date, there are very few circular RNAs reported for humans. In the 1990s, the circRNAs were first identified from DCC transcript study in human cells [7]. The circular RNAs were first experiential in Sry gene of mouse adult testis. This gene plays a vital role in the sex determination of embryo. Many experimental assays were performed including RT-PCR and RNAase protection assays to confirm the presence of circular RNA, and it was concluded that the circular RNAs were highly profuse in testis. These circulars are RNAs differentially expressed at the time of infection or disease. This distinctive characteristic of circular RNA makes them unique and can be used as potential biomarker for the evaluation of human diseases or infections. The high stability and expression of circular RNAs in blood or other body fluids make them unique biomarkers. Many clinical trials have been conducted to identify the role of remarkable circular RNAs in various disease forms, from the clinical serum samples. In this chapter, we will emphasize the differential upregulation and downregulation of circular RNAs at the juncture of different diseases along with molecular identification methods.

2 Role of circRNA as Potential Diagnostic and Prognostic Biomarker

CircRNAs are highly stable and resistant to debranching and exonuclease-mediated degradation in comparison to linear form of RNAs. They arise in nucleus and exported to the cytoplasm via exosomes, where they differentially expressed and function as potential biomarker for the diagnosis of diseased condition. Their presence in mammalian plasma, serum samples, and body fluids (e.g., saliva) secreted from various types of tissues makes them remarkable. In comparison to linear RNAs, the level of circular RNAs is higher in the serum sample, as the canonical splicing-mediated

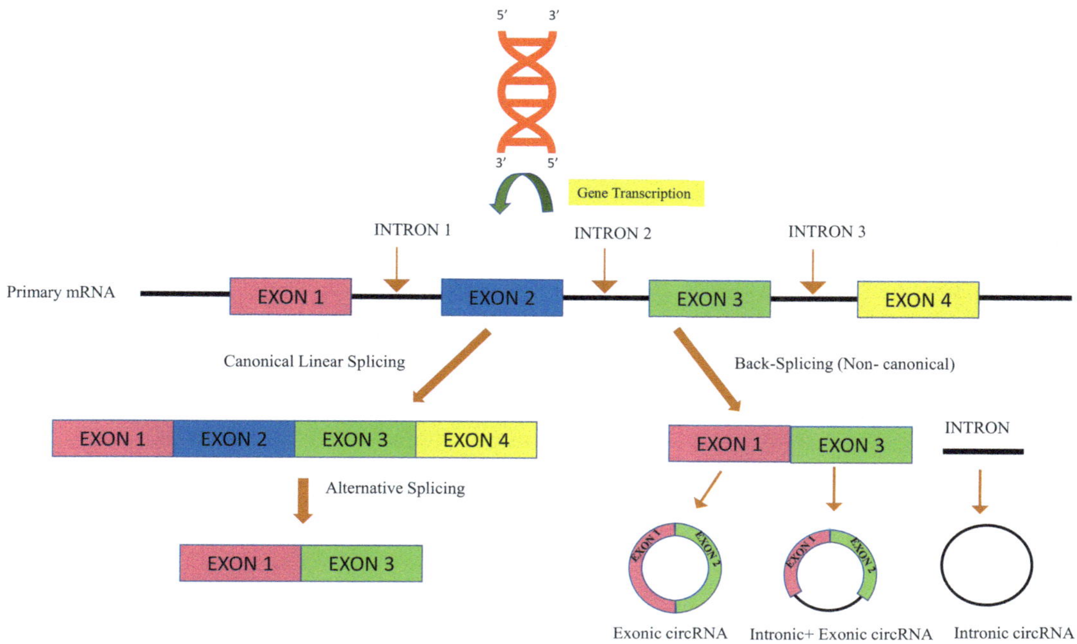

Fig. 12.1 RNA splicing mechanism
Formation of CircRNA isoform diversity. General overview of types of splicing that includes canonical linear splicing and noncanonical splicing. A great diversity of circRNAs is generated from a single genomic locus by noncanonical splicing also known as "back-splicing." Three forms of circular RNAs are formed during splicing – exonic circRNA, intronic RNA, and intronic + exonic circRNA. Colored boxes represent exons; black lines represent introns

linear RNAs got degraded after a short span of time, so, this groovy feature of circular RNA makes them potential diagnostic and prognostic biomarker because of their high stability. With the help of body fluids and serum sampling, the upregulating and downregulating circular RNAs can be easily identified. Even at room temperature, circRNAs are more stable than linear ones, at least for 24 h. In many diseased conditions including cardiovascular diseases, neurological disorders, cancer, and infectious diseases, the differential expression of circRNAs has been observed. For instance, ciRS-7 in Alzheimer's disease was seen to be upregulated which helps in the degradation of amyloid peptides. Secondly, it was observed that the circular RNAs are lavishly present in the heart tissues. During the clinical trials, it was identified that the patients who suffered from ischemia disease had upregulated hsa_circ_0124644 in comparison to healthy individuals. In case of gastric cancer condition, has_circ_0001649 gets downregulated in the serum and tumor tissue samples and is responsible for the development of metastasis and consequently can be used as noninvasive prognostic biomarker [8]. Further, in hepatocellular carcinoma circRNA, hsa_circ_0001649 is downregulated as the tumorigenic condition intensifies. The infectious hepatitis B and cancerous hepatocellular carcinoma (HCC) are interrelated to each other. The circRNA_100338 regulates the expression levels of miR-141-3p which is a disease-relevant miRNA and proliferates the metastasis condition. Thus, the overall differential and tissue-specific expression of circular RNAs in diseased as well as normal condition makes them conventional biomarkers for the diagnosis and prognosis purpose (Fig. 12.2).

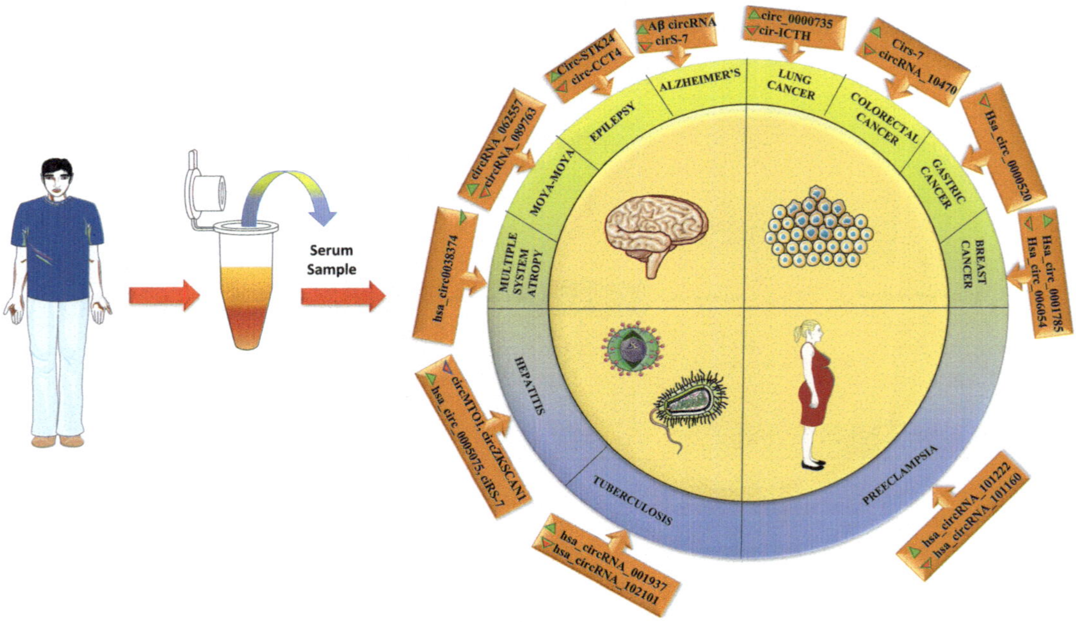

Fig. 12.2 Overview of the differentially expressed circRNAs in various diseases. The upregulated circRNAs are indicated by green color pyramids, while the downregulated ones are shown in red color pyramids

3 Role of circRNA as miRNA Sponge

As it has already been conversed that circular RNAs interact with many molecules, among them, miRNAs are very usual. When the circular RNA interacts with the miRNA, they form miRNA sponges which further suppress the transcript and lead to gene silencing. MiRNAs are a class of small noncoding RNAs which play a vital role in transcriptional silencing. They bind to the target mRNA, due to complementarity among the bases, and perform the process of transcriptional silencing in order to control the gene regulation. Sponges can be either artificial or natural and have multiple miRNA binding sites according to miRNA gene. The two naturally occurring circRNA sponges include SRY and CDR1/ciRS-7, aiming miR-138 and miR-7, respectively. The multiple binding sites present on the sponge are particularly specific for the miRNA and act as a competitive inhibitor, hence overwhelming the binding of miRNA with its target mRNA for posttranscriptional silencing [9]. For instance, the circular RNA ciRS-7 or CDR1 has multiple binding sites for miR-7 and functions as a miRNA sponge which is tremendously expressed during the development of nervous system in mammals [10]. Thereby, it would be more useful to identify the consequences over neuronal development in absence of ciRS-7. So, to decipher the role of ciRS-7 in brain development, people chose zebra fish as animal model because it has lost cdr1 locus, while miR-7 expresses highly in embryonic brain and conserved as well. Furthermore, the researchers identified a tool named Morpholino, which is widely known as Morpholino oligomer. This oligomer can be used for the regulation of gene expression and works as a sponge by base pairing with the target molecules. When this Morpholino type of oligomer synthetically is given as treatment in zebra fish, they observed defects in brain development process. These findings demonstrate that ciRS-7 efficiently interacts with miR-7 and causes loss of function of miR-7 which results in defective midbrain development [11, 12]. This study connotes a noteworthy relationship

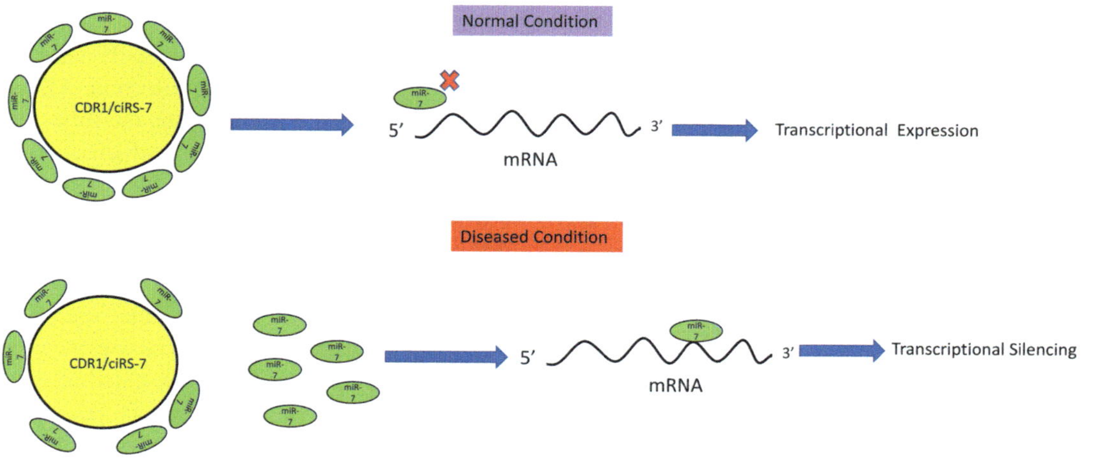

Fig. 12.3 Putative role of circRNAs as sponge CircS-7 acts as sponge or decoy of miR-7. Binding of miRNA-7 to circRNA-7 during the normal condition prevents its binding with mRNA, and this leads to normal transcription of the mRNA. But, in the diseased condition, circRNAs are not able to totally sponge the miR-7 that in turn binds to mRNA leading to silencing of that gene

between miR-7 and ciRS-7. Due to the higher stability, nowadays, synthetically developed circular RNAs are being used as sponge for the regulation of gene expression. During the hepatitis C virus infection, miRNA-122 highly expresses and supports in dissemination of hepatitis infection. Recently, an anti-miRNA drug named Miravirsen has been identified, which plays a vital role in sequestration of miRNA-122 and its activity as circRNA does. Because of the sequestration process, miRNA is unable to bind to the target molecule and hence does not perform the posttranscriptional silencing of gene [13]. As circular RNAs possess multiple binding sites for miRNAs, the artificial circular RNAs as miRNA sponges can be widely used for the sequestration of disease-pertinent miRNAs (Fig. 12.3).

4 Techniques for the Analysis of circRNAs Along with Computational Approaches

The efficient experimental analysis of circular RNA can be achieved by the help of various circular RNA profiling methods. Their detection is important for the diagnosis of various diseases including cancers, neurological disorders, infectious diseases, and many more. The early detection of diseased condition will be advantageous in the treatment of ailment. As it has already been mentioned that circRNAs were initially detected via electron microscopy, this technique was unable to discriminate between linear and circular forms of RNA. After that several analytical methods were recognized for the detection and quantification of circRNAs including microarray technique, RT-PCR followed by PCR (polymerase chain reaction), and Northern blot [14]. But due to low sensitivity and specificity, these techniques are incapable to examine the full sequence of circRNAs. Hence, due to certain drawbacks, these abovementioned techniques cannot be that much reliable. Next method includes RNA-Seq, standard high-throughput RNA sequencing analytical technique, which is widely being used for the detection of circular RNAs, but due to lack of discrimination property between the linear and circular RNA, there will be generation of indeterminate results which is not conventional [15]. To overcome all the complications associated with the aforementioned analytical methods, recently developed RAPD and high-throughput RNA sequencing united with computational approach are used for the high-purity separation of circRNAs on the basis of quantitative and qualitative analyses. The circular RNAs lack 5′ cap and 3′ polyadenylation due to which they become highly stable. The interaction of circRNAs with various ribosome-binding proteins and miRNAs affects the gene regulation and expression. When the RNA sample is treated with exoribonucleases, all the linear forms of RNAs get degraded, and only the circular RNA remnants are left. This RAPD technique determines the circRNA, originated from the back-splicing of transcriptome. Many bioinformatics tools nowadays have been utilizing for the identification of novel back-spliced exonic, intronic, and exo-intronic circular RNAs. The most popular tool associated with the RNA-Seq is find_circ; this analytical tool detects the back-spliced exon reads produced from the RNA-Seq technique. It has ability to read the sequences of more than 100 nucleotides. Next, CIRCexplorer, the solitary annotating tool which is able to identify the back-spliced exonic and intronic circRNAs sequence reads, is obtained from RNA-seq [16]. Subsequently, circRNA_finder is used for determining the proximal splice sites by using RNA-Seq data. These aforesaid profiling methods help in the identification of circular RNAs whose expression levels are associated with the various diseases.

5 Circular RNAs Role in the Diagnosis of Infectious Diseases

5.1 Tuberculosis

Tuberculosis (TB) is an infectious disease caused by bacteria *Mycobacterium tuberculosis* that most often affects the lungs (pulmonary) and sometimes the bones. According to WHO, tuberculosis (TB) is considered to be one of the top ten causes of morbidity and mortality worldwide. They gave an estimate that in the year 2015, approximately 10.4 million people were suffering from TB and 1.8 million people died due to it. Almost every part of world is affected by TB, but in 2016 the largest number of new TB cases occurred in Asia, with 45% of new cases, followed by Africa, with 25% of new case [17]. *Mycobacterium* is aerobic and resides inside the human body easily and can flourish there happily to cause infection. TB is categorized into two types based on the forms of bacteria, the latent pulmonary TB which is considered to be asymptomatic and the active TB which is symptomatic. About one quarter of the world's population is affected from latent form, and this form could develop into symptomatic form as the immune system becomes compromised. Initial symptoms of the disease include extreme cough, fever, night sweats, blood in sputum, excessive body weight loss, loss of appetite, difficulty in breathing, anxiety, and severe chest pains and ultimately lead to lung cancer and death. TB is contagious and can spread through air, sneeze, or spit of the infected patients [17]. Due to the decreasing survival rates, there is urgent need to diagnose the disease condition primarily. This can be achieved with the help of small noncoding circRNAs which can be used as diagnostic and prognostic biomarkers. These circRNAs play important role in gene regulation and contribute to the development of many human disorders. During the diseased (pulmonary) condition, there is a dysregulation in the expression of these circRNAs; some show downregulation, while some show upregulation. A total of 37 circRNAs have shown dysregulated expression in the PBMC cells of infected patients when compared with the healthy cohorts. Among them, 13 circRNAs were upregulated which include 10 exonic, 2 intronic, and 1 sense overlapping, while among 24 downregulated circRNAs, there were 15 exonic, 4 intronic, 3 overlapping, 1 intergenic, and 1 antisense seen. The upregulated circRNAs were hsa_circRNA_001937, hsa_circ_0000414, hsa_circ_0000681, hsa_circ_0002113, hsa_circ_0002362, hsa_circ_0002908, hsa_circ_000879, hsa_circ_0063179hsa_circRNA_009024, hsa_circRNA_005086, hsa_circRNA_103948, hsa_circRNA_003524, hsa_circRNA_015879, hsa_circRNA_009377, hsa_circRNA_103285, hsa_circRNA_406505, and hsa_circRNA_005232. The downregulated circRNAs were hsa_circRNA_102101, hsa_circRNA_104964, hsa_circRNA_104296, hsa_circRNA_003416, hsa_circRNA_002971, hsa_circRNA_007738, hsa_circRNA_000686, hsa_circRNA_048148, hsa_circRNA_092458, and hsa_circRNA_002465. These circRNAs regulate several physiological processes inside the human body such as autophagy, apoptosis, cell cycle regulation, and proliferation [18–21]. Dysregulation of the circRNAs leads to malfunctioning of the physiological processes and ultimately results in several pathological conditions. The pathologies develop due to interruption of many signaling and biological pathways in human PBMCs such as "chemokine signaling pathway," "Fc gamma R-mediated phagocytosis," "neurotrophin signaling pathway," "cytokine-cytokine receptor interaction," and "bacterial invasion of epithelial cells" [22, 23] ultimately lead to development of TB. So, increase or decrease in the number circRNAs during the disease can serve as clinical diagnostic markers and therapeutic targets. Detection of TB is hard to be early diagnosed; several methods for diagnosis include smear microscopy and mycobacterial culture. These are simple but have poor sensitivity, are costly, and are time-consuming. The other method used for the identification of acid fast bacteria *Mycobacterium tuberculosis* is Lowenstein-Jensen culture method, but again it is very time consuming process, as it takes an average of 4–5 weeks to yield the results. The circRNAs are highly nuclease-resistant, more stable than linear transcripts, and may be released into the extracellular space via the exosomes; these peculiar characteristics make them more potent to be used as a biomarker. Therefore, use of these circRNAs as biomarker can be a help in the diagnosis process as they are rapid, sensitive, and cost-effective.

5.2 Hepatitis

An inflammatory ailment of liver leads to hepatitis, which is caused by viral infection. There are a total of five types of hepatitis A, B, C, D, and E. Among them hepatitis B virus (HBV) and hepatitis C virus (HCV) are associated with hepatocellular carcinoma. Hepatocellular carcinoma (HCC) is the cancer of liver tissues; it is considered to be the sixth most common type of cancer worldwide. The major number of HCC has been linked to be caused by chronic hepatitis B. It majorly affects males and is responsible for approximately 600,000 deaths annually across the world. HCV infection is a dominant risk factor in the majority of areas of Asia and sub-Saharan Africa that have a high incidence rate of HCC [24]. The etiological agents that lead to development of HCC are elevated viral loads, having hepatitis virus envelope and surface antigens. Several other factors like obesity, excessive alcohol abuse, metabolic syndrome, diabetes, hereditary hemochromatosis with cirrhosis, etc. are also associated with development of HCC [25, 26]. The pathogenesis of HCC is due to different genetic/epigenetic aberrations and alterations with many signaling pathways that lead to the disease. Various symptoms of the disease include weight loss, pain, emotional distress, sleep disturbance, anemia, nutritional deficiencies, ascites, nausea vomiting, esophageal reflux, pain, peripheral edema, and constipation. Many research studies have been done for analyzing the dysregulated expression circular RNAs during the diseased conditions. Research investigation by Shichang Cui et al. revealed that a total of 24 circRNAs were upregulated, while 23 circRNAs were downregulated. Among these the top five upregulated circRNAs were hsa_circRNA_104351, hsa_circRNA_102814, hsa_circRNA_103489, hsa_circRNA_102109, and hsa_circRNA_100381. Furthermore, the top five downregulated circRNAs were hsa_circRNA_100327, hsa_circRNA_101764, hsa_circRNA_101092, hsa_circRNA_001225, and hsa_circRNA_102904. Whereas, other researchers have detected the dysregulated expression of some other circRNAs. The four circRNAs- circMTO1, hsa_circ_0001649, circZKSCAN1, and hsa_circ_0004018 were downregulated, while three circRNAs named, hsa_circ_0005075, ciRS-7(Cdr1as), and circRNA_100338 got upregulated during HCC condition. CircMTO1 is responsible for repression of HCC progression by sponging miR-9 that in turn leads to increased expression of p21 gene. CiRS-7 targets miR-7 and leads to enhancement of cell proliferation and invasion by promoting CCNE1, and PIK3CD expression CircZKSCAN1 is responsible for inhibiting the HCC cell growth, migration, and invasion by regulating cancer cell signaling pathways.

CircRNAs, hsa_circ_0004018, circRNA_100338, and circRNA_000839, play roles in HCC development [21, 27, 28]. CircRNAs regulate cancer development through a number of mechanisms, including miRNA sponges, modulating epithelial-mesenchymal transition, Wnt signaling pathway, the p53 PIK3CA, and β-catenin gene mutation [29]. Literature survey has suggested that circular RNAs are initially present in viruses as unique noncoding RNA molecules. During the course of infection, these circRNAs are transferred to the host. These CircRNAs are stable structure and have tissue-specific expression and are widely present in the cytoplasm of eukaryotic organisms, in the circular form. They are also responsible for developing the disease by regulating gene expression by competing with the endogenous RNAs of the cells. It modulates the function of miRNAs, by terminating the suppression from their targets, which in turn leads to modulated expression levels of other associated RNA molecules. The interaction between circRNAs and disease-associated miRNAs indicates that circRNAs are important for disease regulation. The different available diagnostics methods are Liver Imaging Reporting and Data System (LIRADS), CT, MRI, contrast imaging, ultrasound, etc. These aforesaid diagnosis procedures are somewhat time-consuming, less sensitive, require validation, and expensive. Because of the drawbacks of above-discussed diagnostic methods, the circular RNAs as biomarkers come into light with unique feature to diagnose the diseased condition primarily.

6 Circular RNAs Role in Diagnosis of Cancers

6.1 Lung Cancer

Lung cancer is the cancer of lung tissues; it is characterized by the abnormal growth of the cells and tissues, alveoli, bronchi, and the surrounding epithelial tissues of the lung. It is the most common cancer among men and ranks third among women in terms of its occurrence and mortality. According to WHO report, in the year 2015 alone, lung cancer caused approximately 1.69 million deaths worldwide. The lung cancer is classified into two categories on the basis of prognosis condition non-small cell lung cancer (NSCLC), which is the major one and small cell lung cancer (SCLC). Approximately 85% of the lung cancers are of non-small cell lung cancer type. Lung adenocarcinoma is a subtype of NSCLC that forms in mucus-secreting glands throughout the lung and is seen to be spreading at very fast rate among the nonsmokers [30, 31]. The various etiological agents that are considered to cause lung cancer include smoking cigarettes, chewing tobacco, snuffing powdered form of tobacco, passive smoking, and occupational exposure to chemicals like arsenic, cadmium, and pollutants, etc. Symptoms include coughing up blood, ache during breathing, persistent breathlessness, loss of appetite, unexplained weight loss, feeling of tiredness, persistent coughing, etc. Differential expression of circRNAs has been reported by some researchers when they compared the tissue sample from NSCLC patients and healthy individuals. Zhang et al. reported a total of ten dysregulated circRNAs, out of which five were upregulated and five were downregulated. The upregulated circRNAs were hsa_circ_0000735, hsa_circ_0016760, hsa_circ_0003645, hsa_circ_0087862, and hsa_circ_0026134, while the downregulated circRNAs were hsa_circ_0005730, hsa_circ_0091000, hsa_circ_0014130, hsa_circ_0071989, and hsa_circ_0092368. Out of these circRNAs, nine were exonic and only one was intronic. Hsa_circ_0014130 was responsible for negatively regulating the KEGG pathway that leads to cancer development [32]. Another group of researchers found that cir-ITCH was downregulated in lung cancer tissues and is responsible for cell proliferation and metastasis, whereas its overexpression inhibits lung cancer cell proliferation. The upregulated cir-ITCH acts as tumor suppressor by acting as sponge of miR-7 and miR-214 as well as it also inhibits the Wnt pathway, leading to reduced lung cancer cell proliferation [33]. CircRNA_100876 upregulated level is related to lymph node metastasis and tumor progression [34]. Luo et al. reported the upregulated level of hsa_circ_0000064 in lung cancer tissues [35]. Circ_0013958/miR-134/cyclin D1 was associated with the development of lung adenocarcinoma [36]. CircRNAFOXO3 (circRNAforkhead box O3 class) has antioncogenic role; it acts as a tumor suppressor by sponging miR-155 during the normal conditions, but during the diseased conditions, it was found to be downregulated [37]. CircPRKC1 was overexpressed in lung adenocarcinoma and hence responsible for cell proliferation and tumorigenesis. It works as sponge for both miR-545 and miR-589 and abrogates the suppression of the pro-tumorigenic transcription factor E2F7 [38]. Circular RNAs (circRNAs) are a class of endogenous noncoding RNAs that have been demonstrated to be potential regulators in the development and progression of lung cancers. These novel circRNAs are differentially expressed during the diseased conditions, and utilization of these circRNAs as biomarkers provides new insights in diagnostic processes.

6.2 Breast Cancer

Breast cancer is a fatal malignant cancerous disease among women, especially in the USA. According to 2011 WHO report 50,8000 women suffered and died because of breast cancer proliferation (http://www.who.int/cancer/detection/breastcancer/en/). The occurrence of breast cancer is very common in developed regions as compared to developing regions. Several studies have been performed on the breast cancer and it was concluded that due to modern lifestyle, various risks are allied with the disease which includes frequent consumption of alcohol, delayed age pregnancy, and extreme usage of estrogen and progesterone hormones. Apart from this, many epigenetic changes such as methylation, acetylation, and phosphorylation are also responsible for the progression of tumorigenic breast cancer. The differential expression of circular RNAs during the breast cancer condition is more concentrating. Circular RNAs came into picture because of their disease diagnosis property, so that the disease condition can be cured in an early stage. A recent study reported the differential expression of circular RNAs by profiling of peripheral blood samples in breast cancer patients. They have reported that during breast cancer, 19 circRNAs were upregulated, while 22 circRNAs were showing the downregulation. Further, it was observed that among the aforementioned number of circular RNAs, hsa_circ_0001785, hsa_circ_00680333, and hsa_circ_0108942 circRNAs were expressed aberrantly. These findings suggested the abnormal expression of hsa_circ_000178 during breast cancer and hence proved the prognostic and diagnostic value of specific circular RNA [39]. Additionally, hsa_circ_103110, hsa_circ_104821, and hsa_circ_104689 were found to be upregulated at the time of cancerous condition [18]. Further, it was identified that these circular RNAs are also associated with different signaling pathways. For instance, an oncogenic protein encoded by circular RNA hsa_circ_103110 is seen to be involved in other cancer development such as ovarian cancer, colorectal cancer, and squamous cancer. The circRNAs expressed during the breast and colon cancer are interlinked, for example, has_circ_104821 associated with PI3K and FAK signaling pathways of breast cancer as well as colon cancer. This was also concluded that in tumorigenic tissues, the expression of hsa_circ_006054, hsa_circ_406697, and hsa_circ_100219 was downregulated. CircFOXO3, a tumor suppressor, is found to be downregulated in tumor tissues, which is another potential biomarker for breast cancer diagnosis [40]. Thus, the promising feature of circRNAs as biomarker for the diagnosis of cancer condition is remarkable.

6.3 Gastric Cancer

Gastric cancer or stomach cancer arises due to the uncontrolled cell growth and is the fifth most common cancerous condition. According to WHO, every year around 723,000 deaths are caused due to stomach cancer, globally. The most common type of gastric cancer condition is adenocarcinoma, which develops from the mucosal cells, produces mucous, and functions as covering or layer of internal organs, especially the stomach. The symptoms of the disease include dysphagia, gastric problems like heartburn and bloating, indigestion, and weight loss. The survival rate is decreasing day by day among people who have developed gastric cancer. So, for early detection of disease, circular RNAs, the potential biomarkers, are used for diagnosis purposes. Circular RNAs are most stable and hence play a vital role in cell cycle regulation. They have the property to act as putative biomarker for the diagnosis of proliferative cancerous condition. By profiling methods, researchers have identified that in gastric cancer condition, the number of downregulated circRNAs is greater than the upregulated circRNAs. Recently, a study was conducted in which it was depicted that hsa_circ_0000520 is notably downregulated in the patients affected from gastric cancer as compared to healthy controls. This hsa_circ_0000520 have showed to be involved in disease progression as it is negatively assisted with TNM stage. TNM staging is a cancer staging system through which the stage of any cancerous condition can be determined. circPVT1, a tumor suppressor, is identified which indicated as good biomarker for the prediction of survival among the gastric cancer patients. Lastly, one more study was conducted between healthy and diseased patients, but it was identified that CNIH4, HIAT1, and KIAA0907 circRNAs were downregulated during gastric cancer condition. The all aforementioned circular RNAs play a dynamic role in the diagnosis of gastric carcinoma.

6.4 Colorectal Cancer

Colorectal cancer (CRC) can be linked to the list of cancers caused by lifestyle factors. It is the cancer that affects the colon or rectum. The third most common type of cancer causes majority of cancer-related deaths globally [21]. As per geographical distribution it was found that the colorectal cancer seems to be highest in south and Midwest part while lowest in the western parts of the world, among them, USA ranked first to encounter the majority of cases. Majority of cases occur in people aged 65 years or older, and relatively men are affected more than women [41]. There are several etiological agents like bacteria, viruses, smoking, and red meat consumption, increased use of aspirin, and physical inactivity. People with inflammatory bowel disease and infected with bacterium *Helicobacter pylori* and *Streptococcus gallolyticus* are at higher risk of developing colorectal cancers in the future [42]. The colon and rectum are parts of the large intestine, which is the lower part of the digestive system. Symptoms are not very particular; they are common as in case of other diseases. They include diarrhea, constipation, narrowing of the stool, rectal bleeding, blood in the stool, cramping or abdominal pain, weakness and fatigue, and unexpected weight loss. The circular RNAs are expressed during the diseased conditions and found to regulate a variety of essential biological functions such as cell proliferation, development, apoptosis, and pathologies. The circRNAs as miRNA sponge are believed to regulate the functions of miRNAs, as they are involved in suppressing translation of the mRNA and sometimes they are even capable of degrading the mRNAs if there is perfect complementarity between them. Bachmayr-Heyda et al. reported the differential expression of several genes between normal colon mucosa and CRC mucosa samples. These genes were responsible for encoding certain mRNAs when further analyzed. These mRNAs had a peculiar characteristic of being circular. The total count revealed that 39 circRNAs are significantly differentially expressed, out of which 11 of them are upregulated and 28 downregulated. Among the upregulated ones, the top circRNAs were circCCDC66, circ-BANP, ciRS-7 (Cdr1as), hsa_circ_0000069, hsa_circ_001569, and hsa_circ_0020397, while the top downregulated ones were hsa_circRNA_103809, hsa_circRNA_10470, hsa_circ_001988, and circRNA0003906. According to Hsiao et al., circCCDC66 and Circ-BANP play an important role in cell proliferation, migration, and invasion that leads to progression of tissues toward cancerous condition. Hsa_circ_ 001569 is expressed at upregulated level and acts as a sponge of miR-

145 that increases colorectal cancer cell proliferation by targeting E2F5, BAG4, and FMNL [43]. Circ-7 (Cdr1as) acts as miR-7 sponge to regulate expression of CDR1 gene. Studies have shown that CDR1 is expressed at high levels in CRC tissues. CDR1 is associated with tumor size, TNM stage, lymph node metastasis, and poor overall survival [44]. The downregulated expression of hsa_-circ_001988 was associated with tumor differentiation and perineural invasion during colorectal cancer [45]. The downregulated expression of hsa_circRNA_103809 and hsa_circRNA_104700 in colorectal cancer tissues was associated with cancer metastasis [46]. Hsa_circ_0020397 promotes TERT and PD-L1 expression by acting as sponge of miR-138 to regulate CRC cell viability, apoptosis, and invasion [47]. One of the few circRNA that have the dual behavior is cir-ITCH; its downregulation inhibits the Wnt/β-catenin pathway and promotes ITCH expression including p63, p73, and Notch1 and is usually associated with tumor formation, whereas upregulated expression of cir-ITCH reduces cell proliferation [48]. These circRNAs are present inside the exosomes in the cancer tissues, plasma, and serum samples and are specifically expressed during colorectal cancers. Thus, they can be used as putative biomarkers for the diagnosis of various oncogenic conditions.

7 Circular RNAs Role in Diagnosis of Neurological Disorders

7.1 Alzheimer's Disease

One of the most common dementia is Alzheimer's disease (AD) which involves rigorous loss of memory, behavior capacities, and thinking with progression in senescence. It is an age-associated neurodegenerative disorder. The neuropathological characterization is marked with the presence of neurofibrillary tangles and neurotic plaques, while clinical characteristics involve the gradual Impairment in cognitive skills. As far as the recent statistical records are concerned, 476,000 people within the age group of 65 and above are suffering from AD only in the USA. Furthermore, there are high chances of these values to undergo a hike of 13.8 million by mid of this century. It is unfortunate enough that there are no effective therapeutics that are victorious in preventing the gradual progression of AD till date [49]. In prior studies, it has been observed that the neurons are enriched with circRNAs, and they get concentrated in high amounts when the brain reaches senescence, which provide hints that there is a possible role of circRNAs in age-related neurodegenerative diseases. In AD neuropathy the accumulation of amyloid-β (Aβ) peptide due to the amyloid precursor protein (APP) proteolytic processes is strongly believed since ages to be one of the key steps. Tau phosphorylation and GSK3β levels are upregulated by Aβ circRNA-derived peptide, which is an Aβ circRNA-encoded expression of an AB-related peptide. Both GSK3β and tau phosphorylation are hallmarks for the AD progression. It has been observed that the mutations that are caused in genes like APP and presenilin which actively participate in APP protein proteolytic processing speed up the accumulation of Aβ peptides rapidly. The accumulation of protein causes the peptides to polymerize into pieces of insoluble amyloid plaques, and toxic forms and drives ahead to tau protein hyperphosphorylation via GSK3β activation, followed by the formation of subsequent tangles of neurofibrils which form inside neurons, ultimately triggering a relay of events that result into neuron death. Alzheimer's associated with this is known as familial Alzheimer's disease (FAD). In the case of sporadic Alzheimer's disease (SAD), Aβ circRNA and Aβ circRNA-DP play a vital role and found within the entire human population [50]. Naturally immune to the exonucleolytically RNA degradation, the circRNAs are abundantly distributed in the brain tissues of mammals. Some noncoding RNAs which are evolutionary conserved such as miRNA-7 are not just overpopulated in human brain but are associated with circRNA (ciRS-7), within the same tissues. ciRS-7 comprises of multiple, tandem anti-miRNA-7 sequences [51]. ciRS-7 behaves as a competing, endogenous, anticomplementary miRNA "sponge" which plays a role in adsorption and thereby quenches normal miRNA-7 functions. When a deficiency in the ciRS-7 and ciRS-7 "sponge activities" takes place, it can be expected to elevate ambient miRNA-7 levels in the brain cells of patients affected with AD which finally results in the downregulation of selective miRNA-7-sensitive messenger RNA (mRNA) goals [52].

7.2 Epilepsy

At an outlook, the epilepsy comprises of a series of brain malfunctioning which ranges from one of those that are much benign to the ones which are life-challenging, severe, and disabling. Epilepsy is well known for disturbing the normal neuronal activity pattern that ultimately results in causing weird sensations, strange emotions and behavior or even muscle spasms, convulsions, and loss of consciousness. The seizures, which are the most consistent emblem for the disease, result due to brain damage, abnormal brain development, incorrect brain wiring, or a disproportionate nerve signaling chemicals widely known as neurotransmitters (https://www.ncbi.nlm.nih.gov/pubmedhealth/PMHT0023036/). Circular RNAs are a category of long noncoding RNAs which have a closed-loop-like structure that participate in regulation of gene expression as microRNA sponges and are highly stable in nature [53]. The circRNAs are the resultant of loop splicing mechanism, are critical regulatory RNA molecules, and are novel. The communication of circRNAs with microRNAs (miRNAs) behaves as miRNA sponge for the regulation of gene expression through circRNAs-miRNA-messenger RNA (mRNA) axes in case of neurological disorders. Again, circRNAs are well known for being evolutionary conserved and for their expression in time, in cell type, and also in gene-specific manners. Initially, it was observed that most circRNAs are normally expressed at very small amounts, but further detailed investigation gradually reported that some circRNAs show tissue type-specific and cell type-specific expression with elevated transcript and copy numbers. Furthermore, synaptic fractions include synaptosomes and neuropil [54]. According to the histological findings, it has been confirmed that the synaptic reorganization of the limbic structures gradually enhances the hyperexcitability of the cortices that may have contributions to the onset of epilepsy. Recent studies have portrayed the significant role of circRNAs in temporal lobe sclerosis (TLS), which is a pathological entity that comes along with temporal lobe epilepsy (TLE). TLS is a type of an abnormal cortical findings associated with epilepsy which serves to study the alterations within the temporal neocortex genes in victims suffering from TLE patients. After a rigorous study, it was found that a significant dysregulation was observed among some of the below-mentioned circRNAs – circ-EFCAB2, circ-STK24, and circ-VPS37C were observed to get notably upregulated within the patients suffering from TLE as in comparison to those with the non-sufferers. On the other hand, circ-DROSHA, circ-CCT4, circ-UBQLN1, circ-USP9Y, and circ-STK17A were all downregulated. The most significant dysregulation was observed in circ-EFCAB2 and circ-DROSHA. Furthermore, the top five miRNAs of circ-EFCAB2 were predicted to be miR-3929, miR-6780a-5p, miR-6884-5p, miR-4739, and miR5787 while for circ-DROSHA were miR-651-3p, miR-1252-5p, miR-548ab, miR-4762-3p, and miR-4446-5p. The interaction between these circRNAs and miRNAs resulted in the formation "miRNA sponges" [55].

7.3 Moyamoya Disease

The moyamoya disease (MMD) is an infrequent cerebrovascular disease that is characterized by gradual occlusion of the internal intracranial carotid artery accompanied with the formation of collateral vessels abnormally at the bottom area of the brain called the basal ganglia [56]. The name "moyamoya" means "puff of smoke" in Japanese language, and it demonstrates the look of tiny vessels tangled together and formed in order to compensate for the blockage caused (https://www.ninds.nih.gov/Disorders/All-

Disorders/Moyamoya-Disease-Information-Page). The disease was first identified in Japan in the 1960s, and then it has been detected in several other countries across the globe. Most of the victims are notably found in Asian countries in comparison to North America or Europe. The disease primarily manifests children but can also make the adults suffer transient ischemic attack, seizures, hemorrhage, and ischemic stroke. In recent studies, it has been revealed that regulatory RNAs such as long noncoding RNAs or microRNA (miRNA) have strong contributions toward the development of MMD. CircRNAs have more number of binding sites and higher levels of expression when compared to other kinds of miRNA sponges and are hence considered much more efficient in terms of sequestering miRNAs and regulating gene expression than linear RNAs [57]. CircRNAs have been linked up with various disorders that involve various neurological disease, and they are correlated with the expression of the miRNAs [58]. It has been found for the first time that 146 circRNAs have been expressed in victims suffering from MMD, and these likely have enough contribution toward the development of this disease. Among these 146 circRNAs, the level of expression varied; 29 of these were upregulated, whereas 117 of them were downregulated. Hsa_circRNA_067130, hsa_circRNA_067209, and hsa_circRNA_062557 were upregulated, while hsa_circRNA_089763, hsa_circRNA_089761, and hsa_circRNA_100914 were downregulated with highest fold variations, which gave enough evidence to state that these circRNAs are potential biomarkers for the diagnosis of moyamoya disease [56].

7.4 Multiple System Atrophy in the Brain

Multiple system atrophy is an infrequent sporadic neurodegenerative disease which comes along with symptoms of slow movement, tremors, postural instability, and muscle rigidity due to malfunctioning of the ataxia and autonomic nervous system. The onset of this disease takes place in adulthood. Around 55% of MSA sufferers are men within the age group of 50s to early 60s [59]. It has been indicated in few researchers that a prion form of alpha-synuclein protein probably causes this disorder [60]. MSA, on a molecular level, is classified as a distinct member from the clan of neurodegenerative disorders named as α-synucleinopathies. A-synuclein is a fibrillary protein that gets fragmented into oligodendrocytes and provides insulation and support to the neuronal cells in the brain [61]. Often there's a tendency to get confused with the symptoms of idiopathic Parkinson's disease and Parkinsonism-predominant MSA. Thus, it's essential to identify potential biomarkers which would facilitate an early and differential diagnosis of multiple system atrophy. CircRNAs are the resultants of back-splicing which is the reciprocal of the canonical splicing of linear RNAs. CircRNAs are immune to hydrolysis by RNA exonucleases as they are covalently closed-loop structures. Furthermore, circRNAs have a half-life time of 48 h, which is very less in case of linear RNAs [62]. Analysis of transcriptome data, which was retrieved from the frontal cortex of the multi-system atrophy brains, identified a significant upregulation of five circular RNAs within the MSA tissue – hsa_circ_0038374, hsa_circ_0005540, hsa_circ_0032891, hsa_circ_0108308, and hsa_circ_00054211. Henceforth, the perceived changes in circRNAs level can be attributed to the premature oligodendritic malfunction which is due to the α-synuclein aggregation [63].

From a bird's eye view, it can be concluded that circRNAs are diverse endogenous noncoding RNAs that occur naturally and regulate gene expressions. CircRNAs are extremely stable molecules and are densely populated within the brain and exosomes. The abundance of circRNAs transcripts in peripheral blood and brain, in comparison to other tissues, has registered these molecules as an attractive proxy for the successful detection and monitoring of neuronal diseases. They are considered as an ideal biomarker because they can successfully transverse the blood-brain barrier, thereby facilitating the studies related to neurological disorders [64]. In particular circRNAs-based monitoring is suitable option for the detection of neurodegenerative diseases.

8 Circular RNAs Role in Diagnosis of Preeclampsia Condition

8.1 Preeclampsia

Preeclampsia (PE) is a pregnancy disorder which is characterized with high blood pressure and frequent presence of protein in the urine of the mother. The disease usually begins manifesting after 5 months of pregnancy. It has been reported as one of the major causes of fetal and maternal morbidity and mortality. Approximately 10% of the pregnant women are victimized with this in developing countries [65, 66]. The route of pathogenesis for this hasn't been unwrapped, yet an early determination is necessary. A combination of biomarkers is required for a prior diagnosis as because PE is accompanied with heterogeneous manifestations. According to the evidences accumulated so far, it has been demonstrated that noncoding RNA have an association with preeclampsia [67]. CircRNAs are a type of noncoding RNA which have the potential to regulate other RNA transcripts by eventually competing for shared miRNAs. This phenomenon has drawn attention toward the fact that they can act as remarkable diagnostic markers. CircRNAs are a unique category of RNA molecules that can be categorized as intronic, intergenic, or exonic. CircRNAs do not possess free 3′ or 5′ ends and can hence bypass the action of RNA exonuclease, thus making it highly stable in nature. It has been found that circRNAs were enriched with miRNA binding sites, thereby behaving as a miRNA sponge within cells, relieving the inhibition of miRNA target genes, and henceforth elevating the level of expression of the targeted gene [12]. The phenomenon is termed as competitive endogenous RNA mechanism. Keeping this context into consideration, it was studied that circRNAs act as an important element in preeclampsia caused during pregnancy [68]. After rigorous studies, it was discovered that 12 circRNAs in blood cells portrayed significant dysregulation in pregnant women affected with PE in comparison with normal pregnant mothers. Hsa_circRNA_101222, hsa_circRNA_101151, hsa_circRNA_104018, hsa_circRNA_101862, hsa_circRNA_100385, hsa_circRNA_105016, hsa_circRNA_103842, and hsa_circRNA_001535 were upregulated; on the other hand, hsa_circRNA_101160, hsa_circRNA_400084, hsa_circRNA_101676, and hsa_circRNA_101218 were downregulated. Hence, it can be used as an early prognosis of pregnant women who are at the verge of developing PE.

9 Conclusion

In a nutshell, here we illuminated the role of circular RNAs associated with different age-old and recently discovered disorders those are victoriously manifesting human life. The areas covered in this chapter include infectious diseases, oncogenic diseases, and neurological disorders. Firstly, in case of pulmonary tuberculosis, the expression of several circRNAs was imbalanced. It was studied that a total of 37 circRNAs were aberrantly expressed in diseased individuals, where 13 were upregulated and 24 were downregulated. Infectious disease further leads to the development of cancer in several cases; one such is hepatitis, widely known as hepatocellular carcinoma which develops due to genetic/epigenetic aberrations and alterations in many signaling pathways. A total of 24 circRNAs were overexpressed, while 23 circRNAs were observed to undergo a lower expression in victims. Among them, circMTO1, hsa_circ_0001649, circZKSCAN1, and hsa_circ_0004018 were found to get downregulated, while three circRNAs hsa_circ_0005075, ciRS-7(Cdr1as), and circRNA_100338 were upregulated. Cancer which is characterized by the uncontrolled growth of cells can also be detected with the help of circular RNAs. In tumorigenic cancerous conditions such as lung cancer, gastric cancer, breast cancer, and colorectal cancer, circular RNA has portrayed itself as a remarkable diagnostic marker. In the colorectal cancer, 39 circRNAs were significantly expressed as compared to the normal adjacent tissues, in which, circCCDC66, circ-BANP, ciRS-7 (Cdr1as), hsa_circ_0000069, hsa_circ_001569, and hsa_circ_0020397 ranked top in the list of the upregulated ones, while hsa_circRNA_103809, hsa_circRNA_10470, hsa_circ_001988, and circRNA0003906 were occupying the initial position in the list of downregulation. Lung cancers are the most common type of cancer worldwide. A total of ten circRNAs were imbalanced, out of which five were upregulated and five were downregulated. The upregulated circRNAs were hsa_circ_0000735 and hsa_circ_0016760 with few others in the list, while the downregulated ones were hsa_circ_0005730, hsa_circ_0091000, hsa_circ_0014130, etc. CircRNA has again proved its potential in determining gastric cancer which is one of the root causes of death globally. Hsa_circ_0000520 was found to get notably downregulated in patients suffering from gastric cancer especially during the TNM stage. Another fatal malignant cancer which is prevalent among women is breast cancer. It was found that a total of 19 circRNAs (e.g., hsa_circ_0001785, etc.) were overexpressed, whereas 22 (e.g., hsa_circ_006054, etc.) of them had a lower expression.

Lastly, one of the most debilitating and untreatable condition which occurs due to degeneration of the neurons is labeled as neurodegenerative disorders. The condition prevails due to the gradual degeneration or eventual death of the nerve cells. This ultimately results into difficulties associated with movement, memory, sensations, tremors, postural defects, seizure, and other mental dysfunction (known as dementias). There are numerous reported neuronal disorders, among which few of them are covered here – Alzheimer's, epilepsy, multiple system atrophy, and moyamoya. Alzheimer is a well-known disease that is associated with rigorous memory loss with age. A huge percentage of the population worldwide is suffering from this disease, and no such remarkable diagnostic or prognostic approaches have been yet developed. This is where circRNAs come into action. CircRNAs being most stable in nature are used as biomarkers, and their dysregulation can significantly help in the detection of the disease. Aβ circRNA and ciRS-7 are the two circRNAs which get upregulated and downregulated, respectively, symbolizing the onset of Alzheimer's disease. Epilepsy is another disorder accompanied with weird sensation and strange emotions. Circ-EFCAB2 along with few others gets upregulated, while circ-DROSHA accompanied with others is observed to get downregulated which behaves like a biomarker for disease detection. A total of 146 circRNAs are imbalanced during moyamoya disease, caused due to a blockage in the carotid artery, for example, Hsa_circRNA_067130 is upregulated, while hsa_circRNA_089763 is downregulated, indicating the prevalence of this disease. A rare sporadic neurodegenerative disease is a multiple system atrophy which has an overexpression of five significant circRNAs that depict that a person showing such upregulation of circRNA, for example, hsa_circ_0038374, hsa_circ_0108308, and so on, is victimized with MSD. Preeclampsia is an emerging disorder reportedly manifesting pregnant mothers. Circular RNAs have played a vital role for the detection of this disease. In recent researches, it was found that 12 circRNAs, namely, hsa_circRNA_101222, hsa_circRNA_105016, and few others, had significant upregulation, strengthening the fact that the level of circRNAs can be used as both prognostic and diagnostic measures.

In each of this, it has been closely scrutinized that circRNAs are playing a significant role as a diagnostic measure to identify the disorder and can also serve as a prognostic measure for the early determination of the disease. With all the evidences accumulated, it can be claimed that circRNAs are functioning as potential biomarkers for all the different fields of diseases that manifest human life.

Acknowledgments RO is thankful to the Central University of Rajasthan for providing UGC fellowship.

Disclosure Statement No potential conflict of interest was reported by authors.

References

1. Zhi-Chun Z, Xiao-Long G, Li X (2017) The novel roles of circular RNAs in metabolic organs. Genes Dis 5(1):16–23
2. Chen Y, Li C, Tan C et al (2016) Circular RNAs: a new frontier in the study of human diseases. J Med Genet 53(6):359–365
3. Memczak S, Papavasileiou P, Peters O et al (2015) Identification and characterization of circular RNAs as a new class of putative biomarkers in human blood. PLoS One 10(10):e0141214
4. Carriero S, Damha MJ (2003) Template-mediated synthesis of lariat RNA and DNA. J Org Chem 68(22):8328–8338
5. Li J, Yang J, Zhou P et al (2015) Circular RNAs in cancer: novel insights into origins, properties, functions and implications. Am J Cancer Res 5(2):472
6. Lasda E, Parker R (2016) Circular RNAs co-precipitate with extracellular vesicles: a possible mechanism for circRNA clearance. PLoS One 11(2):e0148407
7. Jeck WR, Sharpless NE (2014) Detecting and characterizing circular RNAs. Nat Biotechnol 32(5):453
8. Li WH, Song YC, Zhang H (2017) Decreased expression of Hsa_circ_00001649 in gastric cancer and its clinical significance. Dis Markers 2017:4587698
9. Ebert MS, Sharp PA (2010) MicroRNA sponges: progress and possibilities. RNA 16(11):2043–2050
10. Rybak-Wolf A, Stottmeister C, Glažar P et al (2015) Circular RNAs in the mammalian brain are highly abundant, conserved, and dynamically expressed. Mol Cell 58(5):870–885
11. Barrett SP, Salzman J (2016) Circular RNAs: analysis, expression and potential functions. Development 143(11):1838–1847
12. Memczak S, Jens M, Elefsinioti A et al (2013) Circular RNAs are a large class of animal RNAs with regulatory potency. Nature 495(7441):333
13. Jost I, Shalamova LA, Gerresheim GK et al (2018) Functional sequestration of microRNA-122 from Hepatitis C Virus by circular RNA sponges. RNA Biol 28:1–8
14. Panda AC, De S, Grammatikakis I et al (2017) High-purity circular RNA isolation method (RPAD) reveals vast collection of intronic circRNAs. Nucleic Acids Res 45(12):e116–e116
15. Voelkerding KV, Dames SA, Durtschi JD (2009) Next-generation sequencing: from basic research to diagnostics. Clin Chem 55(4):641–658
16. Chen L, Yu Y, Zhang X et al (2016) PcircRNA_finder: a software for circRNA prediction in plants. Bioinformatics 32(22):3528–3529
17. World Health Organization (2016) Global tuberculosis report 2016. World Health Organization, Geneva
18. Lü L, Sun J, Shi P et al (2017) Identification of circular RNAs as a promising new class of diagnostic biomarkers for human breast cancer. Oncotarget 8(27):44096
19. Du WW, Fang L, Yang W et al (2017) Induction of tumor apoptosis through a circular RNA enhancing Foxo3 activity. Cell Death Differ 24(2):357
20. Du WW, Yang W, Liu E et al (2016) Foxo3 circular RNA retards cell cycle progression via forming ternary complexes with p21 and CDK2. Nucleic Acids Res 44(6):2846–2858
21. Bachmayr-Heyda A, Reiner AT, Auer K et al (2015) Correlation of circular RNA abundance with proliferation–exemplified with colorectal and ovarian cancer, idiopathic lung fibrosis, and normal human tissues. Sci Rep 5:8057
22. Qian Z, Liu H, Li M et al (2018) Potential diagnostic power of blood circular RNA expression in active pulmonary tuberculosis. EBioMedicine 27:18–26
23. Huang Z-K, Yao F-Y, Xu J-Q et al (2018) Microarray expression profile of circular RNAs in peripheral blood mononuclear cells from active tuberculosis patients. Cell Physiol Biochem 45(3):1230–1240
24. Zhu RX, Seto W-K, Lai C-L et al (2016) Epidemiology of hepatocellular carcinoma in the Asia-Pacific region. Gut Liver 10(3):332–339
25. Beasley RP (1988) Hepatitis B virus. The major etiology of hepatocellular carcinoma. Cancer 61(10):1942–1956
26. Custer B, Sullivan SD, Hazlet TK et al (2004) Global epidemiology of hepatitis B virus. J Clin Gastroenterol 38(10):S158–S168
27. Yu L, Gong X, Sun L et al (2016) The circular RNA Cdr1as act as an oncogene in hepatocellular carcinoma through targeting miR-7 expression. PLoS One 11(7):e0158347
28. Fu L, Yao T, Chen Q et al (2017) Screening differential circular RNA expression profiles reveals hsa_circ_0004018 is associated with hepatocellular carcinoma. Oncotarget 8(35):58405
29. McKillop IH, Moran DM, Jin X et al (2006) Molecular pathogenesis of hepatocellular carcinoma. J Surg Res 136(1):125–135
30. Wu K, House L, Liu W et al (2012) Personalized targeted therapy for lung cancer. Int J Mol Sci 13(9):11471–11496
31. Ma PC (2012) Personalized targeted therapy in advanced non-small cell lung cancer. Cleve Clin J Med 79:eS56–eS60
32. Zhang S, Zeng X, Ding T et al (2018) Microarray profile of circular RNAs identifies hsa_circ_0014130 as a new circular RNA biomarker in non-small cell lung cancer. Sci Rep 8(1):2878
33. Wan L, Zhang L, Fan K et al (2016) Circular RNA-ITCH suppresses lung cancer proliferation via inhibiting the Wnt/β-catenin pathway. Biomed Res Int 2016(1):1579490
34. Yao J-T, Zhao S-H, Liu Q-P et al (2017) Over-expression of CircRNA_100876 in non-small cell lung cancer and its prognostic value. Pathol Res Pract 213(5):453–456
35. Luo Y-H, Zhu X-Z, Huang K-W et al (2017) Emerging roles of circular RNA hsa_circ_0000064 in the

proliferation and metastasis of lung cancer. Biomed Pharmacother 96:892–898
36. Zhu X, Wang X, Wei S et al (2017) hsa_circ_0013958: a circular RNA and potential novel biomarker for lung adenocarcinoma. FEBS J 284(14):2170–2182
37. Zhang Y, Zhao H, Zhang L (2018) Identification of the tumor-suppressive function of circular RNA FOXO3 in non-small cell lung cancer through sponging miR-155. Mol Med Rep 17(6):7692–7700
38. Ji W, Qiu C, Wang M et al (2018) Hsa_circ_0001649: a circular RNA and potential novel biomarker for colorectal cancer. Biochem Biophys Res Commun 497(1):122–126
39. Yin WB, Yan MG, Fang X, et al (2017) Circulating circular RNA hsa_circ_0001785 acts as a diagnostic biomarker for breast cancer detection. Clin Chim Acta. https://doi.org/10.1016/j.cca.2017.10.011
40. Cocquerelle C, Mascrez B, Hetuin D et al (1993) Mis-splicing yields circular RNA molecules. FASEB J 7(1):155–160
41. Siegel RL, Sahar L, Portier KM et al (2015) Cancer death rates in US congressional districts. CA Cancer J Clin 65(5):339–344
42. Boleij A, van Gelder MM, Swinkels DW et al (2011) Clinical importance of Streptococcus gallolyticus infection among colorectal cancer patients: systematic review and meta-analysis. Clin Infect Dis 53(9):870–878
43. Xie H, Ren X, Xin S et al (2016) Emerging roles of circRNA_001569 targeting miR-145 in the proliferation and invasion of colorectal cancer. Oncotarget 7(18):26680
44. Weng W, Wei Q, Toden S et al (2017) Circular RNA ciRS-7—a promising prognostic biomarker and a potential therapeutic target in colorectal cancer. Clin Cancer Res 23(14):3918–3928
45. Wang X, Zhang Y, Huang L et al (2015) Decreased expression of hsa_circ_001988 in colorectal cancer and its clinical significances. Int J Clin Exp Pathol 8(12):16020
46. Zhang P, Zuo Z, Shang W et al (2017) Identification of differentially expressed circular RNAs in human colorectal cancer. Tumor Biol 39(3):1010428317694546
47. Zhang Y, Zhang Y, Li X et al (2017) Microarray analysis of circular RNA expression patterns in polarized macrophages. Int J Mol Med 39(2):373–379
48. Huang G, Zhu H, Shi Y et al (2015) cir-ITCH plays an inhibitory role in colorectal cancer by regulating the Wnt/β-catenin pathway. PLoS One 10(6):e0131225
49. Huang J-L, Qin M-C, Zhou Y et al (2018) Comprehensive analysis of differentially expressed profiles of Alzheimer's disease associated circular RNAs in an Alzheimer's disease mouse model. Aging (Albany NY) 10(2):253
50. Mo D (2018) The role of Aβ circRNA in Alzheimer's disease. bioRxiv:260968
51. Burmistrova O, Goltsov A, Abramova L et al (2007) MicroRNA in schizophrenia: genetic and expression analysis of miR-130b (22q11). Biochem Mosc 72(5):578–582
52. Cogswell JP, Ward J, Taylor IA et al (2008) Identification of miRNA changes in Alzheimer's disease brain and CSF yields putative biomarkers and insights into disease pathways. J Alzheimers Dis 14(1):27–41
53. Chen L-L, Yang L (2015) Regulation of circRNA biogenesis. RNA Biol 12(4):381–388
54. You X, Vlatkovic I, Babic A et al (2015) Neural circular RNAs are derived from synaptic genes and regulated by development and plasticity. Nat Neurosci 18(4):603
55. Li J, Lin H, Sun Z et al (2018) High-throughput data of circular RNA profiles in human temporal cortex tissue reveals novel insights into temporal lobe epilepsy. Cell Physiol Biochem 45(2):677–691
56. Zhao M, Gao F, Zhang D et al (2017) Altered expression of circular RNAs in Moyamoya disease. J Neurol Sci 381:25–31
57. Wilusz JE, Sharp PA (2013) A circuitous route to noncoding RNA. Science 340(6131):440–441
58. Lin S-P, Ye S, Long Y et al (2016) Circular RNA expression alterations are involved in OGD/R-induced neuron injury. Biochem Biophys Res Commun 471(1):52–56
59. Bensimon G, Ludolph A, Agid Y et al (2008) Riluzole treatment, survival and diagnostic criteria in Parkinson plus disorders: the NNIPPS study. Brain 132(1):156–171
60. Prusiner SB, Woerman AL, Mordes DA et al (2015) Evidence for α-synuclein prions causing multiple system atrophy in humans with parkinsonism. Proc Natl Acad Sci U S A 112(38):E5308–E5317
61. McCann H, Stevens CH, Cartwright H et al (2014) α-Synucleinopathy phenotypes. Parkinsonism Relat Disord 20:S62–S67
62. Enuka Y, Lauriola M, Feldman ME et al (2015) Circular RNAs are long-lived and display only minimal early alterations in response to a growth factor. Nucleic Acids Res 44(3):1370–1383
63. Chen BJ, Mills JD, Takenaka K et al (2016) Characterization of circular RNAs landscape in multiple system atrophy brain. J Neurochem 139(3):485–496
64. Li Y, Zheng Q, Bao C et al (2015) Circular RNA is enriched and stable in exosomes: a promising biomarker for cancer diagnosis. Cell Res 25(8):981
65. Selvaggi L, Loverro G, Schena F et al (1988) Long term follow-up of women with hypertension in pregnancy. Int J Gynaecol Obstet 27(1):45–49
66. Díaz LM, Díaz MPN, Serrano ND et al (2011) The prognosis for children of mothers with preeclampsia. Part 2: long-term effects. Arch Argent Pediatr 109(6):519–524
67. Ji L, Brkić J, Liu M et al (2013) Placental trophoblast cell differentiation: physiological regulation and pathological relevance to preeclampsia. Mol Asp Med 34(5):981–1023
68. Burd CE, Jeck WR, Liu Y et al (2010) Expression of linear and novel circular forms of an INK4/ARF-associated non-coding RNA correlates with atherosclerosis risk. PLoS Genet 6(12):e1001233

Circular RNAs as Novel Biomarkers for Cardiovascular Diseases

13

Qiulian Zhou, Zhongrong Zhang, Yihua Bei, Guoping Li, and Tianhui Wang

Abstract

Cardiovascular diseases are among the most serious diseases, which are a leading cause of death across the world. Early diagnosis and prognosis prediction are keys for treatment and reduction of death rates. Circular RNAs (circRNAs) play a critical role in the physiology and pathology of biological system and participate in the development of diseases. In addition, circRNAs are relative stable and abundant. Therefore, many studies have suggested that circRNAs could be used as biomarkers for diseases, such as neurological diseases, cancers, immune diseases, and digestive diseases. Here we summarize recent studies on circRNAs and compare the characteristics of circRNAs with traditional biomarkers. Finally, we highlight the value of circRNAs as potential biomarkers for cardiovascular diseases, including acute myocardial infarction, heart failure, coronary artery disease, and hypertension. In conclusion, circRNAs may be promising biomarkers for cardiovascular diseases.

Keywords

Circular RNA · Biomarkers · Cardiovascular diseases

Q. Zhou
Shanghai Applied Radiation Institute, School of Environmental and Chemical Engineering, Shanghai University, Shanghai, China

Cardiac Regeneration and Ageing Lab, Institute of Cardiovascular Sciences, School of Life Science, Shanghai University, Shanghai, China

Shanghai Key Laboratory of Bio-Energy Crops, School of Life Sciences, Shanghai University, Shanghai, China

Z. Zhang · Y. Bei · T. Wang (✉)
Cardiac Regeneration and Ageing Lab, Institute of Cardiovascular Sciences, School of Life Science, Shanghai University, Shanghai, China

Shanghai Key Laboratory of Bio-Energy Crops, School of Life Sciences, Shanghai University, Shanghai, China
e-mail: wangth@shu.edu.cn

G. Li
Cardiovascular Division of the Massachusetts General Hospital, Harvard Medical School, Boston, MA, USA

J. Xiao (ed.), *Circular RNAs*, Advances in Experimental Medicine and Biology 1087,
https://doi.org/10.1007/978-981-13-1426-1_13

1 Introduction of Circular RNA

Circular RNAs (circRNAs) are currently a research focus in the field of noncoding RNAs, including microRNAs (miRNAs), long noncoding RNAs (lncRNAs), etc. [1]. Noncoding RNAs are involved in the development of many diseases [2], including cardiovascular diseases. They are recognized as regulators of gene expression. However, in the past, circRNAs were ignored and considered as transcriptional noise [3] or errant by-products of splicing [4], owing to their unique structure and low abundance [4]. Based on advances in bioinformatics and biotechnology, a large number of abundant and significant circRNAs were discovered and identified. The evidence is mounting that circRNAs play essential roles in various physiological and pathological pathways.

2 Biogenesis of circRNAs

CircRNAs are closed circular molecules, which are distinguished from other linear RNA molecules. First circRNA was found by electron microscopy from RNA viruses [5]. Nowadays, more and more circRNAs are being identified by RNA sequencing and bioinformatics. Due to the closed loop characteristic of circRNAs, they can be extracted from the ribosomal RNA and linear RNA molecules in total RNA. RNA sequencing can obtain more circRNAs from specific accumulation of circRNAs samples. The covalently closed loop structure of circRNAs, also called "back-splicing," is formed by joining of 3′ splice site to 5′ splice site [6, 7]. There are three types of circRNAs: exonic circular RNAs (ecircRNAs) [8], intronic circular RNAs (ciRNAs) [9], and exon-intron circular RNAs (EIciRNAs) [10].

Exonic circular RNAs have three formation models: lariat-driven circularization, intron pairing-driven circularization, and resplicing-driven circularization. They form from exon back-splicing and only consist of exonic sequences. The formation of circular intronic RNAs mainly depends on consensus RNA motifs and branchpoint. Circular intronic RNAs only consist of intronic sequences. Exon-intron circular RNAs consist of both exonic and intronic sequences. They predominantly localize in the nucleus and have association with the RNA polymerase II. Therefore, exon-intron circular RNAs are regarded as transcriptional regulators. Three types of circRNAs share the same characteristics, such as stable, abundant, and conservative.

3 Characteristics of circRNAs

Desirable biomarkers are required to be stable against harsh conditions, for example, temperature variation, extreme pH, and thawing cycles [11, 12]. Biomarkers should have high stability in blood and other body fluids. The biggest natural enemy of RNA is RNA exonucleases (or RNase R). CircRNAs are closed circular molecules consist of covalently closed loop structure and therefore resistant to RNA exonucleases that solely digests linear transcripts [13–17]. CircRNAs are more stable compared to linear RNAs [18]. Furthermore, based on their stability in blood and other body fluids [19, 20], circRNAs are suitable biomarkers for disease diagnosis.

Clinical trials must be properly based upon reliable earlier trials and sufficient preclinical studies. CircRNAs are conserved across mammals [21], including human, mouse, and rat [22, 23]. In hearts, RNA-Seq analysis found that plenty of circRNAs could be detected in mammal species: human (16427), mouse (9953), and rat (13086). Among them, about 30% (3171) were conserved between mouse and rat, and about 10% (1288) were conserved in all three species [24]. These circRNAs conserved in different species become abundant resources of promising candidates for biomarkers. In addition to the benefits mentioned above, circRNAs have been demonstrating specificity in expression [25]. Emerging evidence have confirmed thousands of the tissue-specific expression of circRNAs [26], such as the heart, brain, skin, prostate, thyroid, blood, or blood cell. Through biclustering algorithm, studies have already reported thousands of tissue-specific expressed circRNAs in human [26]. Except for tissue-specific expression, circRNAs have specific expression in different diseases. In cardiac development and diseases, the expression of circRNAs is regulated [27, 28]. Through whole transcriptome sequencing, 826 putative circRNAs have been identified in human heart with dilated cardiomyopathy patient and hypertrophic cardiomyopathy [27]. Through microarray expression profile assay, 63 circRNAs were differentially expressed in mouse between heart failure and sham group [28].

In particular, circRNAs (3841) are found unique in different composition and cell types of blood [14], such as plasma (51), serum (37), neutrophils (58), and platelets (2339). Circulating blood markers have enormous potential to be applied to clinical practice. Human peripheral whole blood is an easily accessible body fluid. Thousands of circRNAs have been detected in blood. Except for the blood, distinct circRNAs have also been founded in other body fluids, such as saliva [19]. On the other hand, circRNAs have been found enriched in exosomes [29, 30]. Exosomes are currently regarded as specific secretory membrane vesicles and are involved in intercellular communication. Exosomes are also thought to be the way of transferring miRNA in the circulatory system, which indirectly reflects the changes of miRNAs in diseases. Serum exo-circRNAs may be involved in intercellular communication and can be used as indicators of diseases development. Therefore, exo-circRNAs in the blood of patients are expected to be markers for early diagnosis. According to the characteristics talked above, circRNAs are promising new biomarkers.

4 Functions of circRNAs

As most other noncoding RNAs, most of the circRNAs do not encode proteins. But circRNAs are involved in regulating the expression of genes [31]. Firstly, circRNAs can adsorb some specific miRNAs that prevent mRNA translation, functioning as miRNA sponges [25, 32]. Secondly, circRNAs can bind to RNA-associated proteins and form RNA-protein complexes that function as RNA-binding protein (RBP) sponges [33, 34]. Thirdly, circRNAs can be a regulator of transcription and alternative splicing that directly influence gene expression [35].

For all those reasons and more, circRNAs have been reported to play important roles in physiological and pathological processes [36–38], such as proliferation, apoptosis, and autophagy. CircRNA circHIPK3 regulates the viability and proliferation of endothelial cell viability by inhibiting the activity of miR-30a-3p [36]. CircRNA TTBK2 regulates the cell proliferation and apoptosis of human glioma cells by regulating miR-217/HNF1β/Derlin-1 pathway [37].

In addition, circRNAs can play regulatory function in diseases, such as neurological diseases [39, 40], cancers [41], immune diseases [42], aging [43], diabetes [44], digestive diseases [45], reproductive system diseases, skin, muscles and bones diseases, and especially cardiovascular diseases [46, 47]. CircRNA HRCR can function as an endogenous miR-223 sponge through upregulation of ARC, which attenuate hypertrophic responses [48]. CircRNA Cdr1as can inhibit miR-7a through upregulating PARP and SP1, which induces myocardial apoptosis and aggravates myocardial infarction injuries [49].

Furthermore, circRNAs constitute promising new biomarkers in multiple diseases [25, 50]. CircRNA hsa_circ_0014130 is significantly upregulated in non-small cell lung cancer (NSCLC) tissues that may be a new biomarker of NSCLC [51]. CircRNA hsa_circ_0004277 is downregulated in mononuclear cells from bone marrow of acute myeloid leukemia (AML) patients [52]. CircRNA hsa_circ_0001649 is downregulated in hepatocellular carcinoma (HCC) tissues and is correlated with tumor size and tumor embolus's occurrence [53].

In addition to the above examples, circRNAs have the potential to become biomarkers in many cardiovascular diseases. Here we will briefly introduce the characteristics of circRNAs by comparison with the existing biomarkers of cardiovascular diseases, which expound the value of circRNAs. Then we will briefly summarize recent findings on circRNAs as biomarkers in cardiovascular diseases.

5 Biomarkers of Cardiovascular Diseases

Cardiovascular diseases threaten human health and have become one of the leading diseases which causes death in the world [54]. Nowadays, more and more risk factors are aggravating the development and progression of cardiovascular diseases. Early diagnosis and prognosis prediction are keys for treatment and reducing mortality. There are multiple diagnostic techniques for cardiovascular diseases, such as blood pressure monitoring, dynamic electrocardiogram, cardiovascular ultrasound, magnetic resonance imaging, electronic computed tomography, and so on. These diagnostic techniques require lots of resources and precise instruments. What's more, these diagnostics still cannot ensure the accurate diagnosis of the occurrence and development of cardiovascular diseases.

Some blood markers are commonly regarded as sensitive biomarkers for cardiovascular diseases, such as atrial natriuretic peptide (ANP), creatine kinase (CK), myocardial band isoenzyme (CK-MB), and cardiac troponin (cTn) [55]. Brain natriuretic peptide (BNP) and N-terminal portion of its prohormone (NT-proBNP) are also used as biomarkers for cardiovascular diseases [56]. But there are problems difficult to solve. Firstly, the detection time is uncertain, since markers keep changing over time and exhibit a relative "delayed" release time [57]. Secondly, disease degree and patient's background are varied, while these biomarkers are low in specificity [58]. Thirdly, there are still many kinds of cardiovascular diseases without proper marker, such as coronary artery disease. As a result, there remain many challenges in the field of cardiovascular diseases diagnosis.

There is growing evidence that noncoding RNAs have the potential to become biomarkers in many cardiovascular diseases. Circulating miRNAs are suggested as suitable biomarkers for cardiovascular diseases [59–61], including acute myocardial infarction (miR-1, miR-133, miR-208, and miR-499) [62, 63]; acute coronary syndrome (miR-208a, miR-34a, miR-133a, miR-499) [64, 65]; and heart failure (miR-499, miR-133, miR-423-5p, miR-126) [66–68]. However, only a few of them have been validated, and they are necessary to prove by large-scale clinical studies [69]. Traditional biomarkers are also limited by low stability [70]. Therefore, circRNAs, which are stable, conservative, and specific [71, 72], might open a novel avenue.

6 Circular RNAs as Biomarkers in Cardiovascular Disease

6.1 Acute Myocardial Infarction

Acute myocardial infarction is a myocardial necrosis caused by acute and persistent ischemia. Acute myocardial infarction can be complicated with arrhythmia, shock, or heart failure, which is often life-threatening. It tends to occur in patients with coronary atherosclerosis stenosis and coronary spasm. Acute myocardial infarction carries a high mortality rate and become a global challenge [73, 74]. Reperfusion is the most common treatment strategy for myocardial infarction patient. This treatment strategy will cause myocardial ischemia-reperfusion injury leading to left ventricular remodeling and even heart failure. There are some biomarkers for clinical assessment of the severity, course, and prognosis of acute myocardial infarction. Creatine kinase(CK) and cardiac troponin(cTn) are markers for myocardial injury [75–77]. In skeletal muscles and heart muscles, creatinine kinase facilitates the transfer of energy phosphates between outside and inside mitochondria. The increased concentration of creatine kinase in serum indicates myocardial damage, such as myocardial infarction or ischemia. But it can also be caused by vigorous exercise or other kinds of tissue damage [78]. Cardiac troponin I and T untimely are released into plasma when actin and myosin filaments occur in pathologic degeneration in the heart. Thus, cardiac troponin can be a marker in myocardial infarction. But the abnormal release of cardiac troponin can also occur in myocarditis [75, 76], pulmonary embolism with acute right heart overload [79], and even heart failure [80]. Therefore, new biomarkers are required.

Recently, a study observed that a circRNA called MICRA (myocardial infarction-associated circular RNA) is absent in peripheral blood cells, plasma, and serum samples [81, 82]. The particular circRNA (MICRA) is formed by 874 nucleotides, which is mainly derived from zinc finger protein 609 (ZNF609) gene exon 1 [82, 83]. MICRA is associated with development of heart failure after myocardial infarction and can be used to predict left ventricular remodeling after acute myocardial infarction by multivariable analyses (0.83[0.69–1.01]; p = 0.066) [81]. Lower level of MICRA classifies patients into decreased ejection fraction groups; and high level of MICRA is able to predict a preserved ejection fraction at 4 months. When MICRA is included, a multivariable clinical model, Akaike Information Criteria, becomes more accurate in prediction of preserved ejection fraction at 4 months. CircRNA MICRA is potential to become a novel biomarker in the future prognostication strategies of acute myocardial infarction [81].

In another study, circRNA MICRA was found lower in 642 myocardial infarction patients compared to 86 healthy volunteers [82]. Besides, MICRA was significantly lower in patients who have a low ejection fraction. In particular, both univariate and multivariate analyses proved that circRNA MICRA may be a strong predictor for prognoses of myocardial infarction after 3–4 months. Collectively, circRNA MICRA is suggested to be able to inform the risk of developing left ventricular dysfunction for myocardial infarction patients.

6.2 Heart Failure

Heart failure is a disorder of the systolic function and/or diastolic function of the heart [84–86]. Heart failure is not an independent disease, but a terminal stage in the development of heart diseases [87, 88]. Almost all kinds of cardiovascular diseases can eventually lead to the occurrence of heart failure, such as myocardial infarction, cardiomyopathy, hemodynamic overload, myocardial injury (any cause), and inflammation [89, 90]. All these diseases can lead to alterations of myocardial structure and function. Heart failure is a major cause of morbidity and mortality worldwide [91–93], even though its molecular mechanisms become more and more clear and a large amount of pharmacological interventions come to light. In current guidelines, the plasma levels of a natriuretic peptide (such as brain natriuretic peptide, BNP, and N-terminal pro brain natriuretic peptide, NT-proBNP) are important diagnostic indicators of acute heart failure [94]. Most of biomarkers for the assessment of heart failure are widely acknowledged and used. But there is still limitation that circulating levels of BNP will significantly change with the development of acute myocardial infarction [95]. Therefore, these biomarkers cannot be used in the management of heart failure patients, who have a complicated disease history.

An increasing number of RNA including miRNAs [19, 96] and lncRNAs [83, 97, 98] have been found with enormous potential as biomarkers for heart failure. LncRNA LIPCAR was found downregulated early after myocardial infarction and meanwhile upregulated in plasma of 788 heart failure patients [97]. LncRNA NRON and LncRNA MHRT were found upregulated in plasma of 104 heart failure patients [99].

CircRNAs have great potential to act as biomarker for myocardial infarction and heart failure. By comparing the transcriptome sequencing through bioinformatic analysis, 1163 circRNAs were found changed in hypertrophied myocardial tissues grouping with or without heart failure [28]. Through analyzing by quantitative PCR, 29 upregulated and 34 downregulated circRNAs was validated [28]. Thus, these circRNAs may be potential biomarkers for heart failure. Besides, a circRNA MICRA was downregulated in decreased ejection fraction groups (heart failure). This circRNA might become a biomarker for both myocardial infarction and heart failure.

6.3 Coronary Artery Disease

Coronary artery disease, a big threat to heart health, is the main cause of coronary heart disease, cerebral infarction, and peripheral vascular disease. Coronary artery disease with high morbidity and mortality pose a threat to human health and cause burden to economies [98]. There are some treatments for coronary artery disease, such as medications, percutaneous coronary intervention (PCI), and coronary artery bypass graft surgery (CABG) [100–102]. Minimizing the time between plaque rupture and treatment is critical for reducing the morbidity and mortality [103]. But the diagnosis and prognosis of coronary artery disease is almost blank at present [104, 105].

Early detection of coronary artery disease has the potential to increase survival rates and to improve the life quality of patients. MiR-221 and miR-222 have been reported reduced in carotid plaques after an acute ischemic cerebrovascular syndrome [106]. CircR-284 possesses a miR-221 and miR-222 binding site. Therefore, circR-284 may act as a regulator on miR-221/miR-222 activity. A validation study found that miR-221 is significantly decreased in the urgent carotid atherosclerotic plaques, compared with the asymptomatic control. But perhaps even more interesting is that circR-284 has been reported to be significantly increased in the serum of urgent carotid atherosclerotic plaque patient [103]. Combining the expression of serum circR-284 and miR-221, the ratio of circR-284 and miR-221 was uniquely increased in the urgent group ($P < 0.001$); and this ratio was sensitive (0.76) and specific (0.88) for detecting plaque rupture and stroke [103]. Based on these results, the ratio of circR-284 and miR-221 has the potential to become a diagnostic biomarker in cardiovascular disease for prediction of the risk of plaque rupture.

Identified by microarray analysis, there were 2036 circRNAs in blood differentially expressed between coronary artery disease and control, including 376 upregulated and 1660 downregulated ones [107]. Among them, a circRNA (hsa-circRNA11783-2) was significantly downregulated in coronary artery disease group but also in type 2 diabetes group [107]. In a separate set of experiments, RNA microarray showed 22 circRNAs were differentially expressed in peripheral blood, including 12 upregulated and 10 downregulated. Hsa_circ_0124644 was significantly upregulated in coronary artery disease group, and this circRNA was of high sensitivity (0.867) and specificity (0.767) for coronary artery disease [108]. Therefore, circRNAs are prospective diagnostic biomarkers in coronary artery disease.

6.4 Hypertension

High blood pressure (hypertension) refers to systemic arterial blood pressure (systolic pressure) and/or diastolic blood pressure increase as the main characteristics (systolic blood pressure, 140 mmHg or higher; diastolic blood pressure, 90 mmHg or higher). Hypertension can be associated with the function of the heart, brain, kidney, and other organs or physical damage to the clinical syndrome. Hypertension is the most common chronic disease and the most important risk factor for cardiovascular disease [109–111].

Profiling and bioinformatics identified 13 downregulated and 46 upregulated circRNAs in hypertension patients [112]. Among them, four circRNAs (hsa-circ-0000437, hsa-circ-0008139, hsa-circ-0005870, and hsa-circ-0040809) showed significant difference in hypertensive patients. By further validation, hsa-circ-0005870 was confirmed significantly downregulated in plasma between 49 hypertension patients (systolic blood pressure over 150/90 mmHg) and 49 healthy (systolic blood pressure lower than 140/90 mmHg). In addition, hsa-circ-0005870 has also been found related to many biological processes, such as cellular response to stress. Therefore, this circRNA may be a novel biomarker for hypertension.

In human umbilical vein endothelial cells, 7388 circRNAs have been identified. Under hypoxia in endothelial cells, circRNAs cZNF292, cAFF1, and cDENND4C were found upregulated, whereas cTHSD1 was reduced. Among them, cZNF292 is the highest expressed with proangiogenic activities. These circRNA may be new therapeutic targets and endogenous biomarkers [113].

6.5 Chronic Thromboembolic Pulmonary Hypertension

Chronic thromboembolic pulmonary hypertension (CTEPH) is harmful to human health. Pulmonary hypertension is caused by obstruction of pulmonary arteries and subsequent vascular remodeling [114–116]. Early detection of pulmonary hypertension has the potential to increase physiological function of heart and improve prognosis [117, 118]. In the blood samples of chronic thromboembolic pulmonary hypertension patient, 351 circRNAs showed significant difference, including 122 upregulations and 229 downregulations [119]. Among them, circRNA hsa_circ_0002062 can regulate 761 miRNAs, and circRNA hsa_circ_0022342 can regulate 453 miRNAs. Through miRNAs regulation, these circRNAs may be involved in regulating the pathways in chronic thromboembolic pulmonary hypertension [119–121], such as ErbB signaling pathway. Therefore, these unique circRNAs which are changed and involved in chronic thromboembolic pulmonary hypertension may be novel biomarkers for this disease.

7 Future Perspectives

Cardiovascular diseases are not only manifold but involute. Due to the influence of heredity [122], diet, living habit, and external environment, different people have different age of onset, and some people do not have obvious symptoms in their whole life. Nowadays, many people live a stressful, chronically overloaded, unbalanced life that cause rising morbidity and mortality. In the early stages of cardiovascular diseases, the diagnosis of diseases is important for treatment and prognosis. Traditional nonspecific diagnosis cannot fulfill the needs. A novel biomarker demands (1) noninvasiveness and safety, (2) high sensitivity and specificity, and (3) stability and abundance.

CircRNAs are suitable biomarkers for disease diagnosis. Firstly, circRNAs are closed circular molecules that are resistant to RNA exonucleases or RNase R and hence have stability in blood and other body fluids. Secondly, many circRNAs are reported to be conserved across mammals. Especially, circRNAs have high sensitivity and specificity. CircRNAs can be source for searching new therapeutic targets and candidates in disease diagnosis [123]. Multiple studies have suggested that circRNAs may become potential biomarkers for many cardiovascular diseases, such as heart failure, acute myocardial infarction, coronary artery disease, and hypertension (Table 13.1).

There needs to be more research on exploration of circRNAs as potential biomarkers. In experimental design, patients' physiological conditions need to be more explicit, including criteria such as ages, genders, colors, diet habits, and so on. There are some details we should consider: (1) What kind of sample can be collected, blood sample, serum, or plasma? (2) What can be used for normalization, Cel-mirR-39, GAPDH, or SF3a1? (3) Which test method can be use, SYBR or TaqMan real-time quantitative?

It is important to make strict screening criteria. Although many studies assume that circRNAs have high stability, there are some conditions that need to be investigated, such as sample collection

Table 13.1 Circulating circular RNAs as biomarkers in cardiovascular diseases

Diseases	CircRNAs	Regulation	Normalization	Sample	Ref
Acute myocardial infarction	MICRA	Down	SF3a1	Blood	[81, 82]
Heart failure	MICRA	Down	SF3a1	Blood	[81, 82]
Coronary artery disease	CircR-284/miR-221	Up	–	Serum	[103]
	hsa_circ_0124644	Up	GAPDH	Blood	[108]
Hypertension	Hsa-circ-0005870	Down	GAPDH	Plasma	[112]
Chronic thromboembolic pulmonary hypertension	hsa_circ_0002062	Down	–	Blood	[119]
	hsa_circ_0022342	Down	–	Blood	[119]

and treatment, equipment and procedures, facticity, and reproducible analytics method. It is time to make a universally accepted protocol on identification and quantification of circRNAs, which affects how popular circRNAs will be used as biomarkers for cardiovascular diseases in the future.

Acknowledgments This work was supported by the grants from National Natural Science Foundation of China (81770401 to Y Bei) and National Key Research and Development Program of China (2017YFC1700401 to Y Bei).

Competing Financial Interests The authors declare no competing financial interests.

References

1. Han B, Chao J, Yao H (2018) Circular RNA and its mechanisms in disease: from the bench to the clinic. Pharmacol Ther 187:31. https://doi.org/10.1016/j.pharmthera.2018.01.010
2. Esteller M (2011) Non-coding RNAs in human disease. Nat Rev Genet 12(12):861–874
3. Salzman J (2016) Circular RNA expression: its potential regulation and function. Trends Genet 32(5):309–316
4. Nigro JM, Cho KR, Fearon ER et al (1991) Scrambled exons. Cell 64(3):607–613
5. Sanger HL, Klotz G, Riesner D et al (1976) Viroids are single-stranded covalently closed circular RNA molecules existing as highly base-paired rod-like structures. Proc Natl Acad Sci USA 73(11):3852–3856
6. Tan WL, Lim BT, Anene-Nzelu CG et al (2017) A landscape of circular RNA expression in the human heart. Cardiovasc Res 113(3):298–309
7. Chen LL, Yang L (2015) Regulation of circRNA biogenesis. RNA Biol 12(4):381–388
8. Chen I, Chen CY, Chuang TJ (2015) Biogenesis, identification, and function of exonic circular RNAs. Wiley Interdiscip Rev RNA 6(5):563–579
9. Zhang Y, Zhang XO, Chen T et al (2013) Circular intronic long noncoding RNAs. Mol Cell 51(6):792–806
10. Li Z, Huang C, Bao C et al (2015) Exon-intron circular RNAs regulate transcription in the nucleus. Nat Struct Mol Biol 22(3):256–264
11. Arita T, Ichikawa D, Konishi H et al (2013) Circulating long non-coding RNAs in plasma of patients with gastric cancer. Anticancer Res 33(8):3185–3193
12. Chen X, Ba Y, Ma L et al (2008) Characterization of microRNAs in serum: a novel class of biomarkers for diagnosis of cancer and other diseases. Cell Res 18(10):997–1006
13. Jeck WR, Sharpless NE (2014) Detecting and characterizing circular RNAs. Nat Biotechnol 32(5):453–461
14. Maass PG, Glazar P, Memczak S et al (2017) A map of human circular RNAs in clinically relevant tissues. J Mol Med (Berl) 95(11):1179–1189
15. Alhasan AA, Izuogu OG, Al-Balool HH et al (2016) Circular RNA enrichment in platelets is a signature of transcriptome degradation. Blood 127(9):e1–e11
16. Panda AC, De S, Grammatikakis I et al (2017) High-purity circular RNA isolation method (RPAD) reveals vast collection of intronic circRNAs. Nucleic Acids Res 45(12):e116
17. Suzuki H, Zuo Y, Wang J et al (2006) Characterization of RNase R-digested cellular RNA source that consists of lariat and circular RNAs from pre-mRNA splicing. Nucleic Acids Res 34(8):e63
18. Jeck WR, Sorrentino JA, Wang K et al (2013) Circular RNAs are abundant, conserved, and associated with ALU repeats. RNA 19(2):141–157
19. Bahn JH, Zhang Q, Li F et al (2015) The landscape of microRNA, Piwi-interacting RNA, and circular RNA in human saliva. Clin Chem 61(1):221–230
20. Preusser C, Hung LH, Schneider T et al (2018) Selective release of circRNAs in platelet-derived extracellular vesicles. J Extracell Vesicles 7(1):1424473
21. Rybak-Wolf A, Stottmeister C, Glazar P et al (2015) Circular RNAs in the mammalian brain are highly abundant, conserved, and dynamically expressed. Mol Cell 58(5):870–885
22. Guo JU, Agarwal V, Guo H et al (2014) Expanded identification and characterization of mammalian circular RNAs. Genome Biol 15(7):409
23. Salzman J, Chen RE, Olsen MN et al (2013) Cell-type specific features of circular RNA expression. PLoS Genet 9(9):e1003777
24. Werfel S, Nothjunge S, Schwarzmayr T et al (2016) Characterization of circular RNAs in human, mouse and rat hearts. J Mol Cell Cardiol 98:103–107
25. Kulcheski FR, Christoff AP, Margis R (2016) Circular RNAs are miRNA sponges and can be used as a new class of biomarker. J Biotechnol 238:42–51
26. Liu YC, Chiu YJ, Li JR et al (2018) Biclustering of transcriptome sequencing data reveals human tissue-specific circular RNAs. BMC Genomics 19(Suppl 1):958
27. Khan MA, Reckman YJ, Aufiero S et al (2016) RBM20 regulates circular RNA production from the titin gene. Circ Res 119(9):996–1003
28. Wu HJ, Zhang CY, Zhang S et al (2016) Microarray expression profile of circular RNAs in heart tissue of mice with myocardial infarction-induced heart failure. Cell Physiol Biochem 39(1):205–216
29. Dou Y, Cha DJ, Franklin JL et al (2016) Circular RNAs are down-regulated in KRAS mutant colon

cancer cells and can be transferred to exosomes. Sci Rep 6:37982
30. Li Y, Zheng Q, Bao C et al (2015) Circular RNA is enriched and stable in exosomes: a promising biomarker for cancer diagnosis. Cell Res 25(8):981–984
31. Memczak S, Jens M, Elefsinioti A et al (2013) Circular RNAs are a large class of animal RNAs with regulatory potency. Nature 495(7441):333–338
32. Thomas LF, Saetrom P (2014) Circular RNAs are depleted of polymorphisms at microRNA binding sites. Bioinformatics 30(16):2243–2246
33. Abdelmohsen K, Panda AC, Munk R et al (2017) Identification of HuR target circular RNAs uncovers suppression of PABPN1 translation by CircPABPN1. RNA Biol 14(3):361–369
34. Barbagallo D, Caponnetto A, Cirnigliaro M et al (2018) CircSMARCA5 inhibits migration of glioblastoma Multiforme cells by regulating a molecular Axis involving splicing factors SRSF1/SRSF3/PTB. Int J Mol Sci 19(2)
35. Ashwal-Fluss R, Meyer M, Pamudurti NR et al (2014) circRNA biogenesis competes with pre-mRNA splicing. Mol Cell 56(1):55–66
36. Shan K, Liu C, Liu BH et al (2017) Circular noncoding RNA HIPK3 mediates retinal vascular dysfunction in diabetes mellitus. Circulation 136(17):1629–1642
37. Zheng J, Liu X, Xue Y et al (2017) TTBK2 circular RNA promotes glioma malignancy by regulating miR-217/HNF1beta/Derlin-1 pathway. J Hematol Oncol 10(1):52
38. Du WW, Yang W, Liu E et al (2016) Foxo3 circular RNA retards cell cycle progression via forming ternary complexes with p21 and CDK2. Nucleic Acids Res 44(6):2846–2858
39. Lukiw WJ (2013) Circular RNA (circRNA) in Alzheimer's disease (AD). Front Genet 4:307
40. You X, Vlatkovic I, Babic A et al (2015) Neural circular RNAs are derived from synaptic genes and regulated by development and plasticity. Nat Neurosci 18(4):603–610
41. Tian F, Yu CT, Ye WD et al (2017) Cinnamaldehyde induces cell apoptosis mediated by a novel circular RNA hsa_circ_0043256 in non-small cell lung cancer. Biochem Biophys Res Commun 493(3):1260–1266
42. Wang YH, Yu XH, Luo SS et al (2015) Comprehensive circular RNA profiling reveals that circular RNA100783 is involved in chronic CD28-associated CD8(+)T cell ageing. Immun Ageing 12:17
43. Gruner H, Cortes-Lopez M, Cooper DA et al (2016) CircRNA accumulation in the aging mouse brain. Sci Rep 6:38907
44. Sena CM, Pereira AM, Seica R (2013) Endothelial dysfunction – a major mediator of diabetic vascular disease. Biochim Biophys Acta 1832(12):2216–2231
45. Chen Y, Yuan B, Wu Z et al (2017) Microarray profiling of circular RNAs and the potential regulatory role of hsa_circ_0071410 in the activated human hepatic stellate cell induced by irradiation. Gene 629:35–42
46. Zeng Y, Du WW, Wu Y et al (2017) A circular RNA binds to and activates AKT phosphorylation and nuclear localization reducing apoptosis and enhancing cardiac repair. Theranostics 7(16):3842–3855
47. Holdt LM, Stahringer A, Sass K et al (2016) Circular non-coding RNA ANRIL modulates ribosomal RNA maturation and atherosclerosis in humans. Nat Commun 7:12429
48. Wang K, Long B, Liu F et al (2016) A circular RNA protects the heart from pathological hypertrophy and heart failure by targeting miR-223. Eur Heart J 37(33):2602–2611
49. Geng HH, Li R, Su YM et al (2016) The circular RNA Cdr1as promotes myocardial infarction by mediating the regulation of miR-7a on its target genes expression. PLoS One 11(3):e0151753
50. Meng S, Zhou H, Feng Z et al (2017) CircRNA: functions and properties of a novel potential biomarker for cancer. Mol Cancer 16(1):94
51. Zhang S, Zeng X, Ding T et al (2018) Microarray profile of circular RNAs identifies hsa_circ_0014130 as a new circular RNA biomarker in non-small cell lung cancer. Sci Rep 8(1):2878
52. Li W, Zhong C, Jiao J et al (2017) Characterization of hsa_circ_0004277 as a new biomarker for acute myeloid leukemia via circular RNA profile and bioinformatics analysis. Int J Mol Sci 18(3)
53. Qin M, Liu G, Huo X et al (2016) Hsa_circ_0001649: a circular RNA and potential novel biomarker for hepatocellular carcinoma. Cancer Biomark 16(1):161–169
54. Ronaldson KJ (2017) Cardiovascular disease in clozapine-treated patients: evidence, mechanisms and management. CNS Drugs 31(9):777–795
55. Heil B, Tang WH (2015) Biomarkers: their potential in the diagnosis and treatment of heart failure. Cleve Clin J Med 82(12 Suppl 2):S28–S35
56. Calzetta L, Orlandi A, Page C et al (2016) Brain natriuretic peptide: much more than a biomarker. Int J Cardiol 221:1031–1038
57. Kondkar AA, Abu-Amero KK (2015) Utility of circulating microRNAs as clinical biomarkers for cardiovascular diseases. Biomed Res Int 2015:821823
58. Enroth S, Johansson A, Enroth SB et al (2014) Strong effects of genetic and lifestyle factors on biomarker variation and use of personalized cutoffs. Nat Commun 5:4684
59. Mari-Alexandre J, Sanchez-Izquierdo D, Gilabert-Estelles J et al (2016) miRNAs regulation and its role as biomarkers in endometriosis. Int J Mol Sci 17(1)
60. Van Roosbroeck K, Pollet J, Calin GA (2013) miRNAs and long noncoding RNAs as biomarkers in human diseases. Expert Rev Mol Diagn 13(2):183–204
61. Yang Y, Yu T, Jiang S et al (2017) miRNAs as potential therapeutic targets and diagnostic biomarkers for cardiovascular disease with a particular

focus on WO2010091204. Expert Opin Ther Pat 27(9):1021–1029
62. Zhu H, Fan GC (2013) Whether circulating miRNAs or miRNA-carriers serve as biomarkers for acute myocardial infarction. J Biomark Drug Dev 1(1)
63. Wang F, Long G, Zhao C et al (2014) Atherosclerosis-related circulating miRNAs as novel and sensitive predictors for acute myocardial infarction. PLoS One 9(9):e105734
64. Pleister A, Selemon H, Elton SM et al (2013) Circulating miRNAs: novel biomarkers of acute coronary syndrome? Biomark Med 7(2):287–305
65. Rognoni A, Cavallino C, Lupi A et al (2014) Novel biomarkers in the diagnosis of acute coronary syndromes: the role of circulating miRNAs. Expert Rev Cardiovasc Ther 12(9):1119–1124
66. Ali Sheikh MS, Salma U, Zhang B et al (2016) Diagnostic, prognostic, and therapeutic value of circulating miRNAs in heart failure patients associated with oxidative stress. Oxidative Med Cell Longev 2016:5893064
67. Luscher TF (2016) From heart failure to transplantation: genes, miRNAs, and biomarkers. Eur Heart J 37(33):2561–2563
68. Yan H, Ma F, Zhang Y et al (2017) miRNAs as biomarkers for diagnosis of heart failure: a systematic review and meta-analysis. Medicine (Baltimore) 96(22):e6825
69. Viereck J, Thum T (2017) Circulating noncoding RNAs as biomarkers of cardiovascular disease and injury. Circ Res 120(2):381–399
70. Devaux Y, Creemers EE, Boon RA et al (2017) Circular RNAs in heart failure. Eur J Heart Fail 19(6):701–709
71. Danan M, Schwartz S, Edelheit S et al (2012) Transcriptome-wide discovery of circular RNAs in archaea. Nucleic Acids Res 40(7):3131–3142
72. Hansen TB, Jensen TI, Clausen BH et al (2013) Natural RNA circles function as efficient microRNA sponges. Nature 495(7441):384–388
73. Townsend N, Wilson L, Bhatnagar P, Wickramasinghe K, Rayner M, Nichols M (2016) Cardiovascular disease in Europe 2016: an epidemiological update. Eur Heart J 37(42):3182–3183
74. Benjamin EJ, Blaha MJ, Chiuve SE et al (2017) Heart disease and stroke Statistics-2017 update: a report from the American Heart Association. Circulation 135(10):e146–e603
75. Haggart PC, Adam DJ, Ludman PF et al (2001) Comparison of cardiac troponin I and creatine kinase ratios in the detection of myocardial injury after aortic surgery. Br J Surg 88(9):1196–1200
76. Shyu KG, Lin JL, Chen JJ et al (1996) Use of cardiac troponin T, creatine kinase and its isoform to monitor myocardial injury during radiofrequency ablation for supraventricular tachycardia. Cardiology 87(5):392–395
77. Wang Q, Michiue T, Ishikawa T et al (2011) Combined analyses of creatine kinase MB, cardiac troponin I and myoglobin in pericardial and cerebrospinal fluids to investigate myocardial and skeletal muscle injury in medicolegal autopsy cases. Leg Med (Tokyo) 13(5):226–232
78. Heavens KR, Szivak TK, Hooper DR et al (2014) The effects of high intensity short rest resistance exercise on muscle damage markers in men and women. J Strength Cond Res 28(4):1041–1049
79. Tulevski II, ten Wolde M, van Veldhuisen DJ et al (2007) Combined utility of brain natriuretic peptide and cardiac troponin T may improve rapid triage and risk stratification in normotensive patients with pulmonary embolism. Int J Cardiol 116(2):161–166
80. Stelzle D, Shah ASV, Anand A et al (2018) High-sensitivity cardiac troponin I and risk of heart failure in patients with suspected acute coronary syndrome: a cohort study. Eur Heart J Qual Care Clin Outcomes 4(1):36–42
81. Salgado-Somoza A, Zhang L, Vausort M et al (2017) The circular RNA MICRA for risk stratification after myocardial infarction. Int J Cardiol Heart Vasc 17:33–36
82. Vausort M, Salgado-Somoza A, Zhang L et al (2016) Myocardial infarction-associated circular RNA predicting left ventricular dysfunction. J Am Coll Cardiol 68(11):1247–1248
83. Vausort M, Wagner DR, Devaux Y (2014) Long noncoding RNAs in patients with acute myocardial infarction. Circ Res 115(7):668–677
84. Kasner M, Sinning D, Lober J et al (2015) Heterogeneous responses of systolic and diastolic left ventricular function to exercise in patients with heart failure and preserved ejection fraction. ESC Heart Fail 2(3):121–132
85. Rapezzi C, Milandri A, Lorenzini M (2017) The complex interplay between systolic and diastolic function at rest and during exercise in heart failure: the case of cardiac amyloidosis. Eur J Heart Fail 19(11):1466–1467
86. Rommel KP, von Roeder M, Oberueck C et al (2018) Load-independent systolic and diastolic right ventricular function in heart failure with preserved ejection fraction as assessed by resting and handgrip exercise pressure-volume loops. Circ Heart Fail 11(2):e004121
87. Beltran-Alvarez P, Tarradas A, Chiva C et al (2014) Identification of N-terminal protein acetylation and arginine methylation of the voltage-gated sodium channel in end-stage heart failure human heart. J Mol Cell Cardiol 76:126–129
88. Szczurek W, Szygula-Jurkiewicz B, Zakliczynski M et al (2018) Prognostic utility of the N terminal prohormone of brain natriuretic peptide and the modified model for end stage liver disease in patients with end stage heart failure. Pol Arch Intern Med
89. Callender T, Rahimi K (2015) Heart failure and iron deficiency anaemia: a complex dance. Heart 101(8):579–580
90. Choi KH, Lee GY, Choi JO et al (2018) Outcomes of de novo and acute decompensated heart fail-

ure patients according to ejection fraction. Heart 104(6):525–532
91. Bayes-Genis A, Lanfear DE, de Ronde MWJ et al (2018) Prognostic value of circulating microRNAs on heart failure-related morbidity and mortality in two large diverse cohorts of general heart failure patients. Eur J Heart Fail 20(1):67–75
92. Bhatt AS, Cooper LB, Ambrosy AP et al (2018) Interaction of body mass index on the association between N-terminal-pro-b-type natriuretic peptide and morbidity and mortality in patients with acute heart failure: findings from ASCEND-HF (acute study of clinical effectiveness of Nesiritide in decompensated heart failure). J Am Heart Assoc 7(3)
93. Moser DK, Robinson S, Biddle MJ et al (2015) Health literacy predicts morbidity and mortality in rural patients with heart failure. J Card Fail 21(8):612–618
94. Ponikowski P, Voors AA, Anker SD et al (2016) 2016 ESC guidelines for the diagnosis and treatment of acute and chronic heart failure: the task force for the diagnosis and treatment of acute and chronic heart failure of the European Society of Cardiology (ESC). Developed with the special contribution of the heart failure association (HFA) of the ESC. Eur J Heart Fail 18(8):891–975
95. Talwar S, Squire IB, Downie PF et al (2000) Profile of plasma N-terminal proBNP following acute myocardial infarction; correlation with left ventricular systolic dysfunction. Eur Heart J 21(18):1514–1521
96. Lin X, Lo HC, Wong DT et al (2015) Noncoding RNAs in human saliva as potential disease biomarkers. Front Genet 6:175
97. Kumarswamy R, Bauters C, Volkmann I et al (2014) Circulating long noncoding RNA, LIPCAR, predicts survival in patients with heart failure. Circ Res 114(10):1569–1575
98. Burd CE, Jeck WR, Liu Y et al (2010) Expression of linear and novel circular forms of an INK4/ARF-associated non-coding RNA correlates with atherosclerosis risk. PLoS Genet 6(12):e1001233
99. Xuan L, Sun L, Zhang Y et al (2017) Circulating long non-coding RNAs NRON and MHRT as novel predictive biomarkers of heart failure. J Cell Mol Med 21(9):1803–1814
100. Levine GN, Bates ER, Bittl JA et al (2016) 2016 ACC/AHA Guideline focused update on duration of dual antiplatelet therapy in patients with coronary artery disease: a report of the American College of Cardiology/American Heart Association Task Force on clinical practice guidelines: an update of the 2011 ACCF/AHA/SCAI guideline for percutaneous coronary intervention, 2011 ACCF/AHA guideline for coronary artery bypass graft surgery, 2012 ACC/AHA/ACP/AATS/PCNA/SCAI/STS guideline for the diagnosis and management of patients with stable ischemic heart disease, 2013 ACCF/AHA guideline for the Management of ST-Elevation myocardial infarction, 2014 AHA/ACC guideline for the management of patients with Non-ST-elevation acute coronary syndromes, and 2014 ACC/AHA guideline on perioperative cardiovascular evaluation and management of patients undergoing noncardiac surgery. Circulation 134(10):e123–e155
101. McGrath BM, Norris CM, Hardwicke-Brown E et al (2017) Quality of life following coronary artery bypass graft surgery vs. percutaneous coronary intervention in diabetics with multivessel disease: a five-year registry study. Eur Heart J Qual Care Clin Outcomes 3(3):216–223
102. Zimmermann FM, De Bruyne B, Pijls NH, et al (2015) Rationale and design of the fractional flow reserve versus angiography for multivessel evaluation (FAME) 3 trial: a comparison of fractional flow reserve-guided percutaneous coronary intervention and coronary artery bypass graft surgery in patients with multivessel coronary artery disease. Am Heart J 170 (4):619–626 e612
103. Bazan HA, Hatfield SA, Brug A et al (2017) Carotid plaque rupture is accompanied by an increase in the ratio of serum circR-284 to miR-221 levels. Circ Cardiovasc Genet 10(4):e001720
104. Sarkisian L, Saaby L, Poulsen TS, et al (2016) Clinical characteristics and outcomes of patients with myocardial infarction, myocardial injury, and nonelevated troponins. Am J Med 129(4):446 e445–446 e421
105. Lee PH, Lee JY, Lee CW et al (2018) Comparison of a simple angiographic approach with a synergy between percutaneous coronary intervention with Taxus and cardiac surgery score-based approach for left main coronary artery stenting: a pooled analysis of serial PRECOMBAT (premier of randomized comparison of bypass surgery versus angioplasty using Sirolimus-eluting stent in patients with left main coronary artery disease) studies. Circ Cardiovasc Interv 11(1):e005374
106. Bazan HA, Hatfield SA, O'Malley CB et al (2015) Acute loss of miR-221 and miR-222 in the atherosclerotic plaque shoulder accompanies plaque rupture. Stroke 46(11):3285–3287
107. Li X, Zhao Z, Jian D et al (2017) Hsa-circRNA11783-2 in peripheral blood is correlated with coronary artery disease and type 2 diabetes mellitus. Diab Vasc Dis Res 14(6):510–515
108. Zhao Z, Li X, Gao C et al (2017) Peripheral blood circular RNA hsa_circ_0124644 can be used as a diagnostic biomarker of coronary artery disease. Sci Rep 7:39918
109. Hamamura K, Yanagida M, Ishikawa H et al (2018) Quantitative measurement of a candidate serum biomarker peptide derived from alpha2-HS-glycoprotein, and a preliminary trial of multidimensional peptide analysis in females with pregnancy-induced hypertension. Ann Clin Biochem 55(2):287–295
110. Richter MJ, Schermuly R, Seeger W et al (2016) Relevance of angiopoietin-2 and soluble P-selectin levels in patients with pulmonary arterial hyperten-

sion receiving combination therapy with oral treprostinil: a FREEDOM-C2 biomarker substudy. Pulm Circ 6(4):516–523
111. Smukowska-Gorynia A, Tomaszewska I, Malaczynska-Rajpold K et al (2017) Red blood cells distribution width as a potential prognostic biomarker in patients with pulmonary arterial hypertension and chronic thromboembolic pulmonary hypertension. Heart Lung Circ doi 27:842. https://doi.org/10.1016/j.hlc.2017.08.007
112. Wu N, Jin L, Cai J (2017) Profiling and bioinformatics analyses reveal differential circular RNA expression in hypertensive patients. Clin Exp Hypertens 39(5):454–459
113. Boeckel JN, Jae N, Heumuller AW et al (2015) Identification and characterization of hypoxia-regulated endothelial circular RNA. Circ Res 117(10):884–890
114. Ghofrani HA, Simonneau G, D'Armini AM et al (2017) Macitentan for the treatment of inoperable chronic thromboembolic pulmonary hypertension (MERIT-1): results from the multicentre, phase 2, randomised, double-blind, placebo-controlled study. Lancet Respir Med 5(10):785–794
115. Hoeper MM, Madani MM, Nakanishi N et al (2014) Chronic thromboembolic pulmonary hypertension. Lancet Respir Med 2(7):573–582
116. Miwa H, Tanabe N, Jujo T et al (2018) Long-term outcome of chronic thromboembolic pulmonary hypertension at a single Japanese pulmonary endarterectomy center. Circ J 82(5):1428–1436
117. Ghofrani HA, D'Armini AM, Grimminger F et al (2013) Riociguat for the treatment of chronic thromboembolic pulmonary hypertension. N Engl J Med 369(4):319–329
118. Sulica R, Fenton R, Cefali F (2015) Early observations on the use of Riociguat in a large, metropolitan pulmonary arterial hypertension/chronic thromboembolic pulmonary hypertension treatment center. Cardiol Ther 4(2):209–218
119. Miao R, Wang Y, Wan J et al (2017) Microarray expression profile of circular RNAs in chronic thromboembolic pulmonary hypertension. Medicine (Baltimore) 96(27):e7354
120. Kamath-Rayne BD, Saal H, Lang S et al (2013) Recurrent severe oligohydramnios and fetal pulmonary hypoplasia associated with ErbB4 mutation. Obstet Gynecol 121(2 Pt 2 Suppl 1):499–501
121. Kao DP, Stevens LM, Hinterberg MA et al (2017) Phenotype-specific Association of Single-Nucleotide Polymorphisms with heart failure and preserved ejection fraction: a genome-wide association analysis of the cardiovascular health study. J Cardiovasc Transl Res 10(3):285–294
122. Mitchell BD, McArdle PF, Shen H et al (2008) The genetic response to short-term interventions affecting cardiovascular function: rationale and design of the heredity and phenotype intervention (HAPI) heart study. Am Heart J 155(5):823–828
123. Gorlach A, Holdenrieder S (2017) Circular RNA maps paving the road to biomarker development? J Mol Med (Berl) 95(11):1137–1141

Circular RNAs as Biomarkers for Cancer

14

Lu Xia, Meiyi Song, Mengxue Sun, Fei Wang, and Changqing Yang

Abstract

As a type of novel noncoding RNAs, circular RNAs (circRNAs) have attracted great interest due to its different characteristics from linear RNAs. They are abundantly and stably present in the transcriptome of eukaryotic cells, with development stage specificity and high conservatism. Because circRNAs are not easily degraded by exonuclease RNase R, they can exist more stably in body fluids than linear RNAs. Based on these unique conditions, circRNAs have great potential value as clinical diagnostic and prognostic markers. As the research deepens, more and more evidences suggest that circRNAs may be closely associated with many diseases, especially cancer. Numerous studies have demonstrated the abnormal expression of circRNAs in cancer, and they can regulate the occurrence and progression of cancer by targeting key genes. Abundant circRNAs in tissues and cells can be released into saliva and blood. It is undeniable that circRNAs are a class of promising future biomarkers for cancer diagnosis and prognosis. Here we summarize the researches on circRNAs and cancer over the past few years. We expect this summary to be a stepping stone to further exploration of possible circRNAs as cancer biomarkers.

Keywords

Circular RNAs (circRNAs) · Cancer · Biomarkers · Diagnosis · Prognosis

1 Introduction

With the development of biotechnology and computer technology, more and more "invisible substances" are exposed in the organism, such as circular RNAs (circRNAs). Although do not encode proteins, they are indispensable in regulatory processes. As the latest research hotspot, circRNAs are a group of newly validated noncoding RNA molecules which form a covalent ring structure instead of having a cap structure at the 5′ end and poly (A) tail at the 3′ end. It was found in RNA viruses as early as the 1970s and once thought to be by-products of aberrant RNA splicing due to low expression levels [1, 2]. With the rapid development of bioinformatics and sequencing technologies, large-scale analysis of transcriptome data becomes possible, and the

These authors contributed equally to this work. Lu Xia and Meiyi Song

L. Xia · M. Song · M. Sun · F. Wang (✉)
C. Yang (✉)
Division of Gastroenterology and Hepatology, Digestive Disease Institute, Shanghai Tongji Hospital, Tongji University School of Medicine, Shanghai, China
e-mail: 1132469@tongji.edu.cn; changqingyang_tj@hotmail.com

J. Xiao (ed.), *Circular RNAs*, Advances in Experimental Medicine and Biology 1087,
https://doi.org/10.1007/978-981-13-1426-1_14

characteristics and functions of circRNAs, a sort of ancient and conservative molecules, are being gradually unveiled. It is well known that circRNAs are richly expressed in various tissues and cells throughout the human body and play vital roles in regulating physiological and pathological processes [3, 4]. CircRNAs are abundant in the transcriptome of eukaryotic cells and conservative in species. They are also stable in expression and exhibit tissue and development stage specificity [5–7]. In 2013, professor William R. Jeck's team detected more than 25,000 circRNAs in human fibroblasts, while in the same year, Memczak et al. identified 1950 human circRNAs and 1903 mice circRNAs through RNA-seq data combined with the human leukocyte database [3]. Unlike linear RNAs, circRNAs are not easily degraded by exonuclease RNase R in body fluids; therefore they have the potential application value as clinical diagnostic and prognostic markers [8].

Cancer is characterized by high mortality, and its incidence is increasing in recent years [9]. It is one of the major diseases threatening human life and social economic development and therefore brings the focus of attention [10, 11]. A survey on the incidence of global cancer in 2012 showed that there are approximately 14.1 million new cancer cases and 8.2 million cancer deaths worldwide each year. Lung cancer and breast cancer are the most common types of cancer [9, 12]. The essence of tumor formation is a kind of genetic disease, of which tissues infiltration and metastasis are the major features [13–15]. Its pathogenesis refers to multiple steps and mechanisms. New approaches for early diagnosis and treatment of cancer have been sought for a long time. Biomarkers for cancer have emerged as a class of molecules closely related to the development and progression of tumors [16]. They play key roles in many aspects such as early diagnosis, therapeutic monitoring, and prognostic evaluation [17].

It is universally known that microRNAs (miRNAs) control a large number of biological processes by direct interaction with their target mRNAs. This regulation can be achieved by inhibiting translation or by triggering the degradation of the target mRNA [18, 19]. Multiple researches have shown that miRNAs play an important regulatory role in the development of tumors [20–22]. Recently independent experiments have demonstrated that circRNAs can serve as sponges for miRNAs, which means that they can bind with miRNAs and thus suppress the function of miRNAs [3, 23]. These findings trigger the exploration of the potential regulatory effect of circRNAs on cancer. Technologies such as circRNAs chip and qRT-PCR have been widely used in the studies related to circRNAs and cancer. So far, many circRNAs have been found to be closely related to cancer and bring new dawn of the early diagnosis and treatment of cancer [24].

2 The Characteristics of CircRNAs

There are numerous types of circRNAs, which are huge in number and widely distributed in the organisms. The circRNAs currently found are mainly divided into three categories according to their origin: exonic circRNAs (ecRNAs), intronic circRNAs (ciRNAs), and exon-intron circular RNAs (EIciRNAs) composed of exons and introns. They are characterized by extremely high stability, strong evolutionary conservation, and unique temporal and spatial expression. Studies have found that the expression levels of circRNAs are ten times higher than their linear isomers [25]. The half-life of circRNAs with the peculiar loop structure is more than 48 h, which is far much longer than the linear mRNAs (10 h in average); this is why the circRNAs can be more stable in tissues, cells, and body fluids [25, 26]. However, the expression of the circRNAs is not dependent on the linear mRNA expression of its parental genes but changes with the life activities such as growth and senescence. Moreover, the type and content of circRNAs in different cells and tissues are also distinct [27, 28]. When the abundance of the circRNAs with conservative nucleotide sequences increases, they can compete with other RNA or miRNAs by competitively binding to the RNA-binding protein (RBP) [3]. These properties give circRNAs a unique advantage as diagnostic and prognostic markers for clinical diseases. CircRNAs are widely involved in the processes of physiological and pathological regulation of human beings. Large amounts of circRNAs in organisms influence basic life activities such as cell proliferation, cycle progression, cell senescence, and apoptosis by regulating gene expression.

The mechanisms of the discovered circRNAs mainly include the following three types:

1. As molecular sponges of miRNAs: this is a relatively common function of circRNAs and the most promising direction in circRNAs researches. CircRNAs themselves contain at least one miRNA binding site and therefore can serve as miRNA sponges to regulate the expression of target genes which are inhibited by miRNAs through ceRNAs [3]. The ceRNAs network in the living body is complex, and any minor changes can affect gene expression and induce tumorigenesis. Therefore, the competitive combination of circRNAs is essential to maintain the balance of the ceRNAs network.
2. Regulate gene transcription and cleavage: the mechanism of this process is diverse. It can not only enhance the expression of parental gene through miRNAs but also exert a positive feedback on its parental genes through interaction with RNA polymerase II. It is also possible to regulate parental gene expression through competitive splicing factors. CircRNAs can regulate parental genes through positive feedback and negative feedback control at different expression levels [5, 29, 30].
3. Interact with RNA-binding proteins: circRNAs can bind stably with RNA-binding proteins such as AGO (argonaute), RNA polymerase II, muscleblind protein (MBL), variable factor QKI (Quaking), and eukaryotic translation initiation factor 4A3 (EIF4A3). They can store and transport RBP to compete with RBP substrate for its binding site and thereby regulate the activity of RBP and interfere with the normal function of a protein in a direct or indirect way [3, 5, 30–32].

Since the first discovery of circRNAs in 1976, more and more mature technologies such as high-throughput sequencing and gene chip have been applied to explore their biological functions. On the other hand, the establishment and enrichment of the database also provide extremely convenient conditions for the in-depth study of the circRNAs. The current databases of circRNAs include circRNABase, circBase, Circ2Traits, and CircNet. Through these databases, we can query circRNAs sequences, genome annotations, expression profiles, miRNA-circRNA interaction networks, and related disease information. With these supports we have made tremendous progress in the study of circRNAs.

3 CircRNAs in Cancer

3.1 Lung Cancer

The changes in lifestyle, dietary pattern, and deteriorating environment have led to a rising trend in the prevalence of cancer patients. Worldwide, the incidence and mortality of lung cancer account for the third most severe human tumor. Meanwhile it is one of the most malignant tumors in the world [33]. According to the World Cancer Report 2014 released by WHO, there are 1.82 million new lung cancer cases in 2012 and 1.59 million deaths caused by lung cancer, accounting for 13% and 19.4% of the incidence and deaths of all malignant tumors [34]. The epidemiological data manifest that the morbidity and mortality rates of lung cancer are both the highest among global malignant tumors. More than 220,000 cases were diagnosed with lung cancer in 2015, and the number of patients who died of lung cancer in the same year exceeded 158,000 [35]. Moreover, about 85% of lung cancers are non-small cell lung cancer, and 25–30% of non-small cell lung cancers are squamous cell carcinomas [36]. At present, the common clinical treatment methods, including surgical resection, radiation therapy, and chemotherapy, cannot reduce the incidence of lung cancer. The 5-year survival rate of lung cancer patients is only 15.9% [37]. Therefore, innovative new treatments are urgently needed.

A variety of circRNAs with high expression in lung cancer have been found. Hsa_circ_0000064 is upregulated in lung cancer tissues and lung cancer cell lines A549 and H1229. The abnormal expression of hsa_circ_0000064 is closely related to the clinical features such as tumor lymph node metastasis and TNM staging [38]. The high expression of circRNA-100876 in non-small cell lung cancer (NSCLC) is closely related to the lymph node metastasis and tumor stage of lung cancer. Additionally, the overall survival time of NSCLC patients with high circ_100876 expression is significantly shorter [39]. Circ-HIPK3 can be detected in the NSCLC cell lines H1299, H827, H1975, H2170, H520, and H1650. Experiments have shown that circ-HIPK3 can regulate the expression level of insulin-like growth factor1(IGF1) and promote cell proliferation by binding to miR-379 in NCI-H1299 and NCI-H2170 cells [40]. Hsa_circ_0013958 is significantly upregulated in histiocytes and plasma of lung adenocarcinoma patients and shows statistically significant correlation with lymph node metastasis and tumor staging [41]. By applying high-throughput sequencing on tumor and adjacent tissue from four cases with nonsmoking early lung adenocarcinoma, Zhao et al. detected more than 300 circRNAs differentially expressed in tumor tissues and later verified 5 of them by RT-qPCR. Consistent with the results of the chip, it provided potential targets for early diagnosis and treatment of early-stage lung adenocarcinoma [42]. Cdr1as (ciRS7) contains more than 70 selective binding sites for miRNAs, which can strongly inhibit the activity of miR-7 and hence activate the miR-7-regulated genes. It has a regulatory effect on many diseases [23]. Importantly, similar to liver cancer, Cdr1as can also inhibit tumors in lung cancer by binding to miR-7 [43]. The circ-ITCH is significantly reduced and can act as miR-7 and miR-214 sponge in lung cancer. It can inhibit the activation of Wnt/β-catenin signaling pathway by enhancing the expression of the ITCH gene and thereby suppresses the proliferation of lung cancer cells [44]. A new mechanism of cinnamaldehyde intervention on NSCLC through hsa_circ_0043256/miR-1252/ITCH axis was proposed in another work. The gene expression of ITCH is found positively correlated with hsa_circ_0043256. As a miR-1252 sponge, hsa_circ_0043256 is significantly downregulated, weakening the inhibitory effect of cinnamaldehyde, when cinnamaldehyde is used to block the Wnt/β-catenin signaling pathway [45].

3.2 Gastrointestinal Cancer

Gastrointestinal cancer is one of the most common malignancies which seriously threatens the life and health of human beings [46]. The gastric and colorectal cancers are the most common in this kind of diseases. Studies have shown that the incidence and mortality of gastric cancer rank second in China, which is second only to lung cancer, while colorectal cancer ranks fifth among all malignancies [47]. The 5-year survival rate of advanced gastric cancer is less than 30% [48]. Tumor markers including AFP, CEA, CA19-9, and CA50 are commonly used to assist the clinical diagnosis and prognosis of gastrointestinal tumors. In order to compensate for the lack of tumor detection markers, attempts are made to find circular RNAs with diagnostic and prognostic values.

3.2.1 Gastric Cancer

A study of circRNAs in gastric cancer (GC) tissues and paracancerous tissues identified 467 differentially expressed circRNAs, among which expression of 214 circRNAs were significantly increased and expression of 253 were significantly decreased. Most of the circRNAs could be detected with corresponding miRNA binding sites [49]. Circ-PVT1 is a highly expressed circRNA in GC and a potential independent index for evaluating the prognosis of GC. By analyzing clinical data and tumor tissues of 187 patients with GC, it was found that the survival rate of patients with high circ-PVT1 expression was markedly higher. The promotive effect of circ-PVT1 on GC may act through attenuating the inhibitory effect of miR-125 on cell proliferation by combining with it. The experiment also demonstrated that reducing the expression of circ-PVT1 could inhibit the proliferation of GC cells [50]. In contrast, hsa_circ_0000096 was found obviously downregulated in GC cell lines and tissues compared with normal gastric epithelial cells and paired adjacent non-tumor tissues. It is supposed that hsa_circ_0000096 might interact with miRNAs through endogenous competition and thereby affect GC cell growth and migration by interfering with cell cycle and expression of migration-related protein [51]. Screened by database and verified by qRT-PCR, hsa_circ_002059 was found significantly downregulated in GC tissue compared with the adjacent non-tumor tissue by Li et al. The authors noticed that 10 days after GC tissue resection, the expression of plasma has_circ_002059 was detected to be higher than that before surgery. Low expression level of has_circ_002059 was significantly associated with distant metastases, TNM staging, gender, and age [52]. The combination of hsa_circ_0000096 and hsa_circ_002509 can considerably improve the diagnosis of gastric cancer [51]. Zhang et al. established a prediction system of early recurrence for patients with stage III GC based on four selected circRNAs: hsa_circ_101308, has_circ_104423, hsa_circ_104916, and hsa_circ_100269. The area under the curve (AUC) could reach 0.763 and 0.711 by the test of two centers and could rise to 0.866 and 0.818 by joining TNM staging, which indicated that this circRNA-based predictive model is an effective assessment of the risk of early recurrence after radical gastrectomy for GC. However, this system still needs to be further verified by prospective and multicenter research [53]. Other circRNAs with abnormal expression profiles in GC include has_circ_0001649 (the expression level is significantly downregulated in GC tissue and upregulated in serum after operation) [54], has_circ_0044516 (the expression level is significantly upregulated in GC tissue) [55], has_circ_0014717, and hsa_circ_0000190 (the expression level is significantly downregulated in GC tissues and considered related to distant metastasis, tumor staging, CA19-9) [56, 57].

3.2.2 Colorectal Cancer

Dietmar Pils's team compared circRNAs expression levels between fibrotic lung and normal lung tissue in patients with idiopathic pulmonary fibrosis and ovarian tumor cells and normal ovarian epithelial cells and confirmed that circRNA abundance was negatively correlated with cell proliferation. The expression of circRNAs in colorectal cancer (CRC) tissues were significantly lower than that in normal tissues, and the expression level were also lower in CRC cell lines. The investigators tested four circRNAs (circ_0817, circ_3204, circ_6229, circ_7374) low-expressed in clinical specimens in colon cancer cell lines together with their corresponding linear RNAs. The expression ratio of the four circRNAs to their parental linear RNAs was lower. Finally, the researchers verified the relationship between the expression of circRNAs/corresponding linear RNAs and cell proliferation, confirming that the cell proliferation rate was negatively correlated with the expression circRNAs [58]. As a circRNA generated from the exon 5–11 of BANP gene, circ-BANP is highly expressed in CRC. Knocking down of circ-BANP can significantly reduce the proliferation of CRC cells. Circ-BANP may play an important regulatory role in CRC cells and may serve as a marker for prognosis and treatment of CRC [59]. Recently, Guo et al. discovered a novel abnormal circRNA, hsa_circ_0000069, by employing unsupervised hierarchical clustering analysis. They determined that hsa_circ_0000069 is highly expressed in CRC through quantitative PCR analysis of 30 paired CRC tissues and adjacent noncancerous tissues and is closely related to the patient's age, tumor size, lymph node metastasis, and TNM staging. Functional analysis by using specifically designed siRNA in CRC cells confirms that knocking down hsa_circ_0000069 markedly inhibits cell proliferation, migration, and invasion and induces G0/G1 arrest in vitro [60]. Another study has confirmed that hsa_circ_001569 plays a positive regulatory role in cell proliferation and invasion of CRC. The authors found that the expression of has_circ_001569 is elevated in CRC tissues and correlated with tumor volume, TNM staging, and prognosis. It was further confirmed that hsa_circ_001569 can also be used as a "sponge" to adsorb miR-145. This is why the expression of has_circ_001569 and miR-145 are negatively correlated in CRC tissue. Upregulated circ_001569 increases the expression of miR-145 target genes E2F5, BAG4, and FMNL2 and hence enhances the proliferation and invasion of CRC cells, which in turn promotes the progression of colorectal cancer [61]. Circ-CCDC66 is a newly discovered circRNA which is encoded by the CCDC66 gene. Hsiao et al. found that the expression of circ-CCDC66 was increased in polyps and CRC and was negatively correlated with the prognosis. Inhibition of circ-CCDC66 expression in vitro significantly reduces the tumor volume in nude mice, suggesting that circ-CCDC66 can regulate multiple pathological processes including cell proliferation, invasion, migration, and anchorage-independent growth. This function is achieved by acting as the sponge of miR-33b and miR-93. Experiments showed that knocking down circ-CCDC66 can impede tumor proliferation and invasion in vivo. All of these findings indicate that circ-CCDC66 plays an important role in progression and metastasis of CRC [62]. Other circRNAs such as circ-ITCH can also exert the function of miRNA sponge to suppress tumor proliferation. A research based on 45 specimens of CRC found that circ-ITCH expression is abnormally lower than that of paracancerous tissue. The bioinformatics analysis predicted that circ-ITCH and its parent gene ITCH possess the same miR-214, miR-7, and miR-20a binding sites, and the firefly luciferase reporter assay in CRC cell lines HCT116 and SW480 confirmed that circ-ITCH can obstruct the Wnt signaling pathway to inhibit cancer cell proliferation by competitively absorbing miR-7 and miR-20a [63].

3.2.3 Esophageal Cancer

Researches on the regulation mechanism of circRNAs in esophageal cancer suggest that many of them can function as miRNAs sponges. The radiation resistance obtained during radiotherapy is considered to be the most important factor that affects the therapeutic effect and stimulates local tumor recurrence. Based on this, Su et al. explored the differentially expressed circRNAs in radioresistant esophageal cancer cells by using expression profiling and bioinformatics analysis, in order to reveal the possible circRNAs involved in the generation of radiation resistance during treatment of esophageal cancer. They initially identified 57 remarkable upregulated and 17 remarkable downregulated circRNAs in 3752 candidate circRNAs (fold change $\geq$ 2.0 and $P < 0.05$), of which 9 were validated by real-time qPCR. Supplemented by gene ontology analysis, the authors confirmed a large number of target genes (including most miRNAs) participating in this biological process. Among them, more than 400 target genes are enriched in the Wnt signaling pathway. Circ_001059 and circRNA_000167 are the two largest nodes of the co-expression network of circRNA/microRNA [64]. Based on the detection of circRNAs expression in 684 cases of esophageal squamous cell carcinoma (ESCC) patients and their matched paracancerous tissues, circ-ITCH was also found to play a role as molecular sponge of miR-7, miR-17, and miR-214 in esophageal cancer. Consistent with the mechanism of action in colorectal cancer, increased ITCH promotes ubiquitin-mediated degradation of Dvl2 and reduces the expression of the oncogene c-myc that inhibits the Wnt signaling pathway and ultimately suppresses the tumor growth [65]. In 51 cases of ESCC patients with different staging, the expression of hsa_circ_0067934 in the tumor tissue is obviously higher than that in paired paracancerous tissues; and the high expression level of hsa_circ_0067934 is associated with the tumor stage ($P = 0.025$).] Higher stage of the tumor tissue is companioned with higher expression of hsa_circ_0067934. Interference of hsa_circ_0067934 expression by siRNA in vitro suppresses the proliferation and migration of ESCC cells and blocks cell cycle progression. Cell component analysis and fluorescence in situ hybridization confirm that the circRNAs are mainly located in the cytoplasm [66].

3.2.4 Pancreatic Cancer

Pancreatic cancer (PC) is a common cancer, but there is still lack of reliable biological markers for early diagnosis and evaluation of prognosis. Current studies on the relationship between circRNAs and PC are insufficient. Recently, researchers have explored the expression profiles of circRNAs in four pancreatic ductal adenocarcinoma (PDAC) samples and matched adjacent normal tissues. The result revealed that a large number of circRNAs are abnormally expressed in PDAC, suggesting that they may be involved in the initiation and progression of PDAC. This discovery provides potential biological targets for the diagnosis and treatment of PDAC [67]. Li et al. analyzed the tissue samples from six patients with PDAC by microarray technology and found that compared to normal pancreatic tissues, abnormal expression of circRNAs cluster was gathered in pancreatic cancer tissues. 209 upregulated and 142 downregulated circRNAs were screened out. Subsequently, 7 circRNAs were analyzed with quantitative reverse transcriptase polymerase chain reaction (qRT-PCR) in 20 groups of PC and paracancerous tissues to confirm that the results were consistent with the microarray. GO analysis and pathway analysis suggested that some of the dysregulated circRNAs are involved in molecular biological processes in pancreatic cancer, including influence on endocytosis and meditation of abnormal expression of VEGF pathway. These results reveal the indispensable role of circRNAs in the malignant biological behavior of PC [68].

3.3 Hepatic Cancer

Hepatic cancer generally refers to malignant tumor which originates in hepatocytes or intrahepatic bile duct epithelium [47]. As one of the common malignancies of the digestive system, hepatic cancer is the fifth most common cancer with the third most common mortality in the world [69]. About 250,000 people worldwide are estimated to lose their lives every year due to hepatic cancer, among which China accounts for about 45% [70, 71]. The 5-year survival rate of advanced hepatic cancer is about 10%. Due to the lack of effective early diagnostic tools, only 30–40% of the patients can be diagnosed and take appropriate treatment in early stage [72, 73].

A series of circRNAs associated with hepatic cancer are gradually discovered. A noteworthy downregulation of hsa_circ_0001649 was found associated with inhibition of cell proliferation [74]. In contrast, the upregulated expression of hsa_circ_0005075 and Cdr1as promotes cell adhesion and proliferation, respectively [75, 76]. As the inhibitor and sponge of miR-7, Cdr1as can indirectly interrupt the PI3K/Akt/mTOR signaling pathway by targeting miR-7 [77]. It has been suggested that Cdr1as may be employed as a prognostic biomarker for hepatic carcinoma and a therapeutic target for microvascular infiltration [77]. Huang et al. detected another potential marker hsa_circ_100338 by applying a circRNA microarray and verified miR-141-3p as its direct downstream target through computer calculation and experimental analysis. High expression level of hsa_circ_100338 indicates the metastasis progression and meanwhile makes an influence on the cumulative survival rate [78]. A circRNA-miRNA-mRNA network is constructed with five upregulation circRNAs in hepatic cancer. This network proposes that high levels of circFUT8 (hsa_circ_101368, hsa_circ_0003028), circ-ZFR (hsa_circ_103809, hsa_circ_10072088), and circ-IOP11 (hsa_circ_103847, hsa_circ_0007915) expression are probably associated with the progression of hepatic cancer [79]. Researchers focusing on the genetics and epidemiology of hepatocellular carcinoma (HCC) discovered that the expression of cir-ITCH in HCC tissues is significantly lower than that in matched paracancerous tissues; and suggested that HCC patients with a relatively high cir-ITCH expression have a better prognosis. Collectively, the results revealed that cir-ITCH has an inhibitory action on HCC and may be a hopeful biomarker for the assessment of susceptibility and prognosis in patients with HCC [80]. Modulating function of hsa_circ_0015756 on hepatoblastoma cell by acting as a miR-1250-3p sponge has also been found based on circRNAs microarray analysis. Silencing hsa_circ_0015756 reduces viability, proliferation, and invasiveness of hepatoblastoma cells [81]. Fu et al. conducted a series of experiments to prove the point that the expression level of hsa_circ_0005986 is related to the tumor size, BCLC staging, and microvascular infiltration. The underlying mechanism of hsa_circ_0005986-regulated carcinogenesis of HCC is by modulating Notch1 expression through interaction with miR-129-5p [82]. They further showed the low expression of hsa_circ_0004018 is associated with AFP level in serum, diameter, and differentiation of tumor and Barcelona clinical stage in HCC. Remarkably, hsa_circ_0004018 has the specific expression characteristics indicative of different HCC stage in various chronic liver diseases [83]. Based on all these findings, there are reasons to believe that circRNAs combined with traditional biomarkers can make a more accurate diagnostic method for hepatic cancer.

3.4 Gynecologic Cancer

Gynecologic cancer is closely related to women mainly including breast cancer, endometrial cancer, cervical cancer, and ovarian cancer. The incidence of gynecological tumors has increased year by year, which causes serious harm to women's physical and mental health. Breast cancer has the highest incidence of women malignancy tumors in the world, which is also the biggest factor leading to the death of female cancer patients [84]. Breast cancer accounts for 23% of the global new female cancer cases and accounts for 14% of the total number of global cancer deaths [84, 85]. Ovarian cancer represents the highest mortality rate among gynecologic cancer with no less than 204,000 new cases and 125,000 deaths cases per year [86]. Only a small proportion of patients can be diagnosed in the early stage due to its concealed characteristics and lack of effective early diagnosis. In fact, more than 70% of the patients are diagnosed in advanced stage; and their 5-year survival rate is lower than 30% [87]. Endometrial cancer originates from the endometrial epithelium. Its mortality is second only to ovarian cancer. It accounts for 7% of all malignant tumors in women, and the incidence of malignant tumors in female reproductive system is as high as 20–30% [88]. Although early-stage endometrial cancer has a good prognosis, the median survival time for advanced or recurrent cases does not exceed 1 year [86]. Clinical diagnosis of gynecologic malignancies is mainly by physical signs, tumor markers, B-ultrasonography, computed tomography (CT), and magnetic resonance imaging (MRI). Common gynecological tumor markers include cancer markers such as cancer antigen 125 and 19-9 (CA-125 and CA19-9), but their specificity is doubtful. Researches have been trying to find high-specificity diagnostic markers among circRNAs, which is of great significance for timely treatment of gynecological cancer, reduction of metastasis, and improvement of prognosis.

3.4.1 Breast Cancer

As previously mentioned, Cdr1as can indirectly regulate the expression of miR-7 target gene, while Cdr1as/miR-7 can affect tumorigenesis and development of tumor through multiple pathways. Early studies have showed that the expression level of endogenous miR-7 is negatively correlated with Pak1 and is positively correlated with homeodomain transcription factor HOXD10. In the transformation process of breast cancer from high invasion phenotype to low invasion phenotype, Pak1 protein levels gradually increase, while miR-7 and HOXD10 gradually decrease. In highly invasive breast cancer cells, miR-7 can inhibit their proliferation activity, invasiveness, and tumorigenic potential. These indicate that the miR-7/Pak1 pathway may play an important role in the development of breast cancer. Therefore, Cdr1as is considered also involved in the regulation of breast cancer [89]. A total of 1155 differentially expressed circRNAs were screened out in 51 breast cancer patients by using the whole genome transcript profile technique, of which 715 were upregulated and 440 were downregulated. The expression levels of hsa_circ_103110, hsa_circ_104689, and hsa_circ_104821 were elevated in breast cancer tissues, while the expression levels of hsa_circ_006054, hsa_circ_100219, and hsa_circ_406697 were downregulated among the selected circRNAs. Further investigation of the circRNAs targeting complementary miRNAs response elements revealed that progesterone receptor (PR)-negative was related to the upregulation of hsa_circ_104689 and hsa_circ_104821 and the downregulation of hsa_circ_406697. The diagnostic accuracy of hsa_circ_100219 was the highest with the AUC of 0.78 (95% CI: 0.69–0.88). Combined hsa_circ_006054, hsa_circ_100219, and hsa_circ_406697 had a higher judgment value for judging breast cancer (AUC: 0.82, 95% CI: 0.73–0.90) [90]. With the purpose to investigate the expression profile and possible regulatory mechanisms of oncogenic circRNAs in breast cancer, Liang et al. used circRNA microarray to screen abnormally expressed circRNAs in breast cancer tissues and found that circ-ABCB10 was highly expressed in breast cancer tissues. The authors then verify the result of the chip by using a large amount of samples. The loss-of-function experiments in vitro demonstrated that subtraction of circ-ABCB10 level in breast cancer cells can inhibit cell proliferation and promote apoptosis. The bioinformatics technique was used to predict the existence of complementary sequences in circ-ABCB10 and miR-1271, which was then verified by luciferase reporter assays. Finally, it was confirmed in breast cancer cells that the inhibition of miR-1271 can restore function of circ-ABCB10, demonstrating the spongy effect of circ-ABCB10 on miR-1271 [91]. In a study investigating whether hypoxia regulates proliferation through circRNAs, a hypoxic model in breast cancer cells was established. The increased expression of circ-DENND4C was detected under hypoxic conditions, whereas knockdown of HIF1α could reduce the expression of circ-DENND4C. This confirmed the correlation between circ-DENND4C and HIF1α in the hypoxic model. It was further found that knockdown circ-DENND4C can inhibit the abnormal proliferation of breast cancer cells in anoxic environment, indicating that circ-DENND4C has the function of promoting breast cancer cell proliferation under hypoxic conditions. Finally, the expression level of circ-DENND4C was found related to the tumor volume, and the larger tumors contained more circ-DENND4C [92]. As a noteworthy downregulated circRNA in breast cancer cells, when the expression of circ_000911 is enhanced, the proliferation, migration, and invasion ability of breast cancer cells are all inhibited, and meanwhile the cell apoptosis accelerates. The miR-449a was identified as a related miRNA to circ_000911 by using a biotin-labeled probe method. Overexpression of cir_000911 in breast cancer could increase the expression of Notch1, which is a functional target of miR-449a. Signal transduction reporter array and western blot analysis confirmed that NF-κB signaling transduction is a functional target of the circ_000911/miR-449a pathway [93].

3.4.2 Reproductive System Tumors

By RNA sequencing of primary ovarian cancer, peritoneal metastases, and lymph node metastases in three patients with ovarian cancer, circRNAs with significant differences in expression in epithelial ovarian cancer were found, including many new genes such as HIPK2/3 and ZKSCAN1. The number of differentially expressed circRNAs is much higher than that of the corresponding linear mRNA in metastatic lesions. In addition, various cancer-associated signaling pathways including NF-κB, PI3K/Akt, and TGF-β have the opposite expression trends in circRNAs and linear mRNAs. Consensus of circRNAs expression provides new candidates for cancer treatment and prognosis [94]. Endometrial cancer (EC) and cervical cancer also belong to the female reproductive system malignancy. Lately, researchers used RNA sequencing technology to identify EC-specific circular transcriptomes. By comparison, the overall abundance of circRNAs in EC (14,707) was lower than that in normal endometrium (21,340). On this basis the researchers identified 120 differentially expressed circRNAs between EC tissues and normal endometrial tissues by collecting and analyzing samples from 6 EC patients, in which unique hotspot genes, such as cancer-specific ESR1 circular isoforms, were regarded with the value of EC diagnosis and progress detection. The circular isoform of DNAH14 may be involved in the regulation of tumor-associated pathways. The DMD and DMBT1 genes undergone significant changes during generation of the circular transcript, suggesting that they may be involved in the pathological changes of EC [95]. The work of Abdelmohsen's team demonstrated an example of a functional model of a protein regulation through circRNAs endogenous binding in cervical cancer. They used the RIP assay to identify circRNAs interacting with HuR in Hela cells; and the most obviously changed candidate was circPABPN1(hsa_circ_0031288). Excess circPABPN1 can prevent the binding of HuR to linear PABPN1 mRNA and hence inhibits the translation of HuR [96].

3.5 Other Cancer

Some circRNAs also have the potential of biological markers in other tumors. For instance, the expression level of has_circ_104912 is significantly downregulated in laryngeal squamous cell carcinoma, while the expression level of has_circ_100855 is significantly increased [97]. circ_100290 can function in oral squamous cell carcinoma as a molecular sponge of miR-29 family [98]; high expression of circ-TTBK2 in glioma tissue may promote the development of glioma [99].

Acute pregranulocyte leukemia chromosomal translocation t(15;17)(q24;q21) leads to the formation of a key oncogenic fusion protein PML-RARα, and the circRNA, f-circRNA, forms after the translocation of this chromosome can also be carcinogenic [100]. Circ-TRIM24 (hsa_circ_0082582) and circ-FAM169A (hsa_circ_0007158) are significantly downregulated in bladder cancer tissues, while circ-BC048201 (hsa_circ_0061265), circ-PTK2 (hsa_circ_0005273), circ-ZFR (hsa_circ_0072088), and circ-TCF25 (hsa_circ_0041103) are significantly upregulated, and the circTCF25-miR-103a-3p/miR-107-CDK6 pathway is suggested as an important regulatory axis in bladder cancer [101]. 23 high- and 48 low-expression circRNAs and their corresponding binding sites for 354 bindable miRNA sequences are identified in basal cell carcinoma [102]. The phenomenon of abnormal expression of these circRNAs in various tumor tissues indicates that circRNAs have an inseparable relationship with the occurrence and development of tumors.

4 Conclusion and Prospect

CircRNAs is a type of closed-circular RNA molecule widely distributed in the transcriptome; they participate in the regulation of numerous biological activities in organisms. CircRNAs are not easily degraded by nucleases and are more stable than linear RNAs, which provide a foundation for them to be novel biomarkers for tumor diagnosis. The regulatory expression mechanisms of circRNAs are diverse. They can serve as “miRNA sponges” to perform posttranscriptional regulation by competitively binding to miRNAs, interact with snRNP or RNA polymerase II in the nucleus to regulate transcription, or bind to transcription factors and competitively regulate classic RNA splicing. CircRNAs accumulate in cells and release into exosomes and plasma. The amount of circRNAs released into exosomes from tumor tissue is three times more than that in tumor tissues [24].

CircRNAs play an irreplaceable regulatory role in the complex life process. Their abnormal expression can induce or impede the occurrence and development of cancer. They are promising biomarkers and even therapeutic targets for cancer. With the continuous development of high-throughput sequencing and bioinformatics technologies, the formation and function of circRNAs and their relationship with cancer have gradually attracted widespread attention in the scientific community and become a hotspot for researches of clinical disease after miRNAs and long noncoding RNAs (LncRNAs). For the moment, Circbase, Circ2Traits, CircNet, and other databases have included information of more than 100,000 circRNAs, which can assist us to predict the regulatory relationship of circRNA-miRNA-mRNA. This tool is extremely convenient for us to study circRNA systematically. At the same time, methods for constructing or interfering with circRNAs have emerged and matured, making it possible to artificially regulate the expression of intracellular circRNAs and being helpful to further explore the role of circRNAs in tumor cells.

By summarizing the researches on the relationship between circRNAs and cancer in recent years, we find that many circRNAs have abnormal expression in the tumor tissue/blood/exosome of tumor patients, among which "star molecules" such as circ-ITCH and Cdr1as have differential expression and play a regulatory role in different types of cancer. Cdr1as has the potential for being a biomarker for gastric cancer, hepatic cancer, colorectal cancer, cervical cancer, and so on. Hsa_circ_0000064 and hsa_circ_0013958 are related to tumor lymph node metastasis and TNM staging in lung cancer. Circ-100876 is associated with the prognosis of lung cancer patients. The reduction of has_circ_002059 in gastric cancer has a predictive role in distant metastasis and TNM staging. The combination of hsa_circ_0000096 and hsa_circ_002509 exhibit a high diagnostic value for gastric cancer. Circ-BANP is a potential diagnostic marker for colorectal cancer, and the elevation of has_circ_001569 in colorectal cancer is proven helpful to the diagnosis and staging of disease. The rise of hsa_circ_0067934 in esophageal cancer suggests the progression of tumor staging. The same phenomenon is found in other cancers. In certain kind of cancer, there are many circRNAs with different expression changes; on the other hand, the same circRNA can correspond to different types of cancer. The summary of cancer-related circRNAs that have been uncovered is shown in Table 14.1.

Table 14.1 Cancer-associated circRNAs with its characteristics and related genes

Cancer type	The name of circRNAs	Change in cancer	Features, related molecules, and pathways
Lung cancer	hsa_circ_0000064	Up	Promote cancer cell proliferation and invasion
	circRNA-100876	Up	High expression level suggests shorter survival
	circ-HIPK3	Up	miR-379
	hsa_circ_0013958	Up	miR-314
	Cdr1as (ciRS-7)	/	miR-7
	circ-ITCH	Up	miR-7 and miR-124
	hsa_circ_0043256	Down	miR-1252
	ciR-Sry [103]	/	miR-138
	circ-ZEB1.5	Down	miR-200a-3p [104]
	circ-ZEB1.19	Down	
	circ-ZEB1.17	Down	
	circ-ZEB1.33	Down	
Gastric cancer	circPVT1	Up	Expression level positively correlated with survival rate; miRNA-125
	hsa_circ_0000096	Down	Promote cancer cell proliferation, cycle, and migration miR-224 and miR-200a
	has_circ_002059	Down	Potential for predictive treatment and associated with distant metastases, TNM staging, gender, and age
	has_circ_0001649	Down	Potential for predictive treatment and correlated with pathological differentiation
	has_circ_0044516	Up	191 miRNAs, COL1A1
	has_circ_0014717	Down	Related to distant metastasis, tumor staging, CA19-9
	hsa_circ_0000190	Down	Related to distant metastasis, tumor staging, CA19-9
	hsa_circ_0076305	Down	PGC
	hsa_circ_0037362	Down	C16orf73
	hsa_circ_0035431	Down	CGNL1
	hsa_circ_0000140 [105]	Down	KIAA0907

(continued)

Table 14.1 (continued)

Cancer type	The name of circRNAs	Change in cancer	Features, related molecules, and pathways
Colorectal cancer	circ-BANP	Up	BANP
	hsa_circ_0000069	Up	Related to age, tumor size, lymph node metastasis and TNM staging
	hsa_circ_001569	Up	Have a positive regulatory role in cancer cell proliferation and invasion; miR-145
	circCCDC66	Up	miR-33b, miR-93, CCDC66
	circ-ITCH	Down	Overexpression can inhibit cancer cell proliferation; miR-7, miR-20a,ITCH
	hsa_circ_001988 [106]	Down	Associated with colon cancer cell differentiation and neurotrophic invasion
	hsa_circ_0001946 [43]	/	CDR1
	hsa_circ_0001141 [107]	Down	ITCH
	hsa_circ_0006229	Down	TNS3
Esophageal cancer	circRNA_001059	Up	LIN52
	circRNA_100385	Up	PRRX1
	circRNA_104983	Up	NHS
	circRNA_101877	Down	RFWD3
	circRNA_102913	Down	ATIC
	circRNA_000167	Down	RPPH1
	circRNA_000695	Down	EEFSEC
	cir-ITCH	Down	miR-7, miR-17, miR214; Wnt/β-catenin signaling pathways
	hsa_circ_0067934	Up	High expression level is associated with the tumor stage
Pancreatic cancer	Ci-sirt7 [108]	/	sirt7
	hsa_circ_0001946 [68]	Up	hsa_circ_0005785-miR-181a/miR-181b-VEGF
Hepatic cancer	hsa_circ_0001649	Down	Suppress cancer cell proliferation
	hsa_circ_0005075	Up	Promote cancer cell adhesion
	Cdr1as	Up	miR-7; PI3K/Akt/mTOR
	hsa_circ_100338	Up	Associated with cumulative survival rate; miR-141-3p
	circFUT8 (hsa_circRNA_101368, hsa_circ_0003028)	Up	Associated with the progression of hepatic cancer
	circZFR (hsa_circRNA_103809,hsa_circ_10072088)	Up	Associated with the progression of hepatic cancer
	circIOP11 (hsa_circRNA_103847, hsa_circ_0007915)	Up	Associated with the progression of hepatic cancer
	cir-ITCH	Down	High expression level is associated with good prognosis
	hsa_circ_0015756	Up	miR-1250-3p
	hsa_circ_0005986	Down	Expression level is related to the tumor size, BCLC staging, and microvascular infiltration miR-129-5p, Notch1
	hsa_circ_0004018	Down	Has the specific prompting for different HCC stage

(continued)

Table 14.1 (continued)

Cancer type	The name of circRNAs	Change in cancer	Features, related molecules, and pathways
Breast cancer	hsa_circ_103110	Up	hsa_miR_339_5p
	hsa_circ_104689	Up	Relates to progesterone receptor negative
	hsa_circ_104821	Up	Relates to progesterone receptor negative
	hsa_circ_100219	Down	The AUC for diagnostic accuracy is 0.78 (95% CI: 0.69-0.88).
	hsa_circ_406697	Down	Relates to progesterone receptor negative
	circ-ABCB10	Up	miR-1271
	circDENND4C	Up	Promotes cancer cell proliferation under hypoxic conditions; HIF1α
	circRNA-000911	Down	miR-449a, Notch1
Ovarian cancer	ciR-Sry [109]	/	miR-138
Endometrial cancer	hsa_circ_0031288(circPABPN1)	Up	PABPN1
Laryngeal squamous cell carcinoma	has_circ_104912	Down	/
	has_circ_100855	Up	/
Oral squamous cell carcinoma	circRNA_100290	Up	miR-29 family, CDK6
	ci-mcm5 [110]	/	mcm5
Skin squamous cell carcinoma	Hsa_circ_103736 [111]	Up	miR-876-5p,miR-192-3p,miR-34b-3p,miR-34c-3p,miR-181b-3p
	Hsa_circ_103737 [111]	Up	miR-877-3p,miR-876-5p,miR-181b-2-3p,miR-181b-3p,miR-627-3p
	Hsa_circ_101555 [111]	Up	miR-644a,miR-485-5p,miR-889-5p,miR-329-5p,miR-148a-5p
Glioma	circ-TTBK2	Up	Promotes the development of glioma
Acute pregranulocyte l eukemia	f-circRNA	/	Forms after the translocation of chromosome, has carcinogenic effects
Bladder cancer	circ-TRIM24 (hsa_circ_0082582)	Down	/
	circFAM169A (hsa_circ_0007158)	Down	/
	circBC048201 (hsa_circ_0061265)	Up	/
	circPTK2 (hsa_circ_0005273)	Up	/
	circZFR (hsa_circ_0072088)	Up	/
	circTCF25 (hsa_circ_0041103)	Up	circTCF25-miR-103a-3p/miR-107-CDK6 and PI3K-Akt pathway
Cholangio carcinoma	ciR-Sry [112]	/	miR-138, RhoC
Basal cell carcinoma	Hsa_circ_0008732 [102]	Up	OncomiR-1, miR-19-92 family
Osteosarcoma	Has_circ_0016347 [113]	Up	miR-214
Clear cell carcinoma of kidney	circHIAT1 [114]	up	miR-195-5p,miR-29a-3p,miR-29c-3p

With the continuous expansion of existing researches, the relationship between circRNAs and cancer networks have become increasingly clear. However, our knowledge on circRNAs is only the tip of the iceberg. There is still a long way to go to uncover the mysterious veil of circRNAs function. We have reason to believe that with the discovery of more and more circRNAs, the diagnosis, progression, and prognosis evaluation system of various cancer based on circRNAs will be mature, which is of great significance for cancer treatment.

Acknowledgments This work was supported by the grants from National Natural Science Foundation of China (81670571 and 81370559 to C. Yang; 81400635 to F. Wang), Joint Projects in Major Diseases funding from Shanghai Municipal Commission of Health and Family Planning (2014ZYJB0201 to C. Yang), Joint Projects for Novel Frontier Technology in Shanghai Municipal Hospital from Shanghai Municipal Commission of Health and Family Planning (SHDC12014122 to C. Yang), Shanghai Medical Guide Project from Shanghai Science and Technology Committee (14411971500 to F. Wang), grants from Chinese Foundation for Hepatitis Prevention and Control (TQGB20140141 to F. Wang), and funds from Shanghai Innovation Program (12431901002 to C. Yang).

Competing Financial Interests The authors declare no competing financial interests.

References

1. Sanger HL, Klotz G, Riesner D et al (1976) Viroids are single-stranded covalently closed circular RNA molecules existing as highly base-paired rod-like structures. Proc Natl Acad Sci USA 73(11):3852–3856
2. Sorber K, Dimon MT, Derisi JL (2011) RNA-Seq analysis of splicing in Plasmodium falciparum uncovers new splice junctions, alternative splicing and splicing of antisense transcripts. Nucleic Acids Res 39(9):3820–3835
3. Memczak S, Jens M, Elefsinioti A et al (2013) Circular RNAs are a large class of animal RNAs with regulatory potency. Nature 495(7441):333
4. Hsiao KY, Sun HS, Tsai SJ (2017) Circular RNA – New member of noncoding RNA with novel functions. Exp Biol Med 242(11):1136
5. Zhang Y, Zhang XO, Chen T et al (2013) Circular intronic long noncoding RNAs. Mol Cell 51(6):792–806
6. Li Z, Huang C, Bao C et al (2015) Exon-intron circular RNAs regulate transcription in the nucleus. Nat Struct Mol Biol 22(3):256
7. Rybakwolf A, Stottmeister C, Glažar P et al (2015) Circular RNAs in the mammalian brain are highly abundant, conserved, and dynamically expressed. Mol Cell 58(5):870
8. Guo JU, Agarwal V, Guo H et al (2014) Expanded identification and characterization of mammalian circular RNAs. Genome Biol 15(7):409
9. Ferlay J, Soerjomataram I, Dikshit R et al (2015) Cancer incidence and mortality worldwide: sources, methods and major patterns in GLOBOCAN 2012. Int J Cancer 136(5):E359–E386
10. Dusek L, Muzik J, Maluskova D et al (2014) Epidemiology of screening-targeted cancers according to new data of the Czech National Cancer Registry. Klin Onkol 27(Suppl 2):19–39
11. Rosenberg AR, Kroon L, Chen L et al (2015) Insurance status and risk of cancer mortality among adolescents and young adults. Cancer 121(8):1279
12. Torre LA, Bray F, Siegel RL et al (2015) Global cancer statistics, 2012. CA Cancer J Clin 65(2):87–108
13. Kirwan A, Utratna M, O'Dwyer ME et al (2015) Glycosylation-based serum biomarkers for cancer diagnostics and prognostics. Biomed Res Int 2015:490531
14. Huang G, Li S, Yang N et al (2017) Recent progress in circular RNAs in human cancers. Cancer Lett 404:8
15. Xin Z, Ma Q, Ren S et al (2017) The understanding of circular RNAs as special triggers in carcinogenesis. Brief Funct Genomics 16(2):80
16. Boutros PC (2015) The path to routine use of genomic biomarkers in the cancer clinic. Genome Res 25(10):1508–1513
17. Kalia M (2015) Biomarkers for personalized oncology: recent advances and future challenges. Metabolism 64(3 Suppl 1):S16–S21
18. Hutvagner G, Simard MJ (2008) Argonaute proteins: key players in RNA silencing. Nat Rev Mol Cell Biol 9(9):22–32
19. Ebert MS, Neilson JR, Sharp PA (2007) MicroRNA sponges: competitive inhibitors of small RNAs in mammalian cells. Nat Methods 4(9):721
20. Rupaimoole R, Calin GA, Lopez-Berestein G et al (2016) miRNA deregulation in cancer cells and the tumor microenvironment. Cancer Discov 6(3):235–246
21. Leva GD, Garofalo M, Croce CM (2014) microRNAs in cancer. Annu Rev Pathol 9(1):287
22. Rupaimoole R, Slack FJ (2017) MicroRNA therapeutics: towards a new era for the management of cancer and other diseases. Nat Rev Drug Discov 16(3):203
23. Hansen TB, Jensen TI, Clausen BH et al (2013) Natural RNA circles function as efficient microRNA sponges. Nature 495(7441):384–388
24. Li Y, Zheng Q, Bao C et al (2015) Circular RNA is enriched and stable in exosomes: a promising biomarker for cancer diagnosis. Cell Res 25(8):981
25. Jeck WR, Sorrentino JA, Wang K et al (2013) Circular RNAs are abundant, conserved, and associated with ALU repeats. Rna-a Publication of the Rna. Society 19(2):141–157
26. Schwanhäusser B, Busse D, Li N et al (2013) Corrigendum: global quantification of mammalian gene expression control. Nature 495(7439):126–127
27. Szabo L, Morey R, Palpant NJ et al (2015) Statistically based splicing detection reveals neural enrichment and tissue-specific induction of circular RNA during human fetal development. Genome Biol 16(1):1–26
28. Rybak-Wolf A, Stottmeister C, Glažar P et al (2015) Circular RNAs in the mammalian brain are highly

abundant, conserved, and dynamically expressed. Mol Cell 58(5):870
29. Li Z, Huang C, Bao C et al (2015) Exon-intron circular RNAs regulate transcription in the nucleus. Nat Struct Mol Biol 22(3):256–264
30. Ashwal-Fluss R, Meyer M, Pamudurti NR et al (2014) circRNA biogenesis competes with Pre-mRNA Splicing. Mol Cell 56(1):55–66
31. Conn S, Pillman K, Toubia J et al (2015) The RNA binding protein quaking regulates formation of circRNAs. Cell 160(6):1125–1134
32. Dudekula DB, Panda AC, Grammatikakis I et al (2016) CircInteractome: a web tool for exploring circular RNAs and their interacting proteins and microRNAs. RNA Biol 13(1):34–42
33. Howlader N, Noone AM, Krapcho M, et al (2012) Previous version: SEER cancer statistics review,1975–2009(Vintage 2009 Populations). National Cancer Institute. http://seer.cancer.gov/csr/1975_2009_pops09/
34. Ferlay J, Soerjomataram I, Dikshit R et al (2015) Cancer incidence and mortality worldwide: sources, methods and major patterns in GLOBOCAN 2012. Int J Cancer 136(5):E359
35. Institute NC (2015) Cancer of the pancreas – SEER stat fact sheets. http://seer.cancer.gov/statfacts/html
36. Maemondo M, Inoue A, Kobayashi K et al (2010) Gefitinib or chemotherapy for non-small-cell lung cancer with mutated EGFR. N Engl J Med 362(25):2380
37. Wang BY, Huang JY, Cheng CY et al (2013) Lung cancer and prognosis in taiwan: a population-based cancer registry. J Thorac Oncol 8(9):1128–1135
38. Luo YH, Zhu XZ, Huang KW et al (2017) Emerging roles of circular RNA hsa_circ_0000064 in the proliferation and metastasis of lung cancer. Biomed Pharmacother 96:892–898
39. Yao JT, Zhao SH, Liu QP et al (2017) Over-expression of CircRNA_100876 in non-small cell lung cancer and its prognostic value. Pathol Res Pract 213(5):453–456
40. Fang T, Yun W, Zhe X et al (2017) Circular RNA CircHIPK3 promotes NCI-H1299 and NCI-H2170 cell proliferation through miR-379 and its target IGF1. Chin J Lung Cancer 20(7):459
41. Zhu X, Wang X, Wei S et al (2017) hsa_circ_0013958: a circular RNA and potential novel biomarker for lung adenocarcinoma. FEBS J 284(14):2170
42. Zhao J, Li L, Wang Q et al (2017) CircRNA expression profile in early-stage lung adenocarcinoma patients. Cell Physiol Biochem 44(6):2138–2146
43. Hansen TB, Kjems J, Damgaard CK (2013) Circular RNA and miR-7 in cancer. Cancer Res 73(18):5609
44. Wan L, Zhang L, Fan K et al (2016) Circular RNA-ITCH suppresses lung cancer proliferation via inhibiting the Wnt/β-Catenin pathway. Biomed Res Int 2016(1):1579490
45. Tian F, Yu CT, Ye WD et al (2017) Cinnamaldehyde induces cell apoptosis mediated by a novel circular RNA hsa_circ_0043256 in non-small cell lung cancer. Biochem Biophys Res Commun 493(3):1260
46. Aoyama T, Nishikawa K, Fujitani K et al (2016) Early results of a randomized two-by-two factorial phase II trial comparing neoadjuvant chemotherapy with two and four courses of cisplatin/S-1 and docetaxel/cisplatin/S-1 as neoadjuvant chemotherapy for locally advanced gastric cancer. Ann Oncol 28(8):1876–1881
47. Chen W, Zheng R, Baade PD et al (2016) Cancer statistics in China, 2015. CA Cancer J Clin 66(2):115
48. Acebron SP, Niehrs C (2016) β-Catenin-independent roles of Wnt/LRP6 signaling. Trends Cell Biol 26(12):956–967
49. Sui W, Shi Z, Xue W et al (2017) Circular RNA and gene expression profiles in gastric cancer based on microarray chip technology. Oncol Rep 37(3):1804–1814
50. Chen J (2017) 250P Circular RNA profile identifies circPVT1 as a proliferative factor and prognostic marker in gastric cancer. Cancer Lett 388:208–219
51. Li P, Chen H, Chen S et al (2017) Circular RNA 0000096 affects cell growth and migration in gastric cancer. Br J Cancer 116(5):626–633
52. Peifei L, Shengcan C, Huilin C et al (2015) Using circular RNA as a novel type of biomarker in the screening of gastric cancer. Clin Chim Acta 444:132
53. Zhang Y, Li J, Yu J et al (2017) Circular RNAs signature predicts the early recurrence of stage III gastric cancer after radical surgery. Oncotarget 8(14):22936–22943
54. Li WH, Song YC, Zhang H et al (2017) Decreased expression of Hsa_circ_00001649 in gastric cancer and its clinical significance. Dis Markers 2017:4587698
55. Huang YS, Jie N, Zou KJ et al (2017) Expression profile of circular RNAs in human gastric cancer tissues. Mol Med Rep 16(3):2469–2476
56. Shao Y, Li J, Lu R et al (2017) Global circular RNA expression profile of human gastric cancer and its clinical significance. Cancer Med 6(6):1173–1180
57. Chen S, Li T, Zhao Q et al (2017) Using circular RNA hsa_circ_0000190 as a new biomarker in the diagnosis of gastric cancer. Clin Chim Acta 466:167–171
58. Bachmayrheyda A, Reiner AT, Auer K et al (2015) Correlation of circular RNA abundance with proliferation – exemplified with colorectal and ovarian cancer, idiopathic lung fibrosis, and normal human tissues. Sci Rep 5(8057):8057
59. Zhu M, Xu Y, Chen Y et al (2017) Circular BANP, an upregulated circular RNA that modulates cell proliferation in colorectal cancer. Biomed Pharmacother 88:138
60. Guo J, Jin L, Zhu C et al (2016) Comprehensive profile of differentially expressed circular RNAs reveals that hsa_circ_0000069 is upregulated and promotes cell proliferation, migration, and invasion in colorectal cancer. Onco Targets Ther 9:7451–7458

61. Xie H, Ren X, Xin S et al (2016) Emerging roles of circRNA_001569 targeting miR-145 in the proliferation and invasion of colorectal cancer. Oncotarget 7(18):26680
62. Hsiao KY, Lin YC, Gupta SK et al (2017) Noncoding effects of circular RNA CCDC66 promote colon cancer growth and metastasis. Cancer Res 77(9):2339
63. Huang G, Zhu H, Shi Y et al (2015) cir-ITCHPlays an inhibitory role in colorectal cancer by regulating the Wnt/β-catenin pathway. PLoS One 10(6):e0131225
64. Su H, Lin F, Xia D et al (2016) Profiling and bioinformatics analyses reveal differential circular RNA expression in radioresistant esophageal cancer cells. J Transl Med 14(1):225
65. Li F, Zhang L, Li W et al (2015) Circular RNA ITCH has inhibitory effect on ESCC by suppressing the Wnt/β-catenin pathway. Oncotarget 6(8):6001
66. Xia W, Qiu M, Chen R et al (2016) Circular RNA has_circ_0067934 is upregulated in esophageal squamous cell carcinoma and promoted proliferation. Sci Rep 6:35576
67. Qu S, Song W, Yang X et al (2015) Microarray expression profile of circular RNAs in human pancreatic ductal adenocarcinoma. Genomics Data 5(C):385
68. Li H, Hao X, Wang H et al (2016) Circular RNA expression profile of pancreatic ductal adenocarcinoma revealed by microarray. Cell Physiol Biochem 40(6):1334
69. Wang FS, Fan JG, Zhang Z et al (2014) The global burden of liver disease: the major impact of China. Hepatology 60(6):2099
70. Nguyen VTT, Law MG, Dore GJ (2009) Hepatitis B-related hepatocellular carcinoma: epidemiological characteristics and disease burden. J Viral Hepat 16(7):453–463
71. Zhang JY (2007) Mini-array of multiple tumor-associated antigens to enhance autoantibody detection for immunodiagnosis of hepatocellular carcinoma. Autoimmun Rev 6(3):143
72. Samuel M, Chow PK, Chan SYE et al (2009) Neoadjuvant and adjuvant therapy for surgical resection of hepatocellular carcinoma. Cochrane Database Syst Rev 1(1):CD001199
73. Crissien AM, Frenette C (2014) Current management of hepatocellular carcinoma. Gastroenterol Hepatol 10(3):153
74. Qin M, Liu G, Huo X et al (2016) Hsa_circ_0001649: a circular RNA and potential novel biomarker for hepatocellular carcinoma. Cancer Biomark 16(1):161–169
75. Shang X, Li G, Liu H et al (2016) Comprehensive circular RNA profiling reveals that hsa_circ_0005075, a new circular RNA biomarker, is involved in hepatocellular crcinoma development. Medicine 95(22):e3811
76. Lei Y, Gong X, Lei S et al (2016) The circular RNA Cdr1as Act as an oncogene in hepatocellular carcinoma through targeting miR-7 expression. PLoS One 11(7):e0158347
77. Xu L, Zhang M, Zheng X et al (2017) The circular RNA ciRS-7 (Cdr1as) acts as a risk factor of hepatic microvascular invasion in hepatocellular carcinoma. J Cancer Res Clin Oncol 143(1):1–11
78. Huang XY, Huang ZL, Xu YH et al (2017) Comprehensive circular RNA profiling reveals the regulatory role of the circRNA-100338/miR-141-3p pathway in hepatitis B-related hepatocellular carcinoma. Sci Rep 7(1):5428
79. Ren S, Xin Z, Xu Y et al (2017) Construction and analysis of circular RNA molecular regulatory networks in liver cancer. Cell Cycle 16(22):2204–2211
80. Guo W, Zhang J, Zhang D et al (2017) Polymorphisms and expression pattern of circular RNA circ-ITCH contributes to the carcinogenesis of hepatocellular carcinoma. Oncotarget 8(29):48169–48177
81. Liu BH, Zhang BB, Liu XQ et al (2018) Expression profiling identifies circular RNA signature in hepatoblastoma. Cell Physiol Biochem 45(2):706
82. Fu L, Chen Q, Yao T et al (2017) Hsa_circ_0005986 inhibits carcinogenesis by acting as a miR-129-5p sponge and is used as a novel biomarker for hepatocellular carcinoma. Oncotarget 8(27):43878–43888
83. Fu L, Yao T, Chen Q et al (2017) Screening differential circular RNA expression profiles reveals hsa_circ_0004018 is associated with hepatocellular carcinoma. Oncotarget 8(35):58405–58416
84. Center MM (2011) Global cancer statistics. CA Cancer J Clin 61(2):69
85. Goto N, Hiyoshi H, Ito I et al (2014) Identification of a novel compound that suppresses breast cancer invasiveness by inhibiting transforming growth factor-β signaling via estrogen receptor α. J Cancer 5(5):336–343
86. Grushko TA, Filiaci VL, Mundt AJ et al (2008) An exploratory analysis of HER-2 amplification and overexpression in advanced endometrial carcinoma: a Gynecologic Oncology Group study. Gynecol Oncol 108(1):3–9
87. Wang H, Mu X, Zhou S et al (2014) NEDD9 overexpression is associated with the progression of and an unfavorable prognosis in epithelial ovarian cancer. Hum Pathol 45(2):401–408
88. Tangjitgamol S, Khunnarong J, Srijaipracharoen S (2014) Medical morbidities in endometrial cancer patients. Int J Gynecol Cancer 24(9):1623–1627
89. Reddy SD, Ohshiro K, Rayala SK et al (2008) MicroRNA-7, a homeobox D10 target, inhibits p21-activated kinase 1 and regulates its functions. Cancer Res 68(20):8195
90. Lü L, Jian S, Shi P et al (2017) Identification of circular RNAs as a promising new class of diagnostic biomarkers for human breast cancer. Oncotarget 8(27):44096–44107
91. Liang HF, Zhang XZ, Liu BG et al (2017) Circular RNA circ-ABCB10 promotes breast cancer proliferation and progression through sponging miR-1271. Am J Cancer Res 7(7):1566–1576

92. Liang G, Liu Z, Tan L et al (2017) HIF1Î±-associated circDENND4C promotes proliferation of breast cancer cells in hypoxic environment. Anticancer Res 37(8):4337–4343
93. Wang H, Xiao Y, Wu L et al (2018) Comprehensive circular RNA profiling reveals the regulatory role of the circRNA-000911/miR-449a pathway in breast carcinogenesis. Int J Oncol 52(3):743–754
94. Ahmed I, Karedath T, Andrews SS et al (2016) Altered expression pattern of circular RNAs in primary and metastatic sites of epithelial ovarian carcinoma. Oncotarget 7(24):36366–36381
95. Chen BJ, Byrne FL, Takenaka K et al (2018) Analysis of the circular RNA transcriptome in endometrial cancer. Oncotarget 9(5):5786–5796
96. Abdelmohsen K, Panda AC, Munk R et al (2017) Identification of HuR target circular RNAs uncovers suppression of PABPN1 translation by CircPABPN1. RNA Biol 14(3):361–369
97. Xuan L, Qu L, Han Z et al (2016) Circular RNA: a novel biomarker for progressive laryngeal cancer. Am J Transl Res 8(2):932
98. Chen L, Zhang S, Wu J et al (2017) circRNA_100290 plays a role in oral cancer by functioning as a sponge of the miR-29 family. Oncogene 36(32):4551–4561
99. Zheng J, Liu X, Xue Y et al (2017) TTBK2 circular RNA promotes glioma malignancy by regulating miR-217/HNF1β/Derlin-1 pathway. J Hematol Oncol 10(1):52
100. Guarnerio J, Bezzi M, Jeong JC et al (2016) Oncogenic role of fusion-circRNAs derived from cancer-associated chromosomal translocations. Cell 165(2):289–302
101. Zhong Z, Lv M, Chen J (2016) Screening differential circular RNA expression profiles reveals the regulatory role of circTCF25-miR-103a-3p/miR-107-CDK6 pathway in bladder carcinoma. Sci Rep 6:30919
102. Sand M, Bechara FG, Sand D et al (2016) Circular RNA expression in basal cell carcinoma. Epigenomics 8(5):619–632
103. Yang H, Luo J, Liu Z et al (2015) MicroRNA-138 regulates DNA damage response in small cell lung cancer cells by directly targeting H2AX. Cancer Investig 33(4):126
104. Liu YC, Li JR, Sun CH et al (2016) CircNet: a database of circular RNAs derived from transcriptome sequencing data. Nucleic Acids Res 44(D1):D209
105. Peifei L, Shengcan C, Huilin C et al (2015) Using circular RNA as a novel type of biomarker in the screening of gastric cancer. Clin Chim Acta 444:132–136
106. Wang X, Zhang Y, Huang L et al (2015) Decreased expression of hsa_circ_001988 in colorectal cancer and its clinical significances. Int J Clin Exp Pathol 8(12):16020–16025
107. Huang G, Hua Z, Shi Y et al (2015) cir-ITCH plays an inhibitory role in colorectal cancer by regulating the Wnt/β-catenin pathway. PLoS One 10(6):e0131225
108. Mcglynn LM, Mccluney S, Jamieson NB et al (2015) SIRT3 & SIRT7: potential novel biomarkers for determining outcome in pancreatic cancer patients. PLoS One 10(6):e0131344
109. Yeh YM, Chuang CM, Chao KC et al (2013) MicroRNA-138 suppresses ovarian cancer cell invasion and metastasis by targeting SOX4 and HIF-1α. Int J Cancer 133(4):867–878
110. Yu SY, Wang YP, Chang JY et al (2014) Increased expression of MCM5 is significantly associated with aggressive progression anda poor prognosis of oral squamous cell carcinoma. J Oral Pathol Med 43(5):344–349
111. Sand M, Bechara FG, Gambichler T et al (2016) Circular RNA expression in cutaneous squamous cell carcinoma. J Dermatol Sci 83(3):210–218
112. Wang Q, Tang H, Yin S et al (2013) Downregulation of microRNA-138 enhances the proliferation, migration and invasion of cholangiocarcinoma cells through the upregulation of RhoC/p-ERK/MMP-2/MMP-9. Oncol Rep 29(5):2046–2052
113. Jin H, Jin X, Zhang H et al (2017) Circular RNA hsa-circ-0016347 promotes proliferation, invasion and metastasis of osteosarcoma cells. Oncotarget 8(15):25571–25581
114. Wang K, Sun Y, Tao W et al (2017) Androgen receptor (AR) promotes clear cell renal cell carcinoma (ccRCC) migration and invasion via altering the circHIAT1/miR-195-5p/29a-3p/29c-3p/CDC42 signals. Cancer Lett 394:1–12

Part VI

Circular RNAs and Human Diseases

Circular RNAs in Cardiovascular Diseases

15

Lijun Wang, Xiangmin Meng, Guoping Li, Qiulian Zhou, and Junjie Xiao

Abstract

Circular RNAs (circRNAs), a group of circular RNA molecules with a 3′,5′-phosphodiester bond at the junction site, are generated by back-splicing of precursor mRNAs. Most of the circular RNAs originate from the exon region of the encoded protein, and some are derived from intron regions, antisense transcripts, or long noncoding RNAs. Circular RNAs are abundantly in eukaryotic transcriptome and participate in various biological processes. It is closely associated with various diseases such as tumors, diabetes, nervous system diseases, and cardiovascular diseases. In cardiovascular system, numerous circRNAs have been identified and involved in important processes of cardiovascular development and diseases. Here we will review the latest research progress of circular RNA in cardiovascular diseases. Also, we will outline the specific examples of circRNAs involved in cardiovascular system regulatory effects, including act as miRNA sponges, interaction with RNA-binding proteins, regulated by RNA-binding proteins and serve as biomarkers. In addition, potential mechanisms underlying the regulatory role of circRNAs in cardiovascular diseases will be discussed.

Keywords

Circular RNA · microRNA · RNA-binding protein · Cardiovascular diseases

1 Introduction

Circular RNAs (circRNAs) were first identified as by-products of abnormal splicing with limited functional potential [1–3]. Until 2012, with the

L. Wang · X. Meng
Cardiac Regeneration and Ageing Lab, Institute of Cardiovascular Sciences, School of Life Science, Shanghai University, Shanghai, China

Shanghai Key Laboratory of Bio-Energy Crops, School of Life Sciences, Shanghai University, Shanghai, China

G. Li
Cardiovascular Division of the Massachusetts General Hospital, Harvard Medical School, Boston, MA, USA

Q. Zhou
Shanghai Applied Radiation Institute, School of Environmental and Chemical Engineering, Shanghai University, Shanghai, China

Cardiac Regeneration and Ageing Lab, Institute of Cardiovascular Sciences, School of Life Science, Shanghai University, Shanghai, China

Shanghai Key Laboratory of Bio-Energy Crops, School of Life Sciences, Shanghai University, Shanghai, China

J. Xiao (✉)
School of Life Science, Institute of Cardiovascular Sciences, Shanghai University, Shanghai, China
e-mail: junjiexiao@shu.edu.cn

J. Xiao (ed.), *Circular RNAs*, Advances in Experimental Medicine and Biology 1087,
https://doi.org/10.1007/978-981-13-1426-1_15

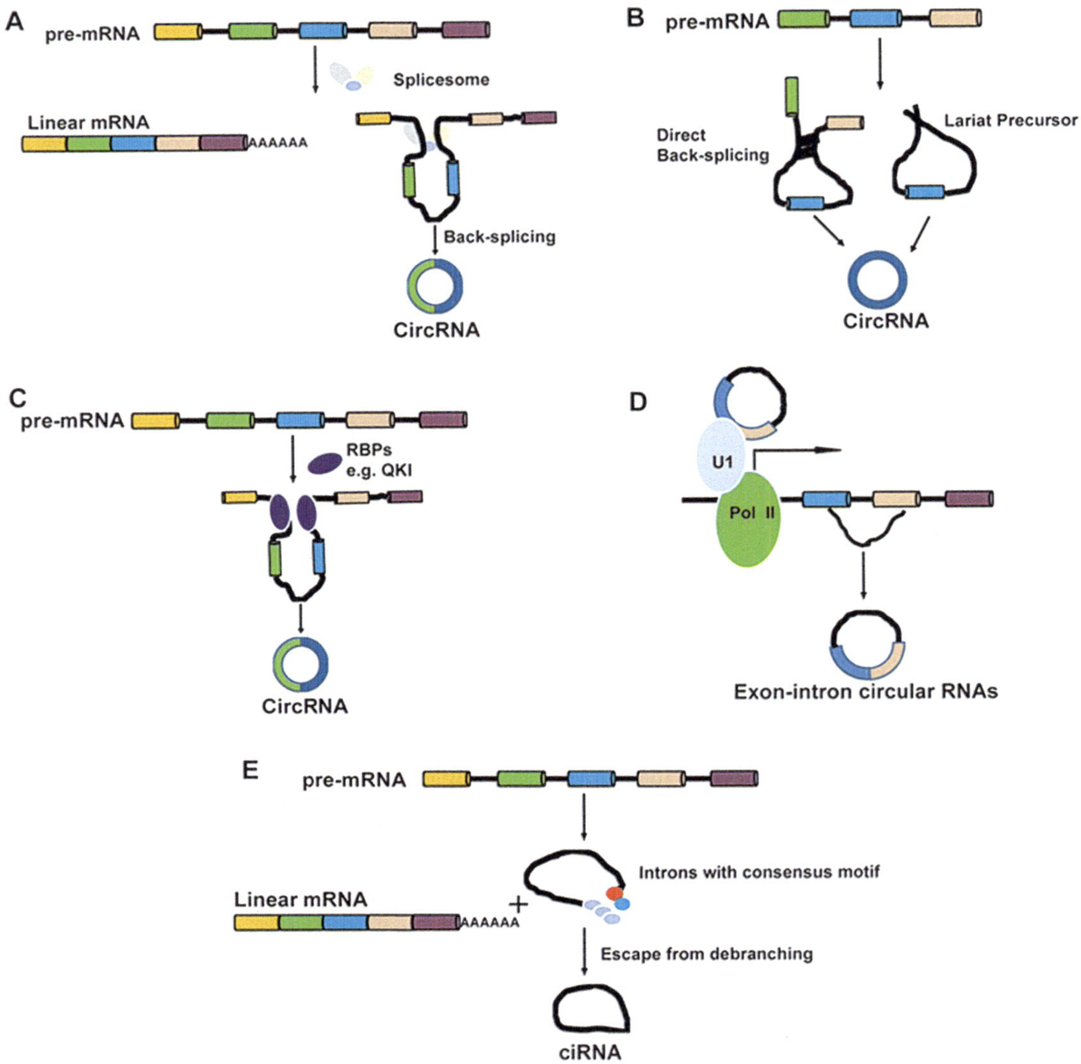

Fig. 15.1 Biogenesis of circular RNA and its regulation. (**a**) Circular RNA is catalyzed by spliceosomal machinery. (**b**) Exon skipping leads to a lariat-driven circularization, and introns contain reverse complementary sequence-driven circularization. (**c**) Circular RNA production is regulated by RNA-binding proteins (RBPs). (**d**) Exon-intron circRNAs (EIciRNAs) contain both exons and introns generated by back-splicing and regulate transcription of its host gene. (**e**) Circular intronic RNAs (ciRNAs) derived from lariat introns during canonical splicing

development of RNA-sequencing technology, widespread and abundant circRNAs have been discovered [4–7]. CircRNAs are produced by the formation of 3′,5′-phosphodiester bond at the junction site [8, 9]. CircRNAs are expressed in various eukaryotic organisms, especially in mammals, and involved in regulating many biological processes. CircRNAs are derived from precursor mRNA back-splicing [10, 11]. Classically, circularization of RNAs requires the assemble of spliceosome at back-splicing sites to catalyze the ligation of 5′ donor sites and 3′ acceptor sites (Fig. 15.1a) [9]. Two models have been proposed for the backsplice formation of circRNAs: Exon skipping leads to a lariat-driven circularization, and introns contain reverse complementary sequence-driven circularization (Fig. 15.1b) [8, 12–14]. Besides, trans-acting factors RNA-binding proteins (RBPs) have been reported to regulate circRNA biogenesis (Fig. 15.1c) [15–

17]. Some circRNAs contain both exons and introns (exon-intron circRNAs, EIciRNAs) (Fig. 15.1d) [18]. Another class of circular RNAs derived from lariat introns during canonical splicing could also lead to the biogenesis of circular intronic RNAs (ciRNAs) (Fig. 15.1e) [19]. Many studies have shown that circRNAs participate in many human diseases, including cancer, diabetes, nervous system diseases, and cardiovascular diseases [20–26]. Numerous excellent reviews have summarized the biogenesis of circular RNAs and the underlying mechanisms [27–36]. Here we will outline the specific examples and discuss potential mechanisms underlying the regulatory role of circRNAs in cardiovascular diseases.

2 Circular RNAs and Cardiovascular Diseases

Cardiovascular diseases are the leading cause of death in human globally, higher than cancer and other diseases [37, 38]. In addition, the increasing direct economic burden of cardiovascular disease has also become a major public health problem. However, the underlying molecular mechanisms of cardiovascular diseases are still less known and remain to be well studied [39–41]. In recent years, in the cardiovascular system, numerous circRNAs have been discovered and involved in important processes of cardiovascular development and diseases [42, 43]. In this following section, we will give a brief review about specific examples of circRNAs involved in the currently reported ways of regulating in the cardiovascular system (Table 15.1).

3 Circular RNA Acts as miRNA Sponge

3.1 Circular RNA HRCR Acts as miR-223 Sponge that Regulates Pathological Cardiac Hypertrophy

Pathological cardiac hypertrophy is a common reaction of the heart against increased hemodynamics, myocardial damage, and neurohormonal stress, manifested as increased myocardial cell volume, increased cardiac fibrosis, and loss of myocardial cells (necrosis or apoptosis) [38, 41]. When the stimulation persists, pathological myocardial hypertrophy is decompensated, resulting in ventricular remodeling and heart failure. Heart failure is the end stage of many cardiovascular diseases and has very poor prognosis. It is the main cause of disability and death. In addition to cardiac transplantation, currently, heart failure

Table 15.1 Circular RNAs in cardiovascular diseases

Actions	CircRNA	Diseases	Regulation	References
Act as miRNA sponges	HRCR	Pathological hypertrophy	↓	[44]
	Cdr1as	Myocardial infarction	↑	[58]
	MFACR	Cardiomyocytes apoptosis	↑	[61]
	CircRNA_000203	Cardiac fibrosis	↑	[63]
	CircRNA_010567	Cardiac fibrosis	↑	[64]
Interaction with RBPs	CircFOXO3	Cardiac senescence	↑	[66]
	CircAmotl1	Dox-induced cardiomyopathy	↓	[69]
	CircANRIL	Atherosclerosis	↓	[71]
Regulated by RBPs	CircTitin	Hypertrophic cardiomyopathy	Unknown	[78]
	CircTitin CircFhod3 CircStrn3	Heart failure	Unknown	[77]
Serve as biomarkers	MICRA	Myocardial infarction	Unknown	[99]
	CircRNA_081881	Myocardial infarction	Unknown	[100]
	has_circ_0124644	Coronary artery disease	Unknown	[101]
Others	cZNF292	Angiogenesis	↑	[80]
	CircSLC8a1	Heart development	Unknown	[43]

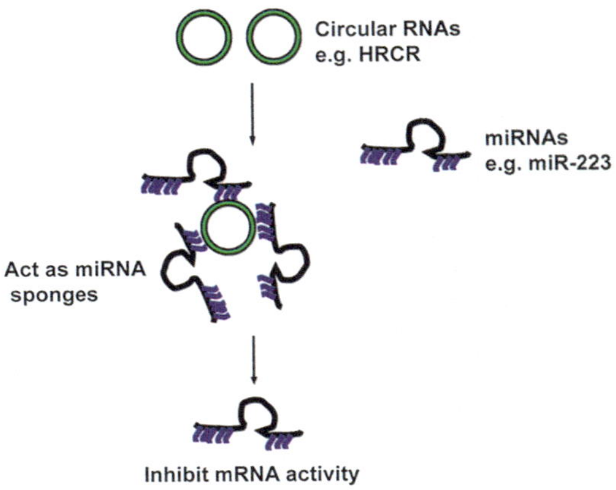

Fig. 15.2 Circular RNAs act as miRNA sponges
For example, circular RNA HRCR binds to miR-223 and acts as an endogenous miR-223 sponge to inhibit miR-223 activity

lacks an effective treatment in clinical practice. Therefore, exploring ways to prevent the progression of pathological cardiac hypertrophy to heart failure remains an important issue in the treatment of cardiovascular diseases and heart failure.

A recent study about circular RNA HRCR (heart-related circRNA) demonstrates that HRCR directly binds to miR-223 and acts as an endogenous miR-223 sponge to inhibit miR-223 activity (Fig. 15.2) [44]. MiR-223 was first identified in myeloid cells in the bone marrow and has been reported to have a prominent role in monocyte/macrophage differentiation and granulocytic differentiation as well as different types of cancers [45–49]. In cardiomyocytes, miR-223 acts as a positive regulator by targeting ARC. Through random screening of 100 circRNAs from circRNA databases as well as the use of the bioinformatics program RNAhybrid, HRCR was found to have six target sites of miR-223. Furthermore, using RNA pulldown and AGO2 immunoprecipitation assays, HRCR was proved to be directly bound to miR-223. The co-localization of HRCR and miR-223 was verified by fluorescence in situ hybridization (FISH). HRCR also regulates cardiomyocytes hypertrophy through targeting miR-223 and apoptosis repressor with CARD domain. Overexpression of HRCR could attenuate cardiomyocyte hypertrophy induced by ISO treatment and represses cardiac hypertrophy as well as heart failure in vivo. This is the first functional research of circRNAs in pathological cardiac hypertrophy.

3.2 Circular RNA Cdr1as Acts as miR-7 Sponge that Promotes Myocardial Infarction

Acute myocardial infarction (MI) remains one of the leading causes of morbidity and mortality globally. Most MI occurs due to coronary artery diseases [50]. During acute MI, cardiomyocyte apoptosis and necrosis trigger the early inflammatory response that clear the wound from dead cells and activate the following reparative response [51]. Early reperfusion is currently the most effective strategy to improve survival rates in patients suffering acute myocardial infarction [50, 52]. However, this reperfusion therapy always leads to myocardial ischemia reperfusion injury (I/RI) and at risk of developing heart failure [52, 53]. Therefore, efforts made to investigate the molecular mechanisms of cardiomyocytes apoptosis are crucial for developing new therapeutic strategies. Cdr1as (antisense to the cerebellar degeneration-related protein 1 transcript), also termed as ciRS-7 (circular RNA sponge for miR-7), has been reported to contain more than 70 conserved miRNA target sites and acts as a miR-7 sponge [54]. In mouse brain, miR-7 inhibition and Cdr1as overexpression demonstrate similar developmental defects [55–57]. In cardiomyocytes, overexpression of miR-7 could reverse Cdr1as-induced apoptosis by targeting PARP and SP1 [58]. In vivo, Cdr1as overexpression aggravates MI injury evidenced by increased cardiac infarct size and upregulation of PARP and SP1, while miR-7 overexpression has opposite effects.

3.3 Circular RNA MFACR Acts as miR-652-3p Sponge that Regulates Mitochondrial Dynamics and Apoptosis in the Heart

Numerous studies have demonstrated that mitochondrial fission dysfunction often exists in cardiovascular disease. Mitochondrial fission process protein 1 (MTFP1) or mitochondrial 18kDa protein (MTP18) is involved in the mitochondrial division by regulating membrane fission [59]. In PC-3 and HaCaT cells, MTP18 knockdown would induce the release of cytochrome c, which activates the caspase cascade and leads to apoptosis [60]. The exact role of MTP18 in heart remains unknown. Interestingly, Wang et al. reported that MTP18 deficiency reduces mitochondrial fission and suppresses cardiomyocytes apoptosis [61]. Using the bioinformatics program RNAhybrid, mitochondrial fission, and apoptosis-related circRNA, MFACR was identified and found to contain 15 binding sites on miR-652-3p. Biotin-coupled RNA pulldown assay in cardiomyocytes was performed to verify this interaction of miR-652-3p and MFACR. AGO2 immunoprecipitation and reverse pulldown assay were further conducted to confirm that MFACR could directly bind to miR-652-3p in vivo. Consistently, overexpression of MFACR increased the protein level of MTP18, while knockdown MFACR resulted in inhibition of MTP18. Besides, inhibition of miR-652-3p would abolish the effect of MFACR knockdown on MTP18. In mice heart with I/R injury, knockdown of MFACR increased miR-652-3p expression and decreased MTP18 protein level. MFACR acts as a miR-652-3p sponge to promote mitochondrial fission and cardiomyocyte apoptosis.

3.4 Circular RNA Acts as miRNA Sponge that Contributes to Cardiac Fibrosis

Cardiac fibrosis is the result of excess deposition of extracellular matrix in the cardiac muscle [62]. It is a main component of adverse ventricular remodeling and heart failure. The strategy of anti-fibrosis for heart failure patients has been raised; however, no effective therapy for cardiac fibrosis was reported. Therefore, understanding underlying mechanism of cardiac fibrosis is important for prevention and treatment of cardiac fibrosis. As newly discovered class of noncoding RNAs, the role of circular RNAs in cardiac fibrosis remains unknown. Recently, circRNA_000203 was reported to upregulate in diabetic mouse myocardium and in AngII-induced mouse cardiac fibroblasts [63]. RNA pulldown and RT-qPCR assay revealed that circRNA_000203 could bind to miR-26b-5p. And duel luciferase assay further showed that miR-26b-5p could specifically interact with the 3′-UTRs of CTGF and Col1a2. Interestingly, this interaction could be abolished by overexpression of circRNA_000203 evidenced by elevated expression of fibrosis-associated gene, such as Col1a2, Col3a1, α-SMA, and CTGF. Therefore, circRNA_000203 could inhibit the anti-fibrotic effect of miR-26b in mouse cardiac fibroblasts. In addition, another circRNA, circRNA_010567, was reported to promote myocardial fibrosis via suppressing miR-141 function [64].

4 Circular RNA Can Interact with RNA-Binding Proteins (RBPs)

4.1 Circular RNA FOXO3 Promotes Cardiac Senescence Through Sequestering RBPs

CircFOXO3 is derived from transcription factor FOXO3 and acts as a scaffold to modulated P21 and CDK2 interaction [65]. In NIH3T3 cell line, circFOXO3 interacts with both P21 and CDK2, repressing cell progression. In the heart, this same circFOXO3 is highly expressed. Interestingly, Du et al. found that the expression levels of circFOXO3 in older hearts are significantly higher than young hearts (Fig. 15.3a) [66]. Therefore, they explored the effects of circFOXO3 on senescence. In mouse embryonic fibroblasts (MEFs), mouse cardiac fibroblasts

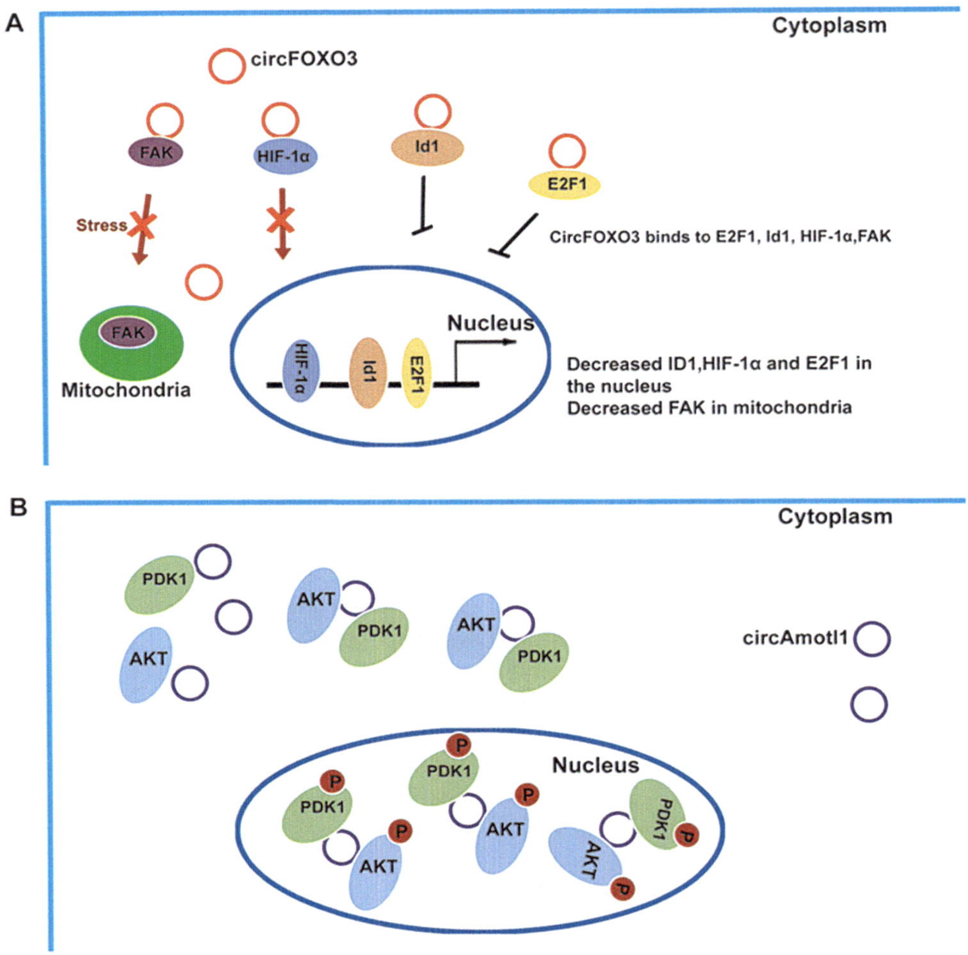

Fig. 15.3 Circular RNA interaction with RNA-binding proteins (RBPs)
(**a**) CircFOXO3 binds to transcription factors E2F1 and Id1 and inhibits their translocation into the nucleus. Also, under stress conditions, circFOXO3 interaction with HIF-1α and FAK in the cytoplasm decreased HIF-1α in the nucleus and FAK in mitochondria. (**b**) CircAmotl1 binds AKT and PDK simultaneously, which increased AKT phosphorylation and pAKT nuclear translocation

(MCFs), NIH3T3 fibroblasts, B16 cells, and primary cardiomyocytes, circFOXO3 was highly expressed in cell lines that underwent cellular senescence. In vivo, in a mouse model of Dox-induced cardiomyopathy, circFOXO3 expression level was found to be associated with the level of tissue apoptosis as evidenced by TUNEL staining. Furthermore, immunoprecipitation and real-time PCR demonstrate that circFOXO3 was pulled down by antibodies against senescence-associated proteins ID1, E2F1, as well as HIF-1α and FAK, but linear FOXO3 mRNA was not. Besides, the expression level of circFOXO3 (whether overexpression or knockdown) did not change the expression of ID1, E2F1, HIF-1α, or FAK. Overexpressed circFOXO3 facilitated the localization of most transcription factors ID1 and E2F1 and anti-stress proteins HIF-1α and FAK in cytoplasm, decreasing the protein levels of ID1 and E2F1, HIF-1α in the nucleus, and FAK in mitochondria. In summary, circFOXO3 sequestering ID1, E2F1, HIF1α, and FAK in the cytoplasm decreases those protein levels in the nucleus and promotes senescence.

4.2 Circular RNA Amotl1 Facilitates Cardioprotective Nuclear Translocation of pAKT

The Ser and Thr kinase AKT plays essential roles in diverse cellular processes. AKT signaling pathways are activated by activation of PI3K and phosphorylation of AKT [67]. In cardiovascular disease, AKT function as a cardioprotective molecule. For the activation of AKT signaling, phosphoinositide-dependent protein kinase 1 (PDK1) was discovered for phosphorylation of AKT1 at T308. During this process, the localization of AKT and PDK1 to membrane sites of PIP3 is required for the access of PDK1 to AKT for phosphorylation [68]. By microarray, Zeng et al. analyzed the expression levels of different circular RNAs in neonatal and mature postnatal human cardiac tissue samples, and circAmotl1 was found to be preferentially expressed in neonate cardiac tissue (Fig. 15.3b) [69]. Overexpression of circAmotl1 could attenuate the Dox-induced cardiomyopathy in mice model. Higher protein levels of pAKT, AKT, PDK1, and pPDK1 were detected in cells transfected with circAmotl1. In the nuclei, pAKT level was also elevated. The authors further investigated the underlying mechanism about how circAmotl1 facilitate nuclear translocation of AKT and PDK1. By conducting AGO2 immunoprecipitation assays, they exclude the possibility that circAmotl1 may function as a sponge for binding miRNAs. In further studies, using computer algorithm, NPDock, and immunoprecipitation, circAmotl1 was found able to bind to PDK1 and AKT simultaneously. The interaction between PDK1 and AKT could be abolished by RNase A indicating that this interaction is mediated by RNA. Based on the docked structure of AKT-circAmotl1-PDK1, antisense oligos complementary to the binding site on circAmotl1 were designed to specifically block interaction between circAmotl1 and AKT or circAmotl1 and PDK1. Interestingly, the expression of AKT, PDK1, and pPDK1 was not affected, but the pAKT and pPKD1 nuclear translocation was reduced. Thus, circAmotl1 binds to AKT and PDK1 directly, induces AKT phosphorylation and pAKT nuclear translocation, leads to activation of AKT signaling pathway, and protects Dox-induced cardiomyopathy.

4.3 Circular RNA ANRIL Regulates Atherosclerosis

ANRIL is a long noncoding RNA called *a*ntisense *n*oncoding *R*NA in the *I*NK4 *l*ocus. This locus on chromosome 9p21.3 has been reported to have strong association with atherosclerotic vascular disease (ASVD). A class of novel circular RNA products from ANRIL locus and their expression are correlated with INK4/ARF transcription and ASVD [70]. Among them, one isoform named circANRIL is well studied [71]. CircANRIL was identified in association with atheroprotection at human 9p21 locus. Overexpression of circANRIL in HEK-293 cells revealed increased apoptosis and decreased proliferation. Using qPCR and RNA immunoprecipitation of AGO2 analysis demonstrates that circANRIL does not regulate 9p21 protein-coding genes and lacks miRNA sponging activity. Further proteomic screen shows that circANRIL potentially binds to proteins. Through RNA pulldown assay and label-free mass spectrometric analysis, 32 proteins were detected with significant enrichment in circANRIL overexpression cells compared to controls. Next, RNA immunoprecipitation was performed to analyze circANRIL-binding proteins, and PES1 was identified to be the strongest binding protein. PES1 is the component of PeBoW complex, which is required for maturation of ribosomal RNAs and formation of the 60S ribosome [72–74]. Northern blot and immunofluorescent staining revealed that overexpression of circANRIL in cells leads to 32S and 36S pre-rRNA accumulation, nucleolar stress, and P53 activation. Taken together, circANRIL overexpression could impair the ribosome biogenesis and P53 activation and further lead to increase in cell apoptosis and decrease in proliferation. Therefore, circANRIL was proposed to promote atheroprotection by suppressing overproliferating cell types in atherosclerotic plaques.

5 RBPs Involved in Regulating the Expression of Circular RNAs

RNA-binding proteins (RBPs) play crucial roles in various cellular processes [75, 76]. It has been reported that RNA-binding protein Quaking regulates circular RNA biogenesis during epithelial to mesenchymal transition [16]. Another splicing factor *muscleblind* (MBL) could specifically bind to circMbl and affect circMbl biosynthesis [15]. In cardiovascular system, several RBPs have been reported as key regulators of cellular function in various cardiac diseases. In doxorubicin-induced cardiotoxicity, Quaking was also found to be downregulated, and overexpression of Quaking could inhibit doxorubicin-induced apoptosis by regulating the biogenesis of cardiac circular RNAs [77]. In addition to Quaking, another report explored circRNAs in cardiac tissue from patients with hypertrophic cardiomyopathy (HCM) or dilated cardiomyopathy (DCM) and from non-diseased individuals [78]. RNA-binding protein 20 (RBM20) is identified as a regulator of mRNA splicing of a subset of genes involved in cardiac development [79]. In this article, RBM20 was found to be essential for the formation of a subset of circRNAs that originate from the titin gene. Therefore, RBPs might play important roles in the formation of circular RNAs and the regulation of circRNA expression.

6 Other Circular RNAs Associated with Cardiovascular Disease

Circular RNAs are widely expressed and involved in many biological processes in cardiovascular system. Boeckel et al. investigated the circRNAs in hypoxia-induced endothelial cells [80]. Through bioinformatics analysis and next-generation sequencing, circular RNA cZNF292 was identified. Circular RNA cZNF292 was one of the highest expressed and significantly hypoxia-regulated circRNAs. The characterization of ZNF292 was confirmed by northern blot and RNase R treatment assay. Further loss-and-gain function studies demonstrated that cZNF292 regulated angiogenic sprouting. Overexpression of circular RNA cZNF292 exhibited enhanced proliferation, while silencing of cZNF292 inhibited spheroid sprouting. The underlying mechanism of cZNF292 was investigated further, and the possibility that cZNF292 acts as miRNA sponge and cis-regulator for its host gene ZNF292 was excluded. As a result, hypoxia-induced endothelial circular RNA cZNF292 is indeed existing and has biological functions.

7 Circular RNA Serves as a Biomarker of Cardiovascular Disease

Circular RNAs are widely present in human pathological tissues, blood, saliva, exosomes, and semen [81–88]. Because of their closed loop structure, they have better stability than linear RNAs. In addition, the expression of circular RNAs has distinct tissue specificity and timing specificity. Recently, circular RNA is expected to be a novel biomarker for human diseases [89–94]. Many studies have been reported that circRNAs acted as biomarkers in some diseases, including cancer, neurological disorders, etc. [95–98]. In cardiovascular diseases, Vausort et al. conducted a study to evaluate circRNA molecules in the peripheral blood with acute myocardial infarction to predict postischemic reperfusion conditions [99]. This study selected myocardial infarction with left ventricular dysfunction and finally identified a circRNA myocardial infarction-associated circular RNA (MICRA), which is an 874-nucleotide-long circRNA formed mainly from exon 1 of the zinc finger protein 609 (ZNF609) gene located on chromosome 15q22 that could predict left ventricular dysfunction 3–4 months after reperfusion. Another study performed by Deng et al. showed that circRNA_081881 can serve as a competitive endogenous RNA molecule of miR-549, thereby regulating PPARγ expression [100]. CircRNA_081881 can be detected in plasma and therefore might serve as a potential marker and therapeutic target for acute myocardial infarction.

Another circular RNA associated with the pathophysiological process of coronary artery disease, has_circ_0124644, has been detected in the peripheral blood and can be used as a potential biomarker [101]. When has_circ_0124644 was introduced as a test marker, both specificity and sensitivity of the diagnoses of cardiovascular disease were significantly increased. Taken together, with the deepening of research on circular RNA, there will be more circular RNAs that are expected to become new biomarkers. It will be greatly helpful for the diagnosis and precise treatment of cardiovascular diseases.

8 Conclusions and Future Perspectives

With the advent of deep sequencing techniques, in recent years, thousands of circular RNA isoforms from numerous tissues and organisms are characterized. A large amount of circular RNAs have been found in the heart, and the expression of many circular RNAs is closely related to the occurrence and development of cardiovascular diseases [42, 43, 102, 103]. In this review, we used various examples to illustrate the roles of circular RNAs in cardiovascular system. For example, circular RNA HRCR, Cdr1as, and MFACR could act as endogenous miRNA sponges to inhibit miRNA activity [44, 58, 61]. Circular RNA FOXO3, Amotl1, and ANRIL have been reported to interact with RBPs [66, 69, 71]. RNA-binding protein Quaking and RBM20 could affect the biogenesis of circular RNAs in the heart [77, 78]. However, it should be noted that many questions underlying the regulatory role of circular RNAs remained; further investigation of circular RNAs is urgently requested.

Although circular RNAs were initially considered to be noncoding RNAs, as the research progressed, some circular RNAs were found to have the potential to encode proteins or polypeptides [104–106]. The presence of internal ribosome entry sites (IRESs) and appropriate open reading frame (ORF) makes the cap-independent translation possible [107]. In addition, N^6-methyladenosine (m6A) modification sites of circular RNA can act as IRESs [108]. The m6A-driven translation could be reduced by m6A demethylase FTO while promoted by adenosine methyltransferase METTL3/14. Circular RNA SHPRH can encode a new protein that inhibits the development of gliomas, and circ-ZNF609 can be translated into proteins that regulate myogenesis [109, 110]. However, no translatable circular RNA has been found to be involved in the regulation of cardiovascular disease. IRES-driven translation is favored under stress conditions. In the event of cardiovascular disease, it is highly likely that the translation of circular RNA into proteins or peptides is involved in the regulation of the occurrence and development of the disease. Due to the strong tissue and spatial specificity of circular RNAs, these circular RNA-derived proteins or peptides may serve as better therapeutic targets. Therefore, it is of great significance to discover the function of these translatable circular RNAs in the heart.

At present, the study of cardiovascular RNAs in circular RNAs has mainly focused on the regulation of circular RNAs in diseases such as myocardial infarction, atherosclerosis, and heart failure [44, 58, 61, 66, 71]. Exercise training is recognized as an effective way to prevent cardiovascular disease [111–114]. A large number of clinical studies have shown that exercise training can improve coronary blood flow by improving vasodilation while reducing myocardial oxidative stress and preventing myocardial cell loss and cardiac fibrosis, thereby improving cardiac function and reducing cardiovascular risk factors [115, 116]. Different from pathological myocardial hypertrophy, physiological cardiac hypertrophy usually occurs after regular exercise [117–123]. Some microRNAs have also been reported to participate in the regulation of exercise-mediated physiological cardiac hypertrophy and have protective functions for the heart [124–127]. However, the function of circular RNAs in exercise-mediated physiological cardiac hypertrophy is still unknown. Therefore, exploring the role of circular RNA in physiological cardiac hypertrophy is also of great significance for the treatment of cardiovascular diseases and prevention of heart failure.

So far, only a few studies have been reported in the study of circular RNAs in cardiovascular system, and much of our understanding comes from a small number of circular RNAs. RNA sequencing data have been demonstrated that circular RNAs are abundant in the heart [42, 43, 103]. However, the underlying regulatory role of most circular RNAs remains unclear. What can be confirmed is that the expression level of circular RNAs has undergone a great change during the occurrence of various cardiovascular diseases. Many circular RNAs may play crucial roles in the development of cardiovascular disease. Therefore, to study the regulatory mechanism of the circular RNAs in the cardiovascular system is of great significance for the diagnosis, intervention, and treatment of cardiovascular diseases.

Acknowledgments The authors thank members of the Cardiac Regeneration and Ageing Lab in Shanghai University for the discussion. Due to space restrictions, the authors cannot cite all the relevant literature in the field. The authors apologize to those colleagues whose work contributed significantly. This work was supported by the grants from the National Natural Science Foundation of China (81722008, 91639101, and 81570362 to JJ Xiao), the Innovation Program of Shanghai Municipal Education Commission (2017-01-07-00-09-E00042 to JJ Xiao), the grant from the Science and Technology Commission of Shanghai Municipality (17010500100 to JJ Xiao), and the development fund for Shanghai talents (to JJ Xiao).

Competing Financial Interests The authors declare no competing financial interests.

References

1. Capel B, Swain A, Nicolis S et al (1993) Circular transcripts of the testis-determining gene Sry in adult mouse testis. Cell 73(5):1019–1030
2. Cocquerelle C, Daubersies P, Majerus MA et al (1992) Splicing with inverted order of exons occurs proximal to large introns. EMBO J 11(3):1095–1098
3. Cocquerelle C, Mascrez B, Hetuin D et al (1993) Mis-splicing yields circular RNA molecules. FASEB J 7(1):155–160
4. Salzman J, Gawad C, Wang PL et al (2012) Circular RNAs are the predominant transcript isoform from hundreds of human genes in diverse cell types. PLoS One 7(2):e30733
5. Lu C, Shi X, Wang AY et al (2018) RNA-Seq profiling of circular RNAs in human laryngeal squamous cell carcinomas. Mol Cancer 17(1):86
6. AbouHaidar MG, Venkataraman S, Golshani A et al (2014) Novel coding, translation, and gene expression of a replicating covalently closed circular RNA of 220 nt. Proc Natl Acad Sci USA 111(40):14542–14547
7. Salzman J, Chen RE, Olsen MN et al (2013) Cell-type specific features of circular RNA expression. PLoS Genet 9(9):e1003777
8. Barrett SP, Wang PL, Salzman J (2015) Circular RNA biogenesis can proceed through an exon-containing lariat precursor. Elife 4:e07540
9. Starke S, Jost I, Rossbach O et al (2015) Exon circularization requires canonical splice signals. Cell Rep 10(1):103–111
10. Wilusz JE (2015) Repetitive elements regulate circular RNA biogenesis. Mob Genet Elements 5(3):1–7
11. Liang D, Tatomer DC, Luo Z et al (2017) The output of protein-coding genes shifts to circular RNAs when the Pre-mRNA processing machinery is limiting. Mol Cell 68(5):940–954 e3
12. Wilusz J (2015) Circular RNA and splicing: skip happens. J Mol Biol 427(15):2411–2413
13. Liang D, Wilusz JE (2014) Short intronic repeat sequences facilitate circular RNA production. Genes Dev 28(20):2233–2247
14. Zhang XO, Wang HB, Zhang Y et al (2014) Complementary sequence-mediated exon circularization. Cell 159(1):134–147
15. Ashwal-Fluss R, Meyer M, Pamudurti NR et al (2014) circRNA biogenesis competes with pre-mRNA splicing. Mol Cell 56(1):55–66
16. Conn SJ, Pillman KA, Toubia J et al (2015) The RNA binding protein quaking regulates formation of circRNAs. Cell 160(6):1125–1134
17. Ivanov A, Memczak S, Wyler E et al (2015) Analysis of intron sequences reveals hallmarks of circular RNA biogenesis in animals. Cell Rep 10(2):170–177
18. Li Z, Huang C, Bao C et al (2015) Exon-intron circular RNAs regulate transcription in the nucleus. Nat Struct Mol Biol 22(3):256–264
19. Zhang Y, Zhang XO, Chen T et al (2013) Circular intronic long noncoding RNAs. Mol Cell 51(6):792–806
20. Yang L, Han B, Zhang Y et al (2018) Engagement of circular RNA HECW2 in the nonautophagic role of ATG5 implicated in the endothelial-mesenchymal transition. Autophagy 14(3):404–418
21. Zheng Q, Bao C, Guo W et al (2016) Circular RNA profiling reveals an abundant circHIPK3 that regulates cell growth by sponging multiple miRNAs. Nat Commun 7:11215
22. Lin SP, Ye S, Long Y et al (2016) Circular RNA expression alterations are involved in OGD/R-induced neuron injury. Biochem Biophys Res Commun 471(1):52–56

23. Veno MT, Hansen TB, Veno ST et al (2015) Spatio-temporal regulation of circular RNA expression during porcine embryonic brain development. Genome Biol 16:245
24. Szabo L, Morey R, Palpant NJ et al (2015) Statistically based splicing detection reveals neural enrichment and tissue-specific induction of circular RNA during human fetal development. Genome Biol 16:126
25. Lukiw WJ (2013) Circular RNA (circRNA) in Alzheimer's disease (AD). Front Genet 4:307
26. You X, Vlatkovic I, Babic A et al (2015) Neural circular RNAs are derived from synaptic genes and regulated by development and plasticity. Nat Neurosci 18(4):603–610
27. Qu S, Zhong Y, Shang R et al (2017) The emerging landscape of circular RNA in life processes. RNA Biol 14(8):992–999
28. Salzman J (2016) Circular RNA expression: its potential regulation and function. Trends Genet 32(5):309–316
29. Qu S, Yang X, Li X et al (2015) Circular RNA: a new star of noncoding RNAs. Cancer Lett 365(2):141–148
30. Chen LL, Yang L (2015) Regulation of circRNA biogenesis. RNA Biol 12(4):381–388
31. Barrett SP, Salzman J (2016) Circular RNAs: analysis, expression and potential functions. Development 143(11):1838–1847
32. Chen LL (2016) The biogenesis and emerging roles of circular RNAs. Nat Rev Mol Cell Biol 17(4):205–211
33. de Bruin RG, Rabelink TJ, van Zonneveld AJ et al (2017) Emerging roles for RNA-binding proteins as effectors and regulators of cardiovascular disease. Eur Heart J 38(18):1380–1388
34. Fischer JW, Leung AK (2017) CircRNAs: a regulator of cellular stress. Crit Rev Biochem Mol Biol 52(2):220–233
35. Viereck J, Thum T (2017) Circulating noncoding RNAs as biomarkers of cardiovascular disease and injury. Circ Res 120(2):381–399
36. Du WW, Zhang C, Yang W et al (2017) Identifying and characterizing circRNA-protein interaction. Theranostics 7(17):4183–4191
37. Cheng S, Vasan RS (2011) Advances in the epidemiology of heart failure and left ventricular remodeling. Circulation 124(20):e516–e519
38. Dorn GW 2nd (2007) The fuzzy logic of physiological cardiac hypertrophy. Hypertension 49(5):962–970
39. Han B, Chao J, Yao H (2018) Circular RNA and its mechanisms in disease: from the bench to the clinic. Pharmacol Ther. https://doi.org/10.1016/j.pharmthera.2018.01.010
40. Viereck J, Thum T (2017) Long noncoding RNAs in pathological cardiac remodeling. Circ Res 120(2):262–264
41. Burchfield JS, Xie M, Hill JA (2013) Pathological ventricular remodeling: mechanisms: part 1 of 2. Circulation 128(4):388–400
42. Tan WL, Lim BT, Anene-Nzelu CG et al (2017) A landscape of circular RNA expression in the human heart. Cardiovasc Res 113(3):298–309
43. Werfel S, Nothjunge S, Schwarzmayr T et al (2016) Characterization of circular RNAs in human, mouse and rat hearts. J Mol Cell Cardiol 98:103–107
44. Wang K, Long B, Liu F et al (2016) A circular RNA protects the heart from pathological hypertrophy and heart failure by targeting miR-223. Eur Heart J 37(33):2602–2611
45. Chen CZ, Li L, Lodish HF et al (2004) MicroRNAs modulate hematopoietic lineage differentiation. Science 303(5654):83–86
46. O'Connell RM, Zhao JL, Rao DS (2011) MicroRNA function in myeloid biology. Blood 118(11):2960–2969
47. Gao Y, Lin L, Li T et al (2017) The role of miRNA-223 in cancer: function, diagnosis and therapy. Gene 616:1–7
48. Johnnidis JB, Harris MH, Wheeler RT et al (2008) Regulation of progenitor cell proliferation and granulocyte function by microRNA-223. Nature 451(7182):1125–1129
49. Lu H, Buchan RJ, Cook SA (2010) MicroRNA-223 regulates Glut4 expression and cardiomyocyte glucose metabolism. Cardiovasc Res 86(3):410–420
50. Parikh NI, Gona P, Larson MG et al (2009) Long-term trends in myocardial infarction incidence and case fatality in the National Heart, Lung, and Blood Institute's Framingham Heart study. Circulation 119(9):1203–1210
51. Saparov A, Ogay V, Nurgozhin T et al (2017) Role of the immune system in cardiac tissue damage and repair following myocardial infarction. Inflamm Res 66(9):739–751
52. Oerlemans MI, Koudstaal S, Chamuleau SA et al (2013) Targeting cell death in the reperfused heart: pharmacological approaches for cardioprotection. Int J Cardiol 165(3):410–422
53. Nahrendorf M, Pittet MJ, Swirski FK (2010) Monocytes: protagonists of infarct inflammation and repair after myocardial infarction. Circulation 121(22):2437–2445
54. Hansen TB, Kjems J, Damgaard CK (2013) Circular RNA and miR-7 in cancer. Cancer Res 73(18):5609–5612
55. Peng L, Yuan XQ, Li GC (2015) The emerging landscape of circular RNA ciRS-7 in cancer (Review). Oncol Rep 33(6):2669–2674
56. Piwecka M, Glazar P, Hernandez-Miranda LR et al (2017) Loss of a mammalian circular RNA locus causes miRNA deregulation and affects brain function. Science 357(6357)
57. Hansen TB, Jensen TI, Clausen BH et al (2013) Natural RNA circles function as efficient microRNA sponges. Nature 495(7441):384–388

58. Geng HH, Li R, Su YM et al (2016) The circular RNA Cdr1as promotes myocardial infarction by mediating the regulation of miR-7a on its target genes expression. PLoS One 11(3):e0151753
59. Tondera D, Czauderna F, Paulick K et al (2005) The mitochondrial protein MTP18 contributes to mitochondrial fission in mammalian cells. J Cell Sci 118(Pt 14):3049–3059
60. Tondera D, Santel A, Schwarzer R et al (2004) Knockdown of MTP18, a novel phosphatidylinositol 3-kinase-dependent protein, affects mitochondrial morphology and induces apoptosis. J Biol Chem 279(30):31544–31555
61. Wang K, Gan TY, Li N et al (2017) Circular RNA mediates cardiomyocyte death via miRNA-dependent upregulation of MTP18 expression. Cell Death Differ 24(6):1111–1120
62. Segura AM, Frazier OH, Buja LM (2014) Fibrosis and heart failure. Heart Fail Rev 19(2):173–185
63. Tang CM, Zhang M, Huang L et al (2017) CircRNA_000203 enhances the expression of fibrosis-associated genes by derepressing targets of miR-26b-5p, Col1a2 and CTGF, in cardiac fibroblasts. Sci Rep 7:40342
64. Zhou B, Yu JW (2017) A novel identified circular RNA, circRNA_010567, promotes myocardial fibrosis via suppressing miR-141 by targeting TGF-beta1. Biochem Biophys Res Commun 487(4):769–775
65. Du WW, Yang W, Liu E et al (2016) Foxo3 circular RNA retards cell cycle progression via forming ternary complexes with p21 and CDK2. Nucleic Acids Res 44(6):2846–2858
66. Du WW, Yang W, Chen Y et al (2017) Foxo3 circular RNA promotes cardiac senescence by modulating multiple factors associated with stress and senescence responses. Eur Heart J 38(18):1402–1412
67. Manning BD, Toker A (2017) AKT/PKB signaling: navigating the network. Cell 169(3):381–405
68. Seong HA, Jung H, Choi HS et al (2005) Regulation of transforming growth factor-beta signaling and PDK1 kinase activity by physical interaction between PDK1 and serine-threonine kinase receptor-associated protein. J Biol Chem 280(52):42897–42908
69. Zeng Y, Du WW, Wu Y et al (2017) A circular RNA binds to and activates AKT phosphorylation and nuclear localization reducing apoptosis and enhancing cardiac repair. Theranostics 7(16):3842–3855
70. Burd CE, Jeck WR, Liu Y et al (2010) Expression of linear and novel circular forms of an INK4/ARF-associated non-coding RNA correlates with atherosclerosis risk. PLoS Genet 6(12):e1001233
71. Holdt LM, Stahringer A, Sass K et al (2016) Circular non-coding RNA ANRIL modulates ribosomal RNA maturation and atherosclerosis in humans. Nat Commun 7:12429
72. Grimm T, Holzel M, Rohrmoser M et al (2006) Dominant-negative Pes1 mutants inhibit ribosomal RNA processing and cell proliferation via incorporation into the PeBoW-complex. Nucleic Acids Res 34(10):3030–3043
73. Rohrmoser M, Holzel M, Grimm T et al (2007) Interdependence of Pes1, Bop1, and WDR12 controls nucleolar localization and assembly of the PeBoW complex required for maturation of the 60S ribosomal subunit. Mol Cell Biol 27(10):3682–3694
74. Holzel M, Grimm T, Rohrmoser M et al (2007) The BRCT domain of mammalian Pes1 is crucial for nucleolar localization and rRNA processing. Nucleic Acids Res 35(3):789–800
75. Errichelli L, Dini Modigliani S, Laneve P et al (2017) FUS affects circular RNA expression in murine embryonic stem cell-derived motor neurons. Nat Commun 8:14741
76. Bayfield OW, Chen CS, Patterson AR et al (2012) Trp RNA-binding attenuation protein: modifying symmetry and stability of a circular oligomer. PLoS One 7(9):e44309
77. Gupta SK, Garg A, Bar C et al (2018) Quaking inhibits doxorubicin-mediated cardiotoxicity through regulation of cardiac circular RNA expression. Circ Res 122(2):246–254
78. Khan MA, Reckman YJ, Aufiero S et al (2016) RBM20 regulates circular RNA production from the Titin gene. Circ Res 119(9):996–1003
79. Guo W, Schafer S, Greaser ML et al (2012) RBM20, a gene for hereditary cardiomyopathy, regulates titin splicing. Nat Med 18(5):766–773
80. Boeckel JN, Jae N, Heumuller AW et al (2015) Identification and characterization of hypoxia-regulated endothelial circular RNA. Circ Res 117(10):884–890
81. Yan L, Feng J, Cheng F et al (2018) Circular RNA expression profiles in placental villi from women with gestational diabetes mellitus. Biochem Biophys Res Commun 498(4):743–750
82. Xu H, Gong Z, Shen Y et al (2018) Circular RNA expression in extracellular vesicles isolated from serum of patients with endometrial cancer. Epigenomics 10(2):187–197
83. Li T, Shao Y, Fu L et al (2018) Plasma circular RNA profiling of patients with gastric cancer and their droplet digital RT-PCR detection. J Mol Med (Berl) 96(1):85–96
84. Zheng F, Yu X, Huang J et al (2017) Circular RNA expression profiles of peripheral blood mononuclear cells in rheumatoid arthritis patients, based on microarray chip technology. Mol Med Rep 16(6):8029–8036
85. Wu N, Jin L, Cai J (2017) Profiling and bioinformatics analyses reveal differential circular RNA expression in hypertensive patients. Clin Exp Hypertens 39(5):454–459
86. Alhasan AA, Izuogu OG, Al-Balool HH et al (2016) Circular RNA enrichment in platelets is a signature of transcriptome degradation. Blood 127(9):e1–e11
87. Li Y, Zheng Q, Bao C et al (2015) Circular RNA is enriched and stable in exosomes: a promising biomarker for cancer diagnosis. Cell Res 25(8):981–984

88. Bahn JH, Zhang Q, Li F et al (2015) The landscape of microRNA, Piwi-interacting RNA, and circular RNA in human saliva. Clin Chem 61(1):221–230
89. Hang D, Zhou J, Qin N et al (2018) A novel plasma circular RNA circFARSA is a potential biomarker for non-small cell lung cancer. Cancer Med. https://doi.org/10.1002/cam4.1514
90. Zhang S, Zeng X, Ding T et al (2018) Microarray profile of circular RNAs identifies hsa_circ_0014130 as a new circular RNA biomarker in non-small cell lung cancer. Sci Rep 8(1):2878
91. Tan S, Gou Q, Pu W et al (2018) Circular RNA F-circEA produced from EML4-ALK fusion gene as a novel liquid biopsy biomarker for non-small cell lung cancer. Cell Res. https://doi.org/10.1038/s41422-018-0033-7
92. Sun H, Tang W, Rong D et al (2018) Hsa_circ_0000520, a potential new circular RNA biomarker, is involved in gastric carcinoma. Cancer Biomark 21(2):299–306
93. Li H, Li K, Lai W et al (2018) Comprehensive circular RNA profiles in plasma reveals that circular RNAs can be used as novel biomarkers for systemic lupus erythematosus. Clin Chim Acta 480:17–25
94. Chen B, Huang S (2018) Circular RNA: an emerging non-coding RNA as a regulator and biomarker in cancer. Cancer Lett 418:41–50
95. Iparraguirre L, Munoz-Culla M, Prada-Luengo I et al (2017) Circular RNA profiling reveals that circular RNAs from ANXA2 can be used as new biomarkers for multiple sclerosis. Hum Mol Genet 26(18):3564–3572
96. Xuan L, Qu L, Zhou H et al (2016) Circular RNA: a novel biomarker for progressive laryngeal cancer. Am J Transl Res 8(2):932–939
97. Qin M, Liu G, Huo X et al (2016) Hsa_circ_0001649: a circular RNA and potential novel biomarker for hepatocellular carcinoma. Cancer Biomark 16(1):161–169
98. Li P, Chen S, Chen H et al (2015) Using circular RNA as a novel type of biomarker in the screening of gastric cancer. Clin Chim Acta 444:132–136
99. Vausort M, Salgado-Somoza A, Zhang L et al (2016) Myocardial infarction-associated circular RNA predicting left ventricular dysfunction. J Am Coll Cardiol 68(11):1247–1248
100. Deng Y-YaZ, Weiping and She, Jianqing and Zhang, Lisa and Chen, Tao and Zhou, Juan and Yuan, Zu-Yi (2016) GW27-e1167 circular RNA related to PPARγ function as ceRNA of microRNA in human acute myocardial infarction. J Am Coll Cardiol 68:C51–C52
101. Zhao Z, Li X, Gao C et al (2017) Peripheral blood circular RNA hsa_circ_0124644 can be used as a diagnostic biomarker of coronary artery disease. Sci Rep 7:39918
102. Jakobi T, Czaja-Hasse LF, Reinhardt R et al (2016) Profiling and validation of the circular RNA repertoire in adult murine hearts. Genomics Proteomics Bioinformatics 14(4):216–223
103. Wu HJ, Zhang CY, Zhang S et al (2016) Microarray expression profile of circular RNAs in heart tissue of mice with myocardial infarction-induced heart failure. Cell Physiol Biochem 39(1):205–216
104. Abe N, Matsumoto K, Nishihara M et al (2015) Rolling circle translation of circular RNA in living human cells. Sci Rep 5:16435
105. Abe N, Hiroshima M, Maruyama H et al (2013) Rolling circle amplification in a prokaryotic translation system using small circular RNA. Angew Chem Int Ed Engl 52(27):7004–7008
106. Wang Y, Wang Z (2015) Efficient backsplicing produces translatable circular mRNAs. RNA 21(2):172–179
107. Pamudurti NR, Bartok O, Jens M et al (2017) Translation of CircRNAs. Mol Cell 66(1):9–21 e27
108. Yang Y, Fan X, Mao M et al (2017) Extensive translation of circular RNAs driven by N(6)-methyladenosine. Cell Res 27(5):626–641
109. Yang Y, Gao X, Zhang M et al (2018) Novel role of FBXW7 circular RNA in repressing Glioma tumorigenesis. J Natl Cancer Inst 110(3)
110. Legnini I, Di Timoteo G, Rossi F et al (2017) Circ-ZNF609 is a circular RNA that can be translated and functions in myogenesis. Mol Cell 66(1):22–37 e29
111. Lenk K, Erbs S, Hollriegel R et al (2012) Exercise training leads to a reduction of elevated myostatin levels in patients with chronic heart failure. Eur J Prev Cardiol 19(3):404–411
112. Wei X, Liu X, Rosenzweig A (2015) What do we know about the cardiac benefits of exercise? Trends Cardiovasc Med 25(6):529–536
113. Fleg JL, Cooper LS, Borlaug BA et al (2015) Exercise training as therapy for heart failure: current status and future directions. Circ Heart Fail 8(1):209–220
114. Lerchenmuller C, Rosenzweig A (2014) Mechanisms of exercise-induced cardiac growth. Drug Discov Today 19(7):1003–1009
115. Fujimoto N, Prasad A, Hastings JL et al (2010) Cardiovascular effects of 1 year of progressive and vigorous exercise training in previously sedentary individuals older than 65 years of age. Circulation 122(18):1797–1805
116. Young DR, Reynolds K, Sidell M et al (2014) Effects of physical activity and sedentary time on the risk of heart failure. Circ Heart Fail 7(1):21–27
117. Weeks KL, McMullen JR (2011) The athlete's heart vs. the failing heart: can signaling explain the two distinct outcomes? Physiology (Bethesda) 26(2):97–105
118. Calvert JW, Condit ME, Aragon JP et al (2011) Exercise protects against myocardial ischemia-reperfusion injury via stimulation of beta(3)-adrenergic receptors and increased nitric oxide signaling: role of nitrite and nitrosothiols. Circ Res 108(12):1448–1458
119. McMullen JR, Amirahmadi F, Woodcock EA et al (2007) Protective effects of exercise and phosphoinositide 3-kinase(p110alpha) signaling in

dilated and hypertrophic cardiomyopathy. Proc Natl Acad Sci USA 104(2):612–617
120. McMullen JR, Shioi T, Zhang L et al (2003) Phosphoinositide 3-kinase(p110alpha) plays a critical role for the induction of physiological, but not pathological, cardiac hypertrophy. Proc Natl Acad Sci USA 100(21):12355–12360
121. DeBosch B, Treskov I, Lupu TS et al (2006) Akt1 is required for physiological cardiac growth. Circulation 113(17):2097–2104
122. Zou J, Li H, Chen X et al (2014) C/EBPbeta knockdown protects cardiomyocytes from hypertrophy via inhibition of p65-NFkappaB. Mol Cell Endocrinol 390(1–2):18–25
123. Bostrom P, Mann N, Wu J et al (2010) C/EBPbeta controls exercise-induced cardiac growth and protects against pathological cardiac remodeling. Cell 143(7):1072–1083
124. Bernardo BC, Ooi JY, Lin RC et al (2015) miRNA therapeutics: a new class of drugs with potential therapeutic applications in the heart. Future Med Chem 7(13):1771–1792
125. Yang L, Li Y, Wang X et al (2016) Overexpression of miR-223 tips the balance of pro- and anti-hypertrophic signaling cascades toward physiologic cardiac hypertrophy. J Biol Chem 291(30):15700–15713
126. Liu X, Xiao J, Zhu H et al (2015) miR-222 is necessary for exercise-induced cardiac growth and protects against pathological cardiac remodeling. Cell Metab 21(4):584–595
127. Shi J, Bei Y, Kong X et al (2017) miR-17-3p contributes to exercise-induced cardiac growth and protects against myocardial ischemia-reperfusion injury. Theranostics 7(3):664–676

Circular RNAs and Neuronal Development

16

Lena Constantin

Abstract

Circular RNAs (circRNAs) are abundant in the brain and are often expressed in complex spatiotemporal patterns that coincide with distinct developmental transitions. This suggests that circRNAs play a significant role in the central nervous system. This book chapter will review research progress into the function of circRNAs during neuronal development. The major themes to be discussed are the enrichment of circRNAs in the synapse and their possible contributions to synaptopathologies, in addition to the findings that neural circRNAs accumulate with age and appear beneficial for neuronal repair. Although more research is needed, some of the possible functions of circRNAs with in the brain are already beginning to come to light.

Keywords

Circular RNA · Brain · Synapse · Ageing · Alzheimer's disease · Ischemic stroke

1 Introduction

High-throughput sequencing (RNA-seq) has greatly expanded our understanding of circRNA biology. The first half of this book chapter will review insights gained from the vast amounts of RNA-seq data, including the general properties for circRNA expression in the brain and their complex spatiotemporal expression patterns. The second half will focus on two potential functions for circRNAs in the brain. The first in synaptic learning, as evidenced by the enrichment (and activity-dependent transcription) of circRNAs in the synapse and their deregulation in some synaptopathologies. Already, a convincing circRNA-associated competing endogenous RNA network has been identified in Alzheimer's disease. The second in the neuroprotection of the ageing brain, on the basis that neuronal circRNAs accumulate with age, are spatiotemporally deregulated with age, and are differentially expressed during the recovery period after stroke or brain injury.

Although many unanswered questions still remain, this book chapter will summarise the current understanding of circRNA function in neuronal development and will put forward two potential biological roles for circRNAs in the brain.

L. Constantin (✉)
Faculty of Medicine, School of Biomedical Sciences, The University of Queensland, St Lucia 4072, QLD, Australia
e-mail: l.constantin@uq.edu.au

J. Xiao (ed.), *Circular RNAs*, Advances in Experimental Medicine and Biology 1087,
https://doi.org/10.1007/978-981-13-1426-1_16

2 The Expression of Circular RNAs in the Brain

2.1 Circular RNAs Are Lowly But Diversely Expressed in the Brain

Circular RNAs are expressed at very low levels in all tissues. The same can be said for the brain. In mouse brain, less than 0.1% of RNA-seq reads from circular junctions can be mapped back to the reference genome [1]. In comparison, 44% of all protein-coding genes RNA-seq reads from the mouse brain are mappable [1]. The majority of the distinct circRNA species that make up this small population of RNAs also appear to be expressed at low levels. For example, 41,027 out of 65,731 circRNA candidates are supported by fewer than 10 RNA-seq reads in the human brain [2]. Similarly, in mouse brain, 10,081 out of 15,849 circRNA candidates are supported by fewer than 10 RNA-seq reads [2]. Further contributing to their low abundance is that only a small percentage of genes are able to synthesise circRNAs. This is also true for the brain, where 21% of expressed genes produce circRNAs in the adult mouse brain [1], 15.8% in the foetal pig cortex [3], and 13% in the human brain [2]. However, the percentage of circRNA-hosting genes in the brain is greater than in other tissues like the heart, liver, and lung, where less than 10% of expressed genes synthesise circRNAs [3]. Therefore, in the brain, a small number of expressed genes are able to synthesise few but a diverse set of circRNAs. This is not unlike in other tissues, although the relative abundance of circRNAs in the brain is much greater.

The brain also stands apart from other tissues in its ability to synthesise the greatest number of distinct circRNAs. This is consistent across species. For example, the most comprehensive analysis of circRNA expression across tissues and developmental staging to date, which involved mining 10 billion RNA-seq reads from 103 fly sample libraries, identified that 90–95% of all circRNAs are expressed in the head [4]. In adult mouse, the brain expresses by far the greatest number of distinct circRNAs: an average of 5925 circRNA candidates (from an average total reads of 19,479,587) compared the second highest tissue, the testis, which expresses an average of 3018 circRNA candidates (out of an average total reads of 20,081,654) [1]. Similarly in the human foetus, the brain produces the greatest variety of circRNAs out of any of the other 14 tissues analysed [5]. Furthermore, one study identified 65,731 distinct circRNAs in the human brain alone [2].

Two factors may help to explain why the brain has a unique capacity to synthesise such a diverse set of circRNA species. One is that the population of circRNA-synthesising genes in the brain, on average, synthesise multiple distinct circRNAs. For example, circRNA-host genes of the adult mouse brain are able to produce an average of 2.4 distinct circRNAs, compared to other tissues like the heart, liver, and lungs, which produce an average of 1.2–1.5 circRNAs per circRNA-host gene [1]. In the human brain, this appears to be even greater, where the average circRNA-producing host genes synthesise 6.4 distinct circRNAs, or a median of three circRNAs per host gene [2]. The second factor is that many circRNA-hosting genes are exclusively expressed in the brain. For example, 225 circRNA-producing genes are exclusively expressed in the brain, relative to 140 in the testis, and fewer than 20 in the heart, liver, and lungs [1]. A similar trend has been reported in adult rat, where 60 circRNA-producing genes are exclusively expressed in the brain, ~35 in the testis, and fewer than 10 in the heart, liver, and lung [1].

Given that circRNA expression is dependent on host gene transcription, it is not unexpected that different tissues should express different subsets of circRNAs. For example, the gene ontologies of linear transcripts derived from the liver are enriched in liver-specific processes like lipoprotein metabolism and extracellular exosomes, while in the brain, the gene ontologies are enriched in brain-specific processes such as protein phosphorylation, postsynaptic density, and protein kinase activity [6]. Likewise, the host genes of tissue-specific circRNAs should also be enriched in pathways specific to that tissue, and indeed this has been shown. For instance, the host genes of brain-specific circRNAs are enriched in pathways specific to neuron development, differentiation, and synaptic transmission, while the host genes of liver-specific circRNAs are enriched in ion transport, proton transport, and caton transport [5]. Therefore, it can be concluded that the brain likely expresses the greatest number of circRNAs out of any tissue simply because of the transcriptional properties of the linear circRNA-synthesising host genes.

2.2 Circular RNA–Hosting Genes May Be Enriched in Long Neuronal Genes

Another point of difference between the linear transcripts of the brain and other tissues is that brain-specific genes tend to have much longer introns. This is particularly obvious when studying topoisomerases, enzymes that catalyse the winding and unwinding of supercoiled DNA strands during transcription or cell replication. Understandably, DNA topoisomerases are particularly important for long genes. This is exemplified by the strong negative correlation between the length of long genes (>67 kb) and their expression levels when DNA topoisomerases are inhibited [7]. Topoisomerases have recently been shown to regulate synaptic genes [7] and to maintain normal synaptic functions [8]. More specifically, the inhibition of topoisomerases at excitatory synapses reduces the number of synapses, while inhibition at inhibitory synapses interferes with the membrane trafficking of GABAA receptor subunits. Therefore, neuronal genes, particularly those involved in synaptic processes, tend to be long (and therefore dependent on DNA topoisomerases).

Circular RNAs tend to be bracketed by longer introns. The flanking introns of circRNAs in human forebrain neurons are on average five times longer than randomly selected introns [9]. Furthermore, a comprehensive analysis of 10 billion RNA-seq reads in a fly identified that the median length of a fly intron is 96 bps, the median length of introns longer than 200 bps is 1009 bp, while the median lengths of introns upstream and downstream of circRNAs are 4662 and 2962 bps, respectively [4]. Moreover, splice sites that are involved in the biogenesis of two or more circRNA isoforms tend to be flanked by even longer introns than splice sites that drive a single circRNA isoform [3]. Given that circRNA-synthesising genes are flanked by much longer introns, neuronal genes (particularly those related to the synapse) may have greater circRNA-synthesising capabilities simply because of their increased length.

2.3 Circular RNAs Are Actively Regulated in the Brain

Although circRNAs are dependent on their host gene for the initiating of transcription, hundreds of neuronal circRNAs are expressed several times higher than their host genes [1–3, 10]. In cases where circRNA-host genes are equivalently expressed in the brain and other tissues, the numbers of circRNAs produced from the brain host genes are significantly higher [1]. This suggests that preferentially expressed circRNAs may have biological roles independent from their linear host genes. CircRNAs are not only more relatively abundantly and diversely expressed in the brain, they are also differentially expressed in the different regions and cell types of the brain. For example, the adult mouse cortex and hippocampus share 4030 out of 6231 circRNA candidates; however, 2201 circRNAs are differentially expressed [11]. Some specific examples of circRNAs with region- or cell-specific expression profiles are *circRims2* and *circDym*, which are expressed at greater than 50% in the mouse adult cerebellum versus the striatum, prefrontal cortex, olfactory bulb, midbrain, and hippocampus [2]. Another example is *circPlxnd1*, which is predominantly expressed in the prefrontal cortex (<60%) versus the other aforementioned brain regions [2].

Many circRNA isoforms, which are derived from the same host gene splice acceptor or donor sites, also display divergent expression profiles in the brain [3]. An intriguing example in human cells are the *circStau2a* (containing exons 2–5) and *circStau2b* (containing exons 2–3) isoforms, which display inverse expression patterns. *circStau2b* is highly expressed in the adult brain relative to almost all other tissues, while the longer *circStau2a* isoform is highly expressed in the adult lung and relatively lowly expressed elsewhere including the brain [2]. The divergent expression profiles of circRNAs in the brain suggests that cis-regulatory elements and brain-specific trans-acting factors may regulate these processes.

There are many possibilities as to why some circRNAs are preferentially upregulated in the brain. First, the brain may be enriched for neuron-specific splicing factors and/or RNA-binding proteins that regulate circRNA biogenesis. The divergent expression profiles and levels of circRNA isoforms from identical splice acceptor or donor sites in particular support this. Second, as already discussed, circularised exons require longer introns, and neuronal genes tend to fulfil this requirement. Third, circRNAs have a half-life almost five times longer than their host transcripts [10], and in quiescent and postmitotic tissues like neurons, this allows for circRNA to accumulate. An accumulation of circRNAs relative to their linear isoforms has already been documented in the ageing fly brain [4]; conversely, circRNAs tend to be reduced in cancers [12]. Finally, highly polarised cells, such as neurons, must regulate multiple cellular functions and translate different combinations of proteins within their different cellular compartments, and this is often mediated by RNA-dependent mechanisms, not unlike circRNAs. One exciting possibility is that circRNAs may form an additional layer of localised posttranscriptional regulation.

Circular RNAs are also temporally regulated in the brain. Differentiating human primary cortical neurons upregulate 1926 and downregulate 797 (out of 5265) circRNA candidates [2]. Similarly, differentiating mouse embryonic carcinoma cells upregulate 1116 and downregulate 238 (out of 2735) circRNAs [2]. Furthermore, metabolic tagging of differentiating human forebrain neuron progenitor cells revealed that 785 (out of 11,185) circRNA candidates are upregulated over 26 days of *in vitro* differentiation (27068474). The temporal expression patterns of some circRNAs in the brain appear highly coordinated and complex. For example, the 200 most highly expressed circRNAs of the foetal pig cortex can be clustered into seven categories based on their temporal expression patterns. For instance, 29 circRNAs are expressed highly in the early half of gestation, 11 circRNAs in the late half of gestation, and 130 at a specific developmental stage during mid-gestation [3]. The tight temporal regulation of circRNAs in the brain, which coincides with distinct developmental transitions, strongly supports a biological function.

3 Circular RNAs Appear Important at the Synapse

3.1 Circular RNAs Are Greatly Enriched in the Synapse

Given that the expression of a particular circRNA is dependent on the transcription of its host gene (although, as already discussed, the relative abundance of circular and linear host transcripts can be regulated divergently), valuable insights can be gained from analysing the molecular function and biological process ontologies of host genes. Intriguing, the synapse is consistently one of the most significantly enriched ontologies for circRNA-associated genes from the brain. This is true for the set of circRNAs that are consistently upregulated in mouse hippocampal cells during postnatal development [1] and for the set of circRNAs most highly elevated with age in the fly brain [4]. Further to this, circRNAs are highly expressed in synaptic subcellular fractions. The majority of circRNAs in the adult mouse brain (1117) are enriched in the synaptoneurosomes, relative to the corresponding cytoplasmic (709) and whole brain (847) fractions

[2]. Similarly, in the mouse and rat hippocampus, circRNAs are enriched in synaptoneurosomes and/or synaptic neuropils (which exhibit robust synaptic plasticity), compared to corresponding cell body or whole hippocampal homogenates [1]. The enrichment of circRNAs in synaptic compartments has also been validated visually using high-resolution *in situ* hybridisation, where at least eight circRNAs derived from synapse-related host genes were located throughout the dendritic arbours of the mouse hippocampus [1].

Evidence is also beginning to suggest that circRNAs may play a direct role in homeostatic scaling. Homeostatic adaptation, or scaling, is the ability of neurons to maintain excitability while the brain is adjusting to environmental change. It involves a cell-wide increase or decrease in postsynaptic α-amino-3-hydroxy-5-methyl-4-isoxazolepropionic acid (AMPA) receptor in excitatory synapses; this scales all synapses by the same multiplicative factor—to become stronger or weaker—while maintaining their relative strengths. Bicuculline, a gamma-aminobutyric acid (GABA)A-receptor antagonist, can also be used to induce homeostatic plasticity. Recently, 37 circRNAs (versus 7 host genes) were transcribed in response to bicuculline-induced homeostatic plasticity, while 5 circRNAs (versus 3 host genes) were downregulated [1]. These circRNAs are examples of synaptic activity-dependent transcription, which suggests a role in neural plasticity. The most dramatically upregulated transcripts were *circHomer1_a* and its host gene, *homer homolog 1 (Drosophila) (Homer1)* [1].

Homer1 can be translated into three different protein products at the postsynaptic density: Homer1a, Homer1b, and Homer1c. Homer1b/c is constitutively expressed, while Homer1a is transiently upregulated during increases to network activity, such as those created by long-term bicuculline treatment [13]. Under neutral network conditions, Homer1b/c interacts with group I metabotropic glutamate receptor (mGluR) at the postsynaptic density; however, when Homer1a is expressed, it interferes with native interactions between mGluR and Homer1b/c [14]. This leads to a cell-wide reorganisation of the postsynaptic density, which activates group I mGluR signalling that in turn initiates homeostatic adaptation [13]. During homeostatic adaptation, *circHomer1_a* is also synthesised. The synthesis of *circHomer1_a* requires the same splice sites as those for *Homer1b/c* transcripts [1]. Therefore, *circHomer1_a* and *Homer1b/c* transcripts cannot be mutually expressed. Circular RNAs are thought to regulate translation by competing with the canonical splicing of the host gene [15]. This may be such an example. That is, the biogenesis of *circHomer1_a* may be actively and purposefully competing with the transcription of *Homer1b/c* transcripts during homeostatic adaptation, with the goal of reducing competition between *Homer1a* and *Homer1b/c* mRNA synthesis. Further experimental validation is required.

3.2 Circular RNAs May Be Linked to Neurodegenerative Synaptopathies

Alzheimer's disease (AD) is the most common form of progressive dementia in the ageing brain. The pathological features of AD are intracellular tau-containing neurofibrillary tangles and extra-cellular amyloid-β plaques, which accumulate in vulnerable regions of the brain such as the cortex and hippocampus. Alzheimer's disease is thought to be a synaptic pathology. Some evidence for this are that subtle alterations in the synaptic efficacy of the hippocampus occur in AD patients prior to the detection of neurofibrillary tangles and amyloid-β plaques [16]. Also, patients within 2–4 years of the clinical onset of AD have reduced numbers of spines per neuron in layers II–III (38%) and V (30%) of the temporal cortex and in layer V (30%) of the frontal cortex [17]. Furthermore, cognitive deficits associated with AD are more strongly correlated to neocortical synapse loss compared to the number of plaques and tangles [18].

Recently, circRNA dysfunction was identified in a sporadic mouse model of AD [19]. More specifically, 94 and 141 circRNAs were, respectively, up- and downregulated (out of 34,096) in the adult brain relative to controls. Based on this, in combination with RNA-seq reads on deregulated miRNAs and linear mRNAs, a circRNA-associated-competing endogenous RNA network was built. This RNA regulatory network did not take into account the number and density of miRNA-binding seed sequences and therefore likely contains a large proportion of false-positive pairings. Nonetheless, the network identified two interactions formally linked to AD involving the genes *deiodinase, iodothyronine, type II* (*Dio2*), and *high-mobility group box 2* (*HMGB2*). *Dio2*, which activates myelination [20] and is reduced in AD [21], putatively associated with *miR-122-5p* and five deregulated circRNAs. *HMGB2*, which activates pathways involved in amyloid-β plaque clearance [22], was paired with *let-7 g-3p* and deregulated 3 circRNAs. These circRNAs may be competitively modulating the activity of *miR-122-5p* and *let-7 g-3p*, to affect the expression of *Dio2* and *HMGB2*, in a process commonly described as 'miRNA sponging' [23]. This study did not identify the most convincing circRNA-mRNA pairing involved in human AD, as would be expected: *ciRS-7* and *ubiquitin-conjugating enzyme E2A, RAD6 homolog (S. cerevisiae)* (*UBE2A*).

The circRNA *ciRS-7* is produced from the antisense of the *cerebellar degeneration-related protein 1* (*CDR1 as*) gene. *ciRS-7* contains 74 tandem seed matches to *miR-7*, and 63 of these are conserved from annelids to humans [24]. Unsurprisingly, *ciRS-7* is a potent negative regulator of *miR-7*. For example, based on RNA-seq data, a single HEK293 cell is estimated to contain ~1400 *ciRS-7* molecules which can sequester up to 20,000 *miR-7* molecules [24]. In patients with sporadic AD, *ciRS-7* is downregulated by 5.4-fold in the hippocampal CA1 region [25, 26] and is significantly reduced by more than fivefold in the superior temporal lobe neocortex (Brodmann area 22). A decrease of *ciRS-7* in hippocampal CA1 and Brodmann area 22 would enable the excess accumulation of *miR-7* in these regions. Indeed, *miR-7* is upregulated in the brain of AD patients by an average of threefold [26]. Excesses of *miR-7* would in turn repress *UBE2A*, which is essential for the proteolytic clearance of amyloid-β peptides in Alzheimer's disease. Indeed, *UBE2A* is also downregulated by 3.7-fold in the hippocampal CA1 region and by 2.8-fold in Brodmann area 22 [26]. Thus, a deficiency in *ciRS-7* 'miRNA sponging' would enable *miR-7* to potentially and efficiently downregulate target genes essential for the clearance of amyloid-β plaques. In a similar fashion, *ciRS-7* could be protective against Parkinson's disease by preventing *miR-7* from silencing *epidermal growth factor receptor* (*EGFR*), *alpha-synuclein* (*SNCA*), and *insulin receptor substrate 2* (*IRS2*) [27].

The fervent sponging capacity of *ciRS-7* is not, however, a general feature of circRNAs. The majority of circRNAs expressed in the brain have comparable numbers of miRNA-binding sites as linear mRNAs and therefore would not make for strong 'sponges' [1, 23, 28]. Also, organisms completely devoid of RNA interference, such as the yeast *Saccharomyces cerevisiae* [29], clearly produce circRNAs [30]. Hence, the function of the majority of circRNAs must extend beyond being competitive-binding moderators of the RNA interference pathway.

4 Circular RNAs Appear Important for the Ageing Brain

4.1 Circular RNAs Accumulate in the Ageing Brain

Circular RNAs accumulate in the ageing brain. This is substantiated by the most comprehensive survey of circRNA expression to date that mined 103 fly tissues from various developmental stages [4]. More specifically, the total circRNA levels were found to increase across fly embryo development and dramatically increase in the adult head relative to earlier time points from the head or all other adult tissues [4]. Similarly in the mouse, the global levels of circRNAs—measured by transcript per kilobase million—significantly increase from 1 to 22 months in the cortex and hippocampus, but not in the heart [11]. CircRNAs not only accumulate in the brain with time, but a subset is differentially upregulated with age independently from host genes. In a fly, 262 (out of 2513) circRNA candidates are significantly upregulated by more than twofold in 20-day-old heads compared to 1-day-old heads [4]. Similarly in the ageing mouse hippocampus, 250 (out of 5528) circRNA candidates are significantly upregulated at 22 months compared to 1 month, and in the ageing cortex, 258 (out of 4733) circRNA candidates are significantly upregulated at 22 months compared to 1 month [11]. In the mouse cortex, the functional ontologies of host genes synthesising age-upregulated circRNAs are enriched for synapse assembly, synapse organisation, neurotransmitter secretion, and neurotransmitter transport [11]. Alternatively, in the mouse hippocampus, the host genes of age-upregulated circRNAs are enriched for protein and chromatin modifications [11].

Given that linear RNA expression does not change with age in the mouse brain [11] and that age-upregulated circRNAs were largely independent to the expression level of the host gene [4, 11], the mechanisms that drive circRNA age-accumulation are not host-dependent. One mechanism may be that circRNAs are especially stable in the quiescent and postmitotic cells of the ageing brain, which allow a greater proportion of circRNAs to accumulate over time. Another mechanism may be related to the phenomenon that more than one-third of genes expressed in the ageing human brain undergo changes to alternative splicing that encourage back-splicing [31]. The next steps would be to determine whether the accumulation of circRNAs is innocuous or serves a protective or detrimental function to the brain. Recent findings that circRNAs are deregulated following stroke provide insights into the potential role of age-upregulated circRNAs.

4.2 Circular RNAs Are Linked to Neural Repair

Stroke is the third leading cause of death in the United States. It is an acute neurological event that leads to the death of neural tissues. The majority of strokes result from vascular occlusions, called ischemic strokes. For some ischemic stroke patients, the restoration of blood flow exacerbates the initial injury, producing a so-called cerebral reperfusion injury. The specific role of circRNAs during 'cerebral reperfusion injury' has been investigated by three different laboratories, using either the intraluminal middle cerebral artery occlusion model in the mouse [32, 33] or oxygen-glucose deprivation and then reoxygenation in cell culture [34]. The most comprehensive analyses to date identified 283 deregulated (out of 1064) circRNA candidates over a 6-, 12-, and/or 24-h time course, with 239 significantly altered at 6 h [32]. This great peak of circRNA deregulation at 6 h, but not at 12 and 24 h, is suggestive of complex temporal regulatory processes taking place. Another study profiled circRNAs at 48 h after artery occlusion and identified over a thousand deregulated circRNAs [33]. Only modest changes in the expression of 15 circRNAs were found at 24 h in the cell culture model [34]. These three studies lack consistencies in the particular circRNAs involved. For example, none of the circRNAs altered post-stroke were shared between the cell culture [34] and intraluminal middle cerebral artery occlusion model [32] at 24 h. Also, there was a far greater number of circRNAs deregulated at 48 h [33] compared to before 24 h [32], perhaps highlighting differences in the models used and the brain regions assayed.

The host genes of stroke-responsive circRNAs across these studies were, however, enriched for repair processes. For example, at 48 h post-stroke, host genes were most significantly enriched for the cell survival and proliferation pathways of Rap1 and Hippo signalling [33]. Similarly, the host genes of all stroke-responsive circRNAs over the entire 24-h time course were most significantly enriched for mitogen-activated kinase signalling, cell cycle and actin cytoskeletal regulators, and focal adhesions molecules that are related to cell growth, proliferation, and death [32]. In a rat model of traumatic injury to the hippocampus, the host genes of injury-responsive circRNAs were most significantly enriched for neurogenesis, neuronal differentiation and development, and in cellular components related to the synapse (5 out of the top 10) [35]. Therefore, circRNAs that are deregulated following brain injury are tightly associated with neuronal repair processes. This tentatively implies that age-upregulated circRNAs may serve as a biological function during recovery in the injured brain, although much more research is required.

Acknowledgement I would like to thank Professor Brandon J. Wainwright for encouraging me to pursue noncoding RNAs as a field of research.

Competing Financial Interests The author declares no competing financial interests.

References

1. You X, Vlatkovic I, Babic A et al (2015) Neural circular RNAs are derived from synaptic genes and regulated by development and plasticity. Nat Neurosci 18(4):603–610
2. Rybak-Wolf A, Stottmeister C, Glazar P et al (2015) Circular RNAs in the mammalian brain are highly abundant, conserved, and dynamically expressed. gs. Mol Cell 58(5):870–885
3. Veno MT, Hansen TB, Veno ST et al (2015) Spatio-temporal regulation of circular RNA expression during porcine embryonic brain development. Genome Biol 16:245
4. Westholm JO, Miura P, Olson S et al (2014) Genome-wide analysis of drosophila circular RNAs reveals their structural and sequence properties and age-dependent neural accumulation. Cell Rep 9(5):1966–1980
5. Xia S, Feng J, Lei L et al (2017) Comprehensive characterization of tissue-specific circular RNAs in the human and mouse genomes. Brief Bioinform 18(6):984–992
6. Li L, Zheng YC, Kayani MUR et al (2017) Comprehensive analysis of circRNA expression profiles in humans by RAISE. Int J Oncol 51(6):1625–1638
7. King IF, Yandava CN, Mabb AM et al (2013) Topoisomerases facilitate transcription of long genes linked to autism. Nature 501(7465):58–62
8. Mabb AM, Kullmann PH, Twomey MA et al (2014) Topoisomerase 1 inhibition reversibly impairs synaptic function. Proc Natl Acad Sci U S A 111(48):17290–17295
9. Zhang XO, Wang HB, Zhang Y et al (2014) Complementary sequence-mediated exon circularization. Cell 159(1):134–147
10. Jeck WR, Sorrentino JA, Wang K et al (2013) Circular RNAs are abundant, conserved, and associated with ALU repeats. RNA 19(2):141–157
11. Gruner H, Cortes-Lopez M, Cooper DA et al (2016) CircRNA accumulation in the aging mouse brain. Sci Rep 6:38907
12. Bachmayr-Heyda A, Reiner AT, Auer K et al (2015) Correlation of circular RNA abundance with proliferation—exemplified with colorectal and ovarian cancer, idiopathic lung fibrosis, and normal human tissues. Sci Rep 5:8057
13. Hu JH, Park JM, Park S et al (2010) Homeostatic scaling requires group I mGluR activation mediated by Homer1a. Neuron 68(6):1128–1142
14. Tu JC, Xiao B, Yuan JP et al (1998) Homer binds a novel proline-rich motif and links group 1 metabotropic glutamate receptors with IP3 receptors. Neuron 21 (4):717–726; Brakeman PR, Lanahan AA, O'Brien R et al (1997) Homer: a protein that selectively binds metabotropic glutamate receptors. Nature 386 (6622):284–288
15. Ashwal-Fluss R, Meyer M, Pamudurti NR et al (2014) circRNA biogenesis competes with pre-mRNA splicing. Mol Cell 56(1):55–66
16. Selkoe DJ (2002) Alzheimer's disease is a synaptic failure. Science 298(5594):789–791
17. Davies CA, Mann DM, Sumpter PQ et al (1987) A quantitative morphometric analysis of the neuronal and synaptic content of the frontal and temporal cortex in patients with Alzheimer's disease. J Neurol Sci 78(2):151–164
18. Terry RD, Masliah E, Salmon DP et al (1991) Physical basis of cognitive alterations in Alzheimer's disease: synapse loss is the major correlate of cognitive impairment. Ann Neurol 30(4):572–580
19. Zhang S, Zhu D, Li H et al (2017) Characterization of circRNA-associated-ceRNA networks in a senescence-accelerated mouse prone 8 brain. Mol Ther 25(9):2053–2061
20. Calza L, Fernandez M, Giuliani A et al (2002) Thyroid hormone activates oligodendrocyte precursors and increases a myelin-forming protein and NGF content in the spinal cord during experimental allergic encephalomyelitis. Proc Natl Acad Sci U S A 99(5):3258–3263
21. Humphries CE, Kohli MA, Nathanson L et al (2015) Integrated whole transcriptome and DNA methylation analysis identifies gene networks specific to late-onset Alzheimer's disease. J Alzheimers Dis 44(3):977–987
22. Yamanaka Y, Faghihi MA, Magistri M et al (2015) Antisense RNA controls LRP1 sense transcript expression through interaction with a chromatin-associated protein, HMGB2. Cell Rep 11(6):967–976
23. Guo JU, Agarwal V, Guo H et al (2014) Expanded identification and characterization of mammalian circular RNAs. Genome Biol 15(7):409
24. Memczak S, Jens M, Elefsinioti A et al (2013) Circular RNAs are a large class of animal RNAs with regulatory potency. Nature 495(7441):333–338

25. Lukiw WJ (2013) Circular RNA (circRNA) in Alzheimer's disease (AD). Front Genet 4:307
26. Zhao Y, Alexandrov PN, Jaber V et al (2016) Deficiency in the ubiquitin conjugating enzyme UBE2A in Alzheimer's disease (AD) is linked to deficits in a natural circular miRNA-7 sponge (circRNA; ciRS-7). Genes (Basel) 7(12):116
27. Hansen TB, Jensen TI, Clausen BH et al (2013) Natural RNA circles function as efficient microRNA sponges. Nature 495 (7441):384–388; Junn E, Lee KW, Jeong BS et al (2009) Repression of alpha-synuclein expression and toxicity by microRNA-7. Proc Natl Acad Sci U S A 106 (31):13052–13057
28. Guo J (2014) Transcription: the epicenter of gene expression. J Zhejiang Univ Sci B 15(5):409–411
29. Drinnenberg IA, Fink GR, Bartel DP (2011) Compatibility with killer explains the rise of RNAi-deficient fungi. Science 333(6049):1592
30. Wang PL, Bao Y, Yee MC et al (2014) Circular RNA is expressed across the eukaryotic tree of life. PLoS One 9(6):e90859
31. Mazin P, Xiong J, Liu X et al (2013) Widespread splicing changes in human brain development and aging. Mol Syst Biol 9:633; Harries LW, Hernandez D, Henley W et al (2011) Human aging is characterized by focused changes in gene expression and deregulation of alternative splicing. Aging Cell 10 (5):868–878
32. Mehta SL, Pandi G, Vemuganti R (2017) Circular RNA expression profiles alter significantly in mouse brain after transient focal ischemia. Stroke 48(9):2541–2548
33. Liu C, Zhang C, Yang J et al (2017) Screening circular RNA expression patterns following focal cerebral ischemia in mice. Oncotarget 8(49):86535–86547
34. Lin SP, Ye S, Long Y et al (2016) Circular RNA expression alterations are involved in OGD/R-induced neuron injury. Biochem Biophys Res Commun 471(1):52–56
35. Xie B, Wang Y, Lin Y et al (2018) Circular RNA expression profiles alter significantly after traumatic brain injury in rats. J Neurotrauma. https://doi.org/10.1089/neu.2017.5468

Circular RNAs in Cancer

17

Susanne Lux and Lars Bullinger

Abstract

Circular RNAs (circRNAs) constitute a class of RNAs that only recently have come into the focus of the scientific cancer community after it was revealed that they are very abundant, highly conserved across species and show tissue- and developmental stage-specific expression. This tightly regulated, dynamic circRNA expression, in line with expression of messenger RNAs, microRNAs, and long noncoding RNAs, is altered in both solid tumors and hematologic malignancies and most likely contributes to tumorigenesis. In this chapter, we will review cancer-associated and cancer-specific circRNAs, some of which have oncogenic or tumor-suppressive potential. We will specifically focus on circRNAs for which the role in cancer has been studied in more detail, and we will discuss the opportunity to use circRNAs as biomarkers and potential therapeutic targets in cancer.

S. Lux
Department of Internal Medicine III, University Hospital of Ulm, Ulm, Germany

L. Bullinger (✉)
Department of Internal Medicine III, University Hospital of Ulm, Ulm, Germany

Department of Hematology, Oncology and Tumorimmunology, Charité – Universitätsmedizin, Berlin, Germany
e-mail: lars.bullinger@charite.de

Keywords

CircRNA · MicroRNA · Cancer · Solid tumor · Hematologic malignancies · Oncogene · Tumor suppressor

1 Introduction

Despite the fact that in some cancers, e.g., in leukemias, usually only few gene mutations affecting protein-coding sequences can be detected, cancer is in general a disease accompanied by a global deregulation of gene expression compared to normal tissue. This leads to the assumption that in addition to genomic mutations, other changes, such as deregulated epigenetic mechanisms, contribute to deregulated gene expression, which by itself might contribute to tumorigenesis. In addition to aberrant expression of genes in cancer, the expression of alternative splice variants as well as noncoding RNAs (ncRNAs) is involved in the global deregulation of gene expression in cancer and might also play an important role.

Recent transcriptome studies have uncovered that, while only 2% of the genome encodes for proteins, more than 60% of the genome is transcribed and can be detected as RNA transcripts [1, 2]. This implies an important biological function of the noncoding transcriptome, and in accordance, several studies have pointed to an important role of noncoding RNAs such as

J. Xiao (ed.), *Circular RNAs*, Advances in Experimental Medicine and Biology 1087,
https://doi.org/10.1007/978-981-13-1426-1_17

microRNAs (miRNA) and long noncoding RNAs (lncRNA) in cancer [3–6].

MiRNAs represent a well-studied subgroup of small ncRNAs that is implicated in several pathologies. Aberrant miRNA expression profiles are found in various types of cancer and can be correlated with clinical parameters like overall survival [7]. In addition, selected miRNAs have been shown to influence apoptosis, proliferation, and differentiation of cancer cells [8, 9]. Furthermore, lncRNAs do also show cell type-specific expression [10] and are regulated during development [11], and selected lncRNAs have been proven as regulators of self-renewal and pluripotency in stem cells [12]. Not surprisingly, some lncRNAs are also implicated in cancer, e.g., lncRNA URHC regulates cell proliferation and apoptosis in hepatocellular carcinoma [13], lncRNA HIS-1 is overexpressed in murine myeloid leukemia [14], and the oncogenic lncRNA HOTAIR is upregulated in breast cancer and promotes tumor metastasis [15]. Moreover, lncRNAs can regulate DNA methylation [16] and coactivate proteins involved in transcriptional regulation [17], thereby possibly contributing to the global transcriptional dysregulation in cancer.

Like the other ncRNAs, circular RNAs (circRNAs) are a class of RNAs that has long been dismissed as rare splicing errors or even experimental artifacts. Only recently it was revealed that they are very abundant, are highly conserved across species, and show tissue- and developmental stage-specific expression [18–22]. With regard to circRNA function, some circRNAs have been shown to serve as miRNA sponges [23], and recent studies show that at least some of the circRNAs can be translated into proteins [24, 25] which adds another level of complexity on this matter and challenges the classification of circRNAs as "noncoding" RNAs. Due to their structure, circRNAs are very stable and protected from exonuclease degradation, and they can be detected in serum and saliva [26–28] which comprises an important criterion for a potential biomarker.

This chapter aims to give an overview of circRNAs that have been implicated in cancer, both in solid tumors and in hematologic malignancies, with focus on cancer-associated and cancer-specific circRNAs and variants for which the functional impact on carcinogenesis has been more comprehensively studied.

2 CircRNAs in Cancer

So far, studies have focused both on the characterization of the circular RNAome in hematopoietic malignancies and in solid tumors. While many studies are mainly descriptive, listing circRNAs differentially expressed in tumor versus healthy tissue, others investigated candidate circRNAs in detail, further elucidating the function of the respective "mysterious" circles in the cell.

In general, studies followed mainly two approaches to screen the circular RNAome of cancer cells: (1) several groups performed ribosomal RNA-depleted RNA sequencing (RNAseq), comparing cancer samples with healthy cells or different subsets of patients. This approach allows the simultaneous quantification of linear parental gene expression and detection of novel, so far unknown circRNAs; (2) the second approach makes use of the Arraystar Human Circular RNA Microarray which can detect 13,617 circRNAs that have been previously found in other studies.

Then, expression of differentially expressed circRNAs was often correlated with clinical parameters like patient survival, cancer grading, and metastasis. Moreover, the impact of candidate circRNAs on the hallmarks of cancer [29], including proliferation, apoptosis, self-renewal, metastasis, and angiogenesis, was investigated.

This chapter provides an overview of current important findings on circRNAs in hematologic malignancies (2.1) and solid tumors (2.2). Furthermore, this chapter about circRNAs in cancer aims to give a summary of putative oncogenic circRNAs (Table 17.1) as well as putative tumor-suppressive circRNAs (Table 17.2).

Table 17.1 Conceivably oncogenic circRNAs and their potential functions in cancer

Parental gene	CircBase [30] ID	Cancer	Expr.	Function	miRNA	References
ABCB10	hsa_circ_0008717	BRCA	High	miRNA sponge, proliferation ↑, cancer progression ↑	miR-1271	[31]
ACP6	hsa_circ_0013958	LUAD	High	miRNA sponge, apoptosis ↓, proliferation ↑, invasion ↑	miR-134	[32]
AMOTL1	hsa_circ_0004214	BRCA	High	Tumorigenesis ↑, c-myc nuclear translocation ↑		[33]
CCDC66		CRC	High	miRNA sponge, proliferation ↑, cancer growth ↑, metastasis ↑	miR-33b, miR-93, miR-185	[34]
CDR1AS / ciRS-7	hsa_circ_0001946	HCC	High	miRNA sponge, EGFR and RAF1 levels↑	miR-7	[35]
		CRC	High			[36]
		GBM	Low	miR-671-5p-mediated degradation		[37]
FLT3	hsa_circ_0100163 hsa_circ_0100164	AML	High			[38]
HIAT1[a]	hsa_circ_0000096	GC	Low	Proliferation ↑, migration ↑, cyclin D1↑, CDK6 ↑, MMP-2 ↑, MMP-9 ↑		[39]
HIPK3[a]	hsa_circ_0000284	CRC	High	miRNA sponge, apoptosis ↓, proliferation ↑, cancer growth ↑, metastasis ↑	miR-7	[40]
		HCC	High	miRNA sponge, cell growth ↑	miR-124	[41]
IKBKB	hsa_circ_0001793	CRC	High			[42]
KCNH1	hsa_circ_0016347	OSA	High	miRNA sponge, proliferation ↑, invasion ↑, metastasis ↑	miR-214	[43]
LARP1B	hsa_circ_0070933	CSCC	High			[44]
	hsa_circ_0070934					
LINC00340	hsa_circ_0075825	BCC	High			[45]
	hsa_circ_0075828					
MLL/AF9		AML with t(9;11)	High	Proliferation ↑, progression ↑, therapy resistance ↑, MAPK and AKT1 signaling ↑		[46]
MYLK	hsa_circ_0002768	Bladder	High	miRNA sponge, proliferation ↑, migration ↑, progression ↑ through VEGFA/VEGFR2, epithelial-mesenchymal transition ↑	miR-29a	[47]
NPM1	hsa_circ_0075001	AML	High	TLR signaling ↓		[38]
PML/RARA		APL with t(15;17)	High	Proliferation ↑		[46]
PRKCI	hsa_circ_0067934	ESCC	High	Proliferation ↑		[48]
PVT1		GC	High	miRNA sponge, proliferation ↑	miR-125	[49]
		MM	High	therapy resistance ↑, proliferation ↑, apoptosis ↓		[50]
SLC30A7	hsa_circ_0013339	OSCC	High	miRNA sponge, CDK6 ↑, proliferation ↑	miR-29	[51]

(continued)

Table 17.1 (continued)

Parental gene	CircBase [30] ID	Cancer	Expr.	Function	miRNA	References
TCF25	hsa_circ_0041103	Bladder	High	miRNA sponge, CDK6 ↑, proliferation ↑, migration ↑	miR-103a-3p miR-107	[52]
TTBK2	hsa_circ_0000594	Glioma	High	miRNA sponge, HNF1β ↑, proliferation ↑, migration ↑, apoptosis ↓,	miR-217	[53]
UBAP2		OSA	High	miRNA sponge, cancer progression ↑	miR-143	[54]
VCAN		Glioma	High			[55]

Abbreviations: *AML* Acute Myeloid Leukemia, *APL* Acute Promyelocytic Leukemia, *BCC* Basal Cell Carcinoma, *BRCA* Breast Invasive Carcinoma, *CRC* Colorectal Carcinoma, *CSCC* Cutaneous Squamous Cell Carcinoma, *ESCC* Esophageal Squamous Cell Carcinoma, *Expr.* Expression, *GBM* Glioblastoma Multiforme, *GC* Gastric Cancer, *HCC* Hepatocellular Carcinoma, *LUAD* Lung Adenocarcinoma, *miRNA* microRNA, *MM* Multiple Myeloma, *OSA* Osteosarcoma, *OSCC* Oral Squamous Cell Carcinoma, *Ref.* References, *TLR* Toll-like Receptor

[a]Oncogenic/tumor-suppressive role is context-specific/inconclusive; ↑ increase; ↓ decrease

Table 17.2 Conceivably tumor-suppressive circRNAs and their potential functions in cancer

Parental gene	CircBase [30] ID	Cancer	Expr.	Function	miRNA	References
FADS2	hsa_circ_00223883	BCC	Low			[45]
		CSCC	Low			[44]
FBXW7		GBM	Low	Proliferation ↓, cell cycle ↓, Encoding FBXW7-185aa, USP28-induced c-Myc stabilization ↓		[56]
FOXO3		BRCA	Low	miRNA sponge, tumor growth ↓, proliferation ↓, cancer cell survival ↓, FOXO3 translation ↑	miR-22, miR-136, miR-138, miR-149, miR-433, miR-762, miR-3614, miR-3622	[57]
				Cell cycle progression ↓ via forming a complex with p21 and CDK2		[58]
		BRCA	Low	Tumor growth ↓, MDM2-induced p53 degradation ↑, PUMA-mediated apoptosis ↑		[59]
HIPK3[a]	hsa_circ_0000284	Bladder	Low	miRNA sponge, heparanase expression ↓, migration ↓, invasion ↓, angiogenesis ↓	miR-588	[60]
ITCH		ESCC	Low	miRNA sponge, ITCH levels ↑, WNT/beta-catenin pathway ↓	miR-7, miR-17, miR-214	[61]
		CRC	Low			[62]
		HCC	Low			[63]
METTL3	hsa_circ_0000523	CRC	Low			[42]
	hsa_circ_006229					
MTO1	hsa_circ_0007874	HCC	Low	miRNA sponge, cancer progression ↓	miR-9	[64]
RNF13	hsa_circ_0001346	CRC	Low			[42]
SMARCA5	hsa_circ_0001445	HCC	Low	miRNA sponge, growth ↓, metastasis ↓, expression of tumor suppressor TIMP3 ↑	miR-17-3p	[65]
					miR-181b-5p	
SMYD4	hsa_circ_0004018	HCC	Low			[66]
WDR37	hsa_circ_0004277	AML	Low			[67]
ZKSCAN1	hsa_circ_0001727	HCC	Low	Growth ↓, migration ↓, invasion ↓		[68]

Abbreviations: *AML* Acute Myeloid Leukemia, *BCC* Basal Cell Carcinoma, *BRCA* Breast Invasive Carcinoma, *CRC* Colorectal Carcinoma, *CSCC* Cutaneous Squamous Cell Carcinoma, *ESCC* Esophageal Squamous Cell Carcinoma, *Expr.* Expression, *GBM* Glioblastoma Multiforme, *GC* Gastric Cancer, *HCC* Hepatocellular Carcinoma, *miRNA* microRNA, *MM* Multiple Myeloma, *OSA* Osteosarcoma, *Ref.* References

[a]Oncogenic/tumor suppressive role is context-specific/inconclusive; ↑ increase; ↓ decrease

2.1 The Role of CircRNAs in the Pathology of Hematologic Malignancies

Expression of circRNAs is known to be differentiation stage-specific [19] in many tissues, and changes in the circRNA repertoire have been extensively studied during neuronal development [69] and epithelial-mesenchymal transition [70]. Blood cells express circRNAs at levels comparable to that in the cerebellum, a tissue that is known to be rich in circRNAs [26]. In the hematopoietic system, we examined the circular RNAome during myeloid differentiation and detected changes upon leukemic transformation [38] using ribosomal RNA-depleted RNA-Seq. The circRNA repertoire of mature myeloid cells including metamyelocytes and neutrophils differed from circRNAs expressed in more immature cells including myeloblasts and promyelocytes, which is in line with findings in other tissue types (Fig. 17.1).

2.1.1 Acute Leukemias

CircRNA expression was found altered in leukemic blasts of acute myeloid leukemia (AML) patients, which constitute a very immature cell type. Nevertheless, the leukemic circular RNAome does not merely resemble that of healthy immature cells but is distinct and different AML subtypes, e.g., patients with mutation of the *nucleophosmin* (*NPM1*) gene could be distinguished from healthy hematopoietic cells (Fig. 17.1). In healthy blood cells, for 44% of the highly expressed genes (~5000 out of 11,000), circRNAs were detectable, whereas in AML the percentage was slightly higher (47%).

In an analysis of a relatively small sample cohort comparing healthy and leukemic samples, 27 genes were associated with differentially expressed circRNAs between the healthy and leukemic samples, with 14 circRNAs being higher in AML than in healthy cells, among them is circ*FLT3* transcribed from the *fms-related tyrosine kinase 3* (*FLT3*) gene that is commonly mutated in AML. Similarly, *NPM1,* which is commonly mutated in AML, was linked to a variety of different circRNAs, both in leukemic and healthy cells. One of them, hsa_circ_0075001, showed differential expression between different AML cell lines and AML patients, independent of the *NPM1* mutational status [38]. Its expression correlated with a distinct gene expression signature characterized by a downregulation of Toll-like receptor (TLR) signaling.

Similar to *NPM1* mutations, chromosomal translocations represent another pathomechanism of leukemic transformation. These translocations result in oncogenic fusion proteins, i.e., the *KMT2A-MLLT3* (alias MLL-AF9) fusion in t(9;11) AML or the *PML-RARA* fusion in acute promyelocytic leukemia (APL) with t(15;17). Guarnerio and colleagues found that also fusion circRNAs (f-circRNAs) are transcribed from these fusion genes which can promote proliferation, viability, and transformation of the cells and contribute to therapy resistance in vivo [46]. The circular *KMT2A-MLLT3* transcript exerted its oncogenic activity via mitogen-activated protein kinase (MAPK) and RAC-alpha serine/threonine-protein kinase (AKT1) signaling.

Salzman and colleagues intended to investigate intragenic rearrangements in childhood acute lymphoblastic leukemia (ALL). To their surprise, they found thousands of genes that produced transcripts with a scrambled exon order and estimated that for many genes, around 10% of the transcripts were circular-derived [18]. However, the most abundant circRNAs they detected and validated were also present in healthy peripheral blood cells and H9 embryonic stem cells and thus not leukemia-specific.

2.1.2 Other Hematological Malignancies

In multiple myeloma (MM), a hematological malignancy that develops in plasma cells, circ*PVT1* transcribed from the nonprotein-coding *PVT1* oncogene locus was shown to promote therapy resistance to glucocorticoid treatment, and knockdown of circ*PVT1* resulted in enhanced apoptosis and reduced proliferation of resistant myeloma cells in vitro and in vivo [50].

Considering the current evidence and the general splicing deregulation in leukemia and myeloma [71–75], it is reasonable to speculate

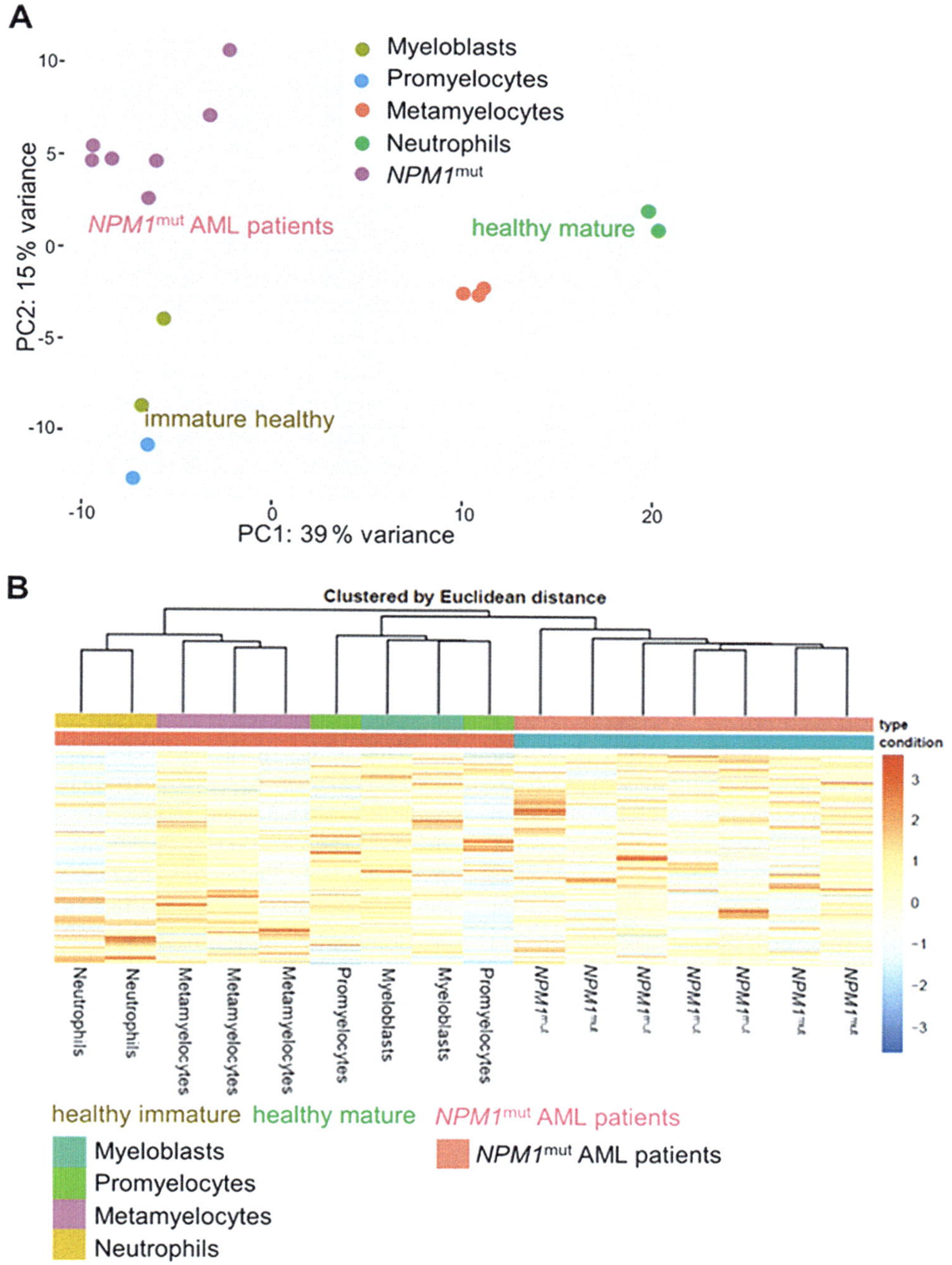

Fig. 17.1 Distinct circRNA expression changes during hematopoietic differentiation and leukemic transformation. (**a**) Principal component analysis (PCA) and (**b**) unsupervised hierarchical clustering based on circRNA expression data derived from RNA-Seq analysis of n = 9 FACS-sorted healthy control samples (including different myeloid differentiation stages as indicated) and n = 7 *NPM1*mut AML patients [38].

that aberrant circRNA expression, in line with deregulation of mRNAs, miRNAs, and lncRNAs, contributes to the pathogenesis of hematologic malignancies. Currently, many additional studies focus on the characterization of circRNAs in hematological malignancies including chronic leukemias, myeloproliferative diseases, and non-Hodgkin lymphoma (NHL).

2.2 The Role of CircRNAs in the Pathology of Solid Tumors

In solid tumors, first studies have shown that based on their function, circRNAs can be categorized into (i) circRNAs that were experimentally proven to serve as miRNA sponges (Sect. 2.2.1) and (ii) circRNAs that have other miRNA-independent functions or were not associated to interact with miRNAs yet (Sect. 2.2.1.1). Moreover, in this chapter we further grouped all circRNAs according to their putative oncogenic or tumor-suppressive properties.

2.2.1 CircRNAs Functioning as MiRNA Sponges in Solid Tumors

MiRNAs represent a well-studied subgroup of small ncRNAs that is implicated in several pathologies, including cancer. They target mRNA transcripts in a sequence-specific manner and are able to block mRNA translation or trigger its degradation. Aberrant miRNA expression profiles were found in various types of cancer [3, 4, 76]. In addition to an impact on clinical parameters, miRNAs were also shown to have the potential to influence proliferation and differentiation of the cancer cells. For example, the oncogenic miR-155 is overexpressed in many solid tumors, in lymphoma, and in AML patients with *FLT3*-ITD mutations, where it can induce myeloid proliferation [7, 77]. In AML patients carrying *NPM1* mutations, elevated levels of miR-10a/b and miR-196a/b were found [8, 77], which led to enhanced proliferation and a block in differentiation of hematopoietic progenitor cells. Conversely, the tumor-suppressive miR-29 family is downregulated in high-risk CLL, lung cancer, and invasive breast cancer [6, 78].

Today, some cytoplasmic circRNAs have been shown to modulate miRNA activity by functioning as a miRNA sponge. A prime example is circRNA CDR1as/ciRS-7 that contains 74 evolutionary conserved miR-7-binding sites and that can be cleaved by AGO proteins [79]. Interestingly, CDR1as overexpression resulted in a phenotype similar to that of miR-7 knockdown and led to impaired midbrain development [19]. Similarly, binding miR-138 the circular sex-determining gene (Sry) transcript can also function as such a "miRNA sponge" [23]. In undifferentiated human embryonic stem cells, circ*BIRC6*, a circRNA of the *BIR repeat-containing ubiquitin-conjugating enzyme* (*BIRC6*) gene, was found to bind miR-34a and miR-145, two miRNAs that target genes maintaining the pluripotent state. By sponging these miRNAs, circ*BIRC6* was able to suppress differentiation [80].

However, sequence analysis of over 7000 newly annotated circRNAs showed that a large part of known circRNAs contains only few, between zero to eight, miRNA-binding sites [81, 82]. Nevertheless, miRNA-binding sites in circRNAs are often depleted of polymorphisms, hinting at an underlying selective pressure conserving functional binding sites [83]. Today, already many studies have investigated the interaction of circRNAs with miRNAs, often in the context of cancer as outlined below.

CircRNAs with Oncogenic Properties

CircRNAs Targeting miR-7 In hepatocellular carcinoma (HCC), CDR1as/ciRS-7 (hsa_circ_0001946), which is probably the most well-studied circRNA that serves as a miRNA sponge, was highly expressed [35, 84] like in colorectal cancer (CRC) [36]. In this study, ciRS-7 increased EGFR and RAF1 expression levels in part via binding of the tumor-suppressive miR-7. In glioblastoma (GBM), however, ciRS-7 levels were low as the circRNA might be targeted by miR-671-5p, which can result in ciRS-7 degradation [37].

In CRC, another circRNA, circ*HIPK3* (hsa_circ_000284) from the *homeodomain-interacting protein kinase 3* (*HIPK3*) gene, was also reported to be highly expressed and associated with metastasis and poor patient survival [40]. Knockdown of circ*HIPK3* decreased proliferation and migration and increased apoptosis in CRC cells in vitro and could suppress cancer growth and metastasis in vivo. This could also be mimicked in vitro by overexpression of miR-7, a miRNA that is sequestered by circ*HIPK3*. Vice versa, overexpression of circ*HIPK3* led to increased levels of proto-oncogene targets of miR-7, including FAK, IGF1R, EGFR, and YY1. Moreover, expression of circ*HIPK3* was also high in HCC where it promoted cell growth, and the circRNA could bind also to other miRNAs such as e.g. miR-124 [41].

Thus, while miR-7 seems to be a prominent target for circRNAs in cancer, these first studies clearly demonstrate that the function of circRNAs is complex and that several different miRNAs can be sponged by individual circRNAs suggesting that many other miRNAs are deregulated by this mechanism in cancer. A summary of circRNAs with conceivably oncogenic function is given in Table 17.1.

CircRNAs Targeting Other miRNAs Similar to its elevated expression in multiple myeloma (see above), circ*PVT1* was highly expressed in gastric cancer where it could sponge miR-125 and increase proliferation [49]. In a different context, circ*PVT1* expression was found to be low in senescent fibroblasts, and senescence could be triggered in proliferating fibroblasts by knocking down circ*PVT1* [85]. In proliferating fibroblasts, circ*PVT1* could sponge miRNA let-7, thereby targeting several proliferative proteins including IGF2BP1, KRAS, and HMGA2. In breast cancer, circ*ABCB10* (hsa_circ_0008717) transcribed from the *ATP-binding cassette subfamily B member 10* (*ABCB10*) gene was highly expressed, and it induced proliferation and cancer progression by sponging miR-1271 [31]. Similarly, in lung adenocarcinoma (LUAC), circ*ACP6* (hsa_circ_0013958) transcribed from the lysophosphatidic *acid phosphatase 6* (*ACP6*) gene was found highly expressed, which in turn induced proliferation and invasion and reduced apoptosis of the cancer cells. As a sponge for miR-134, circ*ACP6* could increase levels of the oncogenic cyclin D1 [32]. In CRC, circ*CCDC66* of the *coiled-coil domain containing 66* (*CCDC66*) locus, but not the linear mRNA, was highly expressed and associated with poor patient survival. Circ*CCDC66* promoted cancer growth and metastasis by sequestering tumor-suppressive miRNAs, including miR-33b, miR-93, and miR-185 [34]. This led to an upregulation of several oncogenes that were otherwise repressed by these miRNAs; among them are DNMT3B, EZH2, and MYC.

In bladder cancer, circ*MYLK* (hsa_circ_0002768), transcribed from the *myosin light chain kinase* (*MYLK*) gene, served as a sponge for miR-29a, thereby increasing proliferation, migration, and tumor progression through VEGFA/VEGFR2 deregulation [47]. Another circRNA (hsa_circ_0013339) sponging members of the tumor-suppressive miR-29 family is highly expressed in oral squamous cell carcinoma (OSCC). Transcribed from the *solute carrier family 30 member 7* (*SLC30A7*) gene expression of this circRNA increased levels of CDK6 and promoted proliferation in OSCC [51]. Similarly, circ*TCF25* (hsa_circ_00411103), a circRNA of the *transcription factor 25* (*TCF25*) gene, was highly expressed in bladder cancer and increased migration, proliferation, and CDK6 levels in cancer cells by sponging miR-103a-3p and miR-107 [52].

In addition to carcinoma, circRNA-mediated deregulation of miRNAs does also play a role in mesenchymal tumors such as glioma and sarcoma. For example, a circRNA of the *Tau tubulin kinase 2* (*TTBK2*) gene was found highly expressed in glioma, in which circ*TTBK2* (hsa_circ_0000594) increased proliferation and decreased apoptosis [53]. As a sponge for miR-217, it increased levels of HNF1β. In osteosarcoma (OSA), a circRNA of the *potassium voltage-gated channel subfamily H member 1* (*KCNH1*) gene, hsa_circ_0016347, was found highly expressed and increased proliferation, invasion, and metastasis of OSA cells by sponging miR-214 [43]. Similarly, circ*UBAP2* transcribed from the *ubiquitin-associated protein 2* (*UBAP2*) gene was highly expressed and promoted OSA progression as a sponge of miR-143 [54].

CircRNAs with Tumor-Suppressive Properties

While for the abovementioned circRNAs, high expression was associated with tumor growth; there have been also several circRNAs reported which due to their tumor suppressive properties are often downregulated in cancer (Table 17.2).

A prime example is a circRNA of the *forkhead box O3* (*FOXO3*) gene that can act as a sponge for multiple miRNAs that regulate *FOXO3* mRNA expression, including miR-22, miR-136, miR-138, miR-149, miR-433, miR-762, miR-3614, and miR-3622 [57]. High expression of circ*FOXO3* leads to increased translation of the FOXO3 protein by taking away the miRNA burden of the *FOXO3* mRNA, resulting in decreased cellular growth, proliferation, and survival. Moreover, ectopic expression of circ*FOXO3* in cancer or non-cancer cell lines blocked cell cycle progression by forming a complex with the cyclin-dependent kinase 2 (CDK2) and the cyclin-dependent kinase inhibitor 1 (p21), thus preventing CDK2 from interaction with cyclin A and cyclin E resulting in a block of G1 to S transition [58]. In accordance, breast cancer circ*FOXO3* expression was reported to be very low [59] and through protein-binding circ*FOXO3* was shown to increase MDM2-induced p53 ubiquitination and subsequent degradation [59]. However, it also increased stress-induced apoptosis since higher levels of FOXO3 protein resulted in upregulation of PUMA, a pro-apoptotic protein that can induce apoptosis independent of p53 [86].

In line, circ*MTO1* (hsa_circ_0007874), a circRNA of the *mitochondrial tRNA translation optimization 1* (*MTO1*) gene, was shown to bind the oncogenic miR-9 which targets p21 [64]. Silencing of circ*MTO1* promoted HCC cell proliferation and invasion via a miR-9-mediated downregulation of p21, and low circ*MTO1* levels were associated with inferior survival of HCC patients. Similarly, circRNA (hsa_circ_0001445) derived from the *SWI/SNF-related matrix-associated actin-dependent regulator of chromatin subfamily A member 5* (*SMARCA5*) gene is lowly expressed in HCC [65]. As a sponge for miR-17-3p and miR-181b-5p, circ*SMARCA5* increased levels of the tumor suppressor TIMP3 and inhibited tumor growth and metastasis.

Additional "tumor suppressive" circRNAs have been found in bladder cancer, in which circ*HIPK3* (hsa_circ_0000284) serves as a sponge for miR-588 [60], a miRNA associated with invasion and angiogenesis [87]. Low circ*HIPK3* was associated with high-grade bladder cancer characterized by vascular invasion, and overexpression of circ*HIPK3* could inhibit bladder cancer growth and metastasis. Thus, by sponging miR-588 circ*HIPK3* exhibited tumor-suppressive properties in bladder cancer. However, circ*HIPK3* can also serve as a sponge for miR-7 in CRC and miR-124 in HCC in which it then promotes tumor growth. This clearly demonstrates that the function of circRNAs is cell context-specific and that circRNAs, like many genes, can have both tumor-suppressive and oncogenic function.

Another potential double-edged sword is circRNA from the *itchy E3 ubiquitin protein ligase* (*ITCH*) gene that was not only expressed at low levels in HCC but also in esophageal squamous cell carcinoma (ESCC) and CRC [61–63]. Circ*ITCH* could sequester miR-7, which is usually involved in tumor progression (see above) but also miR-17 and miR-214 and increase ITCH levels itself. This results in increased ubiquitination and degradation of phosphorylated disheveled segment polarity protein 2 (Dvl2) which resulted in inhibition of the WNT/beta-catenin pathway. This pathway plays a role in tumorigenesis and progression, as well as metastasis [88].

2.2.2 Other CircRNAs Impacting Solid Tumor Pathology

CircRNAs with Oncogenic Properties

In addition to the potential miRNA sponge function, there have been several other mechanisms of action assigned to circRNAs (Table 17.1). For example, highly expressed in breast cancer circ*AMOTL1* (hsa_circ_0004214) transcribed from the *angiomotin like 1* (*AMOTL1*) gene can promote tumorigenesis by increasing the nuclear translocation of the MYC proto-oncogene [33]. This led to increased stability of c-MYC and higher affinity to its target promoters. However, for other oncogenic circRNAs, the exact mechanism of action was not entirely resolved, and an involvement of miRNAs or other ncRNAs cannot be excluded. In gastric cancer, knockdown of circ*HIAT1* (hsa_circ_0000096), a circRNA derived from the *hippocampus abundant transcript 1* (*HIAT1*) gene inhibited migration and proliferation of cancer cells in vitro and in vivo by decreasing protein levels of matrix metalloproteinase-2 (MMP-2), MMP-9, cyclin D1, and cyclin-dependent kinase 6 (CDK6) [39].

A circRNA of the *inhibitor of nuclear factor kappa B kinase subunit beta (IKBKB)* gene (hsa_circ_0001793) was shown to be highly expressed in colorectal cancer; however, the circRNA's effect on NFκB signaling was not investigated [42]. Two circRNAs of *La ribonucleoprotein domain family member 1B* (*LARP1B*), i.e., hsa_circ_0070933 and hsa_circ_0070934), were highly expressed in cutaneous squamous cell carcinoma (CSCC) [44], and the expression of two circRNAs (hsa_circ_0075825, hsa_circ_0075828) from the *LINC00340* locus, also known as the nonprotein-coding *cancer susceptibility 15* locus, was high in basal cell carcinoma (BCC) [45]. The gene also produces a long noncoding RNA that is thought to promote cell proliferation in HCC. Expression of circ*PRKCI* (hsa_circ_0067934) from the *protein kinase C Iota* (*PRKCI*) gene was high in ESCC and could increase proliferation of the cancer cells [48]. In glioma, a highly expressed circRNA derived from the *versican* (*VCAN*) gene, a gene that is known to be involved in processes like proliferation, migration, and angiogenesis [55].

CircRNAs with Tumor-Suppressive Properties

A protein-coding circRNA of the *F-box and WD repeat domain containing 7* (*FBXW7*) gene encoded a novel 185 amino acid (aa) 21-kDa protein that inhibited proliferation and cell cycle progression in cancer cells [56]. By preventing stabilization of c-MYC through USP28, FBXW7-185aa was able to reduce the half-life of c-MYC. In GBM, circ*FBXW7* and FBXW7-185aa levels were decreased compared with healthy tissue.

For other circRNAs, the tumor-suppressive mode of action is less well-studied (Table 17.2). A circRNA of the *zinc finger with KRAB and SCAN domains 1* (*ZKSCAN1*) gene, hsa_circ_0001727, was shown to inhibit growth, migration, and invasion and was expressed at low levels in HCC [68]. Overexpression of circ*ZKSCAN1* could inhibit HCC progression in vivo, but the exact mechanism of action was not investigated. Moreover, hsa_circ_0004018 levels, a circRNA of the *SET and MYND domain containing 4* (*SMYD4*) gene, were also low in HCC tissues and cell lines [66].

Other circRNAs also showed low expression in different cancer types when compared to healthy tissue (Table 17.2). However, since the function of these circRNAs was not investigated in detail, it can only be assumed that they have tumor-suppressive properties. Expression of hsa_circ_00223883, a circRNA of the *fatty acid desaturase 2* (*FADS2*) gene, was low in BCC and CSCC [44, 45], and expression of two circRNAs of the *methyltransferase-like 3* (*METTL3*) gene (hsa_circ_0000523, hsa_circ_0006229), as well as hsa_circ_0001346 of the *ring finger protein 13* (*RNF13*) gene, was low in CRC [42]. In AML, low expression of hsa_circ_0004277 of the *WD repeat domain 37* (*WDR37*) gene was detected [67].

2.3 Methylation and Translation of CircRNAs

CircRNAs have long been classified as noncoding, but it was recently found that at least some of the circRNAs can be translated into proteins [24, 25], such as circ*FBXW7* mentioned above, which adds another level of complexity and challenges the classification of circRNAs as "non-coding" RNAs. The prerequisite for circRNA translation is the presence of an internal ribosomal entry site (IRES) [89]. To date, the only circRNA-encoded protein that was more extensively studied in the context of cancer is indeed FBXW7-185aa which seems to have tumor-suppressive properties, and thus, it was found downregulated in GBM [56].

Yang and colleagues have shown that the translation of circular RNAs can be promoted by methylation of adenosine at position 6, N^6-methyladenosine (m^6A), which is the most common posttranscriptional internal modification found in eukaryotic mRNAs [90]. The methylation pattern of circRNAs was found to be cell-type-specific and also distinct from the respective mRNA [91], and m^6A-circRNAs interacted with YTHDF1/YTHDF2 m^6A-reader proteins. However, for m^6A-circRNAs, unlike mRNAs, interaction with YTHDF2 did not promote RNA degradation. In comparison to circRNAs without methylation, regions flanking m^6A-circRNAs were enriched for transposable elements, and it is known that the methylation writer complex METTL3/14 binds to transposable elements (TE) [92]. TEs in flanking introns represent one mechanism known to promote exon circularization [93], and TEs are enriched in flanking regions of m^6A-circRNAs.

The *METTL3* gene itself also produces a circular RNA that was downregulated in colorectal cancer while linear METTL3 levels remained unchanged [42]. Interestingly, m^6A-methylation is known to play a role in the pathogenesis of cancer. To give some examples, components of the m^6A methyltransferase complex are differentially regulated throughout healthy hematopoiesis, and their expression is altered in AML. In particular, METTL14 is highly expressed in AML, where it is required for the development and maintenance of the disease by conferring self-renewal properties to the leukemia stem cells [94]. METTL3 was shown to increase translation of oncogenes in lung cancer [95]. Changes of m^6A-methylation in the pathogenesis of cancer have not been linked to altered translation of circRNAs yet, but this connection should be investigated in the future given the importance of altered epigenetic and epitranscriptomic mechanisms in tumorigenesis.

3 Conclusion and Outlook

In line with the transcriptome, the circular RNAome of cancer cells is distinct from that of healthy cells. Moreover, the circRNA repertoire can vary in different tumor subtypes and might change with cancer stage and the development of therapy resistance. Differentially expressed circRNAs have been detected in several types of cancer; however, only few studies comprehensively characterized the possible functions of circRNA candidates in tumorigenesis.

One potential pitfall in studies detecting differentially expressed circRNAs between cancerous and healthy tissue is the fact that cancer cells often bear properties of immature, undifferentiated cells. However, only few groups have explicitly investigated differentiation-independent changes in the tumor cells when compared to healthy tissue. Thus, it is possible that the expression of some of these "tumor-associated" circRNAs simply reflects the differentiation status of the cell, and in accordance the expression of the respective circRNAs do not necessarily have to be functionally relevant to tumorigenesis in these cases.

Nevertheless, the functional characterization of single circRNA candidates has shown that circRNAs can affect more or less all cancer-related processes including proliferation, apoptosis, self-renewal, metastasis, and angiogenesis. In some instances, only few miRNA-binding sites within a circRNA were sufficient to impact miRNA function and to affect its downstream targets. As a consequence, deregulation of a single circRNA can impact the expression of a broad range of genes within a cell, thereby having the potential to play an important role in tumorigenesis.

3.1 Diagnostic and Prognostic Potential of CircRNAs in Cancer

Many groups are examining the potential of circRNAs as biomarkers for diagnosis and prognosis of cancer. Due to their circular structure, circRNAs are protected from exonuclease degradation, and their half-life is much longer than that of the parental mRNA, i.e., around 48 h compared to 10 h for mRNAs [82]. Moreover, they are stably detected in body fluids like saliva and blood [26, 27], fluids that could be accessed in a noninvasive way for diagnostic purposes, so-called liquid biopsies. CircRNAs could further be detected in serum exosomes, and analysis of the circRNA content in these vesicles could distinguish colon cancer patients from healthy individuals [28]. Some circRNAs have already been shown to be of prognostic value, either alone or synergistically with known cancer markers [36, 49]. Especially for cases lacking traditional biomarkers, circRNAs could be of value, but the reproducibility and specificity of circRNA detection have yet to be confirmed in larger studies before circRNAs might make their way into cancer diagnostics and might serve as variables that could enter prognostic scores.

3.2 CircRNAs as Therapeutic Targets in Cancer

Both oncogenic and tumor-suppressive circRNAs have the potential to serve as therapeutic targets in cancer. The backsplice junction sequence is unique for a certain circRNA and makes it possible to specifically target a particular circRNA without affecting the parental mRNA. Oncogenic circRNAs could therefore be targeted by siRNAs and subjected to Ago2-mediated degradation [79]. Vice versa, ectopic expression of tumor-suppressive circRNAs could be achieved by using expression vectors with long reverse complementary sequences which facilitate circularization of the enclosed sequence [96]. Since mammalian cells are able to discriminate "self-"circRNAs from foreign ones [97], the use of in vitro-generated circRNAs should be avoided to circumvent activation of the innate immune system. The fact that not all circRNAs are noncoding adds another level of complexity to this field, and the first circRNA-derived protein with tumor-suppressive properties has already been identified [56]. Thus, another therapeutic strategy could be to make use of translatable circRNAs. CircRNAs containing an IRES can be translated using the normal ribosome machinery of the cell. CircRNAs with an infinite open-reading frame without stop codon were able to produce a repeating sequence of peptides in a "rolling circle" manner [98]. This could be exploited to produce peptide sequences that bind and disable oncogene proteins.

In summary, first studies on circRNAs in cancer have shown promising results and ongoing comprehensive analyses will improve our understanding of the impact of deregulated circRNA expression on tumorigenesis, which in turn will also increase our general understanding of the role of circRNAs. In the long term, circRNAs might also become part of the routine clinical work-up and contribute to improved patient management by offering new perspectives to improve diagnosis, to individualize outcome prediction, and to provide targets for innovative therapeutic approaches, both in solid tumors and hematologic malignancies.

Acknowledgments This work was supported by the Deutsche Forschungsgemeinschaft (SFB 1074 project B3 to LB) and the Studienstiftung des Deutschen Volkes (doctoral grant to SL).

Competing Financial Interests The authors declare no competing financial interests.

References

1. Djebali S, Davis CA, Merkel A et al (2012) Landscape of transcription in human cells. Nature 489(7414):101
2. ENCODE Project Consortium (2012) An integrated encyclopedia of DNA elements in the human genome. Nature 489(7414):57
3. Rücker FG, Russ AC, Cocciardi S et al (2013) Altered miRNA and gene expression in acute myeloid leukemia with complex karyotype identify networks of prognostic relevance. Leukemia 27(2):353–361
4. Russ AC, Sander S, Luck SC et al (2011) Integrative nucleophosmin mutation-associated microRNA and gene expression pattern analysis identifies novel microRNA-target gene interactions in acute myeloid leukemia. Haematologica 96(12):1783–1791
5. Garzon R, Volinia S, Papaioannou D et al (2014) Expression and prognostic impact of lncRNAs in acute myeloid leukemia. Proc Natl Acad Sci U S A 111(52):18679–18684
6. Yanaihara N, Caplen N, Bowman E et al (2006) Unique microRNA molecular profiles in lung cancer diagnosis and prognosis. Cancer Cell 9(3):189–198
7. Garzon R, Volinia S, Liu C-G et al (2008) MicroRNA signatures associated with cytogenetics and prognosis in acute myeloid leukemia. Blood 111(6):3183–3189
8. Bryant A, Palma CA, Jayaswal V et al (2012) miR-10a is aberrantly overexpressed in Nucleophosmin1

mutated acute myeloid leukaemia and its suppression induces cell death. Mol Cancer 11(1):8
9. Mott JL, Kobayashi S, Bronk SF et al (2007) mir-29 regulates Mcl-1 protein expression and apoptosis. Oncogene 26(42):6133
10. Mercer TR, Dinger ME, Sunkin SM et al (2008) Specific expression of long noncoding RNAs in the mouse brain. Proc Natl Acad Sci U S A 105(2):716–721
11. Dinger ME, Amaral PP, Mercer TR et al (2008) Long noncoding RNAs in mouse embryonic stem cell pluripotency and differentiation. Genome Res 18(9):1433–1445
12. Guttman M, Amit I, Garber M et al (2009) Chromatin signature reveals over a thousand highly conserved large non-coding RNAs in mammals. Nature 458(7235):223
13. Xu W-H, Zhang J-B, Dang Z et al (2014) Long noncoding RNA URHC regulates cell proliferation and apoptosis via ZAK through the ERK/MAPK signaling pathway in hepatocellular carcinoma. Int J Biol Sci 10(7):664
14. Askew DS, Li J, Ihle JN (1994) Retroviral insertions in the murine His-1 locus activate the expression of a novel RNA that lacks an extensive open reading frame. Mol Cell Biol 14(3):1743–1751
15. Gupta RA, Shah N, Wang KC et al (2010) Long noncoding RNA HOTAIR reprograms chromatin state to promote cancer metastasis. Nature 464(7291):1071
16. Imamura T, Yamamoto S, Ohgane J et al (2004) Non-coding RNA directed DNA demethylation of Sphk1 CpG island. Biochem Biophys Res Commun 322(2):593–600
17. Lanz RB, McKenna NJ, Onate SA et al (1999) A steroid receptor coactivator, SRA, functions as an RNA and is present in an SRC-1 complex. Cell 97(1):17–27
18. Salzman J, Gawad C, Wang PL et al (2012) Circular RNAs are the predominant transcript isoform from hundreds of human genes in diverse cell types. PLoS One 7(2):e30733
19. Memczak S, Jens M, Elefsinioti A et al (2013) Circular RNAs are a large class of animal RNAs with regulatory potency. Nature 495(7441):333
20. Jeck WR, Sorrentino JA, Wang K et al (2013) Circular RNAs are abundant, conserved, and associated with ALU repeats. RNA 19(2):141–157
21. You X, Vlatkovic I, Babic A et al (2015) Neural circular RNAs are derived from synaptic genes and regulated by development and plasticity. Nat Neurosci 18(4):603
22. Gruner H, Cortés-López M, Cooper DA et al (2016) CircRNA accumulation in the aging mouse brain. Sci Rep 6:38907
23. Hansen TB, Jensen TI, Clausen BH et al (2013) Natural RNA circles function as efficient microRNA sponges. Nature 495(7441):384
24. Legnini I, Di Timoteo G, Rossi F et al (2017) Circ-ZNF609 is a circular RNA that can be translated and functions in myogenesis. Mol Cell 66(1):22–37. e9
25. Pamudurti NR, Bartok O, Jens M et al (2017) Translation of circRNAs. Mol Cell 66(1):9–21. e7
26. Memczak S, Papavasileiou P, Peters O et al (2015) Identification and characterization of circular RNAs as a new class of putative biomarkers in human blood. PLoS One 10(10):e0141214
27. Bahn JH, Zhang Q, Li F et al (2015) The landscape of microRNA, Piwi-interacting RNA, and circular RNA in human saliva. Clin Chem 61(1):221–230
28. Li Y, Zheng Q, Bao C et al (2015) Circular RNA is enriched and stable in exosomes: a promising biomarker for cancer diagnosis. Cell Res 25(8):981
29. Hanahan D, Weinberg RA (2011) Hallmarks of cancer: the next generation. Cell 144(5):646–674
30. Glažar P, Papavasileiou P, Rajewsky N (2014) circBase: a database for circular RNAs. RNA 20(11):1666–1670
31. Liang H-F, Zhang X-Z, Liu B-G et al (2017) Circular RNA circ-ABCB10 promotes breast cancer proliferation and progression through sponging miR-1271. Am J Cancer Res 7(7):1566
32. Zhu X, Wang X, Wei S et al (2017) hsa_circ_0013958: a circular RNA and potential novel biomarker for lung adenocarcinoma. FEBS J 284(14):2170–2182
33. Yang Q, Du WW, Wu N et al (2017) A circular RNA promotes tumorigenesis by inducing c-myc nuclear translocation. Cell Death Differ 24(9):1609
34. Hsiao K-Y, Lin Y-C, Gupta SK et al (2017) Noncoding effects of circular RNA CCDC66 promote colon cancer growth and metastasis. Cancer Res 77(9):2339–2350
35. Yu L, Gong X, Sun L et al (2016) The circular RNA Cdr1as act as an oncogene in hepatocellular carcinoma through targeting miR-7 expression. PLoS One 11(7):e0158347
36. Weng W, Wei Q, Toden S et al (2017) Circular RNA ciRS-7—a promising prognostic biomarker and a potential therapeutic target in colorectal cancer. Clin Cancer Res 23(14):3918–3928
37. Barbagallo D, Condorelli A, Ragusa M et al (2016) Dysregulated miR-671-5p/CDR1-AS/CDR1/VSNL1 axis is involved in glioblastoma multiforme. Oncotarget 7(4):4746
38. Hirsch S, Blätte TJ, Grasedieck S et al (2017) Circular RNAs of the nucleophosmin (NPM1) gene in acute myeloid leukemia. Haematologica 102:2039–2047
39. Li P, Chen H, Chen S et al (2017) Circular RNA 0000096 affects cell growth and migration in gastric cancer. Br J Cancer 116(5):626
40. Zeng K, Chen X, Xu M et al (2018) CircHIPK3 promotes colorectal cancer growth and metastasis by sponging miR-7. Cell Death Dis 9(4):417
41. Zheng Q, Bao C, Guo W et al (2016) Circular RNA profiling reveals an abundant circHIPK3 that regulates cell growth by sponging multiple miRNAs. Nat Commun 7:11215
42. Bachmayr-Heyda A, Reiner AT, Auer K et al (2015) Correlation of circular RNA abundance with proliferation–exemplified with colorectal and ovarian cancer,

idiopathic lung fibrosis, and normal human tissues. Sci Rep 5:8057
43. Jin H, Jin X, Zhang H et al (2017) Circular RNA hsa-circ-0016347 promotes proliferation, invasion and metastasis of osteosarcoma cells. Oncotarget 8(15):25571
44. Sand M, Bechara FG, Gambichler T et al (2016) Circular RNA expression in cutaneous squamous cell carcinoma. J Dermatol Sci 83(3):210–218
45. Sand M, Bechara FG, Sand D et al (2016) Circular RNA expression in basal cell carcinoma. Epigenomics 8(5):619–632
46. Guarnerio J, Bezzi M, Jeong JC et al (2016) Oncogenic role of fusion-circRNAs derived from cancer-associated chromosomal translocations. Cell 165(2):289–302
47. Zhong Z, Huang M, Lv M et al (2017) Circular RNA MYLK as a competing endogenous RNA promotes bladder cancer progression through modulating VEGFA/VEGFR2 signaling pathway. Cancer Lett 403:305–317
48. Xia W, Qiu M, Chen R et al (2016) Circular RNA has_circ_0067934 is upregulated in esophageal squamous cell carcinoma and promoted proliferation. Sci Rep 6:35576
49. Chen J, Li Y, Zheng Q et al (2017) Circular RNA profile identifies circPVT1 as a proliferative factor and prognostic marker in gastric cancer. Cancer Lett 388:208–219
50. Wan X-Y, Chu Z-B, Hu Y et al (2017) CircPVT1 inhibit apoptosis and enhance drug resistance in multiple myeloma. Blood 130:3085
51. Chen L, Zhang S, Wu J et al (2017) circRNA_100290 plays a role in oral cancer by functioning as a sponge of the miR-29 family. Oncogene 36(32):4551
52. Zhong Z, Lv M, Chen J (2016) Screening differential circular RNA expression profiles reveals the regulatory role of circTCF25-miR-103a-3p/miR-107-CDK6 pathway in bladder carcinoma. Sci Rep 6:30919
53. Zheng J, Liu X, Xue Y et al (2017) TTBK2 circular RNA promotes glioma malignancy by regulating miR-217/HNF1β/Derlin-1 pathway. J Hematol Oncol 10(1):52
54. Zhang H, Wang G, Ding C et al (2017) Increased circular RNA UBAP2 acts as a sponge of miR-143 to promote osteosarcoma progression. Oncotarget 8(37):61687
55. Song X, Zhang N, Han P et al (2016) Circular RNA profile in gliomas revealed by identification tool UROBORUS. Nucleic Acids Res 44(9):e87–e87
56. Yang Y, Gao X, Zhang M et al (2018) Novel role of FBXW7 circular RNA in repressing glioma tumorigenesis. JNCI: J Natl Cancer Inst 110(3)
57. Yang W, Du WW, Li X et al (2016) Foxo3 activity promoted by non-coding effects of circular RNA and Foxo3 pseudogene in the inhibition of tumor growth and angiogenesis. Oncogene 35(30):3919
58. Du WW, Yang W, Liu E et al (2016) Foxo3 circular RNA retards cell cycle progression via forming ternary complexes with p21 and CDK2. Nucleic Acids Res 44(6):2846–2858
59. Du WW, Fang L, Yang W et al (2017) Induction of tumor apoptosis through a circular RNA enhancing Foxo3 activity. Cell Death Differ 24(2):357
60. Li Y, Zheng F, Xiao X et al (2017) CircHIPK3 sponges miR-558 to suppress heparanase expression in bladder cancer cells. EMBO Rep 18:e201643581
61. Li F, Zhang L, Li W et al (2015) Circular RNA ITCH has inhibitory effect on ESCC by suppressing the Wnt/β-catenin pathway. Oncotarget 6(8):6001
62. Huang G, Zhu H, Shi Y et al (2015) cir-ITCH plays an inhibitory role in colorectal cancer by regulating the Wnt/β-catenin pathway. PloS One 10(6):e0131225
63. Guo W, Zhang J, Zhang D et al (2017) Polymorphisms and expression pattern of circular RNA circ-ITCH contributes to the carcinogenesis of hepatocellular carcinoma. Oncotarget 8(29):48169
64. Han D, Li J, Wang H et al (2017) Circular RNA MTO1 acts as the sponge of miR-9 to suppress hepatocellular carcinoma progression. Hepatology 66(4):1151
65. Yu J, Xu Q-g, Wang Z-g et al (2018) Circular RNA cSMARCA5 inhibits growth and metastasis in hepatocellular carcinoma. J Hepatol 68:1214
66. Fu L, Yao T, Chen Q et al (2017) Screening differential circular RNA expression profiles reveals hsa_circ_0004018 is associated with hepatocellular carcinoma. Oncotarget 8(35):58405
67. Li W, Zhong C, Jiao J et al (2017) Characterization of hsa_circ_0004277 as a new biomarker for acute myeloid leukemia via circular RNA profile and bioinformatics analysis. Int J Mol Sci 18(3):597
68. Yao Z, Luo J, Hu K et al (2017) ZKSCAN1 gene and its related circular RNA (circZKSCAN1) both inhibit hepatocellular carcinoma cell growth, migration, and invasion but through different signaling pathways. Mol Oncol 11(4):422–437
69. Rybak-Wolf A, Stottmeister C, Glažar P et al (2015) Circular RNAs in the mammalian brain are highly abundant, conserved, and dynamically expressed. Mol Cell 58(5):870–885
70. Conn SJ, Pillman KA, Toubia J et al (2015) The RNA binding protein quaking regulates formation of circRNAs. Cell 160(6):1125–1134
71. Le Gouill S, Podar K, Harousseau J-L et al (2004) Mcl-1 regulation and its role in multiple myeloma. Cell Cycle 3(10):1259–1262
72. Dolnik A, Engelmann JC, Scharfenberger-Schmeer M et al (2012) Commonly altered genomic regions in acute myeloid leukemia are enriched for somatic mutations involved in chromatin remodeling and splicing. Blood 120(18):e83–e92
73. Moalli PA, Pillay S, Krett NL et al (1993) Alternatively spliced glucocorticoid receptor messenger RNAs in glucocorticoid-resistant human multiple myeloma cells. Cancer Res 53(17):3877–3879
74. Sun L, Goodman PA, Wood CM et al (1999) Expression of aberrantly spliced oncogenic ikaros isoforms in childhood acute lymphoblastic leukemia. J Clin Oncol 17(12):3753–3766

75. Janssen JW, Ludwig W-D, Borkhardt A et al (1994) Pre-pre-B acute lymphoblastic leukemia: high frequency of alternatively spliced ALL1-AF4 transcripts and absence of minimal residual disease during complete remission. Blood 84(11):3835–3842
76. Trang P, Weidhaas JB, Slack FJ (2017) MicroRNAs and cancer. In: The molecular basis of human cancer. Springer, New York, pp 277–286
77. Jongen-Lavrencic M, Sun SM, Dijkstra MK et al (2008) MicroRNA expression profiling in relation to the genetic heterogeneity of acute myeloid leukemia. Blood 111(10):5078–5085
78. Calin GA, Ferracin M, Cimmino A et al (2005) A MicroRNA signature associated with prognosis and progression in chronic lymphocytic leukemia. New Engl J Med 353(17):1793–1801
79. Hansen TB, Wiklund ED, Bramsen JB et al (2011) miRNA-dependent gene silencing involving Ago2-mediated cleavage of a circular antisense RNA. EMBO J 30(21):4414–4422
80. Yu C-Y, Li T-C, Wu Y-Y et al (2017) The circular RNA circBIRC6 participates in the molecular circuitry controlling human pluripotency. Nat Commun 8(1):1149
81. Guo JU, Agarwal V, Guo H et al (2014) Expanded identification and characterization of mammalian circular RNAs. Genome Biol 15(7):409
82. Jeck WR, Sharpless NE (2014) Detecting and characterizing circular RNAs. Nat Biotechnol 32(5):453
83. Thomas LF, Sætrom P (2014) Circular RNAs are depleted of polymorphisms at microRNA binding sites. Bioinformatics 30(16):2243–2246
84. Xu L, Zhang M, Zheng X et al (2017) The circular RNA ciRS-7 (Cdr1as) acts as a risk factor of hepatic microvascular invasion in hepatocellular carcinoma. J Cancer Res Clin Oncol 143(1):17–27
85. Panda AC, Grammatikakis I, Kim KM et al (2016) Identification of senescence-associated circular RNAs (SAC-RNAs) reveals senescence suppressor CircPVT1. Nucleic Acids Res 45(7):4021–4035
86. Yu J, Zhang L (2009) PUMA, a potent killer with or without p53. Oncogene 27(S1):S71
87. Qu H, Zheng L, Pu J et al (2015) miRNA-558 promotes tumorigenesis and aggressiveness of neuroblastoma cells through activating the transcription of heparanase. Hum Mol Genet 24(9):2539–2551
88. Polakis P (2000) Wnt signaling and cancer. Genes Dev 14(15):1837–1851
89. Chen C-Y, Sarnow P (1995) Initiation of protein synthesis by the eukaryotic translational apparatus on circular RNAs. Science 268(5209):415–417
90. Yang Y, Fan X, Mao M et al (2017) Extensive translation of circular RNAs driven by N 6-methyladenosine. Cell Res 27(5):626
91. Zhou C, Molinie B, Daneshvar K et al (2017) Genome-wide maps of m6A circRNAs identify widespread and cell-type-specific methylation patterns that are distinct from mRNAs. Cell Rep 20(9):2262–2276
92. Kelley DR, Hendrickson DG, Tenen D et al (2014) Transposable elements modulate human RNA abundance and splicing via specific RNA-protein interactions. Genome Biol 15(12):537
93. Ivanov A, Memczak S, Wyler E et al (2015) Analysis of intron sequences reveals hallmarks of circular RNA biogenesis in animals. Cell Rep 10(2):170–177
94. Shi H, Skibbe J, Shen C et al (2018) METTL14 inhibits its hematopoietic stem/progenitor differentiation and promotes Leukemogenesis via mRNA m 6 A modification. Cell Stem Cell 22:1–15
95. Lin S, Choe J, Du P et al (2016) The m 6 A methyltransferase METTL3 promotes translation in human cancer cells. Mol Cell 62(3):335–345
96. Zhang X-O, Wang H-B, Zhang Y et al (2014) Complementary sequence-mediated exon circularization. Cell 159(1):134–147
97. Chen YG, Kim MV, Chen X et al (2017) Sensing self and foreign circular RNAs by intron identity. Mol Cell 67(2):228–238. e5
98. Abe N, Matsumoto K, Nishihara M et al (2015) Rolling circle translation of circular RNA in living human cells. Sci Rep 5:16435

Circular RNAs in Brain Physiology and Disease

18

S. Gokul and G. K. Rajanikant

Abstract

Circular RNAs (circRNAs) are endogenously expressed non-coding RNAs discovered in the early 1990s as a transcriptional by-product of little importance. It was only recently that they were identified as a key player in regulating the gene expression by targeting and modulating the functions of microRNA, a process known as microRNA sponging. They are distributed throughout the system in a tissue-specific manner showing abundant enrichment in neuronal tissue. Their physiological functions in the brain such as neuronal maturation, differentiation, etc. as well as their implications in numerous brain-related disorders have made its entry into the spotlight. Yet the wider scope and molecular mechanism of circRNAs still remain elusive. In this chapter, we describe in detail the functional aspects and importance of circRNAs in the human brain and how it is associated with various neurological diseases.

Keywords

Circular RNA · Brain · Ischemic stroke · Neurodegenerative diseases · Brain tumour

S. Gokul · G. K. Rajanikant (✉)
School of Biotechnology, National Institute of Technology Calicut, Calicut, India
e-mail: rajanikant@nitc.ac.in

Abbreviations

circRNAs	Circular RNAs
GBM	Glioblastoma multiforme
MDD	Major depressive disorder
microRNAs	MicroRNAs
mRNA	Messenger RNA
MSA	Multiple system atrophy
ncRNA	Non-coding RNA
pre-mRNA	Precursor mRNA
SRSF1	Serine- and arginine-rich splicing factor 1
TBI	Traumatic brain injury

1 Introduction

The widespread notion that the circular RNAs (circRNAs) are simply cellular artefacts or transcriptional noises has changed recently, in the light of several studies of late portraying their diverse physiological functions. Circular RNAs are one of the emerging classes of non-coding RNAs, with a widespread expression across the evolutionary tree of life [1]. The biogenesis of circRNA follows a back-splicing mechanism, wherein a covalent bond is formed between 5′ and 3′ splice sites of a pre-mRNA [2, 3]. This unique structural conformation and higher stability compared to other RNA species have raised an immense interest to characterize and

J. Xiao (ed.), *Circular RNAs*, Advances in Experimental Medicine and Biology 1087,
https://doi.org/10.1007/978-981-13-1426-1_18

understand its mechanism within the scientific community. The expression of these RNAs in humans is said to be in a tissue type and developmental stage-specific manner, with a relatively high abundance in the brain, especially in neuronal cells [4, 5]. Here we outline how circRNAs modulate and regulate the normal physiology and development of brain and how their dysregulation is associated with various neurological disorders. We conclude this chapter with a brief look into the future of circRNA research in the context of neurological diseases.

2 circRNAs in Brain Physiology

2.1 Enriched circRNAs in the Brain

The complex morphological nature of brain demands tight orchestrated regulation and control of gene expression throughout the entire lifespan of an organism. This tight regulation can be seen at various developmental stages such as synaptogenesis, neurite outgrowth, neuronal maturation, synaptic plasticity etc. Non-coding RNAs such as microRNAs, long non-coding RNAs, small interfering RNAs have immense roles in regulating these developmental processes. Recently, circRNAs have emerged as a novel regulatory ncRNA with a higher level of expression in the brain as well-documented by several in vitro and in vivo studies [5–7]. Even though 20% of the protein-coding genes in the brain produce circRNAs, their expression pattern is not uniformly distributed throughout the brain, as some regions are more enriched than the others [8]. For example, circRims2, circEl2 and circDym are highly expressed in the cerebellum, whereas circPlxnd1 is enriched in the cortex. Other regions of the brain showing high expression patterns are the striatum, olfactory bulb and hippocampus [5]. However, the expression level of circRNAs does not always correlate with the expression of their linear isoforms. circRNAs such as circMyst4, circKlhl2 and circAagab whose expressions are significantly upregulated during synaptogenesis had their mRNA level downregulated [7], whilst the mRNA transcript of circZfp609 was expressed at a constant level. Intriguingly, the abundant expression of circRNAs in the brain has been attributed to a number of reasons. First, there are a number of genes that are hosts to circRNAs such as cevircRims2, circTulp4, circElf2, circPhf21a and circMyst4, and most of these host genes are expressed exclusively in the brain [8]. Secondly, there is a high resemblance in the intronic regions of both circRNAs and neuronal genes, wherein, the latter often have long intronic regions, while they are highly conserved during evolution in the former. Finally, the regulatory elements of circRNA biogenesis: cis- and trans-factors [9].

It is previously known that evolutionarily conserved genes display a striking functional similarity and physiological significance. In this regard, circRNAs present a well-conserved expression pattern in both mouse and human neuronal cells [7, 10, 11]. Besides, there is an extremely conserved exonic sequence around the head-to-tail junctions of both mouse and rat [7]. All these evidences point to the fact that circRNAs might have a potential role in normal physiological functions of the brain.

2.2 Neuronal Distribution of circRNAs

The structure of neurons is unique in its own way with a long axon having protective covering (myelin sheath), large cell body and cellular projections (dendrites). The synapses or neuronal junction, where neurons communicate with each other, is considered as a circRNA-enriched region, and it tends to accumulate with age [12]. Since the synaptic density in the human brain is higher than in mouse, this is speculated to be one of the reasons for the higher expression of circRNAs in humans compared to the mouse brain [13]. This suggests that these highly enriched circRNAs might have some functional significance in the neurons. Some of these circRNAs such as CDR1as/CiRS-7, circRTN4, circTULP4 and circRIMS2 were found to be involved in neuronal differentiation and developmental processes

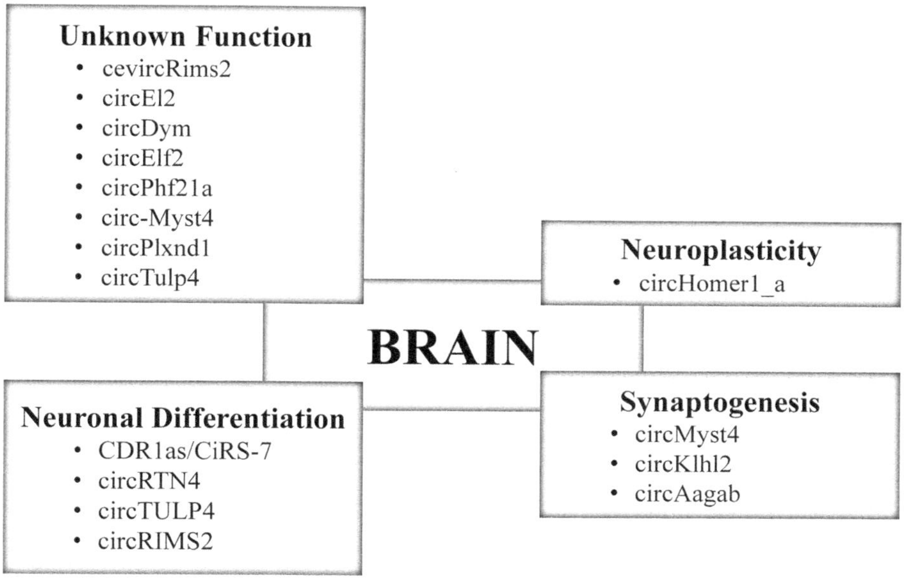

Fig. 18.1 Circular RNAs in the brain and their functional roles

whilst some in synaptogenesis, and they tend to show an increase in their expression level [5, 7].

Neuroplasticity is the most salient feature of the human brain, and the neural networks are kept tightly under control and regulated throughout the lifespan of humans. circRNAs are one of the regulatory RNAs involved in this process as evidenced by the latest report stating a dynamic change in their expression level after neural plasticity. The circRNA which is linked to this process was found to be circHomer1_a [7]. The summarized details of all the circRNAs involved in brain physiology is given in (Fig. 18.1).

2.3 Physiological Role of circRNAs in the Brain

The biological processes in which circRNAs are involved have been mentioned in the previous section. Besides, most of the genes that host circRNAs upon gene ontology analysis were found to be enriched in the process related to nervous system development, neurogenesis and differentiation. The most important role of circRNAs so far documented is to modulate the function of microRNAs (miRNAs), a process known as microRNA sponging. circRNAs have multiple miRNA-binding sites; for instance, one of the highly abundant circRNA in neurons CDR1as/ciRS-7 harbour 74 binding sites for miR-7 by which they competitively bind and supress their function. CDR1as/ciRS-7 is one of the best studied circRNA where they bind to the miRNA effector protein Argonaute in an miR-7-dependent manner [14]. Besides, there are some reports regarding circRNAs' ability to regulate the transcriptional activity in neurons and also in protein and/or RNA transport within the cell [15]. Even though thousands of circRNAs were identified, the functional role of only a handful of them is known.

3 CircRNAs and CNS Disease

3.1 Cerebrovascular Diseases

Cerebrovascular diseases are group of events that affect the blood flow to the brain resulting in permanent brain damage or even cell death. Ischemic stroke is one of the most common cerebrovascular diseases and a leading cause of mortality and disability, affecting millions of people worldwide.

Table 18.1 Circular RNAs in major brain-related disorders

circRNAs	Disease	Regulation	References
mmu_circRNA_40001 mmu_circRNA_013120	Ischemic stroke	Upregulated	[16]
mmu_circRNA_40806		Downregulated	
mmu_circRNA-015947	Ischemic stroke	Upregulated	[17]
circ_008018 circ_015350 circ_016128	Ischemic stroke	Upregulated	[18]
circ_011137 circ_001729 circ_006696	Ischemic stroke	Downregulated	[18]
circDLGAP4	Ischemic stroke	Upregulated	[19]
circ-TTBK2	Glioma	Upregulated	[20]
circSMARCA5	GBM	Downregulated	[24]
circBRAF	Glioma	Downregulated	[23]
cZNF292	Glioma	Upregulated	[21]
circ-FBXW7	Glioma	Downregulated	[22]
IQCK MAP4K3 EFCAB11 DTNA MCTP1	MSA	Upregulated	[28]
circzip-2	Parkinson's disease	Downregulated	[27]
ciRS-7	Alzheimer's disease	Downregulated	[25, 26]
circ_0005402 circ_0035560	MS	Downregulated	[30, 31]
hsa_circRNA_103636	MDD	Downregulated	[29]

The landscape of stroke research has expanded in the last decade, allowing researchers to understand the molecular mechanism underlying its pathology. An extensive research relating non-coding RNAs role in stroke pathology can be seen lately, with circRNAs in particular gaining wider attention amongst all of them. Association of circRNAs with ischemic stroke pathology has recently been revealed and was found to be significantly altered in the ischemic brain (Table 18.1), indicating its significance as potential biomarker (mmu_circRNA_40001, mmu_circRNA_013120 and mmu_circRNA_40806) for stroke. Furthermore, their gene targets were involved in various signalling pathways such as cell survival, death and neuroinflammation [16]. Mouse circRNA-015947 and circ_016423 were significantly altered during ischaemia-reperfusion injury indicating its impact in post-stroke pathology. circRNA-015947 could interact with five microRNA targets, thereby enhancing their target gene expression, whilst circ_016423 showed 625 miRNA-binding sites that can bind to 521 different miRNAs [17, 18]. circRNA DLGAP4 (circDLGAP4) upregulation significantly reduced infarction size and blood-brain barrier damage in a mouse model for ischemic stroke. They act as a microRNA-143 sponge (Table 18.2) thereby inhibiting endothelial-mesenchymal transition. This suggests a potential therapeutic role of circDLGAP4 for acute ischemic injury [19].

3.2 Cancer

Cancer has been in the forefront of research for more than two decades, due to its complex molecular interactions and unique ability to escape and survive the normal homeostatic control. circRNAs being one of the regulatory RNAs have been linked to various cancer-related events like cellular proliferation, migration and invasion. For example, circ-TTBK2 in glioma tissue acts as miR-217 sponge, thereby inhibiting its function in a sequence specific manner, resulting in increased level of HNF1β (a direct target of miR-217) that has an oncogenic role [20]. In contrast, circRNAs like cZNF292 and circ-FBXW7 exert an inhibitory role preventing cell cycle progression and cell proliferation in glioma cells [21, 22]. circBRAF identified as a negatively downregulated circRNA in glioblastoma multiforme (GBM) has been associated with

Table 18.2 circRNAs and their known miRNA targets

S.No.	Circular RNA	miRNA	Reference
1	ciRS-7	miR-7a	[25]
2	circ-TTBK2	miR-217	[20]
3	mmu-circRNA-015947[a]	mmu-miR-188-3p mmu-miR-329-5p mmu-miR-3057-3p mmu-miR-5098 mmu-miR-683	[17]
4	circDLGAP4	miR-143	[19]
5	mmu_circRNA_40806[a]	miR-149-5p mmu-miR-346-3p mmu-miR-20a-3p	[16]
6	circSMARCA5	RBP SRSF1	[24]
7	circzip-2[a]	miR-60-3p	[27]
8	hsa_circRNA_103636[a]	hsa-miR-890 hsa-miR-617 hsamiR-520a hsa-miR-15b-3p hsa-miR-103a-2–5p	[29]

[a]Predicted targets

tumorigenesis and development of GBM, thus making it a potential biomarker to assess the prognosis in glioma patients [23]. circSMARCA5 overexpression has been linked to decreased migration of cells in glioma cell lines. They have a higher binding affinity to serine- and arginine-rich splicing factor 1 (SRSF1), a positive controller of cell migration often overexpressed in GBM. Thus, targeting circSMARCA5 may be a viable therapeutic choice in GBM [24].

3.3 Neurodegenerative Diseases

Neurodegenerative diseases are a group of conditions characterized by progressive degeneration of structure and functions of neurons in the brain. The most common of them are Alzheimer's disease, Parkinson's disease and Huntington's disease. Some of the important circRNAs linked to neurodegenerative diseases are CDR1as/ciRS-7 in Alzheimer's [25, 26], circzip-2 in Parkinson's [27] and IQCK, MAP4K3, EFCAB11, DTNA and MCTP1 circRNAs in multiple system atrophy (MSA) [28]. These studies suggest a strong connection between circRNAs and neurodegenerative diseases, making them a potential therapeutic target and a disease biomarker.

3.4 Other Maladies

Circular RNAs have been linked to numerous other nervous system-related disease such as traumatic brain injury (TBI), major depressive disorder (MDD) and multiple sclerosis (MS). MDD or major depressive disorder is one of the most common problem affecting millions of people worldwide. It is characterized by low mood swing and causes a serious threat to the health of a person. Microarray analysis reports a considerable number of circRNAs to be differentially regulated, in which hsa_circRNA_103636 has been considered as a potential biomarker for MDD [29]. Similarly circ_0005402 and circ_0035560 are considered as potential biomarker for multiple sclerosis, a debilitating autoimmune disorder that destroys the myelin sheath in neurons [30, 31]. Astrocyte activation can be seen in numerous neurological diseases and potentially worsen the inflammatory reactions and neuronal tissue damage. Therapeutic strategies to alleviate astrocyte activation have been shown to have a significant impact on experimental neuroinflammatory models. circHIPK2 has recently been proven to inhibit astrocyte activation by targeting MIR124-2HG via the regulation of autophagy and endoplasmic reticulum (ER) stress, thus acting as a potential therapeutic target in the context of neuroinflammatory disorders and drug abuse [32].

4 Conclusion

Circular RNAs are one of the regulatory ncRNAs that have been in the spotlight of research as of late yet still in its preliminary stages. To the best extent, what we know so far is the functional importance of circRNAs in developmental stages of the human brain and in disease pathology. The wider picture lurks clarity, as more experimental studies are needed to identify their target and its mechanism of regulation. By understanding the complex interlink and regulatory actions of these RNAs, better therapeutic approaches can be taken towards nervous system-related diseases.

Competing Financial Interests The authors declare no competing financial interests.

References

1. Wang PL, Bao Y, Yee MC et al (2014) Circular RNA is expressed across the eukaryotic tree of life. PLoS One 9(6):e90859
2. Chen LL (2016) The biogenesis and emerging roles of circular RNAs. Nat Rev Mol Cell Biol 17(4):205–211
3. Greene J, Baird AM, Brady L et al (2017) Circular RNAs: biogenesis, function and role in human diseases. Front Mol Biosci 4:38
4. Salzman J, Chen RE, Olsen MN et al (2013) Cell-type specific features of circular RNA expression. PLoS Genet 9(9):e1003777
5. Rybak-Wolf A, Stottmeister C, Glazar P et al (2015) Circular RNAs in the mammalian brain are highly abundant, conserved, and dynamically expressed. Mol Cell 58(5):870–885
6. Westholm JO, Miura P, Olson S et al (2014) Genome-wide analysis of drosophila circular RNAs reveals their structural and sequence properties and age-dependent neural accumulation. Cell Rep 9(5):1966–1980
7. You X, Vlatkovic I, Babic A et al (2015) Neural circular RNAs are derived from synaptic genes and regulated by development and plasticity. Nat Neurosci 18(4):603–610
8. Hanan M, Soreq H, Kadener S (2017) CircRNAs in the brain. RNA Biol 14(8):1028–1034
9. Chen W, Schuman E (2016) Circular RNAs in brain and other tissues: a functional enigma. Trends Neurosci 39(9):597–604
10. Chen BJ, Yang B, Janitz M (2018) Region-specific expression of circular RNAs in the mouse brain. Neurosci Lett 666:44–47
11. Reddy AS, O'Brien D, Pisat N et al (2017) A comprehensive analysis of cell type-specific nuclear RNA from neurons and glia of the brain. Biol Psychiatry 81(3):252–264
12. Gruner H, Cortes-Lopez M, Cooper DA et al (2016) CircRNA accumulation in the aging mouse brain. Sci Rep 6:38907
13. Daniel C, Silberberg G, Behm M et al (2014) Alu elements shape the primate transcriptome by cis-regulation of RNA editing. Genome Biol 15(2):R28
14. Hansen TB, Jensen TI, Clausen BH et al (2013) Natural RNA circles function as efficient microRNA sponges. Nature 495(7441):384–388
15. van Rossum D, Verheijen BM, Pasterkamp RJ (2016) Circular RNAs: novel regulators of neuronal development. Front Mol Neurosci 9:74
16. Liu C, Zhang C, Yang J et al (2017) Screening circular RNA expression patterns following focal cerebral ischemia in mice. Oncotarget 8(49):86535–86547
17. Lin SP, Ye S, Long Y et al (2016) Circular RNA expression alterations are involved in OGD/R-induced neuron injury. Biochem Biophys Res Commun 471(1):52–56
18. Mehta SL, Pandi G, Vemuganti R (2017) Circular RNA Expression Profiles Alter Significantly in Mouse Brain After Transient Focal Ischemia. Stroke 48(9):2541–2548
19. Bai Y, Zhang Y, Han B et al (2018) Circular RNA DLGAP4 ameliorates ischemic stroke outcomes by targeting miR-143 to regulate endothelial-mesenchymal transition associated with blood-brain barrier integrity. J Neurosci 38(1):32–50
20. Zheng J, Liu X, Xue Y et al (2017) TTBK2 circular RNA promotes glioma malignancy by regulating miR-217/HNF1beta/Derlin-1 pathway. J Hematol Oncol 10(1):52
21. Yang P, Qiu Z, Jiang Y et al (2016) Silencing of cZNF292 circular RNA suppresses human glioma tube formation via the Wnt/beta-catenin signaling pathway. Oncotarget 7(39):63449–63455
22. Yang Y, Gao X, Zhang M et al (2018) Novel role of FBXW7 circular RNA in repressing glioma tumorigenesis. J Natl Cancer Inst 110(3)
23. Zhu J, Ye J, Zhang L et al (2017) Differential expression of circular RNAs in glioblastoma multiforme and its correlation with prognosis. Transl Oncol 10(2):271–279
24. Barbagallo D, Caponnetto A, Cirnigliaro M et al (2018) CircSMARCA5 inhibits migration of glioblastoma multiforme cells by regulating a molecular axis involving splicing factors SRSF1/SRSF3/PTB. Int J Mol Sci 19(2)
25. Shi Z, Chen T, Yao Q et al (2017) The circular RNA ciRS-7 promotes APP and BACE1 degradation in an NF-kappaB-dependent manner. FEBS J 284(7):1096–1109

26. Zhao Y, Alexandrov PN, Jaber V et al (2016) Deficiency in the ubiquitin conjugating enzyme UBE2A in Alzheimer's Disease (AD) is linked to deficits in a natural circular miRNA-7 sponge (circRNA; ciRS-7). Genes (Basel) 7(12):116
27. Kumar L, Shamsuzzama, Jadiya P et al (2018) Functional characterization of novel circular RNA molecule, circzip-2 and its synthesizing gene zip-2 in C. elegans model of Parkinson's disease. Mol Neurobiol. https://doi.org/10.1007/s12035-018-0903-5
28. Chen BJ, Mills JD, Takenaka K et al (2016) Characterization of circular RNAs landscape in multiple system atrophy brain. J Neurochem 139(3):485–496
29. Cui X, Niu W, Kong L et al (2016) hsa_circRNA_103636: potential novel diagnostic and therapeutic biomarker in Major depressive disorder. Biomark Med 10(9):943–952
30. Cardamone G, Paraboschi EM, Rimoldi V et al (2017) The characterization of GSDMB splicing and backsplicing profiles identifies novel isoforms and a circular RNA that are dysregulated in multiple sclerosis. Int J Mol Sci 18(3)
31. Iparraguirre L, Munoz-Culla M, Prada-Luengo I et al (2017) Circular RNA profiling reveals that circular RNAs from ANXA2 can be used as new biomarkers for multiple sclerosis. Hum Mol Genet 26(18):3564–3572
32. Huang R, Zhang Y, Han B et al (2017) Circular RNA HIPK2 regulates astrocyte activation via cooperation of autophagy and ER stress by targeting MIR124-2HG. Autophagy 13(10):1722–1741

Circular RNA and Alzheimer's Disease

19

Rumana Akhter

Abstract

Circular RNAs (circRNAs) represent a special group of noncoding single-stranded highly stable ribonucleic acid entities abundant in the eukaryotic transcriptome. These circular forms of RNAs are significantly enriched in human brain and retinal tissues. However, the biological evolution and function of these circRNAs are poorly understood. Recent reports showed circRNA to be an important player in the development of neurodegenerative diseases like Alzheimer's disease. With the progression of age, circRNA level increases in the brain and also in age-associated neurological disorder like Alzheimer's disease (AD), Parkinson's disease, inflammatory neuropathy, nervous system neoplasms, and prion diseases. One highly represented circRNA in the human brain and retina is a ciRS-7 (CDR1as) which acts as an endogenous, anticomplementary miRNA inhibitor or "sponge" to quench the normal functioning of miRNA-7. Low CDR1as level can lead to increase in miR-7 expression which downregulates the activity of ubiquitin protein ligase A (UBE2A), an important AD target, functionally involved in clearing toxic amyloid peptides from AD brain. This chapter focuses on the functional relationship of circRNA with AD and interplay of miRNA-mRNA-mediated genetic regulatory networks. Our conceptual understanding also suggests that circRNA can be considered as a potential biomarker and therapeutic target in AD diagnosis and treatment.

Keywords

circRNA · Alzheimer's disease · Amyloid · CDR1as miR-7 · UBE2A

1 Introduction

Alzheimer's disease (AD) is a progressive neurodegenerative disorder and the most common cause of dementia in aging population. AD is clinically apparent as the insidious impairment of higher intellectual functions that ultimately leads to death from complete brain failure. There are regions of the brain that are specifically affected sites of neuropathology in Alzheimer's disease, and they include the hippocampus, the amygdala, the temporal cortex, and the frontal cortex. Complex multifactorial interactions among genetic, epigenetic, and environmental components are responsible for causation of AD. Recent research has come up with an interesting entity of RNA, an endogenous noncoding circular RNA (circRNA) abundantly expressed in eukaryotes,

R. Akhter (✉)
Cleveland Clinic Lerner Research Institute,
Cleveland, OH, USA
e-mail: akhterr@ccf.org

J. Xiao (ed.), *Circular RNAs*, Advances in Experimental Medicine and Biology 1087,
https://doi.org/10.1007/978-981-13-1426-1_19

which has important functions in many disease-associated gene regulations including AD. These structures have the 3′ and 5′ ends joined together by covalent bonds giving a circular appearance which is unlike linear RNA. These molecules which are evolutionarily conserved had been dismissed as a rare, exotic RNA species for decades.

2 circRNAs in Neural Development

circRNAs are enriched in the brain. These neural circRNAs are derived from synaptic genes named *Dscam* and *Homer1* or from genes with important functions in early neural development, such as genes involved in axon guidance, Wnt and TGF-β signaling [1, 2]. These are localized mostly in neuronal cell bodies and neuropil [2] and found to be highly enriched in hippocampal synaptosomes. The developmental role of expression of circRNAs in the brain is determined by profiling the circRNA population in the hippocampus over several stages: embryonic (E18), early postnatal (P1), postnatal at the beginning of synapse formation (P10), and late postnatal hippocampus following the establishment of mature neural circuits (P30) [2]. The existence of circRNA in vivo as well as in primary neuron cultures and cell lines suggests a diverse distribution of circRNAs within neurons. This also cites an important role for circRNAs in neuronal development and plasticity [2, 3].

circRNAs are flanked by introns. Neuronal genes usually have long (> 10 kb) introns which are highly conserved in evolution. So the circRNAs flanked by long introns are mostly evolutionarily conserved [4]. Alternative explanation for conserved circRNA could be that introns in these genes are long for other reasons and that circRNA is produced due to recursive splicing, i.e., circRNAs are produced as by-product of the complex splicing [5].

The function of circRNA in mammalian brain remains to be defined. Since circRNAs are highly stable, they could serve as topologically complex platforms for protein or RNA transportation. The prominence of stable circRNAs in the synapse provides both the stability and flexibility of neuronal networks which are vital to all behavior, including learning and memory. Future functional research should be directed in understanding the effect of genetic perturbation of specific circRNAs followed by phenotypic examination which will address circRNA function in the nervous system in respect to molecular memory.

3 Role of circRNA in Alzheimer's Disease

circRNAs tend to accumulate during normal process of brain aging and thus make susceptible to age-related neurodegenerative diseases like AD. This disease is the most common cause of dementia in elderly population characterized by the presence of neurotoxic senile amyloid plaques, hyper-phosphorylated tau tangles, massive neuron death, and neuro-inflammation. According to the amyloid hypothesis, accumulation of Aβ in the brain is central to AD pathogenesis [6]. Amyloid is a general term for protein fragments of albuminoid proteinaceous material that the body produces normally. β-amyloid is a 36–43 amino acid peptide fragment clipped from amyloid precursor protein (APP). Most cases of Alzheimer's belong to the late-onset category, which occurs after age 60. The reasons of late-onset Alzheimer's are not yet completely elucidated, but they include a combination of genetic, environmental, and lifestyle factors that have an important influence on disease susceptibility of a person. The single-gene mutations are mostly directly responsible for early-onset Alzheimer's disease but do not seem to be involved in late-onset Alzheimer's, and thus a specific gene mutation does not cause the late onset of the disease complex multifactorial interactions among genetic, epigenetic, and environmental components which is responsible for causation of AD. Familial AD is characterized by the genetic mutations involved in Aβ peptide biogenesis which consists of four well-studied AD genes, the *APP*, *PS1*, *PS2*, and

APOE. These genes exhibit mutations that enhance the relative rate of generation of Aβ42, the longer form of the peptide that is much more prone to oligomerization and fibrillation than Aβ40 [7].The spatial and temporal patterns of senile plaques consisting of fibrillar Aβ do not equate very well with the degree of dementia in AD, and thus the traditional amyloid hypothesis remains debatable. In contrast, cognitive malfunctioning displays a profound relationship with most common type of sporadic AD which remains largely unknown. circRNA expressed in the human brain might play a causative role in AD and other neurodegenerative conditions. Interestingly, circRNA could emerge as a potential therapeutic target in AD diagnosis and treatment.

Although circRNA has been reported in many diseases, their role in Alzheimer's disease remains unclear. Interestingly, evolutionarily conserved microRNA-7 which is highly abundant in human brain is associated with a circRNA for miRNA-7 (ciRS-7, also known as CDR1as). ciRS-7 contains multiple, tandem anti-miRNA-7 sequences that thereby act as an endogenous, anticomplementary miRNA "sponge" to adsorb and hence quench normal miRNA-7 functions [8, 9]. In the hippocampal CA1 region of sporadic AD patients, miR-7 circRNA system is dysregulated which is confirmed by Northern blot hybridization techniques and the circularity-sensitive circRNA probe RNase R [8]. Downregulation of ciRS-7 and ciRS-7 "sponging activities" might increase endogenous miRNA-7 levels in AD [10]. The elevated miRNA-7, due to inhibition in ciRS-7 "sponging" effects, can downregulate AD-associated targets like ubiquitin protein ligase, UBE2A, and an autophagic, phagocytic protein essential in the clearance of amyloid peptides in AD brain [11, 12]. Such miRNA-mRNA regulatory systems may represent another crucial aspect of epigenetic control over gene expression in health and disease. Inhibition of "miRNA sponging systems" and increase of specific inducible miRNAs might be a reason for downregulation of important genes related to sporadic AD brain [8, 13].

4 circRNA in Other Neuropathies

circRNA and Parkinson's disease Parkinson's disease (PD) is a neurodegenerative disorder that affects mainly dopamine-producing neurons in a specific area of the brain, substantia nigra pars compacta. CDR1as downregulates miR-7 [14]. It has already been reported that miR-7 can inhibit the expression of α-synuclein, a crucial constituent of Lewy bodies in the PD brain. α-Synuclein protein is expressed highly in the diseased brain and considered as hallmark feature in PD pathogenesis. Moreover, downregulation of α-synuclein by miR-7 protects cells against oxidative stress [15]. miRNA-7 can provide protection against neuron death caused by 1-methyl-4-phenylpyridinium (MPP+) by targeting the nuclear factor (NF)-κB signaling pathway [14, 16].

circRNA and Neoplasms High expression of CDR1as is evident in the brain cerebrum. CDR1as is highly expressed in neuroblastomas and astrocytoma [17]. miR-7 was found to be downregulated in astrocytoma and neuroblastoma compared to other brain tissue. Another study indicated that miR-7 could suppress EGFR expression in a glioblastoma cell line and downregulate IRS-1 and IRS-2 expression by repressing protein kinase B [18]. CDR1as acts as negative regulator of miR-7 [9, 14]. These evidences indicate possible role of circRNAs in the pathophysiology of nervous system neoplasms.

circRNA and Neuro-inflammation Virus-associated miRNA binding sites are present in some circRNA which plays vital role in immunoregulation as, for instance, hsa-circRNA 2149 contains 13 unique, head-to-tail spanning reads. Hsa-circRNA 2149 is present in CD19+ leukocytes but not in CD341 leukocytes or neutrophils. Another circRNA, circRNA100783, has implication in chronic CD28-associated CD8(+)T cell aging which could be utilized as an important biomarker for this disease [14, 19]. circRNA from SRY can repress miR-138 activity. miR-138 downregulates runt-related transcription factor 3 (RUNX3) which plays an essential role in the regulation of T helper cells [14, 20]. These studies provide indication of association of circRNA with neuro-inflammation.

circRNA and Prion Diseases Progressive neurodegeneration is evident in Prion diseases with neuronal loss and a failure to induce inflammatory response. These diseases include Creutzfeldt–Jakob disease (CJD), Gerstmann-Straussler syndrome (GSS), and also fatal familial insomnia. Prion diseases are mainly caused by alteration of a normal cell-surface glycoprotein (PrP^{C}) into a modified isoform (PrP^{Sc}) that renders infectious nature to PrP in the absence of nucleic acid. Reports show PrPC overexpression induces CDR1as expression [21]. Thus, CDR1as might have some implication in the etiology of prion diseases.

5 circRNA Detection in the Brain

Numerous strategies are employed to detect genome-wide circRNA expression over the past few years, but little is known to find out the accuracy of these approaches. Experimental and bioinformatic tools along with accurate statistical approaches can help address these objectives of circRNA detection. Recently, scientists have shown that circular RNA is associated with brain functions. When CDR1as, an RNA molecule highly expressed in human and mouse brain, was deleted from the genome of mice, the animal brain failed to retain important information and disregard the unnecessary ones like in other mental disorders [22]. Copious circRNAs are highly abundant in mammalian brain expression with conserved expression. The well-known CDR1as is strongly bound by miR-7 and miR-671 in the human and mouse brain. Expression of these two microRNAs was posttranscriptionally dysregulated in all brain regions. Early genes such as Fos, a direct miR-7 target, were enhanced in CDR1as-deficient brains, indicating a possible molecular link to the behavioral phenotype [22].

The most popular technology microarray was a preferred way for global RNA expression analysis before the arrival of next-generation sequencing (NGS), but it was not convenient to screen circRNAs or its expression from linear counterparts. The high-throughput NGS technology has provided a competent way to detect circRNAs. Many software packages came to the rescue to decipher circRNAs from RNA-Seq data. Common RNA-seq protocols have limitations as it may introduce technical artifacts that can result in wrong identification of circRNA isoforms as exonucleases might act upon some circRNA and inhibit their expression [1, 23]. In mouse, deep sequencing of multiple organs reveals significantly greater fraction of circRNA junctional reads. Similar reports were found in human tissues [3, 24]. circRNAs show different patterns of expression respective to brain areas which include the striatum, prefrontal cortex, olfactory cortex, cerebellum, and hippocampus. A gene ontology analysis of the transcripts generating circRNAs reveals synaptic genes encoding pre- and postsynaptic functional groups are fortified as circRNA host genes and thus provide an important reasoning for the abundance of circRNA in the brain. Along with that, many circRNAs are highly distributed in synaptic fractions and synaptosomes [24]. Rare circRNA localization in cell body and dendrites of cultured hippocampal neurons and hippocampal slices can be targeted by high-resolution in situ hybridization technique [24].

6 Conclusion

circRNA function and their relationships with Alzheimer's disease and other neuropathies remain to be fully elucidated. circRNAs are usually abundant and found to be stable in vivo, which might attribute to their importance in molecular diagnostics. Until then, the role of circRNA in gene regulation may be utilized as imperative treatment option. Importantly, the potential role of circRNAs as miRNA sponges can be utilized as an innovative approach to regulate gene expression. Further research on circRNA will enhance our understanding in relation to neuropathies like AD and lead to new diagnostic biomarkers and promising therapeutic options.

References

1. Veno MT, Hansen TB, Veno ST et al (2015) Spatio-temporal regulation of circular RNA expression during porcine embryonic brain development. Genome Biol 16:245
2. You X, Vlatkovic I, Babic A et al (2015) Neural circular RNAs are derived from synaptic genes and regulated by development and plasticity. Nat Neurosci 18(4):603–610
3. Rybak-Wolf A, Stottmeister C, Glazar P et al (2015) Circular RNAs in the mammalian brain are highly abundant, conserved, and dynamically expressed. Mol Cell 58(5):870–885
4. Ivanov A, Memczak S, Wyler E et al (2015) Analysis of intron sequences reveals hallmarks of circular RNA biogenesis in animals. Cell Rep 10(2):170–177; Jeck WR, Sorrentino JA, Wang K et al (2013) Circular RNAs are abundant, conserved, and associated with ALU repeats. RNA 19(2):141–157
5. Burnette JM, Miyamoto-Sato E, Schaub MA et al (2005) Subdivision of large introns in Drosophila by recursive splicing at nonexonic elements. Genetics 170(2):661–674
6. Hardy J, Selkoe DJ (2002) The amyloid hypothesis of Alzheimer's disease: progress and problems on the road to therapeutics. Science 297(5580):353–356
7. Bertram L, Tanzi RE (2008) Thirty years of Alzheimer's disease genetics: the implications of systematic meta-analyses. Nat Rev Neurosci 9(10):768–778
8. Lukiw WJ (2013) Circular RNA (circRNA) in Alzheimer's disease (AD). Front Genet 4:307
9. Hansen TB, Jensen TI, Clausen BH et al (2013) Natural RNA circles function as efficient microRNA sponges. Nature 495(7441):384–388
10. Cogswell JP, Ward J, Taylor IA et al (2008) Identification of miRNA changes in Alzheimer's disease brain and CSF yields putative biomarkers and insights into disease pathways. J Alzheimers Dis 14(1):27–41
11. Bingol B, Sheng M (2011) Deconstruction for reconstruction: the role of proteolysis in neural plasticity and disease. Neuron 69(1):22–32
12. Lonskaya I, Shekoyan AR, Hebron ML et al (2013) Diminished parkin solubility and co-localization with intraneuronal amyloid-beta are associated with autophagic defects in Alzheimer's disease. J Alzheimers Dis 33(1):231–247
13. Ginsberg SD, Alldred MJ, Che S (2012) Gene expression levels assessed by CA1 pyramidal neuron and regional hippocampal dissections in Alzheimer's disease. Neurobiol Dis 45(1):99–107
14. Shao Y, Chen Y (2016) Roles of circular RNAs in neurologic disease. Front Mol Neurosci 9:25
15. Junn E, Lee KW, Jeong BS et al (2009) Repression of alpha-synuclein expression and toxicity by microRNA-7. Proc Natl Acad Sci USA 106(31):13052–13057
16. Choi DC, Chae YJ, Kabaria S et al (2014) MicroRNA-7 protects against 1-methyl-4-phenylpyridinium-induced cell death by targeting RelA. J Neurosci 34(38):12725–12737
17. Chen YT, Rettig WJ, Yenamandra AK et al (1990) Cerebellar degeneration-related antigen: a highly conserved neuroectodermal marker mapped to chromosomes X in human and mouse. Proc Natl Acad Sci USA 87(8):3077–3081; Dropcho EJ, Chen YT, Posner JB et al (1987) Cloning of a brain protein identified by autoantibodies from a patient with paraneoplastic cerebellar degeneration. Proc Natl Acad Sci USA 84(13):4552–4556
18. Liu Z, Jiang Z, Huang J et al (2014) miR-7 inhibits glioblastoma growth by simultaneously interfering with the PI3K/ATK and Raf/MEK/ERK pathways. Int J Oncol 44(5):1571–1580
19. Wang YH, Yu XH, Luo SS et al (2015) Comprehensive circular RNA profiling reveals that circular RNA100783 is involved in chronic CD28-associated CD8(+)T cell ageing. Immun Ageing 12:17
20. Fu D, Yu W, Li M et al (2015) MicroRNA-138 regulates the balance of Th1/Th2 via targeting RUNX3 in psoriasis. Immunol Lett 166(1):55–62
21. Satoh J, Obayashi S, Misawa T et al (2009) Protein microarray analysis identifies human cellular prion protein interactors. Neuropathol Appl Neurobiol 35(1):16–35; Satoh J, Yamamura T (2004) Gene expression profile following stable expression of the cellular prion protein. Cell Mol Neurobiol 24(6):793–814
22. Piwecka M, Glazar P, Hernandez-Miranda LR et al (2017) Loss of a mammalian circular RNA locus causes miRNA deregulation and affects brain function. Science 357(6357)
23. Guo JU, Agarwal V, Guo H et al (2014) Expanded identification and characterization of mammalian circular RNAs. Genome Biol 15(7):409
24. Chen W, Schuman E (2016) Circular RNAs in brain and other tissues: a functional enigma. Trends Neurosci 39(9):597–604

Circular RNA in Liver: Health and Diseases

20

Meiyi Song, Lu Xia, Mengxue Sun, Changqing Yang, and Fei Wang

Abstract

Circular RNA (circRNA) is an important class of noncoding RNA characterized by covalently closed continuous loop structures. In recent years, the various functions of circRNAs have been continuously documented, including effects on cell proliferation and apoptosis and nutrient metabolism. The liver is the largest solid organ in mammals, and it also performs many functions in the body, which is considered to be the busiest organ in the body. At the same time, the liver is vulnerable to multiple pathogenic factors, causing various acute and chronic liver diseases. The pathogenesis of liver disease is still not fully understood. As a rising star for the past few years, circRNAs have been proven involved in the regulation of liver homeostasis and disease. This chapter will explain the role of circRNAs in liver health and diseases and sort out the confusion in the present study.

Keywords

Circular RNA · Liver · MicroRNA · Noncoding RNA

1 Introduction

Circular RNAs (CircRNAs) are a class of noncoding RNAs (ncRNAs) that forms covalently closed continuous loop structures, lacking the terminal 5′ and 3′ ends [1, 2]. For quite a long time, circRNA was excluded from the scope of research for being regarded as low abundant RNA and a splicing error without any function. In a variety of organisms, including plants, nematodes, drosophila, mice, and human, there are abundant circRNA expression [3–7]. Most of human circRNAs are transcribed from single or multiple exons, known as exonic circRNAs. However, many studies have shown that the intricate mechanisms of circRNA splicing, including inverted repeated ALU pairs [8–10], reverse complementary sequences and exon skipping [11–14]. CircRNAs can be produced by multiple genetic structures, while the same location of a gene can produce different types of circRNAs. Recent researches have shown that circular RNAs mainly function in four ways: (1) circRNAs may function as miRNA sponges; (2) circular RNAs are involved in transcription and translation regulation; (3) circRNAs may inhibit the splicing of linear RNA; and (4) circRNAs regulate gene expression as sponges of molecules (RBPs), such as RNA-binding protein components [15–17]. The latest research also suggests that circRNAs can also encode proteins, which raises questions about whether they are still classified as noncod-

Authors Meiyi Song and Lu Xia have been contributed equally to this chapter.

M. Song · L. Xia · M. Sun · C. Yang (✉) · F. Wang (✉)
Division of Gastroenterology and Hepatology, Digestive Disease Institute, Shanghai Tongji Hospital, Tongji University School of Medicine, Shanghai, China
e-mail: 1132469@tongji.edu.cn

J. Xiao (ed.), *Circular RNAs*, Advances in Experimental Medicine and Biology 1087,
https://doi.org/10.1007/978-981-13-1426-1_20

ing RNAs [18]. CircRNAs function in the growth or development of tissues or organs and regulation of cell cycle [19, 20], cell apoptosis [21, 22], cell stress [23, 24], cellular senescence [25, 26], and inflammatory responses [22, 27].

The liver is the largest solid organ of the human body, with the functions of immunity, detoxification, biosynthesis, and metabolism. It is responsible for the metabolism of lipids, sugars, proteins, and vitamins in the body, which is a veritable metabolism factory [28–30]. In addition, the liver is the only organ with regenerative function in the body, less than 25% of the normal liver can form a complete liver. But even then, the liver is susceptible to various factors that cause acute or chronic liver injury, and chronic liver disease can lead to liver failure or liver cancer, which is a serious threat to human health. There is a lot of evidence to explain the relationship between circRNAs and the liver homeostasis and diseases; here we review the role of circRNAs in the liver.

2 Identification of Circular RNAs in the Liver

The existence of circRNA was reported as early as the 1990s [31, 32]. The latest research on circRNA expression of six adult and fetal normal tissues by RNA-seq found that 36.97–50.04% of circRNAs detected were organ-specific; meanwhile, 33 circRNAs were universally expressed in almost all tested tissues. Six hundred sixty-eight circRNAs were specifically expressed in liver tissues. Further analysis of the regulatory network found that 17 circRNAs, 22 miRNAs, and 90 mRNAs constitute a tight network, forming a circRNA-miRNA-mRNA interaction network. In contrast with fatal tissue, the particular circRNA is higher expressed in the same tissue of adults. Lin Li et al. used RAISE to detect circRNAs in RNA-seq data. They identified RAISE circRNA candidates in 61 human liver rRNA-depleted samples. Finally, 8270 circRNA candidates for advanced analysis were authenticated, and 59,128 circRNA candidates were identified in HCC and adjacent non-tumor tissues. However, the authors still believe that the expression of circRNAs needed to be confirmed in a larger number of human samples. Although there have not been many studies describing the expression pattern of the liver under a physiological state no matter in mice or human, the rich expression of circRNAs in the liver has been basically revealed, which indicates that circRNAs may be closely related to liver function.

3 Circular RNAs in Liver Damage and Repair

Many pathogenic factors including drugs, alcohol, and virus can lead to liver injuries [33–35]. Since the liver itself has a great capacity to repair after damage, the process of regeneration can be initiated immediately since the attack of the pathogenic factors. Damage in the liver results in injury-related diseases. Poor or insufficient regeneration in the liver will eventually lead to liver fibrosis, irreversible liver cirrhosis, and even liver cancer, so-called chronic liver diseases. CircRNAs are a group of important members of the noncoding RNAs involved in both the damage and the repair process in the liver. The role of circRNAs is gradually revealed (Fig. 20.1).

3.1 Nonalcoholic Fatty Liver Disease (NAFLD)

Nonalcoholic fatty liver disease (NAFLD) is the most common liver disease worldwide. Between 75 million and 100 million individuals are estimated to suffer from NAFLD in America [36–38]. Similar to alcoholic liver disease and viral liver disease, liver inflammation usually occurs in NAFLD, which can lead to liver cirrhosis, liver cancer, and even liver failure [39]. NAFLD is divided into nonalcoholic fatty liver and nonalcoholic steatohepatitis (NASH), which is associated with histological feature of liver fibrosis/cirrhosis from mild to severe [40, 41]. The typical pathological change of nonalcoholic fatty liver includes mild nonspecific inflammation in the liver and isolated hepatic steatosis [42–44]. When it comes

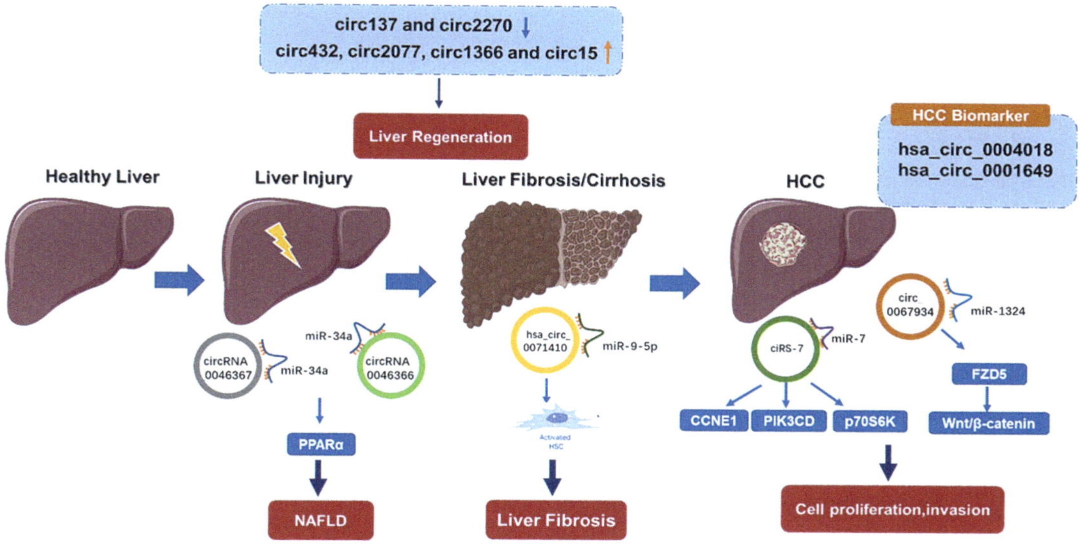

Fig. 20.1 CircRNAs in liver diseases. *HCC* hepatocellular carcinoma, *NAFLD* nonalcoholic fatty liver disease, *PPARα* peroxisome proliferators-activated receptors α

to the stage of nonalcoholic steatohepatitis, liver inflammation will progress with or without fibrosis [45]. Hepatic steatosis with inflammation has a distinct and more progressive natural history than isolated hepatic steatosis as the evidence from clinical and experimental research [46, 47].

NAFLD/NASH is a multifactor comprehensive disease, including metabolic, genetic, environmental, gut microbial [48], and epigenetic factors. At present, scholars have established the theory of "multiple-hit," instead of the previous "two-hit" pathogenesis [49], which is more appropriate to describe the molecular and metabolic changes in the occurrence of NAFLD [50, 51]. The hypothesis proposes that multiple attacks mentioned above are genetically predisposed to NAFLD and provide a more accurate explanation for the pathogenesis of NAFLD [52].

Hepatic triglyceride accumulation is one of the most significant features in NASH. Epigenetics regulation plays a key role in the development of NAFLD. The esterification of glycerol and free fatty acids (FFAs) forms triglycerides [53]. The synthetic triglycerides enter the storage or secretion pool, with varying degrees of lost [54–56]. FFAs derived from dietary or adipose tissue through fat decomposition and/or hepatic DNL undergo acetylation and form fatty acyl-CoA and then participate in the esterification reaction or β-oxidation [57]. Genetic and epigenetic regulations are equally important in process of NAFLD [58–60].

Fatty acid synthesis and degradation [61, 62], insulin resistance (IR) [63], oxidative stress [64, 65], and hepatofibrogenesis [66–68] are regulated by the epigenetics regulation in hepatocyte. Noncoding RNAs play pivotal roles through epigenetics regulation in NAFLD. Recent researches have drawn attention to a novel class of noncoding RNA in NAFLD. The expression profile of circRNA in NAFLD was first determined in 2016 [69]. The authors used MCD diet-induced NASH mice model and preformed circRNA and mRNA microarray at the same time. Sixty-nine up- and 63 down-regulated circRNAs were found; meanwhile, 2760 mRNAs up and 2465 mRNAs down regulated in the same sample. They further searched for the relationship between circRNAs and mRNAs changes based on circRNA-miRNA-mRNA network. Four pathways were built, including circ_002581-miR-122-Slc1a5, circ_002581-miR-122-Plp2, circ_002581-miR-122-Cpeb1, and circ_007585-miR-326-UCP2 based on the downstream miR-

NAs analysis, pathway analysis, and PCR verification. Since circRNAs may function as competitors of pre-mRNA, they also described some possible circRNA-mRNA pairs, such as circ_011775-Rn45s and circ_004300-Malat1. Another research used metformin to treat high-fat diet (HFD)-induced NAFLD mice and then identified related noncoding RNAs. Focusing on circRNA response to HFD and metformin, they revealed 396 (231 up- and 165 downregulated) circRNAs and 222 (126 up- and 96 downregulated) circRNAs respond to HFD and metformin, respectively [70]. Unfortunately, they did not observe a direct association between circRNAs and metabolism in liver diseases, whereas they suggested the extensive interaction between some of the miRNAs and circRNAs was involved in the development of disease and required future investigation. The function of circRNAs has been uncovered in hepatocytes steatosis. The regulation effect of circ_0046367 on NAFLD was revealed by functional experiment in human hepatocytes. Results showed decreased expression of circ_0046367 in FFA-induced hepatocytes steatosis. To explore the underlying mechanism, they searched for the potential miRNAs that regulated hepatocytes steatosis and found that miR-34a could bind to circ_0046367 and hence mediated the effect of the circRNA. The inhibition of miR-34a on its target gene PPARα was also reduced by circ_0046367. In addition, lipid metabolism-related genes which were associated with PPARα were significantly activated by circ_0046367 resulting in alleviated intracellular lipid accumulation, lipid peroxidation, and even mitochondrial dysfunction. In patient with hepatic steatosis, changes in expression levels of circ_0046367 and miR-34a are in the opposite directions which confirm the interaction of each other. Thereafter the same research group identified another circRNA, circ_0046366 acting as miR-34a sponge [71]. Circ_0046366 decreasing is a significant feature in FFA-induced steatosis of hepatocytes. The increasing of miR-34a and corresponding inhibition of its target genes as a result of circ_0046366 deficiency may lead to the transcriptional inhibition of lipometabolic genes. This is the possible mechanism for involvement of circ_0046366 in the progress of NAFLD.

Besides, the functions of other circRNAs are worthwhile to mention here. The circular RNA ciRS-7 can increase both insulin content in islet cells and promote its secretion [72]. These processes may affect by miR-7 and its targets, including Myrip. It suggests that ciRS-7 may play a considerable role in regulating blood glucose and further orchestrate the occurrence of NAFLD.

3.2 Liver Regeneration

The potential of the liver to repair damage or recover weight loss is marvelous. This process is always defective under fat stimulation or some serious lesions, leading to fibrosis/cirrhosis of the liver or other undesirable phenotypes. It is often thought that growth and differentiation are mutually exclusive processes; as an exception, the liver maintains vital, highly differentiated functions throughout its growth. The majority of present studies are based on the in vivo model of liver cell proliferation post-partial hepatectomy. However, liver cell damage and death and the characteristics of nonsurgical resection of liver cell regeneration are also involved in the process of liver regeneration (LR). The partial hepatectomy is the most commonly used model to study the liver regeneration process. The process of liver damage and repair is often interwoven, and many factors can act double-edged simultaneously, such as tumor necrosis factor-α (TNF-α). When liver resection is performed, the whole organ enters a transient regeneration process, without a stage of chronic injury. The process of liver regeneration can be divided into three main phases: priming phase, proliferation phase, and termination phase [73]. Various cytokines or hormones appear to serve as modulators of liver regeneration. Insulin and glucagon are involved in energy metabolism. TNF-α, interleukin-6 (IL-6), and other cytokines begin to be secreted in priming phase [74, 75]. Most quiescent hepatocytes enter back into the cell cycle rapidly from G0 phase and initiate the liver regeneration process. Within approximate 4 h after the injury, the liver began to initiate its regeneration process in mice [76]. During the proliferation phase, remaining hepatocytes start to mitose, under the stimulation of growth factors and metabolic signaling. The primary stage of liver weight gain occurs within 3 days after the partial hepatectomy, and the liver can basically restore to its original weight between 5 and 7 days after surgery. Finally, various regulatory factors are applied to the liver to modulate the weight of the liver and liver regeneration in termination phase. The priming phase drew closer attention because in this phase the normally quiescent hepatocytes reenter the cell cycle to proliferate in response to injury or infection, determining the fate of the liver. Lifei Li et al. described the expression pattern of circRNAs during priming phase of rat LR (0, 2, and 6 h after surgery). Using high-throughput RNA sequencing technology, 2412 circRNAs were identified. Three hundred of the circRNAs were detected at all three time points, among which 15 circRNAs were downregulated and 28 circRNAs were upregulated significantly in both 2 and 6 h after PH [77]. The following functional analysis of circRNAs and their host genes was performed using gene ontology (GO) and Kyoto Encyclopedia of Genes and Genomes (KEGG) biological pathway analysis to predict possible pathway involved. The results indicated that steroid hormone biosynthesis, inflammatory mediator regulation of TRP channels were involved in. According to expression and structure verification, they finally verified two down-regulated circRNAs (circ_137 and circ_2270) and four upregulated circRNAs (circ_432, circ_2077, circ_1366, and circ_15) played a crucial role during the process of liver regeneration. They also highlighted the co-expression of miRNAs with selected circRNAs.

3.3 Liver Fibrosis/Cirrhosis

Liver fibrosis is the end-stage hepatopathy resulting from chronic liver injury due to any common etiology. It can progress and develop irreversible cirrhosis with a survival less than 50% in 10 years. Extracellular matrix (ECM) accumulation leads to structural modification and dysfunction of the liver. Activated hepatic stellate cells (HSCs) make great contributions to the synthesis of ECM. In addition, the role of activation of hepatic Kupffer cells and lack of liver sinusoidal endothelial cells(LSECs) should be mentioned [78]. Inflammation and activation of Kupffer cells prevent ECM secretion and deposition in HSCs [79]. Different from Kupffer cells, LSECs are regarded as a gatekeeper in the fibrotic process [80]. LSEC differentiation occurs during capillarization before the fibrosis development, which has the function of allowing and promoting the activation of HSC [78]. Therefore, the anfractuous interaction between non-parenchymal cells participates in the process of liver fibrosis via affecting the activation of HSC.

In 1876, HSCs were observed and described for the first time by von Kupffer, between liver sinusoidal endothelial cells (LSECs) and hepatocytes in the subendothelial space of Disse, which represent about 10% of all resident liver cells [81, 82]. HSC can transdifferentiate from quiescent, vitamin-A-storing cells to the myofibroblast acquiring the function of proliferation, contractility, fibrogenesis, altered matrix degradation, and chemotaxis. TGF-β1 is essential for the activation of HSCs, which result in expression of α-smooth muscle actin (α-SMA) and secretion of ECM proteins, mostly collagen Type I (ColI) [83, 84]. Targeting HSC has been the research focus to regress liver fibrosis during the last half-century [85]. Noncoding RNA is an especially important regulator of HSC activation, since regulatory activities must occur quickly response to wounding stimulates, and directly activate or inhibit target genes or function through post-transcription regulation. MicroRNA and lncRNA are reported to be involved in the activation of HSC and thereby participate in the process of liver fibrosis/cirrhosis [86]. Knockdown of Lnc-LFAR1, a liver-enriched lncRNA, can impair the activation of HSC, protect TGF-β-induced apoptosis of hepatocytes, and attenuate CCl_4- and bile duct ligation-induced liver fibrosis in vivo by activating TGF-β and Notch pathways. As another member of the noncoding RNA, circRNAs are predicted to be associated with the HSC activation and may be a promising target for treatment of liver fibrosis [87]. Compared to quiescent HSCs, 179 circRNAs were found to be upregulated, and 630 circRNAs were downregulated during irradiation-induced activation [88]. A pathway analysis was performed on the downregulated circRNAs, and eight significantly associated pathways were revealed, including glycosaminoglycan degradation, pentose phosphate pathway, and phosphatidylinositol signaling system. They detected that three circRNAs (hsa_circ_0072765, hsa_circ_0071410, hsa_circ_0054345) were significantly upregulated and three circRNAs (hsa_circ_0070963, hsa_circ_0061893, hsa_circ_0013255) were significantly downregulated which was confirmed by PCR results. The research further explored the function of hsa_circ_0071410 in HSC activation; three putative binding sites of miR-9-5p were predicted. Suppression of miR-9-5p reversed the effect of hsa_circ_0071410 knockdown on promoting HSC activation. These preliminarily prompted that circRNAs may be involved in HSC activation and thereby affect the process of liver fibrosis development. Other circRNAs such as circ_000203 and circ_010567 have also been reported to be regulators implicated in the heart fibrogenesis [89, 90], but whether they are also involved in liver fibrosis is worth further investigation.

4 Circular RNAs in Liver Neoplasm

4.1 The Roles of CircRNAs in Liver Carcinogenesis

Liver cancer is the fifth most common cancer and the third leading cause of cancer death in the world [91]. Proteins encoded by different genes have been shown to play a role in the development and progression of liver cancer. These genes directly or indirectly regulate the cell cycle, cell apoptosis, DNA damage response, and cell survival signaling [92–94]. The function of circular RNAs in cancers has been confirmed in various types of cancer, including digestive system malignancy [95]. circ_100269 shows an inhibition effect on gastric cancer cell proliferation by targeting miR-630 [96]. Investigation of hepatocellular carcinoma (HCC)-associated circRNAs is undergoing. CiRS-7 (also named as CDR1) is one of the famous intensive-studied circRNAs, which function as a sponge of miR-7 [72, 97–99]. Yu et al. enrolled 35 HCC patients in the cohort and found an increased expression of ciRS-7 in human HCC tissue corresponding with downregulated miR-7 expression levels [100]. In addition, they found that ciRS-7 accelerated the proliferation and invasion of liver cancer cell by decreasing the expression of miR-7 and relieving its inhibitory effect on its target gene CCNE1 and PIK3CD. In another cohort consisting of 108 HCC patients, the upregulation of ciRS-7 was detected compared to the adjacent non-tumor tissues [101]. Meanwhile, increased expression of ciRS-7 in HCC tissues was significantly associated with serum AFP level ($P < 0.01$), hepatic MVI ($P = 0.03$), and those with early onset ($P = 0.02$) and the deterioration of HCC ($P = 0.08$). The consistent expression change of ciRS-7 and three target genes of miR-7, PIK3CD, p70S6K, and mTOR were detected in the tissue as well. However, the median time of disease-free survival (DFS) had no significant difference in ciRS-7 higher- or lower-expressed patients. Decreased circ-MTO1 (mitochondrial translation optimization 1 homologue) was detected in the human tumor tissues, which was also named as hsa_circ_0007874/hsa_circ_104135 and related to poor prognosis in patients [102]. Furthermore, inhibition of circMTO1 accelerated hepatocellular cancer cell proliferation by targeting miR-9 and abolished its inhibition on downstream target genes P21 in vivo and in vitro. These all indicate potential therapeutic effect of circRNA on HCC. Both linear and circular (cir-cZKSCAN1) forms of ZKSCAN1 RNA were detected downregulated during HCC development [103]. The downregulation of cir-cZKSCAN1 was correlated with histological grade of tumor (National Comprehensive Cancer Network Clinical Practice Oncology Guidelines), tumor numbers, and underlying cirrhosis, vascular, invasion, or microscopic vascular invasion. In vivo and in vitro experiment showed that inhibition of both ZKSCAN1 mRNA and cir-cZKSCAN1 accelerated tumor progression by facilitating cell proliferation, migration, and invasion. Interestingly, linear and circular RNAs function through entirely different mechanisms without interference on expression of HCC. Cir-cZKSCAN1 affected cancer-related signaling pathways including TGF-β1, ITGB4, and CXCR4, while linear ZKSCAN1 mRNA was found to be associated with cell metabolism. On the other hand, circRNA also reported to be a target regulated by other protein of noncoding RNAs during HCC genesis. The function of cluster of differentiation (CD)90 (Thy-1) on liver cancer cell viability, migration, and invasive abilities was revealed. After that, two specific circRNAs, hsa_circ_0067531 and hsa_circ_0057096, were identified to be differently expressed in CD90 expression or deletion cells [104]. Additionally, the expression level of hsa_circ_0067531 was decreased in human HCC tissue, while hsa_circ_0057096 had no significant difference in expression. But the regulatory effect and diagnosis value of the circRNAs should be further studied. The interactions between in RhoA, circ_000839, and miR-200b have been expounded. miR-200b occupies the core position in the regulation network, which inhibits the migration and invasion of HCC cell in vitro [105]. The expression of circ_000839 has been found negatively related to that of miR-200b and positively related to that of its target gene RhoA, encouraging us to study the function of circ_000839 in detail in the future. Another circRNA has been reported to be involved in HCC. Increased expression of circ_0067934 has been detected in HCC tissue [106]. Inhibition of circ_0067934 can repress the proliferation, migration, and invasion of liver cancer cell. Besides, the abovementioned effect of circ_0067934 was exerted as a sponge of miR-1324 and constitutes a circ_0067934/miR-1324/FZD5/Wnt/β-catenin axis for regulating liver cancer.

Except HCC, the function of circRNAs in other types of liver cancer has also been reported. Intrahepatic cholangiocarcinoma (ICC) is the primary malignant tumor in the liver after HCC with an increasing incidence in recent years [107, 108]. It accounts for about 10–15% of the primary malignant tumor of the liver. A TGF-β-induced long noncoding RNA (TLINC) was identified in ICC, also known as cancer susceptibility candidate 15 (CASC15) [109]. Long and short TLINC isoforms existed and were related to the epithelial to mesenchymal transition phenotype. Circular isoforms of TLINC was verified in TGF-β-induced Huh28 cells and human ICC tissue, which was more stable to be a marker for the diagnosis.

Hepatoblastoma is the most common malignant liver cancer in children [110, 111]. Circ_0015756 was screened from 869 differentially expressed circRNAs in human hepatoblastoma, according to functional analysis of the host genes of above circRNAs [112]. Downregulation of circ_0015756 reduced the cell viability of hepatoblastoma and suppressed their proliferation and invasion in vitro. Functioning as a molecular sponge of miR-1250-3p, circ_0015756 shows a bright prospect of curing hepatoblastoma.

4.2 Circular RNA as Biomarker for Liver Cancer

Searching for efficient biomarkers is the focus of current translational medicine and is the most concerned topic in clinical practice, especially for cancer [113]. On the one hand, the biomarkers in peripheral blood and body fluids are of great significance for the early noninvasive diagnosis of diseases; on the other hand, the discovery of histological biomarkers is important to pathological classification and the prognosis prediction of diseases. Abundant, conserved, and dynamic expression of circRNAs has been reported for many times, which indicate circRNA is an emerging biomarker for cancer and related diseases, including HCC. Liyun Fu et al. established the expression pattern in HCC and explored the association of particular circRNAs and the characteristics of HCC patients [114]. Hsa_circ_0004018 was selected to estimate its value for HCC diagnosis and evaluation. Low expression of hsa_circ_0004018 was correlated with tumor size ($P = 0.045$), degree of tumor differentiation ($P = 0.006$), serum alpha-fetoprotein (AFP) level ($P = 0.027$), Barcelona Clinic Liver Cancer (BCLC) stage ($P = 0.040$), and tumor node metastasis (TNM) stage ($P = 0.029$) in patient with HCC. Further research found that hsa_circ_0004018 was differently expressed in diverse chronic liver disease including chronic hepatitis, cirrhosis, and HCC tissue ($P < 0.001$). It refers as a liver cancer specificity circRNA with an area under receiver operating characteristic curve (ROC) of 0.848 (95% CI = 0.803–0.894, $P < 0.001$). Compared to its adjacent non-tumor tissues, hsa_circ_0001649 is significantly downregulated in the liver cancer with an AUC of 0.63 [115]. Furthermore, hsa_circ_0001649 expression in HCC tissue was related to tumor diameter ($P = 0.045$), the cancer embolus ($P = 0.017$), and metastasis.

5 Challenge of Research in CircRNAs

In terms of what is mentioned above, there have been many researches on the expression and function of circRNAs in liver physiology and different liver pathological conditions. However, current studies still do not have a complete picture of the source of circRNAs and whereabouts in the liver and the underlying mechanisms. There is still a long way towards the practice of applicable researches. First challenge we come across is the complexity of the liver itself, including the complexity of its composition and disease. There are many types of cells in the liver, including parenchymal cells, at least three types, mesenchymal cells, and immune cells [116–118]. They are all involved in multiple physiological and pathological processes of the liver, which plays important roles in maintenance of liver homeostasis and development of liver diseases. These aspects increase the difficulty of the study on circRNAs, because a lot of works are required even after a circRNA is confirmed involved in a certain pathological process in the liver. On the one hand, we need to clarify which cell type circRNAs mostly affect. Sometimes several cells play a synergistic role in mediating the same pathological process. For example, HSC activation is the key process in the pathogenesis of liver fibrosis [119, 120]. Inflammation and activation of Kupffer cells and differentiation of LSECs can indirectly activate HSC [121]. At the same time, hepatocyte also transfers to myofibroblast through epithelial-mesenchymal transition (EMT) [122]. We believe that the identified circRNAs differentially expressed in cirrhosis or fibrosis should be further accessed in specific cell types and clarify its function at the level of individual cell type. On the other hand, it can be prudent to choose the strategy when the molecular mechanism is comprehensively understood. There are four major ways in which circRNAs regulate the biological process, including transcription, translation, and post-transcriptional regulation. Moreover, recent research has revealed the ability of circRNAs in coding protein. Based on up-to-date researches, circRNAs functioning as miRNA sponges have been reported the most [12, 123–126]. It is not only because of the ubiquity of the regulation but also because of the support of bioinformatics [127–129]. Hundreds of binding sites of miRNA have been predicted on circRNAs, providing theoretical basis for further research. High-throughput test, bioinformatics analysis and the expression of linear RNA from the host gene, is still the key means of cracking the puzzle of circRNAs in liver pathology and physiological processes.

The last question is the general problem of circRNA researches. It is well known that circRNAs are characterized as tissue-specific and species-specific [130–134]. The translational value of circRNAs may be greatly limited. The experimental results obtained in mice, rats, and other model organisms have encountered difficulties in the application on human beings. The benefit is that organ specificity makes the treatment of targeting circRNAs less interfered by off-target effect. In terms of the homology of circRNAs, one is through the homology of host genes, and the other is through the homology of circRNAs. In future studies on circRNAs, it is necessary to further address the homologous problems of circRNAs and to establish the bridge of circRNAs between different species.

Acknowledgments This work was supported by the grants from National Natural Science Foundation of China (81670571 and 81370559 C. Yang; 81400635 to F. Wang), Joint Projects in Major Diseases funding from Shanghai Municipal Commission of Health and Family Planning (2014ZYJB0201 to C. Yang), Joint Projects for Novel Frontier Technology in Shanghai Municipal Hospital from Shanghai Municipal Commission of Health and Family Planning (SHDC12014122 to C. Yang), Shanghai Medical Guide Project from Shanghai Science and Technology Committee (14411971500 to F. Wang), grants from Chinese Foundation for Hepatitis Prevention and Control (TQGB20140141 to F. Wang), and funds from Shanghai Innovation Program (12431901002 to C. Yang).

Competing Financial Interests The authors declare no competing financial interests.

References

1. Errichelli L, Dini Modigliani S, Laneve P et al (2017) FUS affects circular RNA expression in murine embryonic stem cell-derived motor neurons. Nat Commun 8:14741
2. Tan WL, Lim BT, Anene-Nzelu CG et al (2017) A landscape of circular RNA expression in the human heart. Cardiovasc Res 113(3):298–309
3. Pan T, Sun X, Liu Y et al (2018) Heat stress alters genome-wide profiles of circular RNAs in Arabidopsis. Plant Mol Biol 96(3):217–229
4. Hansen TB, Jensen TI, Clausen BH et al (2013) Natural RNA circles function as efficient microRNA sponges. Nature 495(7441):384–388
5. Cortes-Lopez M, Gruner MR, Cooper DA et al (2018) Global accumulation of circRNAs during aging in Caenorhabditis elegans. BMC Genomics 19(1):8
6. Kramer MC, Liang D, Tatomer DC et al (2015) Combinatorial control of Drosophila circular RNA expression by intronic repeats, hnRNPs, and SR proteins. Genes Dev 29(20):2168–2182
7. Memczak S, Jens M, Elefsinioti A et al (2013) Circular RNAs are a large class of animal RNAs with regulatory potency. Nature 495(7441):333–338
8. Liang D, Wilusz JE (2014) Short intronic repeat sequences facilitate circular RNA production. Genes Dev 28(20):2233–2247
9. Zhang XO, Wang HB, Zhang Y et al (2014) Complementary sequence-mediated exon circularization. Cell 159(1):134–147
10. Jeck WR, Sorrentino JA, Wang K et al (2013) Circular RNAs are abundant, conserved, and associated with ALU repeats. RNA 19(2):141–157
11. Ivanov A, Memczak S, Wyler E et al (2015) Analysis of intron sequences reveals hallmarks of circular RNA biogenesis in animals. Cell Rep 10(2):170–177
12. Andres-Leon E, Nunez-Torres R, Rojas AM (2016) miARma-Seq: a comprehensive tool for miRNA, mRNA and circRNA analysis. Sci Rep 6:25749
13. Starke S, Jost I, Rossbach O et al (2015) Exon circularization requires canonical splice signals. Cell Rep 10(1):103–111
14. Vicens Q, Westhof E (2014) Biogenesis of circular RNAs. Cell 159(1):13–14
15. Legnini I, Di Timoteo G, Rossi F et al (2017) Circ-ZNF609 is a circular RNA that can be translated and functions in Myogenesis. Mol Cell 66(1):22–37 e29
16. Pamudurti NR, Bartok O, Jens M et al (2017) Translation of CircRNAs. Mol Cell 66(1):9–21 e27
17. Peng L, Yuan XQ, Li GC (2015) The emerging landscape of circular RNA ciRS-7 in cancer (review). Oncol Rep 33(6):2669–2674
18. Han B, Chao J, Yao H (2018) Circular RNA and its mechanisms in disease: from the bench to the clinic. Pharmacol Ther 187:31. https://doi.org/10.1016/j.pharmthera.2018.01.010
19. Du WW, Yang W, Liu E et al (2016) Foxo3 circular RNA retards cell cycle progression via forming ternary complexes with p21 and CDK2. Nucleic Acids Res 44(6):2846–2858
20. Xue J, Liu Y, Luo F et al (2017) Circ100284, via miR-217 regulation of EZH2, is involved in the arsenite-accelerated cell cycle of human keratinocytes in carcinogenesis. Biochim Biophys Acta 1863(3):753–763
21. Zhang R, Xu J, Zhao J et al (2018) Silencing of hsa_circ_0007534 suppresses proliferation and induces apoptosis in colorectal cancer cells. Eur Rev Med Pharmacol Sci 22(1):118–126
22. Cheng X, Zhang L, Zhang K et al (2018) Circular RNA VMA21 protects against intervertebral disc degeneration through targeting miR-200c and X linked inhibitor-of-apoptosis protein. Ann Rheum Dis 77:770. https://doi.org/10.1136/annrheumdis-2017-212056
23. Holdt LM, Stahringer A, Sass K et al (2016) Circular non-coding RNA ANRIL modulates ribosomal RNA maturation and atherosclerosis in humans. Nat Commun 7:12429
24. Place RF, Noonan EJ (2014) Non-coding RNAs turn up the heat: an emerging layer of novel regulators in the mammalian heat shock response. Cell Stress Chaperones 19(2):159–172
25. Huang JL, Qin MC, Zhou Y et al (2018) Comprehensive analysis of differentially expressed profiles of Alzheimer's disease associated circular RNAs in an Alzheimer's disease mouse model. Aging (Albany NY) 10(2):253–265

26. Qu S, Zhong Y, Shang R et al (2017) The emerging landscape of circular RNA in life processes. RNA Biol 14(8):992–999
27. Yang X, Wang J, Zhou Z et al (2018) Silica-induced initiation of circular ZC3H4 RNA/ZC3H4 pathway promotes the pulmonary macrophage activation. FASEB J 32:3264. https://doi.org/10.1096/fj.201701118R:fj201701118R
28. Abd Majid FA, Ab Rahman NS, Abd Wahid E et al (2018) Evaluation of herb-drug interaction of SynacinnTM and individual biomarker through cytochrome 450 inhibition assay. Drug Metab Lett 12. https://doi.org/10.2174/187231281266618031412457
29. Ahmad T, Valentovic MA, Rankin GO (2018) Effects of cytochrome P450 single nucleotide polymorphisms on methadone metabolism and pharmacodynamics. Biochem Pharmacol 153:196. https://doi.org/10.1016/j.bcp.2018.02.020
30. Ahn SB, Jun DW, Jang K et al (2018) Duodenal Niemann-Pick C1-like 1 expression was negatively correlated with liver X receptor expression in non-alcoholic fatty liver disease. Korean J Intern Med. https://doi.org/10.3904/kjim.2017.100
31. Capel B, Swain A, Nicolis S et al (1993) Circular transcripts of the testis-determining gene Sry in adult mouse testis. Cell 73(5):1019–1030
32. Cocquerelle C, Daubersies P, Majerus MA et al (1992) Splicing with inverted order of exons occurs proximal to large introns. EMBO J 11(3):1095–1098
33. Chiu YJ, Chou SC, Chiu CS et al (2018) Hepatoprotective effect of the ethanol extract of Polygonum orientale on carbon tetrachloride-induced acute liver injury in mice. J Food Drug Anal 26(1):369–379
34. George N, Chen M, Yuen N et al (2018) Interplay of gender, age and drug properties on reporting frequency of drug-induced liver injury. Regul Toxicol Pharmacol 94:101–107
35. Su Q, Kumar V, Sud N et al (2018) MicroRNAs in the pathogenesis and treatment of progressive liver injury in NAFLD and liver fibrosis. Adv Drug Deliv Rev 129:54. https://doi.org/10.1016/j.addr.2018.01.009
36. Rahmoune H, Boutrid N, Bioud B (2018) Nonalcoholic steatohepatitis. N Engl J Med 378(8):780–781
37. Aspichueta P (2018) Lipid-rich environment: a key role promoting carcinogenesis in obesity-related non-alcoholic fatty liver disease. Gut:gutjnl-2018-316047. https://doi.org/10.1136/gutjnl-2018-316047
38. Fukushima H, Yamashina S, Arakawa A et al (2018) The formation of p62-positive inclusion body is associated with macrophage polarization in non-alcoholic fatty liver disease. Hepatol Res. https://doi.org/10.1111/hepr.13070
39. Wesolowski SR, Kasmi KC, Jonscher KR et al (2017) Developmental origins of NAFLD: a womb with a clue. Nat Rev Gastroenterol Hepatol 14(2):81–96
40. Del Campo JA, Gallego-Duran R, Gallego P et al (2018) Genetic and epigenetic regulation in nonalcoholic fatty liver disease (NAFLD). Int J Mol Sci 19(3)
41. Krenkel O, Puengel T, Govaere O et al (2017) Therapeutic inhibition of inflammatory monocyte recruitment reduces steatohepatitis and liver fibrosis. Hepatology 67:1270. https://doi.org/10.1002/hep.29544
42. Tarantino G, Caputi A (2011) JNKs, insulin resistance and inflammation: a possible link between NAFLD and coronary artery disease. World J Gastroenterol 17(33):3785–3794
43. Tilg H, Moschen AR (2011) IL-1 cytokine family members and NAFLD: neglected in metabolic liver inflammation. J Hepatol 55(5):960–962
44. Vespasiani-Gentilucci U, Carotti S, Perrone G et al (2015) Hepatic toll-like receptor 4 expression is associated with portal inflammation and fibrosis in patients with NAFLD. Liver Int 35(2):569–581
45. Neuschwander-Tetri BA (2014) NASH: the tribulations of conducting NASH trials. Nat Rev Gastroenterol Hepatol 11(5):274–276
46. Diehl AM, Day C (2017) Cause, pathogenesis, and treatment of nonalcoholic steatohepatitis. N Engl J Med 377(21):2063–2072
47. Caussy C, Hsu C, Lo MT et al (2018) Novel link between gut-microbiome derived metabolite and shared gene-effects with hepatic steatosis and fibrosis in NAFLD. Hepatology. https://doi.org/10.1002/hep.29892
48. Arab JP, Martin-Mateos RM, Shah VH (2018) Gut-liver axis, cirrhosis and portal hypertension: the chicken and the egg. Hepatol Int 12(Suppl 1):24–33
49. Buzzetti E, Pinzani M, Tsochatzis EA (2016) The multiple-hit pathogenesis of non-alcoholic fatty liver disease (NAFLD). Metabolism 65(8):1038–1048
50. Younossi ZM, Loomba R, Anstee QM et al (2017) Diagnostic modalities for non-alcoholic fatty liver disease (NAFLD), non-alcoholic steatohepatitis (NASH) and associated fibrosis. Hepatology. https://doi.org/10.1002/hep.29721
51. Younossi ZM, Loomba R, Rinella ME et al (2017) Current and future therapeutic regimens for non-alcoholic fatty liver disease (NAFLD) and non-alcoholic steatohepatitis (NASH). Hepatology. https://doi.org/10.1002/hep.29724
52. Beier JI, Banales JM (2018) Pyroptosis: an inflammatory link between NAFLD and NASH with potential therapeutic implications. J Hepatol 68:643. https://doi.org/10.1016/j.jhep.2018.01.017
53. Ruiz-Ramirez A, Chavez-Salgado M, Peneda-Flores JA et al (2011) High-sucrose diet increases ROS generation, FFA accumulation, UCP2 level, and proton leak in liver mitochondria. Am J Physiol Endocrinol Metab 301(6):E1198–E1207
54. Kawano Y, Cohen DE (2013) Mechanisms of hepatic triglyceride accumulation in non-alcoholic fatty liver disease. J Gastroenterol 48(4):434–441

55. Bril F, Barb D, Portillo-Sanchez P et al (2017) Metabolic and histological implications of intrahepatic triglyceride content in nonalcoholic fatty liver disease. Hepatology 65(4):1132–1144
56. Alves-Bezerra M, Cohen DE (2017) Triglyceride metabolism in the liver. Compr Physiol 8(1):1–8
57. Ezquerro S, Mendez-Gimenez L, Becerril S et al (2016) Acylated and desacyl ghrelin are associated with hepatic lipogenesis, beta-oxidation and autophagy: role in NAFLD amelioration after sleeve gastrectomy in obese rats. Sci Rep 6:39942
58. Zhang Y, Cheng X, Lu Z et al (2013) Upregulation of miR-15b in NAFLD models and in the serum of patients with fatty liver disease. Diabetes Res Clin Pract 99(3):327–334
59. An X, Yang Z, An Z (2017) MiR-149 compromises the reactions of liver cells to fatty acid via its polymorphism and increases non-alcoholic fatty liver disease (NAFLD) risk by targeting methylene tetrahydrofolate reductase (MTHFR). Med Sci Monit 23:2299–2307
60. Ghareghani P, Shanaki M, Ahmadi S et al (2018) Aerobic endurance training improves nonalcoholic fatty liver disease (NAFLD) features via miR-33 dependent autophagy induction in high fat diet fed mice. Obes Res Clin Pract 12(1S1):80–89
61. Drygalski K, Berk K, Charytoniuk T et al (2017) Does the enterolactone (ENL) affect fatty acid transporters and lipid metabolism in liver? Nutr Metab (Lond) 14:69
62. Cheng J, Liu C, Hu K et al (2017) Ablation of systemic SIRT1 activity promotes nonalcoholic fatty liver disease by affecting liver-mesenteric adipose tissue fatty acid mobilization. Biochim Biophys Acta 1863(11):2783–2790
63. Salgado AL, Carvalho L, Oliveira AC et al (2010) Insulin resistance index (HOMA-IR) in the differentiation of patients with non-alcoholic fatty liver disease and healthy individuals. Arq Gastroenterol 47(2):165–169
64. Klisic A, Isakovic A, Kocic G et al (2017) Relationship between oxidative stress, inflammation and dyslipidemia with fatty liver index in patients with type 2 diabetes mellitus. Exp Clin Endocrinol Diabetes 126:371. https://doi.org/10.1055/s-0043-118667
65. Firidin G (2018) Oxidative stress parameters, induction of lipid peroxidation, and ATPase activity in the liver and kidney of Oreochromis niloticus exposed to lead and mixtures of lead and zinc. Bull Environ Contam Toxicol 100(4):477–484
66. Francque S, De Maeght S, Adler M et al (2011) High prevalence of advanced fibrosis in association with the metabolic syndrome in a Belgian prospective cohort of NAFLD patients with elevated ALT. Results of the Belgian NAFLD registry. Acta Gastroenterol Belg 74(1):9–16
67. Tan TC, Crawford DH, Jaskowski LA et al (2011) Altered lipid metabolism in Hfe-knockout mice promotes severe NAFLD and early fibrosis. Am J Physiol Gastrointest Liver Physiol 301(5):G865–G876
68. Al-Serri A, Anstee QM, Valenti L et al (2012) The SOD2 C47T polymorphism influences NAFLD fibrosis severity: evidence from case-control and intra-familial allele association studies. J Hepatol 56(2):448–454
69. Guo XY, Chen JN (2017) circRNA_0046367 prevents hepatoxicity of lipid peroxidation: an inhibitory role against hepatic steatosis. Oxid Med Cell Longev 2017:3960197
70. Guo J, Zhou Y, Cheng Y et al (2018) Metformin-induced changes of the coding transcriptome and non-coding RNAs in the livers of non-alcoholic fatty liver disease mice. Cell Physiol Biochem 45(4):1487–1505
71. Guo XY, Sun F, Chen JN et al (2018) circRNA_0046366 inhibits hepatocellular steatosis by normalization of PPAR signaling. World J Gastroenterol 24(3):323–337
72. Xu H, Guo S, Li W et al (2015) The circular RNA Cdr1as, via miR-7 and its targets, regulates insulin transcription and secretion in islet cells. Sci Rep 5:12453
73. Argemi J, Kress TR, Chang HCY et al (2017) X-box binding protein 1 regulates unfolded protein, acute-phase, and DNA damage responses during regeneration of mouse liver. Gastroenterology 152(5):1203–1216 e1215
74. Schmidt-Arras D, Rose-John S (2016) IL-6 pathway in the liver: from physiopathology to therapy. J Hepatol 64(6):1403–1415
75. Paschos KA, Bird NC (2010) Liver regeneration and its impact on post-hepatectomy metastatic tumour recurrence. Anticancer Res 30(6):2161–2170
76. Tao Y, Wang M, Chen E et al (2017) Liver regeneration: analysis of the main relevant signaling molecules. Mediat Inflamm 2017:4256352
77. Li L, Guo J, Chen Y et al (2017) Comprehensive CircRNA expression profile and selection of key CircRNAs during priming phase of rat liver regeneration. BMC Genomics 18(1):80
78. DeLeve LD (2015) Liver sinusoidal endothelial cells in hepatic fibrosis. Hepatology 61(5):1740–1746
79. Pradere JP, Kluwe J, De Minicis S et al (2013) Hepatic macrophages but not dendritic cells contribute to liver fibrosis by promoting the survival of activated hepatic stellate cells in mice. Hepatology 58(4):1461–1473
80. Cai C, Zhu X, Li P et al (2017) NLRP3 deletion inhibits the non-alcoholic steatohepatitis development and inflammation in Kupffer cells induced by palmitic acid. Inflammation 40(6):1875–1883
81. Li Z, Ji L, Su S et al (2018) Leptin up-regulates microRNA-27a/b-3p level in hepatic stellate cells. Exp Cell Res 366:63. https://doi.org/10.1016/j.yexcr.2018.03.015
82. Qu C, Zheng D, Li S et al (2018) Tyrosine kinase SYK is a potential therapeutic target for liver

fibrosis. Hepatology. https://doi.org/10.1002/hep.29881
83. Bu FT, Chen Y, Yu HX et al (2018) SENP2 alleviates CCl4-induced liver fibrosis by promoting activated hepatic stellate cell apoptosis and reversion. Toxicol Lett 289:86–98
84. Coppola N, Perna A, Lucariello A et al (2018) Effects of treatment with Maraviroc a CCR5 inhibitor on a human hepatic stellate cell line. J Cell Physiol 233:6224. https://doi.org/10.1002/jcp.26485
85. Kiagiadaki F, Kampa M, Voumvouraki A et al (2018) Activin-a causes hepatic stellate cell activation via the induction of TNFalpha and TGFbeta in Kupffer cells. Biochim Biophys Acta 1864(3):891–899
86. Lan T, Li C, Yang G et al (2018) Sphingosine kinase 1 promotes liver fibrosis by preventing miR-19b-3p-mediated inhibition of CCR2. Hepatology. https://doi.org/10.1002/hep.29885
87. Zhang K, Han X, Zhang Z et al (2017) The liver-enriched lnc-LFAR1 promotes liver fibrosis by activating TGFbeta and Notch pathways. Nat Commun 8(1):144
88. Chen Y, Yuan B, Wu Z et al (2017) Microarray profiling of circular RNAs and the potential regulatory role of hsa_circ_0071410 in the activated human hepatic stellate cell induced by irradiation. Gene 629:35–42
89. Tang CM, Zhang M, Huang L et al (2017) CircRNA_000203 enhances the expression of fibrosis-associated genes by derepressing targets of miR-26b-5p, Col1a2 and CTGF, in cardiac fibroblasts. Sci Rep 7:40342
90. Zhou B, Yu JW (2017) A novel identified circular RNA, circRNA_010567, promotes myocardial fibrosis via suppressing miR-141 by targeting TGF-beta1. Biochem Biophys Res Commun 487(4):769–775
91. Forner A, Reig M, Bruix J (2018) Hepatocellular carcinoma. Lancet (18):30010–30012. https://doi.org/10.1016/S0140-6736
92. Abitbol S, Dahmani R, Coulouarn C et al (2018) AXIN deficiency in human and mouse hepatocytes induces hepatocellular carcinoma in the absence of beta-catenin activation. J Hepatol 68:1203. https://doi.org/10.1016/j.jhep.2017.12.018
93. Cai Z, Qian ZY, Jiang H et al (2018) hPCL3s promotes hepatocellular carcinoma metastasis by activating beta-catenin signaling. Cancer Res 78:2536. https://doi.org/10.1158/0008-5472.CAN-17-0028
94. Buschauer S, Koch A, Wiggermann P et al (2018) Hepatocellular carcinoma cells surviving doxorubicin treatment exhibit increased migratory potential and resistance to doxorubicin re-treatment in vitro. Oncol Lett 15(4):4635–4640
95. Yang F, Liu DY, Guo JT et al (2017) Circular RNA circ-LDLRAD3 as a biomarker in diagnosis of pancreatic cancer. World J Gastroenterol 23(47):8345–8354
96. Zhang Y, Liu H, Li W et al (2017) CircRNA_100269 is downregulated in gastric cancer and suppresses tumor cell growth by targeting miR-630. Aging (Albany NY) 9(6):1585–1594
97. Geng HH, Li R, Su YM et al (2016) The circular RNA Cdr1as promotes myocardial infarction by mediating the regulation of miR-7a on its target genes expression. PLoS One 11(3):e0151753
98. Barrett SP, Parker KR, Horn C et al (2017) ciRS-7 exonic sequence is embedded in a long non-coding RNA locus. PLoS Genet 13(12):e1007114
99. Zhao ZJ, Shen J (2017) Circular RNA participates in the carcinogenesis and the malignant behavior of cancer. RNA Biol 14(5):514–521
100. Yu L, Gong X, Sun L et al (2016) The circular RNA Cdr1as act as an oncogene in hepatocellular carcinoma through targeting miR-7 expression. Hepatol Res Treat 11(7):e0158347
101. Xu L, Zhang M, Zheng X et al (2017) The circular RNA ciRS-7 (Cdr1as) acts as a risk factor of hepatic microvascular invasion in hepatocellular carcinoma. J Cancer Res Clin Oncol 143(1):17–27
102. Han D, Li J, Wang H et al (2017) Circular RNA circMTO1 acts as the sponge of microRNA-9 to suppress hepatocellular carcinoma progression. Hepatology 66(4):1151–1164
103. Yao Z, Luo J, Hu K et al (2017) ZKSCAN1 gene and its related circular RNA (circZKSCAN1) both inhibit hepatocellular carcinoma cell growth, migration, and invasion but through different signaling pathways. Mol Oncol 11(4):422–437
104. Zhang K, Che S, Su Z et al (2018) CD90 promotes cell migration, viability and sphere forming ability of hepatocellular carcinoma cells. Int J Mol Med 41(2):946–954
105. Wang BG, Li JS, Liu YF et al (2017) MicroRNA-200b suppresses the invasion and migration of hepatocellular carcinoma by downregulating RhoA and circRNA_000839. Tumour Biol 39(7):1010428317719577
106. Zhu Q, Lu G, Luo Z et al (2018) CircRNA circ_0067934 promotes tumor growth and metastasis in hepatocellular carcinoma through regulation of miR-1324/FZD5/Wnt/beta-catenin axis. Biochem Biophys Res Commun 497(2):626–632
107. Dong LQ, Shi Y, Ma LJ et al (2018) Spatial and temporal clonal evolution of intrahepatic cholangiocarcinoma. J Hepatol 69:89. https://doi.org/10.1016/j.jhep.2018.02.029
108. Lee DD, Croome KP, Musto KR et al (2018) Caveat emptor, liver transplantation for intrahepatic cholangiocarcinoma – a diagnostic conundrum for 44 patients at a single center. Liver Transpl 24:634. https://doi.org/10.1002/lt.25052
109. Merdrignac A, Angenard G, Allain C et al (2018) A novel transforming growth factor beta-induced long noncoding RNA promotes an inflammatory microenvironment in human intrahepatic cholangiocarcinoma. Hepatol Commun 2(3):254–269
110. Xie L, Onysko J, Morrison H (2018) Childhood cancer incidence in Canada: demographic and

geographic variation of temporal trends (1992-2010). Health Promot Chronic Dis Prev Can 38(3):79–115
111. Zhong S, Zhao Y, Fan C (2018) Hepatoblastoma with pure fetal epithelial differentiation in a 10-year-old boy: a rare case report and review of the literature. Medicine (Baltimore) 97(2):e9647
112. Liu BH, Zhang BB, Liu XQ et al (2018) Expression profiling identifies circular RNA signature in hepatoblastoma. Cell Physiol Biochem 45(2):706–719
113. Li P, Chen S, Chen H et al (2015) Using circular RNA as a novel type of biomarker in the screening of gastric cancer. Clin Chim Acta 444:132–136
114. Fu L, Yao T, Chen Q et al (2017) Screening differential circular RNA expression profiles reveals hsa_circ_0004018 is associated with hepatocellular carcinoma. Oncotarget 8(35):58405–58416
115. Xu Y, Yao Y, Zhong X et al (2018) Downregulated circular RNA hsa_circ_0001649 regulates proliferation, migration and invasion in cholangiocarcinoma cells. Biochem Biophys Res Commun 496(2):455–461
116. Blaya D, Aguilar-Bravo B, Hao F et al (2018) Expression of microRNA-155 in inflammatory cells modulates liver injury. Hepatology. https://doi.org/10.1002/hep.29833
117. Bourgine M, Crabe S, Lobaina Y et al (2018) Nasal route favors the induction of CD4(+) T cell responses in the liver of HBV-carrier mice immunized with a recombinant hepatitis B surface- and core-based therapeutic vaccine. Antivir Res 153:23–32
118. Neumann K, Rudolph C, Neumann C et al (2015) Liver sinusoidal endothelial cells induce immunosuppressive IL-10-producing Th1 cells via the Notch pathway. Eur J Immunol 45(7):2008–2016
119. Zhuang S, Hua X, He K et al (2018) Inhibition of GSK-3beta induces AP-1-mediated osteopontin expression to promote cholestatic liver fibrosis. FASEB J:fj.201701137R. https://doi.org/10.1096/fj.201701137R:fj201701137R
120. Kostallari E, Hirsova P, Prasnicka A et al (2018) Hepatic stellate cell-derived PDGFRalpha-enriched extracellular vesicles promote liver fibrosis in mice through SHP2. Hepatology. https://doi.org/10.1002/hep.29803
121. O'Leary JG, Demetris AJ, Philippe A et al (2017) Non-HLA antibodies impact on C4d staining, stellate cell activation and fibrosis in liver allografts. Transplantation 101(10):2399–2409
122. Li TZ, Kim SM, Hur W et al (2017) Elk-3 contributes to the progression of liver fibrosis by regulating the epithelial-mesenchymal transition. Gut Liver 11(1):102–111
123. Caiment F, Gaj S, Claessen S et al (2015) High-throughput data integration of RNA-miRNA-circRNA reveals novel insights into mechanisms of benzo[a]pyrene-induced carcinogenicity. Nucleic Acids Res 43(5):2525–2534
124. Jin X, Feng CY, Xiang Z et al (2016) CircRNA expression pattern and circRNA-miRNA-mRNA network in the pathogenesis of nonalcoholic steatohepatitis. Oncotarget 7(41):66455–66467
125. Zhao Y, Alexandrov PN, Jaber V et al (2016) Deficiency in the ubiquitin conjugating enzyme UBE2A in Alzheimer's disease (AD) is linked to deficits in a natural circular miRNA-7 sponge (circRNA; ciRS-7). Genes (Basel) 7(12)
126. Andres-Leon E, Nunez-Torres R, Rojas AM (2018) Corrigendum: miARma-Seq: a comprehensive tool for miRNA, mRNA and circRNA analysis. Sci Rep 8:46928
127. Hansen TB (2018) Improved circRNA identification by combining prediction algorithms. Front Cell Dev Biol 6:20
128. Lei W, Feng T, Fang X et al (2018) Signature of circular RNAs in human induced pluripotent stem cells and derived cardiomyocytes. Stem Cell Res Ther 9(1):56
129. Liang J, Wu X, Sun S et al (2018) Circular RNA expression profile analysis of severe acne by RNA-seq and bioinformatics. J Eur Acad Dermatol Venereol. https://doi.org/10.1111/jdv.14948
130. Jakobi T, Dieterich C (2018) Deep computational circular RNA analytics from RNA-seq data. Methods Mol Biol 1724:9–25
131. Quan G, Li J (2018) Circular RNAs: biogenesis, expression and their potential roles in reproduction. J Ovarian Res 11(1):9
132. Gao Y, Wang J, Zheng Y et al (2016) Comprehensive identification of internal structure and alternative splicing events in circular RNAs. Nat Commun 7:12060
133. Zhao J, Li L, Wang Q et al (2017) CircRNA expression profile in early-stage lung adenocarcinoma patients. Cell Physiol Biochem 44(6):2138–2146
134. Xia S, Feng J, Chen K et al (2018) CSCD: a database for cancer-specific circular RNAs. Nucleic Acids Res 46(D1):D925–D929

21 Circular RNAs in Organ Fibrosis

Jianhua Yao, Qiying Dai, Zhuyuan Liu, Lei Zhou, and Jiahong Xu

Abstract

Fibrosis refers to a process involving the accumulation of extracellular matrix components. It could happen in chronic organ injury or during the recovery of acute organ injury. The severity of fibrosis interferes with the function of the organ involved. Numerous studies have been carried out to explore the mechanism of fibrosis, including parenchyma injury, fibrillar ECM accumulation, fibroblast activation, microvasculature rarefaction, and a mononuclear infiltrate. Unfortunately, its underlying mechanism is at largely unknown. The studying of noncoding RNAs has provided novel insight for circRNA-miRNA-mRNA in learning disease progress. Emerging evidence has shown that circRNA is related to fibrosis activity and could potentially be a monitoring factor for fibrosis or, more excitingly, could be a target for treatment. In this chapter, we will first present the basic mechanism of organ fibrosis. Then we will focus on the recent studies about how circRNA dysregulation contributes to organ fibrosis. Finally, the advantages and potential challenges of circRNA-based therapeutics for the treatment of fibroproliferative diseases will be discussed.

Keywords

circRNA · Fibrosis

1 Introduction

Cell damage and tissue repair is a complex biological process [1]. A successful repairment involves at least three distinct phases: (1) an elimination phase, in which damaged or dead cells are eliminated; (2) a regenerative phase, in which the eliminated cells are replaced by cells of the same or different origin; and (3) a fibrogenic phase, in which fibrosis tissue is generated to partially restore the function of the organ [2–7]. The fibrogenic phase will continue until damaged tissue or

Author contributed equally with all other contributors. Jianhua Yao and Qiying Dai

J. Yao
Department of Cardiology, Shanghai Tenth People's Hospital, Tongji University School of Medicine, Shanghai, China

Q. Dai
MetroWest Medical Center, Framingham, MA, USA

Department of Cardiology, First Affiliated Hospital of Nanjing Medical University, Nanjing, China

Z. Liu · L. Zhou (✉)
Department of Cardiology, First Affiliated Hospital of Nanjing Medical University, Nanjing, China

J. Xu (✉)
Department of Cardiology, Shanghai Tongji Hospital, Tongji University School of Medicine, Shanghai, China
e-mail: xujiahong@tongji.edu.cn

J. Xiao (ed.), *Circular RNAs*, Advances in Experimental Medicine and Biology 1087,
https://doi.org/10.1007/978-981-13-1426-1_21

lost cells are regenerated. Ideally, temporary fibrosis tissue will be replaced by fully functional cells. In some cases, like chronic tissue injury or extensive injury, the generation of fibrotic tissue overgrows that of the original tissue. And then, the function of the injured site would be impaired because of the replacement by the nonfunctional fibrosis tissue [8, 9].

It is a common phenomenon in both chronic and acute diseases, and it can occur in nearly all organs including the heart, liver, kidney, lung, and so on [10–14]. Chronic stimuli such as persistent infections and hormonal changes and acute stimuli such as the sudden change of blood dynamics can induce organ fibrosis [15–17]. The progressive organ fibrosis can eventually cause organ dysfunction or even failure [18–20]. A great amount of efforts have been put into exploring the nature of fibrosis with minimal achievement, and no effective management has been available either [21].

The prevalence of non-polyadenylated transcriptome data has prospered the study in the field of circRNAs, which are found to modulate disease progress, including fibrosis. CircRNAs are single-stranded RNAs which are detectable in humans, mice, rats, fungi, and other organisms [22–35]. There are over ten thousands of different circRNAs in metazoans, from insects [22] to mammals [23, 24]. Moreover, over 10% of expressed genes in examined cells and tissues can produce circRNAs [25]. However, only a few number of circRNAs have been found. Considering their low expression, it was hypothesized that they were produced by aberrant splicing [26–30].While approximately 50 genes have circRNAs that are expressed more abundantly than their linear isoforms [31], few numbers of circRNAs have been convinced to be highly expressed in a cell-type-specific or organ-specific manner [32, 33]. The high ubiquitousness of the co-expression of circRNA and its linear isoforms suggests that circRNAs are not simply by-products. Several circRNAs have been found to participate in disease progress. For example, circRNA from sex-determining region Y (circSRY) has a key role in testes development in mouse [34]. CircRNAs from ANRIL positively affect the development of atherosclerosis [35]. CiRS-7 (circular RNA sponge for miR-7) in neuronal tissues [34, 36] is significantly reduced in sporadic Alzheimer disease [37]. Moreover, a lot of other circRNAs were found to be significantly altered during cancer-associated cell proliferation [38].

The organ specific expression makes circRNAs a candidate for disease biomarkers [22, 39, 40]. Unlike other noncoding RNAs, circRNAs are highly expressed and concentrated due to their resistance to RNA exonucleases or RNase R [41, 42]. What's more, circRNA appears to be highly conserved among mammals [43]. In mice, 4% of orthologous genes can generate circRNAs [31], and approximately 5–30% of these circRNAs are completely preserved in human and mice [36, 41, 44]. About 5–10% of human brain circRNAs are also expressed in the porcine brain [43, 45]. Fibrosis, an irreversible damage to organs, has been the focus of studies for years. The discovery of circRNA has created a new path in understanding this process in a different level. Here we will start with the basic mechanism of organ fibrosis and then take a deep look into how circRNAs affect this process.

2 Mechanism of Organ Fibrosis

Fibrosis is the formation of excess fibrous tissue during tissue repair. It can lead to malfunction of the organ after replacing the original cells. Organs like the liver, kidneys, heart, and lungs which contain a lot of parenchymal cells are most vulnerable to fibrosis [55–57]. Other organs might be affected as well. For example, fibrosis involving the skin and joints will impair movement and decrease life quality. Bone marrow fibrosis could lead to cytopenia [46–50]. In addition to normal organs, tumors, especially solid ones, could also be affected by fibrosis [51–53].

2.1 Common Features of Tissue Fibrosis (Fig. 21.1)

The fibrosis in all organs shares similar histomorphology, both macroscopically and microscopically [54]. The fibrosis organs are stiff, pale, and uneven. The accumulation of excessive extracel-

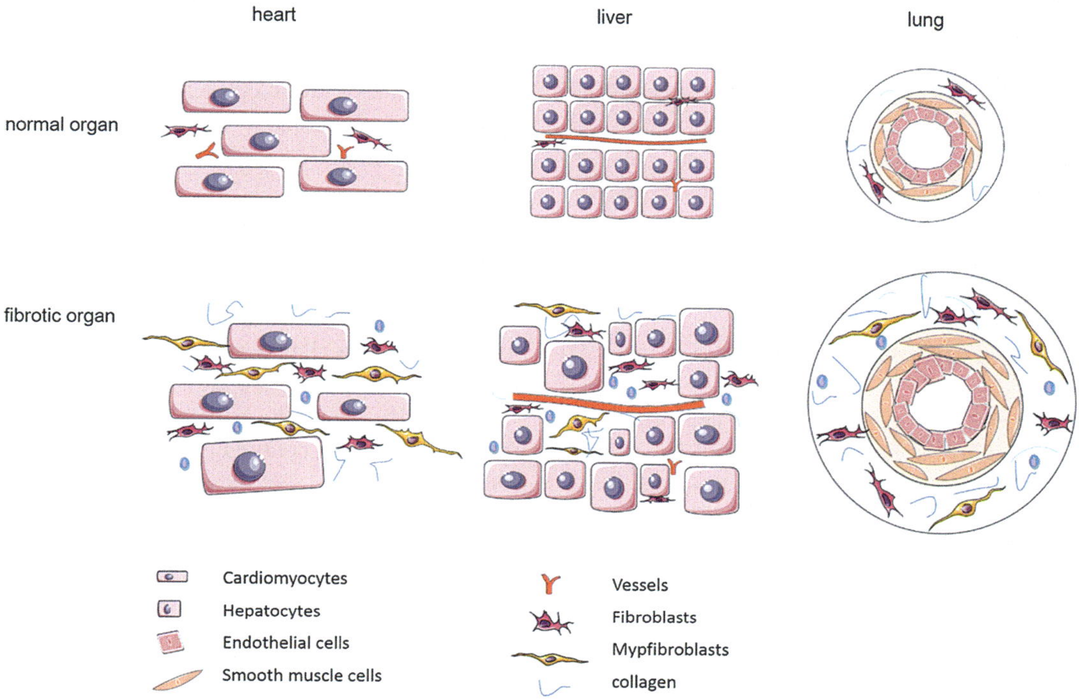

Fig. 21.1 Common features of tissue fibrosis
The fibrosis in all organs shared similar histomorphology, and macroscopically and microscopically fibrosis organs share commonalities. The fibrosis organs are stiff, pale, and uneven. The histopathological analysis reveals that organ fibrosis is associated with the parenchyma injury, fibrillar ECM accumulation, fibroblast activation, microvasculature rarefaction, and a mononuclear infiltrate. The schematic diagram showed the common mechanisms of organ fibrosis in the heart, liver, and lung. Normal organ consists of functional parenchyma, connective tissue, vessels, and fibroblasts. In the heart, the functional parenchyma is cardiomyocytes. In the liver, the functional parenchyma is hepatocytes, and in the lung, the functional parenchymas are endothelial cells and smooth muscle cells. The fibrotic organs have more connective tissue than normal organs. The connective tissue mainly contains fibroblasts, myofibroblasts, collagens, and mononuclear infiltrate

lular matrix (ECM) causes the stiffness [55]. Destruction of the vasculature makes them look pale. Contraction of the fibroblasts contributes to the uneven surface. The histopathological analysis reveals that organ fibrosis is associated with the parenchyma injury, fibrillar ECM accumulation, fibroblast activation, microvasculature rarefaction, and a mononuclear infiltrate [56–59].

2.2 Parenchymal Injury

Each parenchymal organ has its specific parenchymal cells. Different parenchymal cells have different potentials to regenerate. Some parenchymal cells have low regeneration capacity, such as cardiomyocytes, while some have relatively high regenerative capacity such as hepatocytes [60–62]. The balance of regeneration breaks when the injury stimulation exceeds the rate of repair, causing a suppression on the regeneration capacity [53, 63–65]. This in turn facilitates the formation of the fibrous tissue. Meanwhile, the injured parenchymal cells would secrete chemokines like growth factors and other profibrotic metabolites to promote fibrogenesis. The typical pathway is TGF-β signaling pathway. TGF-β was initially identified as a hallmark of malignant transformation in embryonic kidney fibroblasts [66]. It turns out that TGF-β is overexpressed in fibrosis tissues as well. It induces collagen production both in vitro and in vivo. Fibrotic activity was noticed to be suppressed after TGF-β neutralization [67–70].

2.3 Fibrillar ECM Accumulation

Fibroblasts are the major source of extracellular matrix in fibrosis [71, 72]. The main goal of anti-fibrosis therapy is to inhibit fibroblasts, thus suppressing the accumulation of extracellular matrix. Little is known in this field because of the complexity and multiple factors it involves [73].

2.4 Fibroblast Activation

Fibroblast activation is the first step in the process of fibrosis. Fortunately, fibroblasts are easy to culture, making it convenient for us to do in vitro experiment [74, 75]. Fibroblasts are usually found in connective tissue. They have prominent cytoskeleton and endoplasmic reticulum and specific leaflet-shaped cytoplasm [76, 77]. Based on their location and basic structure, we can identify them through the light microscope. Activated fibroblasts and myofibroblasts, which have great potential for biosynthesis and proliferation, consist of the major source of fibrosis. They are characterized by pronounced rough endoplasmic reticulum, stress fibers, and large nucleolus.

Cell biomarkers are frequently used in fibroblast-related researches. Both fibroblasts and activated fibroblasts can be labelled with α-smooth muscle actin (α-SMA) [78–80]. Even so, specific markers are still needed because of the existence of subpopulations. The types of fibroblasts vary among different organs. In addition, there are heterogeneous phenotypes within the same organ. For example, in the kidney, the lipid-laden fibroblasts in medulla are different from the ones in the cortex [81]. Functions of fibroblasts vary according to their origins. Fibroblasts in the kidneys could transform into erythropoietin-producing cells [82, 83]. A part of the cardiac conduction system is atrial fibroblasts [84–87]. Tracking the expression of Hox gene, the original position of dermal fibroblasts can be found [88]. In order to study the functions of these subtypes, additional markers have been developed, like α-SMA, vimentin, fibroblast-specific protein 1 (FSP1), desmin, etc. [89]. One of the biggest challenges that make it difficult to study fibroblasts is their heterogeneity.

Myofibroblasts are the principal mediators of fibrogenesis with unique α-SMA-expressing features. Using FSP1 as a marker, the group of FSP1+ fibroblasts are found to contribute for kidney fibrosis [90]. Due to the large number of different types of myofibroblasts, it is a big challenge to carry on the study on myoblasts. Moreover, the dynamic changes of these cells during aging make it more difficult to study [91–94]. Searching simple markers remains the most significant task.

On the other hand, the origination of activated fibroblasts remains controversial. It was previously acknowledged that these cells were generated from the in situ fibroblasts until bone marrow-derived fibroblasts were discovered [95]. Later on, it was found that epithelial cells, pericytes, and even vascular smooth muscle cells participate in fibrosis [96–99]. Moreover, by using fate-mapping strategies and marker analysis, it was found that mechanisms of fibrosis might be different in different organs and diseases [53, 65, 100–111].

2.5 Microvasculature Rarefaction

Microvasculature rarefaction causes the pale look in the fibrotic tissues by decreasing perfusion chronically. The microvascular system plays an important role in transporting oxygen. The declination of these small vessels creates a chronic hypoxic environment, which promotes fibrosis by activating HIF-α signaling pathway [112–114]. In turn, fibrosis can attenuate the capacity of vessel regeneration through decreasing the expression of proangiogenic molecules and increasing the expression of antiangiogenic molecules. Pro-angiogenic therapy, for example, VEGF administration, has been reported to have cardiac protection through decreasing fibrosis in experimental models [115–118].

Interestingly, cell loss by either apoptosis or necrosis is related to microvascular rarefaction [119]. One of the well-known explanations is endothelial-mesenchymal transition (EndMT)

[109, 120–122]. It was discovered in cardiac development in which mesenchymal cells are found to be derived from the endocardium through a process called EndMT [123–125]. The mesenchymal cells are group of cells which form the atrioventricular cushion, the primordia of valve, and the septa of the heart. During EndMT, endothelial cells lose their own characteristic markers and acquire a mesenchymal phenotype. After the transformation, they will initiate the fibrosis process by producing mesenchymal cell products. Transforming growth factor-β (TGF-β) is one of the well-studied EndMT factors [122]. In the heart and kidney, it was found that inhibition of EndMT could attenuate fibrosis [105]. This provides a potential therapeutic target for fibrosis.

Fibrosis can influence inflammatory process by affecting mononuclear infiltrate. It is common to see the coexistence of inflammation and fibrosis, especially in chronic diseases such as viral hepatitis, schistosomiasis, or bacterial pyelonephritis-induced fibrosis. Whether inflammation is a constituent of fibrogenesis or they are independent of each other is controversial [126–128]. Of note, anti-inflammatory regimens are responsible for anti-fibrogenesis in clinical practice, suggesting that the anti-fibrogenesis could be an independent process.

3 CircRNA in Organ Fibrosis

Fibrosis has different characteristics and clinical effects in different organ systems (Table 21.1). Kidney fibrosis decreases erythropoietin production and causes anemia. Liver fibrosis impairs the metabolism of lots of medication. Electromechanical coupling is debilitated when cardiac fibrosis occurred. It will lead to all kinds of arrhythmias and also affect the heart pump function. Causes of fibrosis vary in these organs too. Infections have been reported the most common cause of fibrosis in the liver. Also, fibrosis in organs like the liver and lung can potentially lead to cancer.

Emerging evidence has suggested that circRNA participates in this process. Here we will discuss the role of circRNA in fibrosis in different organ systems.

Table 21.1 Specific aspects of organ fibrosis

	Kidney	Liver	Heart	Lung	Skin
Etiology	Diabetes mellitus Hypertension Glomerulonephritis	viral hepatitis alcohol-induced steatohepatitis Nonalcoholic steatohepatitis	Hypertension Coronary artery disease Aortic stenosis	Idiopathic pulmonary fibrosis Occupational disease Sarcoidosis	Physical injury Idiopathic scleroderma
Diagnosis	Kidney function(BUN, GFR, serum creatinine) Kidney biopsy	Liver function MRI Liver biopsy	Cardiac function	X-ray	Manifestations (stiffness, elasticity)
Specific features	Kidney fibrosis causes anemia because of cessation of erythropoietin production	Stellate cells participate in liver fibrosis Liver fibrosis has detrimental microvascular shunts which cause porto-venous hypertension	Cardiac fibrosis can result in atrial fibrillation because fibroblasts are required for electromechanical coupling	Survival mean time of IPF patients is <6 years	Skin fibrosis is associated with skin color and keloid formation

3.1 Liver Fibrosis

Liver fibrosis has devastating effects on metabolism and has been a concern worldwide. The major causes of liver fibrosis are alcohol, nonalcoholic steatohepatitis (NASH), and viral hepatitis. The major cells involved in liver fibrosis are stellate cells, which are derived from neurocrest. They can be transdifferentiated into myofibroblasts, which, in turn, can activate common fibroblasts and peripheral fibroblasts. Picrosirius red staining and transmission electron microscopy are often used to identify cross-linked and stable collagen fibers to assess liver fibrosis.

In cancer patients, radiation-induced liver injury accounts for most of liver fibrosis [143–146]. It is recognized as radiation-induced liver fibrosis (RILF). RILF is a complex process. Radiation can facilitate the transformation of quiescent hepatic stellate cells (HSC) into myofibroblast-like cells (activated HSC) which are mediated by TGF-β signaling pathway [129, 130]. In the hepatic microenvironment, quiescent hepatic stellate cells can also be activated via prolonged stimulation from other inflammatory factors, which result in excessive extracellular matrix accumulation. On the whole, the activated HSCs contribute to the progression of liver fibrosis [131].

CircRNA plays an important role in various biological processes through acting as microRNA (miRNA) sponges. Recently, it has been reported that the expression of circRNA was correlated with RILF. By using circRNA microarray, circRNA expression was profiled in HSCs treated with or without irradiation. It was found that the expressions of 179 circRNAs were significantly increased, while the expressions of 630 circRNAs were decreased in irradiated HSCs compared with normal HSCs. Bioinformatic analyses indicated these expression changes might be the response to irradiation and the product of fibrotic process. In another similar study, researcher reported that inhibiting the expression of hsa_circ_0071410 increased the expression of miR-9-5p and attenuated irradiation-induced HSC activation [132].

3.2 Cardiac Fibrosis

Unlike hepatocytes, cardiomyocytes have less capacity to regenerate; fibrosis can occur after a sudden loss of cardiomyocytes. Ischemia is the major trigger of the process. After an episode of acute myocardial infarction, the dead portion will be ultimately replaced by scar tissue.

Apart from the acute causes, chronic causes like aging, hypertension, valvular disease, and cardiotoxic medication also contribute to cardiac fibrosis. They can induce the perivascular and interstitial collagen production and accumulation. Aging is an unpreventable cause for fibrosis. During aging, the cardiac functional cell loss and fibrotic tissue replacement increase. These changes could impair cardiac function. Stimuli which change the hemodynamic balance in the heart could cause more extensive damage. For instance, uncontrolled hypertension and valvular diseases like aortic stenosis generate stimuli on cardiomyocytes by increasing intraventricular pressure. Due to the limited regenerative ability of cardiomyocytes, fibrous tissue grows to compensate for the elevated workload. The major consequence is increased stiffness of the ventricular wall. Heart failure would eventually occur. Chemotherapy with cardiotoxicity could cause heart injury as well.

It is known that metabolic disease like diabetes could induce heart failure independent of any coronary artery disease or hypertension. Fibrosis was observed in both metabolic disease patients and experimental models [133]. In diabetes-related heart fibrosis, hyperglycemia, increased fatty acid metabolism, insulin resistance, and microcirculatory changes all contribute to the cardiac remodeling by fibrosis [134–138]. Myocardial fibrosis increases myocardial stiffness, disturbs relaxation, and causes cardiac dysfunction. It also damages conduction system and results in atrial and ventricular arrhythmias [139].

There are two main types of fibrotic lesions in the heart—perivascular fibrosis and interstitial fibrosis. The latter is also known as "subendocardial fibrosis" or "endomyocardial fibrosis." It is characterized by subendocardial deposition of elastic fibers and is usually found in newborns with congenital heart defects [152–155]. The unique feature of cardiac fibroblasts is that they are able to function as mechanoelectrical transducers [84–87, 140]. Thus, as the most abundant cell type in the cardiac, the cardiac fibroblasts play important physiological roles.

CircRNAs keep the balance of fibrosis by interacting with miRNAs. A good number of miRNAs have been recognized to enhance or suppress the activation of fibroblasts. CircRNAs act as "miRNA sponges" to help fibrosis modulation.

In diabetes mouse model, circRNA_000203 was found to be increased in the myocardium as well as in Ang II-induced cardiac fibroblasts in vivo. Upregulated circRNA_000203 enhanced the expression of fibrosis markers (Col1a2, Col3a1, and α-SMA) in mouse cardiac fibroblasts. In addition, circRNA_000203 further facilitated fibrosis by blocking the effect of miR-26b-5p which is known to downregulate Col1a2, CTGF, Col3a1, and α-SMA expression in cardiac fibroblasts. Consistently, the anti-fibrosis effect of miR-26b-5p could be eliminated by the overexpression of circRNA_000203 in cardiac fibroblasts. In this case, circRNA and its downstream miRNA formed a circRNA_000203/miR-26b-5p axis to modulate cardiac fibrosis [141].

CircRNA_010567 is another circRNA that is found to regulate cardiac fibrosis. MiR-26b-5p and miR-141 are identified to be the downstream miRNAs of circRNA_010567. miR-141 is also an anti-fibrotic factor which directly targets TGF-b1. Subsequent study showed that the block of circRNA_010567 upregulated miR-141 and thus decreased TGF-b1 and other fibrosis-associated protein in CFs, including Col1, Col3, and a-SMA. Taken together, circRNA_010567/miR-141/TGF-b1 axis prohibited cardiac fibrosis and provides a novel insight for circRNA-miRNA-mRNA in cardiac fibrosis [142].

3.3 Lung Fibrosis

Lung fibrosis is a devastating condition which could be resulted from any chronic lung diseases including obstructive pulmonary diseases, emphysema, pneumonia, and so on [108, 143–147]. These are usually considered as secondary pulmonary fibrosis. Primary pulmonary fibrosis, also known as idiopathic pulmonary fibrosis, could be the sole presentation or a part of other systemic diseases [147, 148]. Inflammation is the main cause of pulmonary fibrosis, and anti-inflammatory treatment is the standard treatment in most circumstances. However, in primary pulmonary fibrosis, anti-inflammatory therapy does not lead to favorable outcome [127].

Silicosis is a chronic fibrotic pulmonary disease caused by the inhalation of silica. It is one of the most serious occupational diseases in the world. Silicosis is characterized by chronic progressive pulmonary fibrosis. Diagnosis is usually made when patients enter the late stage. Early diagnosis remains challenging [127, 149, 150]. Unfortunately, there is no specific therapy for this disease. Avoidance of the substance, smoking cessation, and symptomatic treatment with bronchodilators are the major management for mild disease. For more advanced silicosis, lung transplantation is the only way to go [151, 152]. Lung fibrosis complications include but not limited to lung infections, rheumatoid disorders, airway obstruction, and even cancer [153]. Fibrosis in silicosis is initiated with the ingestion of silica dust by macrophages. Stimulated by silica dust, these macrophages will produce cytokines to attract fibroblasts. Collagens recruited by macrophages participate in "eating" silica dust and produce inflammatory reaction. The inflammatory reaction further attracts more collagen and macrophages. Fibrotic nodules are produced in this vicious cycle. On the other hand, EMT also promotes the accumulation of fibroblasts.

Researches on macrophages found that circZC3H4 and its downstream product ZC3H4 were positively related to SiO2 stimulation in macrophages. Fibroblast proliferation and migration can be activated through the circZC3H4 RNA/ZC3H4 pathway. This study suggests that ZC3H4 could be the candidate target for fibrosis therapy.

Study on the EMT in silicosis confirmed the occurrence of the EMT in response to SiO2 exposure both in vivo and in vitro. In addition, circHECTD1/HECTD1 pathway was reported to regulate silica-induced fibrosis. SiO2 modulated the recruitment of cell proliferation and migration by targeting circHECTD1 or upregulating HECTD1 [154].

4 Clinical Applications of circRNAs in Fibrosis

4.1 Implications of circRNA Therapies for Anti-Fibrosis

While the anti-fibrosis studies have been prosperous, investigations for according therapy has been lagged behind. This is due to the challenges of finding the common targets in fibrosis. Organ fibrosis is not a disease; it is a common and progressive process in chronic disease. It involves various complicated interactions. Currently, a great number of strategies have been reported to be useful in experimental fibrosis models in vivo and in vitro. The anti-fibrosis strategies proposed nowadays are majorly based on the following mechanisms: (1) modulation of the accompanying inflammation, (2) pharmacological prevention of ECM deposition, (3) correction of the altered epigenome, and (4) inhibition of profibrotic growth factors, especially TGF-β [155–158]. Among these, direct inhibition of TGF-β is arguable since studies have shown that it might not be safe [21, 159]. Apart from inducing fibrosis, TGF-β performs anti-inflammatory and anticancer activities [160, 161]. TGF-β1 knockout mice were found to have more generalized inflammation. What's more, TGF-β1 mutations are associated with certain gastrointestinal cancers [162–164].

With the emergence of researches on circRNAs, more attention has shifted toward circRNAs. Compared with synthetic molecules (modified chemical drugs, RNA interference constructs), circRNAs might have reduced side effects. The main function of circRNAs is sponging miRNA, which is found to play a significant role in all kinds of diseases, including fibrosis. CircRNAs could modulate protein (especially inflammatory cytokines and other protein signals) expression by interacting with miRNAs, thus interfering disease progress. Ideally, miRNA manipulation could produce the similar effects. However, previous researchers have found that miRNAs and small interfering RNAs have inevitable off-target effects due to their short lengths [165–167]. Unlike miRNAs or small interfering RNAs, circRNAs have low off-target effects and more stable structure.

Circular RNAs provide potential drug targets for fibrosis therapy. However, this approach will base on innovations in drug synthetics to allow clinical applications. Given the rapid progress in these areas, circRNA-based therapeutics will undoubtedly enrich the development of antifibrotic drugs in the coming years.

4.2 Implications of Diagnostic and Prognostic circRNA Markers in Fibrosis

CircRNAs could also be candidate markers for disease. As mentioned above, circRNAs have more stable structures and are resistant to nuclease [168–170]. The half-lives of circRNAs can be as long as 50 h. On average, the half-lives of circRNAs are about 2.5-fold longer than their linear counterparts [42, 171]. Also, the expressions of circRNAs in the blood are generally more stable than linear mRNAs. All the above advantages make circRNAs attractive diagnostic and prognostic tools [172, 173].

CircRNAs have been reported to be potential biomarkers for several types of cancer [174, 175]. In liver cancer, it was found that the expression of hsa_circ_0001649 (circSHPRH) is significantly downregulated. What's more, the level of circSHPRH is associated with the size of the tumor as well as the occurrence of tumor embolus.

The expression levels of circRNAs correlate with fibrosis too. CircRNA expression profiles revealed that expressions of over 800 circRNAs were changed during liver fibrosis. Among these circRNAs, hsa_circ_0071410 performed antifibrosis effects [199].

Determining fibrosis status appears to be more challenging in certain organs, such as the heart, in which case functional test seems more prevalent. However, functional test is not sensitive enough to detect early fibrosis. Progress has been made to identify serum markers in early cardiac fibrosis. The expression of circRNA_010567 and circRNA_000203 changed during cardiac fibrosis. CircRNA_010567 is markedly increased in Ang II-induced cardiac fibrosis through circRNA_010567/miR-141/TGF-b1 axis [200]. Similarly, circRNA_000203/miR-26b-5p axis was found to effect fibroblasts in diabetic myocardium [201]. These findings provide a novel insight for circRNA-miRNA-mRNA in cardiac fibrosis and their potential role in diagnosis and prognosis. However, cardiac fibrosis is still less assessed in the lab setting, compared with other organs. Until now no blood test has been approved for clinical use yet.

5 Conclusions

CircRNAs are a class of noncoding RNAs that regulate the expression of genes by a variety of mechanisms. They might also have the potential roles in regulating protein expression. The functions of circRNAs are so complicated, and the regulation mechanisms have not yet been completely understood. Current studies mainly focus on their expression changes during physiological and pathological process. Generally speaking, circRNA has emerged as a new tool to understand different diseases and their progressions. Efforts have been made to explore their potential use in detecting disease status and gene therapy. The function study of circRNAs in organ fibrosis is a rapid developing field. CircRNAs involved in fibrosis have been discovered, but these studies are still in their initial stage. Difficulties and challenges are on the way. CircRNAs will be important players in genetic regulation in the future.

Acknowledgments This work was supported by the grants from National Natural Science Foundation of China (81370280 and 81570332 to LZ, 81470515 and 81670362 to JH Xu), Shanghai Medical Guide Project from Shanghai Science and Technology Committee (134119a3000 to Jiahong Xu), and the grant from Jiangsu Province's Key Provincial Talents Program (ZDRCA2016019 to LZ).

Competing Financial Interests The authors declare no competing financial interests.

References

1. Murakami M (1991) Study of the mechanism of supporting cells repairing the organ of Corti in terms of cell kinetics–nuclear DNA synthesis of supporting cell of the organ of Corti in the cochlea damaged by nitromin administration. Nihon Jibiinkoka Gakkai Kaiho 94(3):386–395
2. Muller I, Vogl T, Pappritz K et al (2017) Pathogenic role of the damage-associated molecular patterns S100A8 and S100A9 in coxsackievirus B3-induced myocarditis. Circ Heart Fail 10(11)
3. Guerrero-Juarez CF, Plikus MV (2017) Gli-fully halting the progression of fibrosis. Cell Stem Cell 20(6):735–736
4. Nightingale S, Stormon MO, O'Loughlin EV et al (2017) Early posthepatoportoenterostomy predictors of native liver survival in biliary atresia. J Pediatr Gastroenterol Nutr 64(2):203–209
5. Schwab ME (2002) Increasing plasticity and functional recovery of the lesioned spinal cord. Prog Brain Res 137:351–359
6. Tampe B, Steinle U, Tampe D et al (2017) Low-dose hydralazine prevents fibrosis in a murine model of acute kidney injury-to-chronic kidney disease progression. Kidney Int 91(1):157–176
7. Zoubek ME, Trautwein C, Strnad P (2017) Reversal of liver fibrosis: from fiction to reality. Best Pract Res Clin Gastroenterol 31(2):129–141
8. Friedman SL, Sheppard D, Duffield JS et al (2013) Therapy for fibrotic diseases: nearing the starting line. Sci Transl Med 5(167):167sr161
9. Wynn TA (2007) Common and unique mechanisms regulate fibrosis in various fibroproliferative diseases. J Clin Invest 117(3):524–529
10. Fani F, Regolisti G, Delsante M et al (2017) Recent advances in the pathogenetic mechanisms of sepsis-associated acute kidney injury. J Nephrol. https://doi.org/10.1007/s40620-017-0452-4
11. Fischereder M, Schroppel B (2009) The role of chemokines in acute renal allograft rejection and chronic allograft injury. Front Biosci (Landmark Ed) 14:1807–1814
12. Gomez H, Kellum JA, Ronco C (2017) Metabolic reprogramming and tolerance during sepsis-induced AKI. Nat Rev Nephrol 13(3):143–151
13. Kusano KF, Pola R, Murayama T et al (2005) Sonic hedgehog myocardial gene therapy: tissue repair through transient reconstitution of embryonic signaling. Nat Med 11(11):1197–1204
14. Li X, Shu R, Filippatos G et al (2004) Apoptosis in lung injury and remodeling. J Appl Physiol (1985) 97(4):1535–1542

15. Mehrotra P, Collett JA, Gunst SJ et al (2018) Th17 cells contribute to pulmonary fibrosis and inflammation during chronic kidney disease progression after acute ischemia. Am J Physiol Regul Integr Comp Physiol 314(2):R265–R273
16. Niroomand F, Kubler W (1994) Hibernating, stunning and ischemic preconditioning of the myocardium: therapeutic implications. Clin Investig 72(10):731–736
17. Popovic B, Sutic I, Skocibusic N et al (2015) Cholestasis and Inflammation of the Pancreas in family medicine. Acta Med Croatica 69(4):319–326
18. Ranganathan P, Jayakumar C, Ramesh G (2013) Proximal tubule-specific overexpression of netrin-1 suppresses acute kidney injury-induced interstitial fibrosis and glomerulosclerosis through suppression of IL-6/STAT3 signaling. Am J Physiol Renal Physiol 304(8):F1054–F1065
19. Tewes S, Gueler F, Chen R et al (2017) Functional MRI for characterization of renal perfusion impairment and edema formation due to acute kidney injury in different mouse strains. PLoS One 12(3):e0173248
20. Xiao Y, Yang N, Zhang Q et al (2014) Pentraxin 3 inhibits acute renal injury-induced interstitial fibrosis through suppression of IL-6/Stat3 pathway. Inflammation 37(5):1895–1901
21. Wynn TA, Ramalingam TR (2012) Mechanisms of fibrosis: therapeutic translation for fibrotic disease. Nat Med 18(7):1028–1040
22. Westholm JO, Miura P, Olson S et al (2014) Genome-wide analysis of drosophila circular RNAs reveals their structural and sequence properties and age-dependent neural accumulation. Cell Rep 9(5):1966–1980
23. Salzman J, Gawad C, Wang PL et al (2012) Circular RNAs are the predominant transcript isoform from hundreds of human genes in diverse cell types. PLoS One 7(2):e30733
24. Yang L, Zou M, Fu B et al (2013) Genome-wide identification, characterization, and expression analysis of lineage-specific genes within zebrafish. BMC Genomics 14:65
25. Kelly S, Greenman C, Cook PR et al (2015) Exon Skipping Is Correlated with Exon Circularization. J Mol Biol 427(15):2414–2417
26. Nigro JM, Cho KR, Fearon ER et al (1991) Scrambled exons. Cell 64(3):607–613
27. Cocquerelle C, Daubersies P, Majerus MA et al (1992) Splicing with inverted order of exons occurs proximal to large introns. EMBO J 11(3):1095–1098
28. Capel B, Swain A, Nicolis S et al (1993) Circular transcripts of the testis-determining gene Sry in adult mouse testis. Cell 73(5):1019–1030
29. Cocquerelle C, Mascrez B, Hetuin D et al (1993) Mis-splicing yields circular RNA molecules. FASEB J 7(1):155–160
30. Pasman Z, Been MD, Garcia-Blanco MA (1996) Exon circularization in mammalian nuclear extracts. RNA 2(6):603–610
31. Salzman J, Chen RE, Olsen MN et al (2013) Cell-type specific features of circular RNA expression. PLoS Genet 9(9):e1003777
32. Liang D, Wilusz JE (2014) Short intronic repeat sequences facilitate circular RNA production. Genes Dev 28(20):2233–2247
33. Starke S, Jost I, Rossbach O et al (2015) Exon circularization requires canonical splice signals. Cell Rep 10(1):103–111
34. Hansen TB, Jensen TI, Clausen BH et al (2013) Natural RNA circles function as efficient microRNA sponges. Nature 495(7441):384–388
35. Burd CE, Jeck WR, Liu Y et al (2010) Expression of linear and novel circular forms of an INK4/ARF-associated non-coding RNA correlates with atherosclerosis risk. PLoS Genet 6(12):e1001233
36. Memczak S, Jens M, Elefsinioti A et al (2013) Circular RNAs are a large class of animal RNAs with regulatory potency. Nature 495(7441):333–338
37. Lukiw WJ (2013) Circular RNA (circRNA) in Alzheimer's disease (AD). Front Genet 4:307
38. Bachmayr-Heyda A, Reiner AT, Auer K et al (2015) Correlation of circular RNA abundance with proliferation--exemplified with colorectal and ovarian cancer, idiopathic lung fibrosis, and normal human tissues. Sci Rep 5:8057
39. Rybak-Wolf A, Stottmeister C, Glazar P et al (2015) Circular RNAs in the mammalian brain are highly abundant, conserved, and dynamically expressed. Mol Cell 58(5):870–885
40. You X, Vlatkovic I, Babic A et al (2015) Neural circular RNAs are derived from synaptic genes and regulated by development and plasticity. Nat Neurosci 18(4):603–610
41. Jeck WR, Sorrentino JA, Wang K et al (2013) Circular RNAs are abundant, conserved, and associated with ALU repeats. RNA 19(2):141–157
42. Suzuki H, Zuo Y, Wang J et al (2006) Characterization of RNase R-digested cellular RNA source that consists of lariat and circular RNAs from pre-mRNA splicing. Nucleic Acids Res 34(8):e63
43. Barrett SP, Salzman J (2016) Circular RNAs: analysis, expression and potential functions. Development 143(11):1838–1847
44. Guo JU, Agarwal V, Guo H et al (2014) Expanded identification and characterization of mammalian circular RNAs. Genome Biol 15(7):409
45. Veno MT, Hansen TB, Veno ST et al (2015) Spatio-temporal regulation of circular RNA expression during porcine embryonic brain development. Genome Biol 16:245
46. Bellemare J, Roberge CJ, Bergeron D et al (2005) Epidermis promotes dermal fibrosis: role in the pathogenesis of hypertrophic scars. J Pathol 206(1):1–8
47. Holvoet T, Devriese S, Castermans K et al (2017) Treatment of intestinal fibrosis in experimental inflammatory bowel disease by the pleiotropic actions of a local rho kinase inhibitor. Gastroenterology 153(4):1054–1067

48. Alcalde O, Cabrera Gomez S, Valles Gras E et al (2017) Rheumatoid arthritis with severe atrial fibrosis and multiple atrial arrhythmias: chronic atrial myocarditis? Rev Esp Cardiol (Engl Ed). https://doi.org/10.1016/j.rec.2017.04.003
49. Abdul N, Dixon D, Walker A et al (2015) Fibrosis is a common outcome following total knee arthroplasty. Sci Rep 5:16469
50. Ramos F, Robledo C, Izquierdo-Garcia FM et al (2016) Bone marrow fibrosis in myelodysplastic syndromes: a prospective evaluation including mutational analysis. Oncotarget 7(21):30492–30503
51. Dvorak HF, Harvey VS, McDonagh J (1984) Quantitation of fibrinogen influx and fibrin deposition and turnover in line 1 and line 10 guinea pig carcinomas. Cancer Res 44(8):3348–3354
52. Dvorak HF (1986) Tumors: wounds that do not heal. Similarities between tumor stroma generation and wound healing. N Engl J Med 315(26):1650–1659
53. Iwano M, Plieth D, Danoff TM et al (2002) Evidence that fibroblasts derive from epithelium during tissue fibrosis. J Clin Invest 110(3):341–350
54. Jian XC, Liu SF, Shen ZH et al (1988) Histomorphology of oral submucous fibrosis. Report of 24 cases. Chin Med J (Engl) 101(7):505–509
55. Zhang SS, Gong ZJ, Xiong W et al (2016) A rat model of oral submucous fibrosis induced by bleomycin. Oral Surg Oral Med Oral Pathol Oral Radiol 122(2):216–223
56. Harwood IR, Greene LM, Kozakowski-Koch JA et al (1992) New peripherally inserted midline catheter: a better alternative for intravenous antibiotic therapy in patients with cystic fibrosis. Pediatr Pulmonol 12(4):233–239
57. Jiao LR, Seifalian AM, Habib N et al (1999) The effect of mechanically enhancing portal venous inflow on hepatic oxygenation, microcirculation, and function in a rabbit model with extensive hepatic fibrosis. Hepatology 30(1):46–52
58. Perry KA, Banarjee A, Liu J et al (2013) Gastric ischemic conditioning increases neovascularization and reduces inflammation and fibrosis during gastroesophageal anastomotic healing. Surg Endosc 27(3):753–760
59. Tucker RD, Gibbs GE, Christensen MB (1979) Cystic fibrosis serum effect on the short circuit current of rat jejunum. Pediatr Res 13(12):1371–1374
60. Poling J, Gajawada P, Lorchner H et al (2012) The Janus face of OSM-mediated cardiomyocyte dedifferentiation during cardiac repair and disease. Cell Cycle 11(3):439–445
61. Hano H, Takasaki S (2003) Three-dimensional observations on the alterations of lobular architecture in chronic hepatitis with special reference to its angioarchitecture for a better understanding of the formal pathogenesis of liver cirrhosis. Virchows Arch 443(5):655–663
62. Forbes MS, Thornhill BA, Chevalier RL (2011) Proximal tubular injury and rapid formation of atubular glomeruli in mice with unilateral ureteral obstruction: a new look at an old model. Am J Physiol Renal Physiol 301(1):F110–F117
63. Akram KM, Lomas NJ, Spiteri MA et al (2013) Club cells inhibit alveolar epithelial wound repair via TRAIL-dependent apoptosis. Eur Respir J 41(3):683–694
64. Aravinthan A, Scarpini C, Tachtatzis P et al (2013) Hepatocyte senescence predicts progression in non-alcohol-related fatty liver disease. J Hepatol 58(3):549–556
65. Rowe RG, Lin Y, Shimizu-Hirota R et al (2011) Hepatocyte-derived Snail1 propagates liver fibrosis progression. Mol Cell Biol 31(12):2392–2403
66. Roberts AB, Anzano MA, Lamb LC et al (1982) Isolation from murine sarcoma cells of novel transforming growth factors potentiated by EGF. Nature 295(5848):417–419
67. Border WA, Okuda S, Languino LR et al (1990) Suppression of experimental glomerulonephritis by antiserum against transforming growth factor beta 1. Nature 346(6282):371–374
68. Denis M (1994) Neutralization of transforming growth factor-beta 1 in a mouse model of immune-induced lung fibrosis. Immunology 82(4):584–590
69. Kuwahara F, Kai H, Tokuda K et al (2002) Transforming growth factor-beta function blocking prevents myocardial fibrosis and diastolic dysfunction in pressure-overloaded rats. Circulation 106(1):130–135
70. Roberts AB, Sporn MB, Assoian RK et al (1986) Transforming growth factor type beta: rapid induction of fibrosis and angiogenesis in vivo and stimulation of collagen formation in vitro. Proc Natl Acad Sci USA 83(12):4167–4171
71. Crosas-Molist E, Fabregat I (2015) Role of NADPH oxidases in the redox biology of liver fibrosis. Redox Biol 6:106–111
72. Kendall RT, Feghali-Bostwick CA (2014) Fibroblasts in fibrosis: novel roles and mediators. Front Pharmacol 5:123
73. Kalluri R, Zeisberg M (2006) Fibroblasts in cancer. Nat Rev Cancer 6(5):392–401
74. Kalluri R, Zeisberg E (2006) Controlling angiogenesis in heart valves. Nat Med 12(10):1118–1119
75. Kalluri R, Zeisberg M (2003) Exploring the connection between chronic renal fibrosis and bone morphogenic protein-7. Histol Histopathol 18(1):217–224
76. Moore-Morris T, Cattaneo P, Puceat M et al (2016) Origins of cardiac fibroblasts. J Mol Cell Cardiol 91:1–5
77. Zeisberg EM, Kalluri R (2010) Origins of cardiac fibroblasts. Circ Res 107(11):1304–1312
78. Dufourcq P, Louis H, Dandre F et al (1997) Phenotypic modification of arterial smooth muscle cells in response to medial dissection. Coron Artery Dis 8(3–4):163–170
79. Shujia J, Haider HK, Idris NM et al (2008) Stable therapeutic effects of mesenchymal stem cell-based multiple gene delivery for cardiac repair. Cardiovasc Res 77(3):525–533

80. Tanaka Y, Nouchi T, Yamane M et al (1991) Phenotypic modulation in lipocytes in experimental liver fibrosis. J Pathol 164(3):273–278
81. Grupp C, Lottermoser J, Cohen DI et al (1997) Transformation of rat inner medullary fibroblasts to myofibroblasts in vitro. Kidney Int 52(5):1279–1290
82. Maxwell PH, Osmond MK, Pugh CW et al (1993) Identification of the renal erythropoietin-producing cells using transgenic mice. Kidney Int 44(5):1149–1162
83. Paliege A, Rosenberger C, Bondke A et al (2010) Hypoxia-inducible factor-2alpha-expressing interstitial fibroblasts are the only renal cells that express erythropoietin under hypoxia-inducible factor stabilization. Kidney Int 77(4):312–318
84. Camelliti P, Devlin GP, Matthews KG et al (2004) Spatially and temporally distinct expression of fibroblast connexins after sheep ventricular infarction. Cardiovasc Res 62(2):415–425
85. Camelliti P, Green CR, Kohl P (2006) Structural and functional coupling of cardiac myocytes and fibroblasts. Adv Cardiol 42:132–149
86. Camelliti P, Green CR, LeGrice I et al (2004) Fibroblast network in rabbit sinoatrial node: structural and functional identification of homogeneous and heterogeneous cell coupling. Circ Res 94(6):828–835
87. Tveito A, Lines G, Artebrant R et al (2011) Existence of excitation waves for a collection of cardiomyocytes electrically coupled to fibroblasts. Math Biosci 230(2):79–86
88. Chang HY, Chi JT, Dudoit S et al (2002) Diversity, topographic differentiation, and positional memory in human fibroblasts. Proc Natl Acad Sci USA 99(20):12877–12882
89. Sugimoto H, Mundel TM, Kieran MW et al (2006) Identification of fibroblast heterogeneity in the tumor microenvironment. Cancer Biol Ther 5(12):1640–1646
90. Iwano M, Fischer A, Okada H et al (2001) Conditional abatement of tissue fibrosis using nucleoside analogs to selectively corrupt DNA replication in transgenic fibroblasts. Mol Ther 3(2):149–159
91. Bayreuther K, Francz PI, Rodemann HP (1992) Fibroblasts in normal and pathological terminal differentiation, aging, apoptosis and transformation. Arch Gerontol Geriatr 15(Suppl 1):47–74
92. Bayreuther K, Francz PI, Rodemann HP (1995) Fibroblasts in normal and pathological terminal differentiation, aging, apoptosis and transformation. Ontogenez 26(1):22–37
93. Bayreuther K, Rodemann HP, Francz PI et al (1988) Differentiation of fibroblast stem cells. J Cell Sci Suppl 10:115–130
94. Bayreuther K, Rodemann HP, Hommel R et al (1988) Human skin fibroblasts in vitro differentiate along a terminal cell lineage. Proc Natl Acad Sci USA 85(14):5112–5116
95. Bucala R, Spiegel LA, Chesney J et al (1994) Circulating fibrocytes define a new leukocyte subpopulation that mediates tissue repair. Mol Med 1(1):71–81
96. Lin SL, Kisseleva T, Brenner DA et al (2008) Pericytes and perivascular fibroblasts are the primary source of collagen-producing cells in obstructive fibrosis of the kidney. Am J Pathol 173(6):1617–1627
97. Ronnov-Jessen L, Petersen OW (1996) A function for filamentous alpha-smooth muscle actin: retardation of motility in fibroblasts. J Cell Biol 134(1):67–80
98. Ronnov-Jessen L, Petersen OW, Koteliansky VE et al (1995) The origin of the myofibroblasts in breast cancer. Recapitulation of tumor environment in culture unravels diversity and implicates converted fibroblasts and recruited smooth muscle cells. J Clin Invest 95(2):859–873
99. Ronnov-Jessen L, Petersen OW (1993) Induction of alpha-smooth muscle actin by transforming growth factor-beta 1 in quiescent human breast gland fibroblasts. Implications for myofibroblast generation in breast neoplasia. Lab Invest 68(6):696–707
100. Boutet A, De Frutos CA, Maxwell PH et al (2006) Snail activation disrupts tissue homeostasis and induces fibrosis in the adult kidney. EMBO J 25(23):5603–5613
101. Broekema M, Harmsen MC, van Luyn MJ et al (2007) Bone marrow-derived myofibroblasts contribute to the renal interstitial myofibroblast population and produce procollagen I after ischemia/reperfusion in rats. J Am Soc Nephrol 18(1):165–175
102. Humphreys BD, Lin SL, Kobayashi A et al (2010) Fate tracing reveals the pericyte and not epithelial origin of myofibroblasts in kidney fibrosis. Am J Pathol 176(1):85–97
103. Kim KK, Wei Y, Szekeres C et al (2009) Epithelial cell alpha3beta1 integrin links beta-catenin and Smad signaling to promote myofibroblast formation and pulmonary fibrosis. J Clin Invest 119(1):213–224
104. Kisseleva T, Uchinami H, Feirt N et al (2006) Bone marrow-derived fibrocytes participate in pathogenesis of liver fibrosis. J Hepatol 45(3):429–438
105. Li J, Qu X, Yao J et al (2010) Blockade of endothelial-mesenchymal transition by a Smad3 inhibitor delays the early development of streptozotocin-induced diabetic nephropathy. Diabetes 59(10):2612–2624
106. Osterreicher CH, Penz-Osterreicher M, Grivennikov SI et al (2011) Fibroblast-specific protein 1 identifies an inflammatory subpopulation of macrophages in the liver. Proc Natl Acad Sci USA 108(1):308–313
107. Quaggin SE, Kapus A (2011) Scar wars: mapping the fate of epithelial-mesenchymal-myofibroblast transition. Kidney Int 80(1):41–50
108. Rock JR, Barkauskas CE, Cronce MJ et al (2011) Multiple stromal populations contribute to pulmonary fibrosis without evidence for epithelial to mesenchymal transition. Proc Natl Acad Sci USA 108(52):E1475–E1483
109. Zeisberg EM, Potenta SE, Sugimoto H et al (2008) Fibroblasts in kidney fibrosis emerge via

endothelial-to-mesenchymal transition. J Am Soc Nephrol 19(12):2282–2287
110. Zeisberg M, Yang C, Martino M et al (2007) Fibroblasts derive from hepatocytes in liver fibrosis via epithelial to mesenchymal transition. J Biol Chem 282(32):23337–23347
111. Zhou B, von Gise A, Ma Q et al (2010) Genetic fate mapping demonstrates contribution of epicardium-derived cells to the annulus fibrosis of the mammalian heart. Dev Biol 338(2):251–261
112. Distler O, Distler JH, Scheid A et al (2004) Uncontrolled expression of vascular endothelial growth factor and its receptors leads to insufficient skin angiogenesis in patients with systemic sclerosis. Circ Res 95(1):109–116
113. Higgins DF, Kimura K, Bernhardt WM et al (2007) Hypoxia promotes fibrogenesis in vivo via HIF-1 stimulation of epithelial-to-mesenchymal transition. J Clin Invest 117(12):3810–3820
114. Moon JO, Welch TP, Gonzalez FJ et al (2009) Reduced liver fibrosis in hypoxia-inducible factor-1alpha-deficient mice. Am J Physiol Gastrointest Liver Physiol 296(3):G582–G592
115. Corpechot C, Barbu V, Wendum D et al (2002) Hypoxia-induced VEGF and collagen I expressions are associated with angiogenesis and fibrogenesis in experimental cirrhosis. Hepatology 35(5):1010–1021
116. Ioannou M, Pyrpasopoulou A, Simos G et al (2013) Upregulation of VEGF expression is associated with accumulation of HIF-1alpha in the skin of naive scleroderma patients. Mod Rheumatol 23(6):1245–1248
117. Maeshima Y, Makino H (2010) Angiogenesis and chronic kidney disease. Fibrogenesis Tissue Repair 3:13
118. Yoon YS, Uchida S, Masuo O et al (2005) Progressive attenuation of myocardial vascular endothelial growth factor expression is a seminal event in diabetic cardiomyopathy: restoration of microvascular homeostasis and recovery of cardiac function in diabetic cardiomyopathy after replenishment of local vascular endothelial growth factor. Circulation 111(16):2073–2085
119. Worda M, Sgonc R, Dietrich H et al (2003) In vivo analysis of the apoptosis-inducing effect of anti-endothelial cell antibodies in systemic sclerosis by the chorionallantoic membrane assay. Arthritis Rheum 48(9):2605–2614
120. Rieder F, Kessler SP, West GA et al (2011) Inflammation-induced endothelial-to-mesenchymal transition: a novel mechanism of intestinal fibrosis. Am J Pathol 179(5):2660–2673
121. Zeisberg EM, Potenta S, Xie L et al (2007) Discovery of endothelial to mesenchymal transition as a source for carcinoma-associated fibroblasts. Cancer Res 67(21):10123–10128
122. Zeisberg EM, Tarnavski O, Zeisberg M et al (2007) Endothelial-to-mesenchymal transition contributes to cardiac fibrosis. Nat Med 13(8):952–961
123. Eisenberg LM, Markwald RR (1995) Molecular regulation of atrioventricular valvuloseptal morphogenesis. Circ Res 77(1):1–6
124. Liebner S, Cattelino A, Gallini R et al (2004) Beta-catenin is required for endothelial-mesenchymal transformation during heart cushion development in the mouse. J Cell Biol 166(3):359–367
125. Nakajima Y, Yamagishi T, Hokari S et al (2000) Mechanisms involved in valvuloseptal endocardial cushion formation in early cardiogenesis: roles of transforming growth factor (TGF)-beta and bone morphogenetic protein (BMP). Anat Rec 258(2):119–127
126. Murray PJ, Wynn TA (2011) Protective and pathogenic functions of macrophage subsets. Nat Rev Immunol 11(11):723–737
127. Wynn TA (2011) Integrating mechanisms of pulmonary fibrosis. J Exp Med 208(7):1339–1350
128. Wynn TA, Barron L, Thompson RW et al (2011) Quantitative assessment of macrophage functions in repair and fibrosis. Curr Protoc Immunol Chapter 14:Unit14.22
129. Bataller R, Brenner DA (2005) Liver fibrosis. J Clin Invest 115(2):209–218
130. Bataller R, Sancho-Bru P, Gines P et al (2005) Liver fibrogenesis: a new role for the renin-angiotensin system. Antioxid Redox Signal 7(9-10):1346–1355
131. Seki E, Schwabe RF (2015) Hepatic inflammation and fibrosis: functional links and key pathways. Hepatology 61(3):1066–1079
132. Chen Y, Yuan B, Wu Z et al (2017) Microarray profiling of circular RNAs and the potential regulatory role of hsa_circ_0071410 in the activated human hepatic stellate cell induced by irradiation. Gene 629:35–42
133. Kong P, Christia P, Frangogiannis NG (2014) The pathogenesis of cardiac fibrosis. Cell Mol Life Sci 71(4):549–574
134. Goyal BR, Mehta AA (2013) Diabetic cardiomyopathy: pathophysiological mechanisms and cardiac dysfuntion. Hum Exp Toxicol 32(6):571–590
135. Poornima IG, Parikh P, Shannon RP (2006) Diabetic cardiomyopathy: the search for a unifying hypothesis. Circ Res 98(5):596–605
136. Suys BE, Katier N, Rooman RP et al (2004) Female children and adolescents with type 1 diabetes have more pronounced early echocardiographic signs of diabetic cardiomyopathy. Diabetes Care 27(8):1947–1953
137. Aneja A, Tang WH, Bansilal S et al (2008) Diabetic cardiomyopathy: insights into pathogenesis, diagnostic challenges, and therapeutic options. Am J Med 121(9):748–757
138. Tarquini R, Lazzeri C, Pala L et al (2011) The diabetic cardiomyopathy. Acta Diabetol 48(3):173–181
139. Dzeshka MS, Lip GY, Snezhitskiy V et al (2015) Cardiac fibrosis in patients With atrial fibrillation: mechanisms and clinical implications. J Am Coll Cardiol 66(8):943–959

140. Tveito A, Lines GT, Edwards AG et al (2012) Slow Calcium-Depolarization-Calcium waves may initiate fast local depolarization waves in ventricular tissue. Prog Biophys Mol Biol 110(2–3):295–304
141. Tang CM, Zhang M, Huang L et al (2017) CircRNA_000203 enhances the expression of fibrosis-associated genes by derepressing targets of miR-26b-5p, Col1a2 and CTGF, in cardiac fibroblasts. Sci Rep 7:40342
142. Zhou B, Yu JW (2017) A novel identified circular RNA, circRNA_010567, promotes myocardial fibrosis via suppressing miR-141 by targeting TGF-beta1. Biochem Biophys Res Commun 487(4):769–775
143. Barkauskas CE, Noble PW (2014) Cellular mechanisms of tissue fibrosis. 7. New insights into the cellular mechanisms of pulmonary fibrosis. Am J Physiol Cell Physiol 306(11):C987–C996
144. Noble PW, Barkauskas CE, Jiang D (2012) Pulmonary fibrosis: patterns and perpetrators. J Clin Invest 122(8):2756–2762
145. Churg A, Wright JL, Tazelaar HD (2011) Acute exacerbations of fibrotic interstitial lung disease. Histopathology 58(4):525–530
146. Tazelaar HD, Wright JL, Churg A (2011) Desquamative interstitial pneumonia. Histopathology 58(4):509–516
147. Wright JL, Tazelaar HD, Churg A (2011) Fibrosis with emphysema. Histopathology 58(4):517–524
148. Mathai SK, Gulati M, Peng X et al (2010) Circulating monocytes from systemic sclerosis patients with interstitial lung disease show an enhanced profibrotic phenotype. Lab Invest 90(6):812–823
149. Carneiro PJ, Clevelario AL, Padilha GA et al (2017) Bosutinib therapy ameliorates lung inflammation and fibrosis in experimental silicosis. Front Physiol 8:159
150. Miao R, Ding B, Zhang Y et al (2016) Proteomic profiling differences in serum from silicosis and chronic bronchitis patients: a comparative analysis. J Thorac Dis 8(3):439–450
151. Rosengarten D, Fox BD, Fireman E et al (2017) Survival following lung transplantation for artificial stone silicosis relative to idiopathic pulmonary fibrosis. Am J Ind Med 60(3):248–254
152. Zhao MM, Cui JZ, Cui Y et al (2013) Therapeutic effect of exogenous bone marrowderived mesenchymal stem cell transplantation on silicosis via paracrine mechanisms in rats. Mol Med Rep 8(3):741–746
153. Gungen AC, Aydemir Y, Coban H et al (2016) Lung cancer in patients diagnosed with silicosis should be investigated. Respir Med Case Rep 18:93–95
154. Yang X, Wang J, Zhou Z et al (2018) Silica-induced initiation of circular ZC3H4 RNA/ZC3H4 pathway promotes the pulmonary macrophage activation. FASEB J. https://doi.org/10.1096/fj.201701118R:fj201701118R
155. Kawaratani H, Moriya K, Namisaki T et al (2017) Therapeutic strategies for alcoholic liver disease: focusing on inflammation and fibrosis (Review). Int J Mol Med 40(2):263–270
156. Nogueira A, Pires MJ, Oliveira PA (2017) Pathophysiological mechanisms of renal fibrosis: a review of animal models and therapeutic strategies. In Vivo 31(1):1–22
157. Owens GM (2017) Strategies to manage costs in idiopathic pulmonary fibrosis. Am J Manag Care 23(11 Suppl):S191–S196
158. Stasi C, Milani S (2017) Evolving strategies for liver fibrosis staging: non-invasive assessment. World J Gastroenterol 23(2):191–196
159. Friedman SL (2004) Mechanisms of disease: mechanisms of hepatic fibrosis and therapeutic implications. Nat Clin Pract Gastroenterol Hepatol 1(2):98–105
160. Bayomi HS, Elsherbiny NM, El-Gayar AM et al (2013) Evaluation of renal protective effects of inhibiting TGF-beta type I receptor in a cisplatin-induced nephrotoxicity model. Eur Cytokine Netw 24(4):139–147
161. Xu Y, Lou Z, Lee SH (2017) Arctigenin represses TGF-beta-induced epithelial mesenchymal transition in human lung cancer cells. Biochem Biophys Res Commun 493(2):934–939
162. Munjal C, Opoka AM, Osinska H et al (2014) TGF-beta mediates early angiogenesis and latent fibrosis in an Emilin1-deficient mouse model of aortic valve disease. Dis Model Mech 7(8):987–996
163. Katz LH, Likhter M, Jogunoori W et al (2016) TGF-beta signaling in liver and gastrointestinal cancers. Cancer Lett 379(2):166–172
164. Rao S, Zaidi S, Banerjee J et al (2017) Transforming growth factor-beta in liver cancer stem cells and regeneration. Hepatol Commun 1(6):477–493
165. Kiryu H, Terai G, Imamura O et al (2011) A detailed investigation of accessibilities around target sites of siRNAs and miRNAs. Bioinformatics 27(13):1788–1797
166. Samuel P, Pink RC, Brooks SA et al (2016) miRNAs and ovarian cancer: a miRiad of mechanisms to induce cisplatin drug resistance. Expert Rev Anticancer Ther 16(1):57–70
167. Scimeca JC, Verron E (2017) The multiple therapeutic applications of miRNAs for bone regenerative medicine. Drug Discov Today 22(7):1084–1091
168. Chen Y, Li C, Tan C et al (2017) Circular RNA in human disease and their potential clinic significance. Zhonghua Yi Xue Yi Chuan Xue Za Zhi 34(1):133–137
169. Park WS, Miyano-Kurosaki N, Abe T et al (1999) Properties of circular dumbbell RNA/DNA chimeric

oligonucleotides containing antisense phosphodiester oligonucleotides. Nucleic Acids Symp Ser doi 42:225–226

170. Yamakawa H, Abe T, Saito T et al (1998) Properties of nicked and circular dumbbell RNA/DNA chimeric oligonucleotides containing antisense phosphodiester oligodeoxynucleotides. Bioorg Med Chem 6(7):1025–1032
171. Zuo H, Suzuki S, Sotoda M et al (2006) New technique for visualizing cerebral vessels in MR angiographic images using three-dimensional discrete wavelet transform. Igaku Butsuri 26(2):65–74
172. Chen LL (2016) The biogenesis and emerging roles of circular RNAs. Nat Rev Mol Cell Biol 17(4):205–211
173. Greene J, Baird AM, Brady L et al (2017) Circular RNAs: biogenesis, function and role in human diseases. Front Mol Biosci 4:38
174. Qin M, Liu G, Huo X et al (2016) Hsa_circ_0001649: a circular RNA and potential novel biomarker for hepatocellular carcinoma. Cancer Biomark 16(1):161–169
175. Bao C, Lyu D, Huang S (2016) Circular RNA expands its territory. Mol Cell Oncol 3(2):e1084443

Circular RNAs in Metabolic Diseases

22

Tianhui Wang, Wen Pan, Jun Hu,
Zhongrong Zhang, Guoping Li, and Yajun Liang

Abstract

Metabolic diseases include diabetes mellitus (DM), obesity, metabolic syndrome, and non-alcoholic fatty liver disease (NAFLD). Circular RNA is a new type of RNA that is different from traditional linear RNA and has a closed loop structure. However, the function of circular RNA is not yet well elucidated in metabolic diseases. Only a few studies have reported about the relationship between circular RNA and metabolic diseases such as DM and NAFLD. This chapter presents a brief review of epidemiology, pathophysiology, or treatment of DM and NAFLD and then discusses the relationship between circular RNA and DM or NAFLD. Besides, this chapter further provides an updated discussion of the most relevant discoveries regarding circular RNA and their potential applications in molecular diagnostics, nucleic acid therapy, and biomarkers.

Keywords

Circular RNA · Metabolic diseases

1 Introduction

Since the first discovery of circular RNAs in plants in 1976 [1], thousands of circular RNAs have been demonstrated to express in eukaryotic cells [2–5]. Generally, the biological functions of circular RNA mainly include the following six aspects: (1) microRNA sponge, (2) RBP sponge, (3) regulator of transcription, (4) interaction with long non-coding RNAs, (5) interaction with mRNAs, and (6) secreted into exosomes. To date, the function of circular RNAs serving as

Author contributed equally with all other contributors. Tianhui Wang, Wen Pan and Jun Hu

T. Wang · Y. Liang (✉)
Cardiac Regeneration and Ageing Lab, Institute of Cardiovascular Sciences, School of Life Science, Shanghai University, Shanghai, China

Shanghai Key Laboratory of Bio-Energy Crops, School of Life Sciences, Shanghai University, Shanghai, China

W. Pan · Z. Zhang
Cardiac Regeneration and Ageing Lab, Institute of Cardiovascular Sciences, School of Life Science, Shanghai University, Shanghai, China

J. Hu
Department of Pediatric Surgery, The First Affiliated Hospital of Wenzhou Medical University, Wenzhou, China

G. Li
Cardiovascular Division of the Massachusetts General Hospital, Harvard Medical School, Boston, MA, USA

J. Xiao (ed.), *Circular RNAs*, Advances in Experimental Medicine and Biology 1087,
https://doi.org/10.1007/978-981-13-1426-1_22

Table 22.1 Circular RNAs in metabolic diseases

Circular RNA	Biological process	Reference
Diabetes mellitus		
ciRS-7/CDR1as	Sponge for miR-7 and involved in the regulation of insulin secretion	[39]
circRNA HIPK3	Enhance retinal vascular disorders via sponging miR-30a-3p Sponge for miR-7 and involved in the regulation of insulin secretion	[6, 39]
Cdr1as	Increases the insulin secretion via targeting miR-7 in mouse β cells	[40]
hsa_circ_0054633	Capable of predicting prediabetes in peripheral blood	[57]
hsa_circRNA11783-2	Differentially expressed in the peripheral blood of diabetes mellitus patients or control individuals	[58]
circRNA_010567	circRNA_010567/miR-141/TGF-β1 axis plays an important regulatory role in the diabetic mice myocardial fibrosis model	[37]
circRNA_000203	Eliminate the anti-fibrosis effect of miR-26b-5 in the diabetic mice myocardial fibrosis model	[59]
circ_0005015	Facilitate retinal endothelial angiogenic function via sponging miR-519d-3p in retinal endothelial cell	[61]
circRNA WDR77	Regulate proliferation and migration via circRNA WDR77-miR-124-FGF2 axis in high-glucose-induced VSMCs	[62]
Non-alcoholic fatty liver disease		
circRNA_021412	Lead to hepatic steatosis via circRNA_021412/miR-1972/LPIN1 axis in HepG2 cells induced by high-fat mixture	[85]
circRNA_0046366	Lead to hepatocyte steatosis via circRNA_0046366 / miR-34a/PPARα axis in HepG2 cells induced by high-fat mixture	[89]
circRNA_0046367	Facilitate steatosis resolution via circRNA_0046367/miR-34a/PPARα axis in HepG2 cells induced by high-fat mixture	[90]

microRNA sponge is well elucidated in accumulating studies [6–9]. Metabolic diseases, including diabetes mellitus, obesity, metabolic syndrome, and non-alcoholic fatty liver disease (NAFLD), are commonly caused by metabolic homeostasis imbalance such as glucose metabolism, lipid metabolism, insulin signaling, or metabolism-related genes dysregulation [10–12]. Although numerous studies have suggested that microRNAs played vital roles in metabolic diseases [13], the potential roles of circular RNAs in metabolic diseases are less mentioned. Given the close relationships between circular RNAs and microRNAs, researchers need to pay more attention on the functions of circular RNAs in metabolism diseases. In this chapter, we will introduce the functions of microRNAs and circular RNAs in metabolic diseases. And then we will discuss the potential applications of circular RNAs in molecular diagnostics, nucleic acid therapy, and biomarkers (Table 22.1).

2 Diabetes Mellitus (DM)

2.1 The Epidemiology and Pathophysiology of Diabetes Mellitus

2.1.1 Diabetes Mellitus Epidemiology

Recently, diabetes mellitus has emerged as a major threat to human health worldwide. 425 million adults were diagnosed with diabetes all over the world, with an estimated increase to 629 million by 2045. And there are 352 million people who have the risk of developing type 2 diabetes mellitus (T2DM) [14]. The International Diabetes Federation (IDF) has predicted the prevalence of diabetes mellitus in many countries. It is reported that the morbidity and prevalence of diabetes mellitus vary in different geographical regions [14, 15]. The burden of diabetes has grown faster in low-income and middle-income countries than in high-income countries.

About 79% of the people who were diagnosed with diabetes live in low- or middle-income countries [16].

2.1.2 Diabetes Mellitus Etiology, Pathophysiology, and Diagnosis

It is reported that type 1 diabetes mellitus (T1DM) is caused by the absolute lack of insulin which is due to autoimmune-mediated destruction of pancreatic β cells. Ninety percent of diabetes are T2DM which is characterized by relative insulin deficiencies and abnormal hyperglycemia [17, 18]. Long-term dysregulation of metabolism-related substances will lead to damages to organs, including the heart, brain, kidney, liver, etc. [19–21]. The World Health Organization defines the clinical diagnosis of diabetes mellitus based on the clinical symptoms and values of plasma glucose [22].

However, the diagnosis does not provide any clues to the causes or progression of the disease. In addition, almost half of the cases have not been diagnosed, leading to increased risk of organ damages through epigenetic mechanisms. Therefore, identifying novel biomarkers that can indicate the early stages of disease is highly needed.

People who are chronically suffering diabetes mellitus are at high risk of developing severe microvascular or macro-vascular complications [23–25]. If patients are not diagnosed early or treated appropriately, these complications will eventually develop into cardiovascular disease (diabetic cardiomyopathy), kidney disease (diabetic nephropathy), diabetic eye disease (diabetic retinopathy), or diabetic foot [26–28].

Interestingly, some studies proposed a β cell-centric model to clarify that all persistent dysglycemia shares a common feature: the damaged or abnormal β cells [29, 30]. There were several mechanisms that mediated the abnormal glucose metabolism in β cells, such as insulin resistance, insulin signal dysregulation, inflammation dysregulation, and immunity dysregulation [29, 31]. Due to the heterogeneity and complexity of diabetic complications, it is needed to diagnose the disease elaborately and accurately. Studies suggested that new therapeutic strategies should focus on ameliorating the function of β cells [31].

2.2 MicroRNAs and Diabetes Mellitus

Numerous studies have demonstrated the functions of non-coding RNAs in diabetes mellitus. Non-coding RNAs, especially microRNAs, have been reported to play important roles in diabetes mellitus pathophysiology. For example, miR-375 is specifically expressed in pancreas cells and affects insulin secretion [32]. MiR-124a modulates insulin secretion by regulating the expression of Rab27a in pancreatic β cells [33]. Let-7 family members regulate glucose homeostasis and insulin sensitivity by directly targeting several components of insulin-signaling pathway, such as Insr, Igf1r, Pik3ip1, and Irs2 [34]. In addition to the above microRNAs, other microRNAs, including miR-29 family [35], miR-107 [36], miR-33 family [37], Let-7 family [38], miR-221/222 family [39], and miR-223 family [40], are also reported to affect β-cell proliferation, insulin secretion, and exocytosis.

2.3 Circular RNAs and Diabetes Mellitus

2.3.1 Circular RNAs and Diseases

It has been investigated that circular RNAs play an important role of being "super sponge" in other diseases. Circular RNAs-microRNAs-mRNAs interaction network widely exists in multiple pathophysiology processes. For instance, defects in ciRS-7-mediated "sponging events" lead to the dysregulation of ciRS-7-miRNA-7-UBE2A circuit in neocortex and hippocampal CA1, which eventually results in sporadic Alzheimer's disease [6]. In colorectal cancer, hsa_circ_001569, acting as a positive regulator of cell proliferation and invasion, is identified as a sponge of miR-145 [41]. Circular RNA CER affects cartilage-related extracellular matrix degradation by sponging miR-136 in chondrocyte [7]. In vivo and in vitro experiments demonstrate that circMTO1 sponges microRNA-9 to modulate p21 expression and inhibit hepatocellular carcinoma progression [42]. It is reported that circRNA_010567/miR-141/TGF-β1 axis regulates myocardial fibrosis in diabetic mice [43]. On the whole, circular RNAs, as newly discovered non-coding RNAs, are involved in multiple pathophysiological processes of diseases.

2.3.2 Circular RNAs and Pancreatic β Cells

Pancreatic β cells are the only source of insulin, and the secreted insulin is essential for maintaining blood glucose homeostasis. Insulin secretion defection will lead to chronic hyperglycemia and then the development of diabetes [44]. The functions of circular RNAs in β cells are studied by several researchers. CiRS-7 and circHIPK3 are reported to be involved in β-cell function regulation and the development of diabetes [45]. Another study reveals that circular RNA Cdr1as is able to improve insulin secretion by targeting miR-7, Pax6, and Myrip in mouse β cells. The results indicate that Cdr1as might be a potential therapeutic target in diabetes [46].

2.3.3 Circular RNAs as Potential Biomarkers in Diabetes Mellitus

Biomarkers are defined as biological molecules that are the hallmarks of normal or abnormal biological process, or of healthy condition or disease [47]. Circulating circular RNAs derived from blood can indicate the physiological changes of the whole body. Biomarkers function in the following four aspects: (1) identify people at risk of developing diseases, (2) diagnose diabetes or other metabolic diseases, (3) predict the development of complications, and (4) monitor the response to treatments [48]. As we all know, the dynamic interplay of circular RNAs, microRNAs, and long non-coding RNAs is needed to regulate cellular homeostasis [49]. Numerous studies focus on developing microRNAs and long non-coding RNAs as diagnosis or prognosis biomarkers of diseases. For instance, circulating miR-203 associates with poor survival and metastasis in patients with colorectal carcinoma [50]. In a cohort of 1112 patients with CAD, Cox regression analyses suggest that miR-132, miR-140-3p, and miR-210 are correlated with cardiovascular death events [51]. Kaplan-Meier analysis and Cox proportional hazards model reveal that miR-425-5p is an independent prognostic factor for cervical cancer [52].

Circular RNAs have the potential to be disease biomarkers due to the following characteristics: (1) circular RNAs are covalently closed RNAs which are resistant to nucleases; (2) circular RNAs are relatively abundant, stable, and conserved in the blood [2]; (3) circular RNAs are enriched and stable in exosomes [53]; (4) circular RNAs can be detected not only in tumor tissues but also in the blood,cerebral spinal fluid, and saliva [54, 55]; and (s5) compared with linear RNAs, the half-lives of circular RNAs are much longer [56]. Recent studies have demonstrated that circular RNAs could be utilized as potential biomarkers for the diagnosis or prognosis of diseases, such as cancers [57–60], coronary artery disease [61], and central nerve system disease [62]. In addition to the above studies, it is reported that 489 circular RNAs were found to be differentially expressed in the peripheral blood of patients with T2DM. Moreover, hsa_circ_0054633 is reported to be a promising diagnostic biomarker for prediabetes and T2DM [63]. In another study, real-time polymerase chain reaction is used to explore the differentially expressed circular RNAs in the peripheral blood from diabetes mellitus patients and control individuals. They verify that the expression of hsa-circRNA11783-2 correlates with T2DM [64]. However, most patients are asymptomatic in the early stages of T2DM, which makes it difficult to diagnose the disease. A convenient, specific, and sensitive diagnostic method is urgently needed to facilitate the early diagnosis of T2DM.

2.3.4 Circular RNAs and Diabetes Mellitus-Related Complications

The potentials of circular RNAs as biomarkers in predicting diabetes mellitus complications remain in further investigations. Several studies are devoted to identify the relationships between diabetes-related complications and circular RNAs.

Zhou et al. report that circRNA_010567 is upregulated in diabetic mice myocardium. The knockdown of circRNA_010567 suppresses myocardium fibrosis. Further studies demonstrate that circRNA_010567/miR-141/TGF-β1 axis plays an important role in diabetic mice myocardium [43]. Tang et al. discover that circRNA_000203 is upregulated in the diabetic mouse myocardium. Further studies demonstrate that circRNA_000203 can specifically sponge miR-26b-5p, and the overexpression of circRNA_000203 eliminates the anti-fibrosis effects of miR-26b-5p [65].

Diabetic vascular complications are the main cause of disability and mortality in diabetic patients [66]. These complications include diabetic cardiomyopathy, diabetic nephropathy, diabetic retinopathy, diabetic cardiomyopathy, and diabetic foot. Diabetic retinopathy (DR) is a frequent diabetic vascular complication. In a recent study, Zhang et al.. use circular RNA microarrays to identify the differential expression profiles of diabetic retinas and nondiabetic retinas. They find that circ_0005015 is significantly upregulated in the diabetic retina. Functional analysis reveals that circ_0005015 regulates retinal endothelial cell proliferation, migration, and tube formation by sponging miR-519d-3p [67]. Another study suggests that circHIPK3 is a potential therapeutic target for alleviating diabetes-related retinopathy. In mechanism, gain-of-function and loss-of-function assays reveal that circHIPK3 affect diabetic retinopathy by sponging miR-30a-3p and regulating downstream genes VEGFC, FZD4, and WNT2 [8]. In addition, researchers identify that circWDR77-miR-124-FGF2 regulatory pathway plays a role in VSMC proliferation and migration. It provides the theoretical basis for diabetes mellitus-related vasculopathy treatment [68]. In conclusion, more efforts are needed to study the functions of circular RNAs in diabetes-related complications.

3 Non-alcoholic Fatty Liver Disease (NAFLD)

3.1 NAFLD Epidemiology

Non-alcoholic fatty liver disease (NAFLD) has become the most common cause for chronic liver disease, cirrhosis, and hepatocellular carcinoma worldwide [69]. Up to 20% of NAFLD have non-alcoholic steatohepatitis (NASH) with active hepatic necrotizing inflammation and injury, which lead to progressive hepatic fibrosis and complications. It is now believed that NAFLD is the representative liver metabolic syndrome (MS) and affects liver metabolic homeostasis [70]. NAFLD increases liver-related morbidity, liver-related mortality, and the risk of other metabolic complications such as type 2 diabetes [71], obesity [72], and metabolic syndrome [73]. The global prevalence rate of NAFLD in different regions is as follows: Middle East, 32%; the United States, 30%; South America, 30%; Asia, 27%; Europe, 24%; and Africa, 13% [74–76]. 18–33% of NAFLD cases are found to have type 2 diabetes (T2DM), and 66–83% of NAFLD cases are identified with insulin resistance (IR).

3.2 NAFLD Pathogenesis

NAFLD affects many liver-related symptoms such as asymptomatic steatosis, steatohepatitis, fibrosis, cirrhosis, and hepatocellular carcinoma [77]. The pathogenesis of NAFLD is still controversial: according to traditional double-hit theory, the first hit of NAFLD is the abnormal accumulation of triglycerides in liver cells. The second hit is that oxidative stress and inflammatory mediators promote liver steatosis to steatohepatitis, fibrosis, and cirrhosis [75, 78]. However, recent findings suggest that "multiple parallel hits" hypothesis is more reasonable to NAFLD pathogenesis. Multiple pathogenic factors, such as insulin resistance, oxidative stress, mitochondrial dysfunction, lipotoxicity, endoplasmic reticulum stress, adipose tissue dysfunction, altered innate immune regulation, and cytokine secretion, contribute to liver injury [79]. Therefore, the pathogenesis of NAFLD is complex and heterogeneous. The underlying mechanisms of NAFLD are still needed to be elucidated. In addition, the early diagnosis of non-alcoholic steatohepatitis may prevent disease progression and improve patient outcomes [80].

3.3 MicroRNAs and NAFLD

Recently, researchers study the potential roles of microRNAs in NAFLD pathogenesis. It has recently been reported that microRNAs serve as key regulators in NAFLD. For instance, liver-specific miR-122 knockout mice display abnormal triglyceride accumulation in the liver due to the increased triglyceride synthesis and reduced triglyceride secretion [81]. Compared with healthy people, NASH patients have elevated expression of miR-122 [82]. MiR-34a regulates VLDL metabolism, fatty acid β oxidation, cholesterol synthesis, and liver injury by targeting several key transcription factors such as HNF4α, PPARα, Sirt1, and p53 [83, 84]. MiR-155 plays a protective role for NAFLD by targeting LXRα-SREBP-1c pathway [85].

3.4 NAFLD Treatment

Non-alcoholic steatohepatitis is a more aggressive form of NAFLD and is also considered as a first hit to liver. Currently, no Food and Drug Administration (FDA)-approved medicines are applied in the treatment of non-alcoholic steatohepatitis. NAFLD-related medicine development focuses on the following fields: medicines that regulate nuclear transcription factors, medicines that target lipotoxicity and oxidative stress, and medicines that regulate cellular energy homeostasis [86, 87].

In a double-blind clinical study, NGM282, an antidiabetic drug, is reported to reduce hepatic fat contents in patients with non-alcoholic steatohepatitis [88]. In another clinical trial, subcutaneous injection of liraglutide which is a glucagon-like peptide-1 (GLP-1) analogue leads to histological improvement of non-alcoholic steatohepatitis [89]. Given the complexity and heterogeneity of NAFLD, more efforts should be expended on the drug development.

3.5 Circular RNAs and NAFLD

Insulin sensitizers, such as metformin and thiazolidinedione are commonly used for those NAFLD patients who are resistant to insulin. Researchers detect the circular RNA expression changes in metformin-treated or high-fat diet mice livers. They find that the altered circular RNAs by metformin treatment can modulate NAFLD-related pathways through interacting with downstream microRNAs [90]. NAFLD starts with hepatic steatosis symptom and then progresses to steatohepatitis. 5–20% steatosis patients will develop non-alcoholic steatohepatitis (NASH). 10–20% of NASH patients will develop higher-grade fibrosis, and less than 5% of them will progress to full-blown cirrhosis. Therefore, numerous studies focus on uncovering the underlying mechanisms of NAFLD progression.

A recent study reports that 357 circular RNAs are associated with hepatic steatosis. Further study demonstrates that the circRNA_021412-miR-1972-LPIN1 signaling plays an important role in liver metabolism [91]. Peroxisome proliferator-activated receptors (PPARs) are reported as fatty acid regulators which control lipid metabolism and inflammation [92]. PPARα signaling pathway is widely involved in various physiological processes, including glucose metabolism; lipid metabolism; inflammation; cell proliferation, differentiation, and apoptosis; and aging [93]. It is reported that PPARα signaling pathway is inhibited by PPAR1 in NAFLD patients. They suggest that the restoration of PPARα signaling is important for NAFLD treatment [94].

In HepG2 cells, circRNA_0046366 is identified as antagonist of miR-34a. The inhibition of circRNA_0046366 to miR-34a restores the expression of PPARα, which suppresses hepatocytes steatosis. These findings suggest that circRNA_0046366-miR-34a-PPARα pathway plays an important role in hepatocyte steatosis regulation [95]. Another study reveals that circRNA_0046367 abolishes the inhibitory effects of miR-34a on PPARα, which improves the transcriptional activation lipid metabolism associated genes and steatosis [96].

4 Conclusion and Perspective

4.1 Conclusion

In brief, circular RNAs play important roles in the regulation of metabolic diseases. In particular, circular RNAs are involved in the development of diabetes and are closely correlated with the secretion of insulin by β cells. Moreover, circular RNAs take part in the development of diabetes-related complications such as diabetic retinopathy. Circulating circular RNAs can be used as biomarkers for early diabetes assessment.

On the other hand, circular RNAs are involved in NAFLD pathogenesis. The current research mainly focuses on the regulation of circular RNAs to NASH. Circular RNAs regulate NASH pathogenesis mainly through circular RNA/microRNA/mRNA axis. At present, NAFLD is still lack of effective drug therapies. In-depth study of the molecular mechanisms in NAFLD is highly needed.

Circular RNAs are evolutionarily conserved and resistant to nuclease digestion. The half-life of circular RNAs is more than 48 h which makes it more stable than other non-coding RNAs. Based on the above characteristics, circular RNAs are proposed to be the potential diagnostic markers and therapeutic targets in various diseases. Nowadays, only several groups focus on the studies of circular RNAs in metabolic diseases. The functions of circular RNAs in metabolism disease pathogenesis remain to be elucidated [6].

4.2 Perspective

Circular RNAs, acting as microRNA sponges, RBP sponges, or regulators of transcription, are involved in the progression of various diseases. The fact that circular RNAs are detected in saliva [27], exosomes [83], and the blood [86] encourages researchers to develop circular RNAs as diagnostic biomarkers for metabolic diseases. Further studies demonstrate that circular RNAs biomarkers can also be detected in urine and cerebrospinal fluid. Although a growing number of studies focus on the functions of circular RNAs, several issues should be elucidated such as the biosynthesis of circular RNAs, the metabolisms of circular RNAs, and the regulation of circular RNAs to downstream genes.

It is reported that circular RNAs are enriched in exosomes. Exosomes are filled with nucleic acids such as messenger RNAs, microRNAs, circular RNAs, and long non-coding RNAs [53, 97, 98]. They can be detected in serum and serve as mediators for cell or organ communication. The interplays between circular RNAs and exosomes are manifested in the following aspects: (1) exosomes transport circular RNAs from one metabolic organ to another, (2) circular RNAs in exosomes can be a biomarker for some metabolic diseases, and (3) exosomes may help to eliminate circular RNAs. Until now, the researches on circular RNAs still remain in laboratory studies. There is still a long way to go before we apply circular RNAs as diagnostic and prognostic biomarkers.

Nucleic acid treatments were first proposed 30 years ago. RNA interference technology represents a revolutionary breakthrough in the pharmaceutical field and has become a frontier area for drug development [99]. Nucleic acid drugs have many advantages compared with traditional medicines, such as nucleic acid drugs do not produce exogenous proteins and they are safer than conventional drugs. To date, microRNAs are used to target downstream genes in two forms, namely, microRNA antagonists and microRNA mimics [100]. For hepatitis C treatment, miR-122 antagonists have entered phase 1B clinical trials [5, 101]. And for treatment of patients with advanced solid tumors, MRX34, a liposomal miR-34a mimic, have entered phase I clinical trials [102]. Until now, no therapeutic circular RNAs have been produced or approved. However, with the development of high-throughput sequencing, increasing circular RNAs will be discovered and the circular RNA mysteries will finally be uncovered. In the future, circular RNAs will be used in clinical treatment. Based on their important biological functions, circular RNAs are proposed to be the most promising field in nucleic acid therapy.

Competing Financial Interests The authors declare no competing financial interests.

References

1. Sanger HL, Klotz G, Riesner D et al (1976) Viroids are single-stranded covalently closed circular RNA molecules existing as highly base-paired rod-like structures. Proc Natl Acad Sci USA 73(11):3852–3856
2. Jeck WR, Sorrentino JA, Wang K et al (2013) Circular RNAs are abundant, conserved, and associated with ALU repeats. RNA 19(2):141–157
3. Memczak S, Jens M, Elefsinioti A et al (2013) Circular RNAs are a large class of animal RNAs with regulatory potency. Nature 495(7441):333–338
4. Afonina ZA, Myasnikov AG, Shirokov VA et al (2014) Formation of circular polyribosomes on eukaryotic mRNA without cap-structure and poly(A)-tail: a cryo electron tomography study. Nucleic Acids Res 42(14):9461–9469
5. Jopling CL, Schutz S, Sarnow P (2008) Position-dependent function for a tandem microRNA miR-122-binding site located in the hepatitis C virus RNA genome. Cell Host Microbe 4(1):77–85
6. Zhao Y, Alexandrov PN, Jaber V et al (2016) Deficiency in the ubiquitin conjugating enzyme UBE2A in Alzheimer's Disease (AD) is linked to deficits in a natural circular miRNA-7 sponge (circRNA; ciRS-7). Genes (Basel) 7(12)
7. Liu Q, Zhang X, Hu X et al (2016) Circular RNA related to the chondrocyte ECM regulates MMP13 expression by functioning as a MiR-136 'Sponge' in human cartilage degradation. Sci Rep 6:22572
8. Shan K, Liu C, Liu BH et al (2017) Circular noncoding RNA HIPK3 mediates retinal vascular dysfunction in diabetes mellitus. Circulation 136(17):1629–1642
9. Wang K, Long B, Liu F et al (2016) A circular RNA protects the heart from pathological hypertrophy and heart failure by targeting miR-223. Eur Heart J 37(33):2602–2611
10. Rottiers V, Naar AM (2012) MicroRNAs in metabolism and metabolic disorders. Nat Rev Mol Cell Biol 13(4):239–250
11. Perry RJ, Peng L, Barry NA et al (2016) Acetate mediates a microbiome-brain-beta-cell axis to promote metabolic syndrome. Nature 534(7606):213–217
12. Afdhal NH (2012) Management of nonalcoholic fatty liver disease: a 60-year-old man with probable nonalcoholic fatty liver disease: weight reduction, liver biopsy, or both? JAMA 308(6):608–616
13. Fernandez-Hernando C, Ramirez CM, Goedeke L et al (2013) MicroRNAs in metabolic disease. Arterioscler Thromb Vasc Biol 33(2):178–185
14. Cho NH, Shaw JE, Karuranga S et al (2018) IDF diabetes Atlas: global estimates of diabetes prevalence for 2017 and projections for 2045. Diabetes Res Clin Pract 138:271–281
15. Worldwide trends in diabetes since 1980: a pooled analysis of 751 population-based studies with 4·4 million participants (2016). The Lancet 387(10027):1513–1530.
16. Benjamin EJ, Virani SS, Callaway CW et al (2018) Heart disease and stroke statistics-2018 update: a report from the American Heart Association. Circulation 137:e67–e492
17. Chatterjee S, Khunti K, Davies MJ (2017) Type 2 diabetes. The Lancet 389(10085):2239–2251
18. Koekkoek PS, Kappelle LJ, van den Berg E et al (2015) Cognitive function in patients with diabetes mellitus: guidance for daily care. Lancet Neurol 14(3):329–340
19. Collier A, Ghosh S, Hair M et al (2015) Impact of socioeconomic status and gender on glycaemic control, cardiovascular risk factors and diabetes complications in type 1 and 2 diabetes: a population based analysis from a Scottish region. Diabetes Metab 41(2):145–151
20. Augustin K, Khabbush A, Williams S et al (2018) Mechanisms of action for the medium-chain triglyceride ketogenic diet in neurological and metabolic disorders. Lancet Neurol 17(1):84–93
21. Kaczorowska M, Ryterska K, Ossowski P et al (2016) Metabolic risk factors of coronary heart disease in relation to anthropometric measures in nonalcoholic fatty liver disease patients following dietary intervention. Pomeranian J Life Sci 62(2):8–14
22. Organization WH Definition, Diagnosis and classification of diabetes mellitus and its complications part 1: diagnosis and classification of diabetes mellitus provisional report of a WHO consultation
23. Ghaemi N, Hasanabadi H, Ashrafzadeh F et al (2018) Peripheral neuropathy in children and adolescents with insulin-dependent diabetes Mellitus. Iran J Child Neurol 12(2):83–90

24. Packer M (2018) Heart failure: the most important, preventable, and treatable cardiovascular complication of type 2 diabetes. Diabetes Care 41(1):11–13
25. Shanbhogue VV, Hansen S, Frost M et al (2017) Bone disease in diabetes: another manifestation of microvascular disease? Lancet Diabetes Endocrinol 5(10):827–838
26. Kahn SE, Cooper ME, Del Prato S (2014) Pathophysiology and treatment of type 2 diabetes: perspectives on the past, present, and future. Lancet 383(9922):1068–1083
27. Kobayashi M, Zochodne DW (2018) Diabetic neuropathy and the sensory neuron: new aspects of pathogenesis and their treatment implications. J Diabetes Investig.
28. Kanwar YS, Sun L, Xie P et al (2011) A glimpse of various pathogenetic mechanisms of diabetic nephropathy. Annu Rev Pathol 6:395–423
29. Schwartz SS, Epstein S, Corkey BE et al (2016) The time is right for a new classification system for diabetes: rationale and Implications of the beta-cell-centric classification schema. Diabetes Care 39(2):179–186
30. Ashcroft FM, Rorsman P (2012) Diabetes mellitus and the beta cell: the last ten years. Cell 148(6):1160–1171
31. Schwartz SS, Epstein S, Corkey BE et al (2017) A unified pathophysiological construct of diabetes and its complications. Trends Endocrinol Metab 28(9):645–655
32. Poy MN, Eliasson L, Krutzfeldt J et al (2004) A pancreatic islet-specific microRNA regulates insulin secretion. Nature 432(7014):226–230
33. Lovis P, Gattesco S, Regazzi R (2008) Regulation of the expression of components of the exocytotic machinery of insulin-secreting cells by microRNAs. Biol Chem 389(3):305–312
34. Frost RJA, Olson EN (2011) Control of glucose homeostasis and insulin sensitivity by the Let-7 family of microRNAs. Proc. Natl Acad Sci 108(52):21075–21080
35. Pullen TJ, da Silva Xavier G, Kelsey G et al (2011) miR-29a and miR-29b contribute to pancreatic beta-cell-specific silencing of monocarboxylate transporter 1 (Mct1). Mol Cell Biol 31(15):3182–3194
36. Trajkovski M, Hausser J, Soutschek J et al (2011) MicroRNAs 103 and 107 regulate insulin sensitivity. Nature 474(7353):649–653
37. Davalos A, Goedeke L, Smibert P et al (2011) miR-33a/b contribute to the regulation of fatty acid metabolism and insulin signaling. Proc Natl Acad Sci USA 108(22):9232–9237
38. Zhang J, Zhang L, Fan R et al (2013) The polymorphism in the let-7 targeted region of the Lin28 gene is associated with increased risk of type 2 diabetes mellitus. Mol Cell Endocrinol 375(1–2):53–57
39. Lightell DJ Jr, Moss SC, Woods TC (2018) Upregulation of miR-221 and -222 in response to increased extracellular signal-regulated kinases 1/2 activity exacerbates neointimal hyperplasia in diabetes mellitus. Atherosclerosis 269:71–78
40. Long Y, Zhan Q, Yuan M et al (2017) The expression of microRNA-223 and FAM5C in cerebral infarction patients with diabetes mellitus. Cardiovasc Toxicol 17(1):42–48
41. Xie H, Ren X, Xin S et al (2016) Emerging roles of circRNA_001569 targeting miR-145 in the proliferation and invasion of colorectal cancer. Oncotarget 7(18):26680–26691
42. Han D, Li J, Wang H et al (2017) Circular RNA circMTO1 acts as the sponge of microRNA-9 to suppress hepatocellular carcinoma progression. Hepatology 66(4):1151–1164
43. Zhou B, Yu JW (2017) A novel identified circular RNA, circRNA_010567, promotes myocardial fibrosis via suppressing miR-141 by targeting TGF-beta1. Biochem Biophys Res Commun 487(4):769–775
44. Heit JJ, Karnik SK, Kim SK (2006) Intrinsic regulators of pancreatic beta-cell proliferation. Annu Rev Cell Dev Biol 22:311–338
45. Stoll L, Sobel J, Rodriguez-Trejo A et al (2018) Circular RNAs as novel regulators of beta-cell functions in normal and disease conditions. Mol Metab 9:69–83
46. Xu H, Guo S, Li W et al (2015) The circular RNA Cdr1as, via miR-7 and its targets, regulates insulin transcription and secretion in islet cells. Sci Rep 5:12453
47. Henry NL, Hayes DF (2012) Cancer biomarkers. Mol Oncol 6(2):140–146
48. Parrizas M, Novials A (2016) Circulating microRNAs as biomarkers for metabolic disease. Best Pract Res Clin Endocrinol Metab 30(5):591–601
49. Guil S, Esteller M (2015) RNA-RNA interactions in gene regulation: the coding and noncoding players. Trends Biochem Sci 40(5):248–256
50. Hur K, Toiyama Y, Okugawa Y et al (2017) Circulating microRNA-203 predicts prognosis and metastasis in human colorectal cancer. Gut 66(4):654–665
51. Karakas M, Schulte C, Appelbaum S et al (2017) Circulating microRNAs strongly predict cardiovascular death in patients with coronary artery disease-results from the large AtheroGene study. Eur Heart J 38(7):516–523
52. Sun L, Jiang R, Li J et al (2017) MicoRNA-425-5p is a potential prognostic biomarker for cervical cancer. Ann Clin Biochem 54(1):127–133
53. Li Y, Zheng Q, Bao C et al (2015) Circular RNA is enriched and stable in exosomes: a promising biomarker for cancer diagnosis. Cell Res 25(8):981–984
54. Bonizzato A, Gaffo E, Te Kronnie G et al (2016) CircRNAs in hematopoiesis and hematological malignancies. Blood Cancer J 6(10):e483
55. Bahn JH, Zhang Q, Li F et al (2015) The landscape of microRNA, Piwi-interacting RNA, and circular RNA in human saliva. Clin Chem 61(1):221–230
56. Enuka Y, Lauriola M, Feldman ME et al (2016) Circular RNAs are long-lived and display only mini-

mal early alterations in response to a growth factor. Nucleic Acids Res 44(3):1370–1383
57. Yang F, Liu D-Y, Guo J-T et al (2017) Circular RNA circ-LDLRAD3 as a biomarker in diagnosis of pancreatic cancer. World J Gastroenterol 23(47):8345–8354
58. Chen S, Li T, Zhao Q et al (2017) Using circular RNA hsa_circ_0000190 as a new biomarker in the diagnosis of gastric cancer. Clin Chim Acta 466:167–171
59. Xu L, Zhang M, Zheng X et al (2017) The circular RNA ciRS-7 (Cdr1as) acts as a risk factor of hepatic microvascular invasion in hepatocellular carcinoma. J Cancer Res Clin Oncol 143(1):17–27
60. Li P, Chen S, Chen H et al (2015) Using circular RNA as a novel type of biomarker in the screening of gastric cancer. Clin Chim Acta 444:132–136
61. Zhao Z, Li X, Gao C et al (2017) Peripheral blood circular RNA hsa_circ_0124644 can be used as a diagnostic biomarker of coronary artery disease. Sci Rep 7:39918
62. Westholm JO, Miura P, Olson S et al (2014) Genome-wide analysis of drosophila circular RNAs reveals their structural and sequence properties and age-dependent neural accumulation. Cell Rep 9(5):1966–1980
63. Zhao Z, Li X, Jian D et al (2017) Hsa_circ_0054633 in peripheral blood can be used as a diagnostic biomarker of pre-diabetes and type 2 diabetes mellitus. Acta Diabetol 54(3):237–245
64. Li X, Zhao Z, Jian D et al (2017) Hsa-circRNA11783-2 in peripheral blood is correlated with coronary artery disease and type 2 diabetes mellitus. Diab Vasc Dis Res 14(6):510–515
65. Tang CM, Zhang M, Huang L et al (2017) CircRNA_000203 enhances the expression of fibrosis-associated genes by derepressing targets of miR-26b-5p, Col1a2 and CTGF, in cardiac fibroblasts. Sci Rep 7:40342
66. Sena CM, Pereira AM, Seica R (2013) Endothelial dysfunction – a major mediator of diabetic vascular disease. Biochim Biophys Acta 1832(12):2216–2231
67. Zhang SJ, Chen X, Li CP et al (2017) Identification and characterization of circular RNAs as a new class of putative biomarkers in diabetes retinopathy. Invest Ophthalmol Vis Sci 58(14):6500–6509
68. Chen J, Cui L, Yuan J et al (2017) Circular RNA WDR77 target FGF-2 to regulate vascular smooth muscle cells proliferation and migration by sponging miR-124. Biochem Biophys Res Commun 494(1-2):126–132
69. Bhala N, Angulo P, van der Poorten D et al (2011) The natural history of nonalcoholic fatty liver disease with advanced fibrosis or cirrhosis: an international collaborative study. Hepatology 54(4):1208–1216
70. Vanni E, Bugianesi E, Kotronen A et al (2010) From the metabolic syndrome to NAFLD or vice versa? Dig Liver Dis 42(5):320–330
71. Portillo-Sanchez P, Bril F, Maximos M et al (2015) High prevalence of nonalcoholic fatty liver disease in atients with type 2 diabetes Mellitus and normal plasma aminotransferase levels. J Clin Endocrinol Metab 100(6):2231–2238
72. Henao-Mejia J, Elinav E, Jin C et al (2012) Inflammasome-mediated dysbiosis regulates progression of NAFLD and obesity. Nature 482(7384):179–185
73. Adams LA, Waters OR, Knuiman MW et al (2009) NAFLD as a risk factor for the development of diabetes and the metabolic syndrome: an eleven-year follow-up study. Am J Gastroenterol 104(4):861–867
74. Pappachan JM, Babu S, Krishnan B et al (2017) Non-alcoholic fatty liver disease: a clinical update. J Clin Transl Hepatol 5(4):384–393
75. Carr RM, Oranu A, Khungar V (2016) Nonalcoholic fatty liver disease: pathophysiology and management. Gastroenterol Clin North Am 45(4):639–652
76. Vernon G, Baranova A, Younossi ZM (2011) Systematic review: the epidemiology and natural history of non-alcoholic fatty liver disease and non-alcoholic steatohepatitis in adults. Aliment Pharmacol Ther 34(3):274–285
77. Angulo P (2002) Nonalcoholic fatty liver disease. N Engl J Med 346(16):1221–1231
78. Farrell GC, Larter CZ (2006) Nonalcoholic fatty liver disease: from steatosis to cirrhosis. Hepatology 43(2 Suppl 1):S99–s112
79. Birkenfeld AL, Shulman GI (2014) Nonalcoholic fatty liver disease, hepatic insulin resistance, and type 2 diabetes. Hepatology 59(2):713–723
80. Rinella ME (2015) Nonalcoholic fatty liver disease: a systematic review. JAMA 313(22):2263–2273
81. Hsu SH, Wang B, Kota J et al (2012) Essential metabolic, anti-inflammatory, and anti-tumorigenic functions of miR-122 in liver. J Clin Invest 122(8):2871–2883
82. Auguet T, Aragones G, Berlanga A et al (2016) miR33a/miR33b* and miR122 as Possible Contributors to Hepatic Lipid Metabolism in Obese Women with Nonalcoholic Fatty Liver Disease. Int J Mol Sci 17(10)
83. Yamada H, Suzuki K, Ichino N et al (2013) Associations between circulating microRNAs (miR-21, miR-34a, miR-122 and miR-451) and non-alcoholic fatty liver. Clin Chim Acta 424:99–103
84. Xu Y, Zalzala M, Xu J et al (2015) A metabolic stress-inducible miR-34a-HNF4alpha pathway regulates lipid and lipoprotein metabolism. Nat Commun 6:7466
85. Miller AM, Gilchrist DS, Nijjar J et al (2013) MiR-155 has a protective role in the development of non-alcoholic hepatosteatosis in mice. PLoS One 8(8):e72324
86. Gastaldelli A (2017) Insulin resistance and reduced metabolic flexibility: cause or consequence of NAFLD? Clin Sci (Lond) 131(22):2701–2704
87. Musso G, Cassader M, Gambino R (2016) Non-alcoholic steatohepatitis: emerging molecular targets and therapeutic strategies. Nat Rev Drug Discov 15(4):249–274

88. Harrison SA, Rinella ME, Abdelmalek MF et al (2018) NGM282 for treatment of non-alcoholic steatohepatitis: a multicentre, randomised, double-blind, placebo-controlled, phase 2 trial. Lancet 391(10126):1174–1185
89. Armstrong MJ, Gaunt P, Aithal GP et al (2016) Liraglutide safety and efficacy in patients with non-alcoholic steatohepatitis (LEAN): a multicentre, double-blind, randomised, placebo-controlled phase 2 study. Lancet 387(10019):679–690
90. Guo J, Zhou Y, Cheng Y et al (2018) Metformin-induced changes of the coding transcriptome and non-coding RNAs in the livers of non-alcoholic fatty liver disease mice. Cell Physiol Biochem 45(4):1487–1505
91. Guo XY, He CX, Wang YQ et al (2017) Circular RNA profiling and bioinformatic modeling identify its regulatory role in hepatic steatosis. Biomed Res Int 2017:5936171
92. Tanaka N, Aoyama T, Kimura S et al (2017) Targeting nuclear receptors for the treatment of fatty liver disease. Pharmacol Ther 179:142–157
93. Samuel VT, Shulman GI (2018) Nonalcoholic fatty liver disease as a Nexus of metabolic and hepatic diseases. Cell Metab 27(1):22–41
94. Huang K, Du M, Tan X et al (2017) PARP1-mediated PPARalpha poly(ADP-ribosyl)ation suppresses fatty acid oxidation in non-alcoholic fatty liver disease. J Hepatol 66(5):962–977
95. Guo XY, Sun F, Chen JN et al (2018) circRNA_0046366 inhibits hepatocellular steatosis by normalization of PPAR signaling. World J Gastroenterol 24(3):323–337
96. Guo XY, Chen JN, Sun F et al (2017) circRNA_0046367 Prevents hepatoxicity of lipid peroxidation: an inhibitory role against hepatic steatosis. Oxid Med Cell Longev 2017:3960197
97. Zhao R, Zhou J, Dong X et al (2018) Circular RNA expression alteration in exosomes from the brain extracellular space after traumatic brain injury in mice. J Neurotrauma
98. Lasda E, Parker R (2016) Circular RNAs co-precipitate with extracellular vesicles: a possible mechanism for circRNA clearance. PLoS One 11(2):e0148407
99. Sridharan K, Gogtay NJ (2016) Therapeutic nucleic acids: current clinical status. Br J Clin Pharmacol 82(3):659–672
100. Bouchie A (2013) First microRNA mimic enters clinic. Nat Biotechnol 31(7):577
101. van der Ree MH, de Vree JM, Stelma F et al (2017) Safety, tolerability, and antiviral effect of RG-101 in patients with chronic hepatitis C: a phase 1B, double-blind, randomised controlled trial. Lancet 389(10070):709–717
102. Beg MS, Brenner AJ, Sachdev J et al (2017) Phase I study of MRX34, a liposomal miR-34a mimic, administered twice weekly in patients with advanced solid tumors. Invest New Drugs 35(2):180–188

Circular RNAs in Vascular Functions and Diseases

23

Shengguang Ding, Yujiao Zhu, Yajun Liang, Haitao Huang, Yiming Xu, and Chongjun Zhong

Abstract

Vascular disease is one of the top five causes of death and affects a variety of other diseases, such as heart, nervous system, and metabolic disorders. Vascular dysfunction is a hallmark of ischemia, cancer, and inflammatory diseases and can accelerate the progression of diseases. Circular RNAs (circRNAs) are a new type of noncoding RNAs with covalent bond ring structure, which have been reported to be abnormally expressed in many human diseases. circRNAs regulate gene expression through the sponging of microRNAs (miRNAs) and can also be used as disease biomarkers. Here we will summarize the functions of circRNAs in vascular diseases, including vascular dysfunction, atherosclerosis, diabetes mellitus-related retinal vascular dysfunction, chronic thromboembolic pulmonary hypertension, carotid atherosclerotic disease, hepatic vascular invasion in hepatocellular carcinoma, aortic aneurysm, coronary artery disease, and type 2 diabetes mellitus.

Keywords

Circular RNAs · Vascular function · Vascular diseases

1 Introduction

Noncommunicable diseases (NCDs) cause a great deal of disabilities and deaths in the world [1]. Among NCDs, vascular diseases rank on the top of the list and are among the top five causes of patient deaths, which affect many other diseases such as the cardiac, nervous, and metabolic diseases [2, 3]. Of the 57 million global deaths, 60% are caused by NCDs. About 17.5 million NCDs deaths are caused by cardiovascular diseases, including 6.7 million cases of stroke [4]. Vascular disorders affect not only cardiac diseases but also nervous system disorders. Hypertension and elevated serum lipids are classical risk factors for stroke, a disorder that affects approximately 15 million individuals every year and leads to an annual cost of 75 billion dollars in the United States [5]. Other NCDs such as diabetes mellitus (DM) can also lead to neurodegenerative diseases. Moreover, DM significantly

Authors Shengguang Ding, Yujiao Zhu and Yajun Liang have been equally contributed to this chapter.

S. Ding · H. Huang · Y. Xu · C. Zhong (✉)
Department of Thoracic and Cardiovascular Surgery, The Second Affiliated Hospital of Nantong University, Nantong, China

Y. Zhu · Y. Liang
Cardiac Regeneration and Ageing Lab, Institute of Cardiovascular Sciences, School of Life Science, Shanghai University, Shanghai, China

Shanghai Key Laboratory of Bio-Energy Crops, School of Life Sciences, Shanghai University, Shanghai, China

J. Xiao (ed.), *Circular RNAs*, Advances in Experimental Medicine and Biology 1087,
https://doi.org/10.1007/978-981-13-1426-1_23

impacts vascular diseases, atherosclerosis, endothelial cell senescence, endothelial cell injury, endothelial progenitor cell dysfunction, impaired angiogenesis, and cardiovascular disease [6]. In particular, 82% of NCDs deaths occur in low-income or middle-income countries [7].

Blood vessels are conduits for the transport of blood, which can be divided into arteries, veins, and microvessels. Arteries carry blood from the heart to organs [8], while veins carry blood back to the heart. Arteries and veins are connected by self-organization capillaries, which are the main places for material exchanges [9, 10]. The aorta walls are thick and contain rich elastic fibers, which are expansive and elastic [11, 12]. When the left ventricle shoots blood, the pressure in the artery rises, which pushes the blood forward in the artery. After the aortic valve is closed, the expansion of the aorta and artery is still able to maintain the potential energy from the left ventricular systole, so they are called elastic reservoir vessels [13]. As the artery branches, the walls of arteries thin and the elastic fibers are gradually reduced. Smooth muscles become the main components of the arteriole walls. The contraction and relaxation of smooth muscle endow resistance to blood flow [14]. The resistance to blood flow mostly occurs in small arteries, especially the microarteries, which are called resistance vessels [15]. Meanwhile, the resistance of peripheral vascular to blood flow is called peripheral resistance or capillary vascular resistance [16, 17].

In all kinds of blood vessels, blood capillaries have the smallest caliber and the largest number. The total cross-sectional area of blood capillaries is the largest, and the blood flow velocity is the slowest. The blood walls of capillaries consist of monolayer endothelial cells and basement membrane, which create conditions for material exchanges [18]. When capillaries converge into a vein, the vessel wall has a complete smooth muscle layer [19]. Compared with arteries, the veins have larger diameter and thinner tube wall [20]. Usually, veins are in quiescent state, and they contain 60–70% of the circulating blood [21, 22].

In the early 1990s, circRNAs were first found in plants. Since then circRNAs have been found to express in viruses, archaea, and animals [23]. circRNAs, formed by non-sequential back splicing of pre-messenger RNA (pre-mRNA) transcripts, have been widely studied in recent years [24, 25]. Unlike linear RNAs, circRNAs have 5′ caps and 3′ tails. Its 5′ caps and 3′ tails bond together covalently and form a ring. Recent reports reveal that circRNAs can function as competing endogenous RNAs or microRNA sponges to regulate gene expression [26]. With the development of high-throughput RNA sequencing technology, a large number of circRNAs are discovered. circRNAs are found to express endogenously in mammalian cells and be involved in multiple diseases, such as atherosclerosis, neurological disorder, prion disease, and cancer [27].

2 Circular RNAs and Vascular Diseases

2.1 The Roles of circRNAs in Vascular Dysfunction

Endothelium and smooth muscles are essential components for vascular function [28, 29]. Vascular dysfunction is the classical symptom of ischemia, cancer, and inflammatory diseases [30]. Vascular dysfunction is often associated with abnormal gene regulation and endothelial cell dysfunction [31]. Aberrant circRNA expressions have been detected in cardiovascular diseases and cancers, which are usually accompanied with vascular dysfunction [32].

The host gene, ZNF609, is an important member of zinc finger protein family [33]. Zinc finger protein plays an important role in DNA identification, RNA packaging, cell apoptosis regulation, protein folding and assembly, and other biological processes [34]. In endothelium and smooth muscle, circular RNAs cZNF609 and ZNF609 derive from the same transcripts. Researchers report that cZNF609 is one of the top ten abundantly expressed circular RNAs in endothelial cells. CircBase data shows that the sequence of cZNF609 is homologous between mouse and human genome. In vivo experiments demonstrate that cZNF609 silence decreases the loss of retinal vessels and pathologic angiogenesis [35]. Mechanistically, cZNF609 acts as an endogenous miR-615 sponge to sequester and inhibit miR-615 activity, which leads to the increased expression of MEF2A [36]. In summary, cZNF609/miR-615-5p/MEF2A regulatory network plays an important role in vascular regulation [37].

2.2 The Roles of circRNAs in Atherosclerosis

Atherosclerotic lesions start with endothelial dysfunction in arterial vascular system, and then circulating monocytes and lipoprotein particles accumulate in the subendothelial space to block the blood flow [51]. Atherosclerosis is a chronic inflammatory disease and is the leading cause of global vascular deaths [38–40]. The risk factors of atherosclerosis include tobacco use, obesity, and hyperlipidemia. In addition, genetic factors are proved to be associated with atherosclerosis [41]. The main symptoms of atherosclerosis include coronary heart disease (CHD), ischemic stroke, and peripheral artery disease [42, 43]. In the United States, the prevalence of CHD in adults is estimated to be 6.2%, and the annual cost of CHD and strokes is approximately $317 billion [44].

The INK4/ARF locus encodes three tumor suppressor genes, namely, $p16^{INK4a}$, $p15^{INK4b}$, and ARF. In addition, a long noncoding RNA called antisense noncoding RNA in the INK4 locus (ANRIL) is also encoded from INK4/ARF locus [45, 46]. INK4/ARF protein is indispensable in the development of normal mammals, and it plays an important role in inhibiting abnormal proliferation. circANRIL is reported to bind to pescadillo homologue 1 (PES1) which is an essential 60S-preribosomal assembly factor. The binding of circANRIL to PES1 impairs exonuclease-mediated pre-rRNA processing and ribosome biogenesis in vascular smooth muscle cells [44]. In addition, circANRIL is reported to induce p53 activation and promote cell apoptosis. In sum, circANRIL protects atherosclerosis by controlling ribosomal RNA processing or inhibiting cell proliferation.

2.3 The Roles of circRNAs in Diabetes Mellitus-Induced Retinal Vascular Dysfunction

Diabetic retinal dysfunction is a common and severe microvascular complication [47–49]. Hyperglycemia can cause retinal vascular damage and blood–retinal barrier damage, which is the main cause of morbidity and mortality in diabetic patients [50]. The vascular endothelium is composed of a layer of endothelial cells. The fissures between adjacent endothelial cells are closely connected by the transmembrane protein complex, such as occludins, claudins, and zonula occludens [51, 52]. These complexes contribute to the paracellular barriers, such as the blood–brain barrier and blood–retinal barrier [53]. Based on the above information, endothelial cell barrier can be used as therapeutic targets for vascular permeability-associated diabetic retinopathy [54].

Homeodomain-interacting protein kinase 3 (HIPK3) is highly expressed in heart and muscle tissue and localized in nuclear speckles. HIPK3, as a corepressor for homeodomain transcription factor, involves in cell cycle regulation [55, 56]. circHIPK3 derives from the second exon of HIPK3 gene and has long introns on both sides, which include many complementary Alu repeats [24]. circHIPK3 affects the activity, proliferation, migration, and tube formation of retinal endothelial cells [57]. circHIPK3 is found to be upregulated in diabetic retinas [55]. In vivo silence of circHIPK3 alleviates retinal vascular dysfunction symptoms, such as retinal acellular capillary decrease, vascular leakage, and inflammation [58]. Furthermore, circHIPK3 acts as an endogenous miR-30a-3p sponge to sequester and inhibit miR-30a-3p activity, which leads to the increased expression of vascular endothelial growth factor-C, FZD4, and WNT2. Overall, circHIPK3 plays a role in diabetic retinopathy and is a potential target for diabetic retinopathy treatment.

2.4 The Roles of Circular RNAs in Chronic Thromboembolic Pulmonary Hypertension

Chronic thromboembolic pulmonary hypertension (CTEPH) is a rare but debilitating and life-threatening complication of acute pulmonary embolism, which is caused by pulmonary arterial obstruction and progressive vascular remodeling [59–62]. The incidence of CTEPH is about 0.1–9.1% in the 2 years after acute pulmonary embolism [63]. Risk factors for CTEPH include inflammatory bowel disease, splenectomy, and myeloproliferative disease. According to the recent ESC/ERS guidelines for pulmonary hypertension, CTEPH is divided into four types [64, 65]. CTEPH is a life-threatening disease, and if it is not treated appropriately, CTEPH will develop into refractory right ventricular failure [66]. Patients with CTEPH exhibit a poor prognosis unless they receive treatment at an early stage [67]. Thus, the early diagnosis and treatment of CTEPH are important. CTEPH is a pan-vascular disease of the pulmonary arteries. The feature of CTEPH is the nonlinear arterial pulmonary pressure increase caused by major vessels obstruction [67–70]. CTEPH is also the complication of pulmonary hypertension (PH), which is caused by pulmonary endarterectomy (PEA) [71–73].

Twenty years ago, circRNAs were considered as scrambled exons, most of which were misread as splicing errors [74]. Until recently, circRNAs are gradually recognized as regulatory transcripts that derive from protein-coding exons [75]. circRNAs are reported to function in CTEPH mainly by affecting ribonucleotide biosynthesis, the cellular response to stress, the response to DNA damage, and gene expression. Microarray analysis is performed to identify the differential expression of circRNAs between CTEPH patients and control individuals [76]. Statistically, 351 circRNAs are abnormally expressed in CTEPH patients. Among these circRNAs, hsa_circ_0022342 is found to play an important role in CTEPH by targeting hsa-miR-940 [77, 78]. CDKN1A is reported as the downstream gene of hsa-miR-940 in CTEPH. And then, CDKN1A activates the downstream ErbB signaling pathway, which leads to increased cell apoptosis. The hsa_circ_0022342/hsa-miR-940/CRKL/ErbB signaling pathway is proposed to be an important pathway in CTEPH development [79].

2.5 The Roles of Circular RNAs in Carotid Atherosclerotic Disease

Early detection of acute ischemic stroke may reduce morbidity and mortality in patients with advanced carotid atherosclerosis [80]. Currently, there are still no biomarkers for atherosclerotic plaque rupture and stroke [81]. In stable carotid atherosclerotic plaques, the fibrous cap is composed of vascular smooth muscle cells (VSMCs) and collagen-enriched matrix. Here, the fibrous cap is a kind of artery inflammation, which is caused by intimal thickening. In vulnerable plaques, fibrous cap is thin and is composed with decreased VSMCs and increased inflammatory cells [82]. As a result, identifying circular biomarkers for fibrous cap change provides a new strategy for carotid artery prediction.

Recently, it is reported that the expressions of miR-221 and miR-222 are reduced in acute ischemic cerebrovascular-related carotid plaques [83]. miR-221 and miR-222 promote intimal thickening through depressing the expression of $p27^{Kip1}$ and inhibiting VSMC cell cycle progression [83]. Memczak et al. report that there are several circRNAs with miR-221 and miR-222 binding sites. Among them, circRNA-284 is demonstrated to be able to regulate the activities of miR-221 and miR-222, which indicate that circRNA-284 may function in cerebrovascular-related carotid plaques [42].

2.6 The Roles of Circular RNAs in Hepatic Vascular Invasion

Hepatocellular carcinoma (HCC) is the most common malignant tumor and is a major health problem worldwide. Surgical treatment, such as hepatectomy and transplantation, remains the most effective treatment for patients with early liver cancer [84]. However, due to the high incidence of postoperative recurrence, the prognosis of HCC treatment is still not satisfactory [85]. Multiple studies have demonstrated that microvascular infiltration (MVI) is the most important prognostic factor for HCC recurrence and survival [86, 87].

It is reported that cirs-7 can promote HCC cell proliferation [88–90]. In a further study, the expression of cirs-7 is investigated in 108 paired HCC tissues. They find that the expression of cirs-7 is correlated with the clinical pathological parameters of HCC and the deterioration of the disease [91]. Another study analyzes the relationships between cirs-7 and MVI. Univariate and multivariate analyses suggest that the elevated cirs-7 expression is an independent risk factor for MVI in HCC [92–95].

2.7 The Roles of Circular RNAs in Aortic Aneurysm

Aortic aneurysm is one of the leading causes of cardiovascular deaths [96]. Aortic aneurysm is divided into true aortic aneurysm and aortic pseudoaneurysm [97]. In true aortic aneurysm, the pathological aorta dilation is greater than 50% compared with normal blood vessel diameter. In contrast, the pathological aorta dilation in aortic pseudoaneurysm is less than 50% [98–100]. Considerable progress has been made to improve cardiovascular disease survival over the past few decades. However, the 5 year survival rate of cardiovascular disease remains below 35% [101–103]. Circular RNAs have been reported to function in cardiovascular diseases. However, their expression and functions in aortic aneurysm remain elusive. Hsa-circ-000595 is reported to be located on chromosome 14 and regulate the function of miR-19a [104]. In addition, it is indicated that hsa-circ-000595 may take part in the regulation of aortic aneurysm by preventing cell apoptosis [105].

2.8 The Roles of Circular RNAs in Coronary Artery Disease and Type 2 Diabetes Mellitus

In 2015, the mortality rate of coronary artery disease (CAD) was 2.6% higher than that in 2013. The death rate of cardiovascular disease (CVD) rapidly increases with population aging [106, 107]. According to the International Diabetes Federation (IDF) data, there are nearly 410 million diabetic patients (DM) worldwide, and about 46.5% of them have not been diagnosed [108, 109]. A Chinese survey reports the clinical data of 3513 CVD patients from 7 cities across the country. They find that 77% of CVD patients have hyperglycemia [110]. The high morbidity and mortality of CAD and T2DM have brought enormous social and economic burdens worldwide [109].

Hsa-circRNA11783–2 is located in chromosome 18, and its parental gene is ENST00000251081 [111, 112]. Mutations of the parental gene are linked to cardiovascular disease. A recent study reports that hsa-circRNA11783–2 is also correlated with CAD and T2DM. Most circRNAs regulate gene expression by acting as "miRNA sponges." In this context, hsa-circRNA11783–2 might take part in disease regulation by targeting miR-608 or miR-3907 [113–115]. Overall, the specific role of hsa-circRNA11783–2 in CAD and T2DM requires further exploration [116].

2.9 The Roles of Circular RNAs in Essential Hypertension

Essential hypertension (EH) is one of the most common diseases in human cardiovascular system [116–118]. EH is a kind of global disease. In China, it is estimated that a third of Chinese have EH. Since the symptoms of EH are not obvious, it is also known as the silent killer [119]. Risk factors of EH mainly include age, gender, and regional and socioeconomic conditions [120]. Here, we will focus on the function of circular RNAs in EH [121–123].

Table 23.1 Circular RNA associated with vascular function and disease

Vascular diseases	Vascular function	Circular RNA	Target genes	References
Retinopathy of prematurity diabetic retinopathy	Apoptosis	cZNF609	miR-615-5p, MEF2A	[30]
Ischemia-reperfusion injury	Apoptosis	circRNAs Cdr1	miR-7a	[137]
Atherosclerosis	Apoptosis	circANRIL	PES1	[44]
Heart failure	Proliferation	circ-Foxo3	CDK2,p21	[138]
Retinal vascular dysfunction induced by diabetes mellitus	Proliferation	circHIPK3	miR-30a-3p, FZD4, WNT2	[55]
Chronic thromboembolic pulmonary hypertension	Proliferation	hsa_circ_0022342	hsa-miR-940, CRKL, ErbB	[67]
Carotid atherosclerotic disease	Necrosis	circR-284	miR-221,miR-222	[83]
Hepatic vascular invasion in hepatocellular carcinoma	Apoptosis	ciRS-7	miR-7,PIK3CD, p70S6K, mTOR	[93]
Aortic aneurysm	Proliferation	hsa-circ-000595	miR-19a	[100]
Diabetes mellitus	Migration	circWDR77	miR-124, FGF-2	[139]
Hypertension	Migration and invasion	hsa_circ_0037911	miR-637	[116]
Coronary artery disease and type 2 diabetes mellitus	Proliferation	hsa-circRNA11783–2	miR-608	[140]

Hsa_circ_0037911 is located in chromosome 16 and contains four exons [124]. According to the bioinformatic analysis, hsa_circ_0037911 is proposed to be associated with EH. Evidence demonstrates that hsa_circ_0037911 acts as miR-637 sponge to regulate the pathogenesis of EH. In previous studies, miR-637 has been shown to function as a tumor suppressor in hepatocellular carcinoma, breast cancer, and follicular thyroid cancer [125–128]. The upregulation of miR-637 decreases the activity of Akt1 and inhibits the migration and invasion of cancer cells [129, 130].

3 Prospectives

In this chapter, we have introduced the functions of circRNAs in multiple vascular-related diseases. Intervention of certain circRNAs might improve certain vascular diseases [131, 132]. Vascular disease-related deaths have been increasing. Clarifying the relationships between circular RNAs and vascular diseases will provide new methods for vascular disease treatment [133]. At present, the early diagnosis of vascular diseases is urgently needed [134]. It has been reported that some circRNAs are specifically expressed in certain vascular diseases. Identifying circular RNAs as vascular disease biomarkers might help the early diagnosis of diseases [35, 135, 136]. With the rapid development of RNA sequencing technology, more and more circular RNAs are identified. In future, the circular RNAs study will gain more and more attention. In the past decade, a number of circRNAs are reported to be associated with vascular diseases. However, there is still a long way to go before we uncover the underlying mechanisms of circular RNAs in vascular diseases (Table 23.1).

Competing Financial Interests The authors declare no competing financial interests.

References

1. Maiese K (2017) Harnessing the power of SIRT1 and non-coding RNAs in vascular disease. Curr Neurovasc Res 14(1):82–88
2. Arokiasamy P, Uttamacharya KP et al (2017) Chronic noncommunicable diseases in 6 low- and middle-income countries: findings from wave 1 of the World Health Organization's study on global ageing and adult health (SAGE). Am J Epidemiol 185(6):414–428
3. Strazhesko I, Tkacheva O, Boytsov S et al (2015) Association of insulin resistance, arterial stiffness and telomere length in adults free of cardiovascular diseases. PLoS One 10(8):e0136676
4. Minino AM, Murphy SL (2012) Death in the United States, 2010. NCHS Data Brief 99:1–8

5. Turtle JR (2000) The economic burden of insulin resistance. Int J Clin Pract Suppl(113):23–28
6. Mels CM, Schutte AE, Schutte R et al (2013) The link between vascular deterioration and branched chain amino acids in a population with high glycated haemoglobin: the SABPA study. Amino Acids 45(6):1405–1413
7. Yao Z, Luo J, Hu K et al (2017) ZKSCAN1 gene and its related circular RNA (circZKSCAN1) both inhibit hepatocellular carcinoma cell growth, migration, and invasion but through different signaling pathways. Mol Oncol 11(4):422–437
8. Siddiqui R, Saleem S, Khan NA (2016) The effect of peptidic and non-peptidic proteasome inhibitors on the biological properties of Acanthamoeba castellanii belonging to the T4 genotype. Exp Parasitol 168:16–24
9. Christensen HM, Schou M, Goetze JP et al (2013) Body mass index in chronic heart failure: association with biomarkers of neurohormonal activation, inflammation and endothelial dysfunction. BMC Cardiovasc Disord 13:80
10. Boeckel JN, Jae N, Heumuller AW et al (2015) Identification and characterization of hypoxia-regulated endothelial circular RNA. Circ Res 117(10):884–890
11. Health Quality O (2005) Biventricular pacing (cardiac resynchronization therapy): an evidence-based analysis. Ont Health Technol Assess Ser 5(13):1–60
12. Khafaji HA, Bener AB, Rizk NM et al (2012) Elevated serum leptin levels in patients with acute myocardial infarction; correlation with coronary angiographic and echocardiographic findings. BMC Res Notes 5:262
13. Kiuchi MG, Andrea BR, da Silva GR et al (2015) Pulmonary artery ablation to treat pulmonary arterial hypertension: a case report. J Med Case Rep 9:284
14. Fang S, Guo H, Cheng Y et al (2018) circHECTD1 promotes the silica-induced pulmonary endothelial-mesenchymal transition via HECTD1. Cell Death Dis 9(3):396
15. Koide T, Nakanishi A, Ito I et al (1975) Hemodynamic and clinical significances of atrial fibrillation, pulmonary vascular resistance and left ventricular function in rheumatic mitral stenosis. Jpn Heart J 16(2):122–142
16. Lindsay DC, Holdright DR, Clarke D et al (1996) Endothelial control of lower limb blood flow in chronic heart failure. Heart 75(5):469–476
17. Pluta A, Strozecki P, Krintus M et al (2015) Left ventricular remodeling and arterial remodeling in patients with chronic kidney disease stage 1-3. Ren Fail 37(7):1105–1110
18. Radauskaite G, Vaskelyte J, Jurkevicius R (2007) Progression of mitral regurgitation following ischemic mitral valve repair. Medicina (Kaunas) 43(8):619–622
19. Rees W, Schuler S, Hummel M et al (1993) Heart transplantation in patients with muscular dystrophy associated with end-stage cardiomyopathy. J Heart Lung Transplant 12(5):804–807
20. Scherbakov N, Sandek A, Martens-Lobenhoffer J et al (2012) Endothelial dysfunction of the peripheral vascular bed in the acute phase after ischemic stroke. Cerebrovasc Dis 33(1):37–46
21. Song MG, Yang HS, Lee DH et al (2014) Mid-term results in patients having tricuspidization of the quadricuspid aortic valve. J Cardiothorac Surg 9:29
22. Zeng ZY, Gui C, Li L et al (2016) Effects of percutaneous coronary intervention on serum Angiopoietin-2 in patients with coronary heart disease. Chin Med J 129(6):631–635
23. Chen Y, Li C, Tan C et al (2017) Circular RNA in human disease and their potential clinic significance. Zhonghua Yi Xue Yi Chuan Xue Za Zhi 34(1):133–137
24. Zheng Q, Bao C, Guo W et al (2016) Circular RNA profiling reveals an abundant circHIPK3 that regulates cell growth by sponging multiple miRNAs. Nat Commun 7:11215
25. Militello G, Weirick T, John D et al (2017) Screening and validation of lncRNAs and circRNAs as miRNA sponges. Brief Bioinform 18(5):780–788
26. Haque S, Harries LW (2017) Circular RNAs (circRNAs) in health and disease. Genes (Basel) 8(12):353
27. Qu S, Yang X, Li X et al (2015) Circular RNA: a new star of noncoding RNAs. Cancer Lett 365(2):141–148
28. Mazzuca MQ, Wlodek ME, Dragomir NM et al (2010) Uteroplacental insufficiency programs regional vascular dysfunction and alters arterial stiffness in female offspring. J Physiol 588(Pt 11):1997–2010
29. Maass PG, Glazar P, Memczak S et al (2017) A map of human circular RNAs in clinically relevant tissues. J Mol Med (Berl) 95(11):1179–1189
30. Liu C, Yao MD, Li CP et al (2017) Silencing of circular RNA-ZNF609 ameliorates vascular endothelial dysfunction. Theranostics 7(11):2863–2877
31. Puro DG, Kohmoto R, Fujita Y et al (2016) Bioelectric impact of pathological angiogenesis on vascular function. Proc Natl Acad Sci U S A 113(35):9934–9939
32. Carqueja IM, Sousa J, Mansilha A (2018) Vascular malformations: classification, diagnosis and treatment. Int Angiol 37(2):127. https://doi.org/10.23736/S0392-9590.18.03961-5
33. Yang P, Qiu Z, Jiang Y et al (2016) Silencing of cZNF292 circular RNA suppresses human glioma tube formation via the Wnt/beta-catenin signaling pathway. Oncotarget 7(39):63449–63455
34. Laity JH, Lee BM, Wright PE (2001) Zinc finger proteins: new insights into structural and functional diversity. Curr Opin Struct Biol 11(1):39–46
35. Ding HX, Lv Z, Yuan Y et al (2018) The expression of circRNAs as a promising biomarker in the diagnosis and prognosis of human cancers: a

systematic review and meta-analysis. Oncotarget 9(14):11824–11836
36. Peng L, Chen G, Zhu Z et al (2017) Circular RNA ZNF609 functions as a competitive endogenous RNA to regulate AKT3 expression by sponging miR-150-5p in Hirschsprung's disease. Oncotarget 8(1):808–818
37. Sena CM, Pereira AM, Seica R (2013) Endothelial dysfunction - a major mediator of diabetic vascular disease. Biochim Biophys Acta 1832(12):2216–2231
38. Burd CE, Jeck WR, Liu Y et al (2010) Expression of linear and novel circular forms of an INK4/ARF-associated non-coding RNA correlates with atherosclerosis risk. PLoS Genet 6(12):e1001233
39. Kumar L, Shamsuzzama HR et al (2017) Circular RNAs: the emerging class of non-coding RNAs and their potential role in human neurodegenerative diseases. Mol Neurobiol 54(9):7224–7234
40. Li Y, Dong Y, Huang Z et al (2017) Computational identifying and characterizing circular RNAs and their associated genes in hepatocellular carcinoma. PLoS One 12(3):e0174436
41. Zhao L, Varghese Z, Moorhead JF et al (2018) CD36 and lipid metabolism in the evolution of atherosclerosis. Br Med Bull 126:101. https://doi.org/10.1093/bmb/ldy006
42. Maiese K (2016) Disease onset and aging in the world of circular RNAs. J Transl Sci 2(6):327–329
43. Traylor M, Makela KM, Kilarski LL et al (2014) A novel MMP12 locus is associated with large artery atherosclerotic stroke using a genome-wide age-at-onset informed approach. PLoS Genet 10(7):e1004469
44. Holdt LM, Stahringer A, Sass K et al (2016) Circular non-coding RNA ANRIL modulates ribosomal RNA maturation and atherosclerosis in humans. Nat Commun 7:12429
45. Li H, Li K, Lai W et al (2018) Comprehensive circular RNA profiles in plasma reveals that circular RNAs can be used as novel biomarkers for systemic lupus erythematosus. Clin Chim Acta 480:17–25
46. Holdt LM, Teupser D (2012) Recent studies of the human chromosome 9p21 locus, which is associated with atherosclerosis in human populations. Arterioscler Thromb Vasc Biol 32(2):196–206
47. Yang LP, Sun HL, Wu LM et al (2009) Baicalein reduces inflammatory process in a rodent model of diabetic retinopathy. Invest Ophthalmol Vis Sci 50(5):2319–2327
48. Wei L, Yin Z, Yuan Y et al (2010) A PKC-beta inhibitor treatment reverses cardiac microvascular barrier dysfunction in diabetic rats. Microvasc Res 80(1):158–165
49. Takeda M, Mori F, Yoshida A et al (2001) Constitutive nitric oxide synthase is associated with retinal vascular permeability in early diabetic rats. Diabetologia 44(8):1043–1050
50. Kim J, Jo K, Kim CS et al (2017) Aster koraiensis extract prevents diabetes-induced retinal vascular dysfunction in spontaneously diabetic Torii rats. BMC Complement Altern Med 17(1):497
51. Su W, Dai DZ, Liu HR et al (2007) Upregulated endothelin system in diabetic vascular dysfunction and early retinopathy is reversed by CPU0213 and total triterpene acids from Fructus Corni. Clin Exp Pharmacol Physiol 34(12):1228–1233
52. Wondemagegn AT, Bizuayehu HM, Abie DD et al (2017) Undiagnosed diabetes mellitus and related factors in east Gojjam (NW Ethiopia) in 2016: a community-based study. J Public Health Res 6(1):834
53. Herbreteau D, Brunereau L, Cottier J et al (1997) Hemangiomas and superficial vascular malformations of the head and neck. Classification, diagnosis, treatment. J Neuroradiol 24(4):274–290
54. Shan K, Li CP, Liu C et al (2017) RNCR3: a regulator of diabetes mellitus-related retinal microvascular dysfunction. Biochem Biophys Res Commun 482(4):777–783
55. Shan K, Liu C, Liu BH et al (2017) Circular noncoding RNA HIPK3 mediates retinal vascular dysfunction in diabetes mellitus. Circulation 136(17):1629–1642
56. Chen G, Shi Y, Liu M et al (2018) circHIPK3 regulates cell proliferation and migration by sponging miR-124 and regulating AQP3 expression in hepatocellular carcinoma. Cell Death Dis 9(2):175
57. Li Y, Zheng F, Xiao X et al (2017) CircHIPK3 sponges miR-558 to suppress heparanase expression in bladder cancer cells. EMBO Rep 18(9):1646–1659
58. Tian F, Wang Y, Xiao Z et al (2017) Circular RNA CircHIPK3 promotes NCI-H1299 and NCI-H2170 cell proliferation through miR-379 and its target IGF1. Zhongguo Fei Ai Za Zhi 20(7):459–467
59. Pesavento R, Prandoni P (2018) Prevention and treatment of the chronic thromboembolic pulmonary hypertension. Thromb Res 164:150. https://doi.org/10.1016/j.thromres.2018.02.149
60. Aldabbous L, Abdul-Salam V, McKinnon T et al (2016) Neutrophil extracellular traps promote angiogenesis: evidence from vascular pathology in pulmonary hypertension. Arterioscler Thromb Vasc Biol 36(10):2078–2087
61. Bonderman D, Jakowitsch J, Redwan B et al (2008) Role for staphylococci in misguided thrombus resolution of chronic thromboembolic pulmonary hypertension. Arterioscler Thromb Vasc Biol 28(4):678–684
62. Chen Z, Nakajima T, Tanabe N et al (2010) Susceptibility to chronic thromboembolic pulmonary hypertension may be conferred by miR-759 via its targeted interaction with polymorphic fibrinogen alpha gene. Hum Genet 128(4):443–452
63. Pesavento R, Prandoni P (2015) Prevention and treatment of the post-thrombotic syndrome and of the chronic thromboembolic pulmonary hypertension. Expert Rev Cardiovasc Ther 13(2):193–207
64. Gu S, Li G, Zhang X et al (2015) Aberrant expression of long noncoding RNAs in chronic

thromboembolic pulmonary hypertension. Mol Med Rep 11(4):2631–2643
65. Guo L, Yang Y, Liu J et al (2014) Differentially expressed plasma microRNAs and the potential regulatory function of let-7b in chronic thromboembolic pulmonary hypertension. PLoS One 9(6):e101055
66. Mercier O, Sage E, Izziki M et al (2010) Endothelin a receptor blockade improves regression of flow-induced pulmonary vasculopathy in piglets. J Thorac Cardiovasc Surg 140(3):677–683
67. Miao R, Wang Y, Wan J et al (2017) Microarray expression profile of circular RNAs in chronic thromboembolic pulmonary hypertension. Medicine (Baltimore) 96(27):e7354
68. Miyata M, Ito M, Sasajima T et al (2001) Effect of a serotonin receptor antagonist on interleukin-6-induced pulmonary hypertension in rats. Chest 119(2):554–561
69. Ogawa A, Firth AL, Yao W et al (2009) Inhibition of mTOR attenuates store-operated Ca2+ entry in cells from endarterectomized tissues of patients with chronic thromboembolic pulmonary hypertension. Am J Physiol Lung Cell Mol Physiol 297(4):L666–L676
70. Wang L, Guo LJ, Liu J et al (2013) MicroRNA expression profile of pulmonary artery smooth muscle cells and the effect of let-7d in chronic thromboembolic pulmonary hypertension. Pulm Circ 3(3):654–664
71. Wynants M, Vengethasamy L, Ronisz A et al (2013) NF-kappaB pathway is involved in CRP-induced effects on pulmonary arterial endothelial cells in chronic thromboembolic pulmonary hypertension. Am J Physiol Lung Cell Mol Physiol 305(12):L934–L942
72. Yang M, Deng C, Wu D et al (2016) The role of mononuclear cell tissue factor and inflammatory cytokines in patients with chronic thromboembolic pulmonary hypertension. J Thromb Thrombolysis 42(1):38–45
73. Zhang YM, Liu WJ, Shi YZ et al (2008) Expressions of urotensin II and its receptor in pulmonary arteries in rats with chronic thromboembolic pulmonary hypertension. Zhonghua Jie He He Hu Xi Za Zhi 31(1):37–41
74. Qu S, Zhong Y, Shang R et al (2017) The emerging landscape of circular RNA in life processes. RNA Biol 14(8):992–999
75. Fu C, Lv R, Xu G et al (2017) Circular RNA profile of infantile hemangioma by microarray analysis. PLoS One 12(11):e0187581
76. Zhou Z, Du D, Chen A et al (2018) Circular RNA expression profile of articular chondrocytes in an IL-1beta-induced mouse model of osteoarthritis. Gene 644:20–26
77. Bhajun R, Guyon L, Pitaval A et al (2015) A statistically inferred microRNA network identifies breast cancer target miR-940 as an actin cytoskeleton regulator. Sci Rep 5:8336
78. Ding D, Zhang Y, Yang R et al (2016) miR-940 suppresses tumor cell invasion and migration via regulation of CXCR2 in hepatocellular carcinoma. Biomed Res Int 2016:7618342
79. Zhang EY, Cristofanilli M, Robertson F et al (2013) Genome wide proteomics of ERBB2 and EGFR and other oncogenic pathways in inflammatory breast cancer. J Proteome Res 12(6):2805–2817
80. Jovanovic I, Zivkovic M, Jovanovic J et al (2014) The co-inertia approach in identification of specific microRNA in early and advanced atherosclerosis plaque. Med Hypotheses 83(1):11–15
81. Rootman J, Heran MK, Graeb DA (2014) Vascular malformations of the orbit: classification and the role of imaging in diagnosis and treatment strategies*. Ophthal Plast Reconstr Surg 30(2):91–104
82. Raitoharju E, Lyytikainen LP, Levula M et al (2011) miR-21, miR-210, miR-34a, and miR-146a/b are up-regulated in human atherosclerotic plaques in the tampere vascular study. Atherosclerosis 219(1):211–217
83. Bazan HA, Hatfield SA, Brug A et al (2017) Carotid plaque rupture is accompanied by an increase in the ratio of serum circR-284 to miR-221 levels. Circ Cardiovasc Genet 10(4):e001720
84. Werner JA, Dunne AA, Folz BJ et al (2001) Current concepts in the classification, diagnosis and treatment of hemangiomas and vascular malformations of the head and neck. Eur Arch Otorhinolaryngol 258(3):141–149
85. Altendorf-Hofmann A, Scheele J (2002) Indication and surgical outcome in hepatocellular carcinoma with infiltration of blood vessels, bile ducts and lymph nodes. Kongressbd Dtsch Ges Chir Kongr 119:635–641
86. Yamashita Y, Shirabe K, Aishima S et al (2015) Predictors of microvascular invasion in hepatocellular carcinoma. Dig Dis 33(5):655–660
87. Aravindhan V, Mohan V, Arunkumar N et al (2015) Chronic endotoxemia in subjects with type-1 diabetes is seen much before the onset of microvascular complications. PLoS One 10(9):e0137618
88. Xu L, Zhang M, Zheng X et al (2017) The circular RNA ciRS-7 (Cdr1as) acts as a risk factor of hepatic microvascular invasion in hepatocellular carcinoma. J Cancer Res Clin Oncol 143(1):17–27
89. Shi Z, Chen T, Yao Q et al (2017) The circular RNA ciRS-7 promotes APP and BACE1 degradation in an NF-kappaB-dependent manner. FEBS J 284(7):1096–1109
90. Su C, Han Y, Zhang H et al (2018) CiRS-7 targeting miR-7 modulates the progression of non-small cell lung cancer in a manner dependent on NF-kappaB signalling. J Cell Mol Med 22:3097. https://doi.org/10.1111/jcmm.13587
91. Barrett SP, Parker KR, Horn C et al (2017) ciRS-7 exonic sequence is embedded in a long non-coding RNA locus. PLoS Genet 13(12):e1007114
92. Geng HH, Li R, Su YM et al (2016) The circular RNA Cdr1as promotes myocardial infarction by

mediating the regulation of miR-7a on its target genes expression. PLoS One 11(3):e0151753
93. Pan H, Li T, Jiang Y et al (2018) Overexpression of circular RNA ciRS-7 abrogates the tumor suppressive effect of miR-7 on gastric cancer via PTEN/PI3K/AKT signaling pathway. J Cell Biochem 119(1):440–446
94. Peng L, Yuan XQ, Li GC (2015) The emerging landscape of circular RNA ciRS-7 in cancer (review). Oncol Rep 33(6):2669–2674
95. Xu H, Guo S, Li W et al (2015) The circular RNA Cdr1as, via miR-7 and its targets, regulates insulin transcription and secretion in islet cells. Sci Rep 5:12453
96. Woo KI, Kim YD (2014) Re: "vascular malformations of the orbit: classification and the role of imaging in diagnosis and treatment strategies". Ophthal Plast Reconstr Surg 30(6):532
97. Guo X, Chang Q, Pei H et al (2017) Long non-coding RNA-mRNA correlation analysis reveals the potential role of HOTAIR in pathogenesis of sporadic thoracic aortic aneurysm. Eur J Vasc Endovasc Surg 54(3):303–314
98. He Q, Tan J, Yu B et al (2015) Long noncoding RNA HIF1A-AS1A reduces apoptosis of vascular smooth muscle cells: implications for the pathogenesis of thoracoabdominal aorta aneurysm. Pharmazie 70(5):310–315
99. Li Y, Yang N (2018) Microarray expression profile analysis of long non-coding RNAs in thoracic aortic aneurysm. Kaohsiung J Med Sci 34(1):34–42
100. Liu G, Huang Y, Lu X et al (2010) Identification and characteristics of microRNAs with altered expression patterns in a rat model of abdominal aortic aneurysms. Tohoku J Exp Med 222(3):187–193
101. Iyer V, Rowbotham S, Biros E et al (2017) A systematic review investigating the association of microRNAs with human abdominal aortic aneurysms. Atherosclerosis 261:78–89
102. Joviliano EE, Ribeiro MS, Tenorio EJR (2017) MicroRNAs and current concepts on the pathogenesis of abdominal aortic aneurysm. Braz J Cardiovasc Surg 32(3):215–224
103. Li Y, Maegdefessel L (2017) Non-coding RNA contribution to thoracic and abdominal aortic aneurysm disease development and progression. Front Physiol 8:429
104. Zheng C, Niu H, Li M et al (2015) Cyclic RNA hsacirc000595 regulates apoptosis of aortic smooth muscle cells. Mol Med Rep 12(5):6656–6662
105. Hu ZY, Luo JF, Zhong SL et al (2012) MicroRNAs expression in normal and dissected aortic tissue. Zhonghua Xin Xue Guan Bing Za Zhi 40(5):406–410
106. Kolopp-Sarda MN, Miossec P (2018) Cryoglobulins: an update on detection, mechanisms and clinical contribution. Autoimmun Rev 17:457. https://doi.org/10.1016/j.autrev.2017.11.035
107. Howson JMM, Zhao W, Barnes DR et al (2017) Fifteen new risk loci for coronary artery disease highlight arterial-wall-specific mechanisms. Nat Genet 49(7):1113–1119
108. Li C, Zhao L, Jiang W et al (2018) Correct microarray analysis approaches in 'Hsa-circRNA11783-2 in peripheral blood is correlated with coronary artery disease and type 2 diabetes mellitus'. Diab Vasc Dis Res 15(1):92–93
109. Bunnag SC (2006) Implications of microcirculation-research based information on prevention and treatment of diabetes mellitus type 2: a perspective. Clin Hemorheol Microcirc 34(1–2):43–50
110. Zhao Z, Li X, Li M (2018) Response to letter to the editor entitled 'Correct microarray analysis approaches in "Hsa-circRNA11783-2 in peripheral blood is correlated with coronary artery disease and type 2 diabetes mellitus"'. Diab Vasc Dis Res 15(2):169–170
111. Bhavadharini B, Mahalakshmi MM, Maheswari K et al (2016) Use of capillary blood glucose for screening for gestational diabetes mellitus in resource-constrained settings. Acta Diabetol 53(1):91–97
112. Li CY, Ma L, Yu B (2017) Circular RNA hsa_circ_0003575 regulates oxLDL induced vascular endothelial cells proliferation and angiogenesis. Biomed Pharmacother 95:1514–1519
113. Dai ZM, Kang HF, Zhang WG et al (2016) The associations of single nucleotide polymorphisms in miR196a2, miR-499, and miR-608 with breast cancer susceptibility: a STROBE-compliant observational study. Medicine (Baltimore) 95(7):e2826
114. Hashemi M, Sanaei S, Rezaei M et al (2016) miR-608 rs4919510 C>G polymorphism decreased the risk of breast cancer in an Iranian subpopulation. Exp Oncol 38(1):57–59
115. Kupcinskas J, Bruzaite I, Juzenas S et al (2014) Lack of association between miR-27a, miR-146a, miR-196a-2, miR-492 and miR-608 gene polymorphisms and colorectal cancer. Sci Rep 4:5993
116. Bao X, Zheng S, Mao S et al (2018) A potential risk factor of essential hypertension in case-control study: circular RNA hsa_circ_0037911. Biochem Biophys Res Commun 498:789. https://doi.org/10.1016/j.bbrc.2018.03.059
117. Wise IA, Charchar FJ (2016) Epigenetic modifications in essential hypertension. Int J Mol Sci 17(4):451
118. Zhu HY, Wang SW, Martin LJ et al (2009) The role of mitochondrial genome in essential hypertension in a Chinese Han population. Eur J Hum Genet 17(11):1501–1506
119. Dangwal S, Schimmel K, Foinquinos A et al (2017) Noncoding RNAs in heart failure. Handb Exp Pharmacol 243:423–445
120. Hoebeeck LI, Rietzschel ER, Langlois M et al (2011) The relationship between diet and subclinical atherosclerosis: results from the Asklepios study. Eur J Clin Nutr 65(5):606–613
121. Marques FZ, Booth SA, Charchar FJ (2015) The emerging role of non-coding RNA in essential

hypertension and blood pressure regulation. J Hum Hypertens 29(8):459–467
122. Marques FZ, Charchar FJ (2015) microRNAs in essential hypertension and blood pressure regulation. Adv Exp Med Biol 888:215–235
123. Marques FZ, Morris BJ (2012) Neurogenic hypertension: revelations from genome-wide gene expression profiling. Curr Hypertens Rep 14(6):485–491
124. Zou M, Huang C, Li X et al (2017) Circular RNA expression profile and potential function of hsa_circRNA_101238 in human thoracic aortic dissection. Oncotarget 8(47):81825–81837
125. Stokowy T, Wojtas B, Fujarewicz K et al (2014) miRNAs with the potential to distinguish follicular thyroid carcinomas from benign follicular thyroid tumors: results of a meta-analysis. Horm Metab Res 46(3):171–180
126. Xu RL, He W, Tang J et al (2018) Primate-specific miRNA-637 inhibited tumorigenesis in human pancreatic ductal adenocarcinoma cells by suppressing Akt1 expression. Exp Cell Res 363:310. https://doi.org/10.1016/j.yexcr.2018.01.026
127. Zhang JF, He ML, Fu WM et al (2011) Primate-specific microRNA-637 inhibits tumorigenesis in hepatocellular carcinoma by disrupting signal transducer and activator of transcription 3 signaling. Hepatology 54(6):2137–2148
128. Li YM, Liu XY (2016) Molecular mechanisms underlying application of serum procalcitonin and stool miR-637 in prognosis of acute ischemic stroke. Am J Transl Res 8(10):4242–4249
129. Que T, Song Y, Liu Z et al (2015) Decreased miRNA-637 is an unfavorable prognosis marker and promotes glioma cell growth, migration and invasion via direct targeting Akt1. Oncogene 34(38):4952–4963
130. Sang HY, Jin YL, Zhang WQ et al (2016) Downregulation of microRNA-637 increases risk of hypoxia-induced pulmonary hypertension by modulating expression of Cyclin Dependent Kinase 6 (CDK6) in pulmonary smooth muscle cells. Med Sci Monit 22:4066–4072
131. Salem ESB, Fan GC (2017) Pathological effects of exosomes in mediating diabetic cardiomyopathy. Adv Exp Med Biol 998:113–138
132. Wu HJ, Zhang CY, Zhang S et al (2016) Microarray expression profile of circular RNAs in heart tissue of mice with myocardial infarction-induced heart failure. Cell Physiol Biochem 39(1):205–216
133. Xu T, Wu J, Han P et al (2017) Circular RNA expression profiles and features in human tissues: a study using RNA-seq data. BMC Genomics 18(Suppl 6):680
134. Muller S, Appel B (2017) In vitro circularization of RNA. RNA Biol 14(8):1018–1027
135. Zhang HD, Jiang LH, Sun DW et al (2018) CircRNA: a novel type of biomarker for cancer. Breast Cancer 25(1):1–7
136. Xin Z, Ma Q, Ren S et al (2017) The understanding of circular RNAs as special triggers in carcinogenesis. Brief Funct Genomics 16(2):80–86
137. Memczak S, Jens M, Elefsinioti A et al (2013) Circular RNAs are a large class of animal RNAs with regulatory potency. Nature 495(7441):333–338
138. Du WW, Yang W, Chen Y et al (2017) Foxo3 circular RNA promotes cardiac senescence by modulating multiple factors associated with stress and senescence responses. Eur Heart J 38(18):1402–1412
139. Chen J, Cui L, Yuan J et al (2017) Circular RNA WDR77 target FGF-2 to regulate vascular smooth muscle cells proliferation and migration by sponging miR-124. Biochem Biophys Res Commun 494(1–2):126–132
140. Li X, Zhao Z, Jian D et al (2017) Hsa-circRNA11783-2 in peripheral blood is correlated with coronary artery disease and type 2 diabetes mellitus. Diab Vasc Dis Res 14(6):510–515

Functional Role of Circular RNA in Regenerative Medicine

24

Richard Y. Cao, Qiying Dai, Qing Li, and Jian Yang

Abstract

Every year, millions of people around the world suffer from different forms of tissue trauma. Regenerative medicine refers to therapy that replaces the injured organ or cells. Stem cells are the frontiers and hotspots of current regenerative medicine research. Circular RNAs (circRNAs) are essential for the early development of many species. It was found that they could guide stem cell differentiation through interacting with certain microRNAs (miRNAs). Based on this concept, it is meaningful to look into how circRNAs influence stem cells and its role in regenerative medicine. In this chapter we will discuss the functional roles of circRNAs in the prevention, repair, or progression of chronic diseases, through the communication between stem cells.

Keywords

Circular RNA · Regenerative medicine · Stem cell

R. Y. Cao · Q. Li · J. Yang (✉)
Zhongshan-Xuhui Hospital, Fudan University/Shanghai Clinical Research Center, Chinese Academy of Sciences, Shanghai, China

Q. Dai
MetroWest Medical Center, Framingham, MA, USA

Department of Cardiology, First Affiliated Hospital of Nanjing Medical University, Nanjing, China

1 Introduction

In recent years, chronic diseases like cardiovascular diseases, cancer, diabetes, and chronic lung diseases account for more than 70% of patient deaths in the world [1]. Regenerative medicine refers to therapies that aim at creating replacements for the tissue or organ function loss due to trauma or diseases. Researches on regenerative medicine mainly focus on human cells, including somatic cells, stem cells, and embryo-derived cells. Unlike conventional medical technology, regenerative medicine avoids the immune response by using patient's own cells [2].

At present, various approaches of regenerative medicine have been found to repair damaged tissues or cells [3]. It brings hope for many incurable diseases [4]. Besides, the application of regenerative medicine opens up a new era of health care and promotes radical innovations in patient management. One of the approaches is the use of embryonic stem cells (ESCs) in regenerative medicine [5]. ESCs play an important role in the tissue regeneration in various diseases due to their ability of self-renewal [6]. They are widely studied in stem cell therapy and tissue engineering [7].

Circular RNA (circRNA) is a group of RNA which has a covalently closed structure. Usually circRNAs do not code proteins, but they have been found to have significant regulatory function in protein expressions. In addition, recent

J. Xiao (ed.), *Circular RNAs*, Advances in Experimental Medicine and Biology 1087,
https://doi.org/10.1007/978-981-13-1426-1_24

studies have found that they were essential for the early growth of various species. With its interaction with certain microRNAs, somatic cells could be transformed back to their pluripotent state [8–10]. circRNAs are prevalently expressed in undifferentiated human embryonic stem cells (hESCs) [11]. circRNAs regulate gene transcription mainly by acting as a "microRNA sponge." MicroRNAs are another noncoding RNA group that regulates gene expression. ESCs differentiate under the interaction of microRNAs and circRNAs [12]. Because of the abundant number of noncoding RNAs and the complexity in their interactions, few circRNAs have been well studied.

In this chapter, we will discuss the impact of circRNAs on ESCs and their use in regenerative medicine.

2 The Status of Regenerative Medicine

Human life has been greatly extended with the improvement of medical care in the past decades. Along with this privilege, aging-associated disease and aging itself have become a problem [13]. Aging-associated disease includes atherosclerosis, cardiovascular disease, cancer, arthritis, etc. Aging is also an independent risk factor for many diseases. These health problems greatly affect patients' life quality. Each year, about 100 million people suffer from different forms of tissue and organ damage in the world [14].

Transplantation has been the only therapy for permanent organ damage. However, the source of organs hardly ever met the demand. Not to mention the huge cost of the transplantation, the other huge concern is the immune rejection it brings. Patients have to take immunosuppressant after the transplantation, which further increases other complications such as infection [15, 16]. The appearance of regenerative medicine has brought significant changes in treating aging-associated diseases. As part of regenerative medicine, stem cell therapy has achieved great success in recent years [17, 18].

Stem cells have long been studied to explore disease pathophysiology, drug screening, and safety evaluation [17]. The application of stem cells in clinical practice has been increasing, and the benefit it brings is tremendous [18]. The advantage of stem cells or ancestral/precursor cells is that they can be quickly generated and have the potential of differentiation [19]. Combining bioengineering techniques and new biomaterials, it is possible to create human organ [20, 21].

Hematopoietic stem cell transplantation has been successfully used in treating leukemia, lymphoma, and multiple myeloma [22]. It reveals the next frontier of medical treatment, with the aim of achieving structural or functional repair [23]. Indeed, hematopoietic stem cell transplant can also be applied in other diseases including but not limited to metabolic storage diseases and extracellular matrix disorders like epidermolysis bullosa [24].

Apart from saving life, regenerative medicine could improve patients' quality of life [25, 26]. Based on the continuous medical and social needs of regenerative science, there is an urgent need to design, implement, and validate feasible models [27]. Regenerative medicine will benefit to both patients and stakeholders.

2.1 The Definition and Source of Stem Cells

Stem cells are a class of cells with the ability of self-renewal and multi-differentiation. They include embryonic stem cells isolated from in vivo embryos, pluripotent stem cells induced in vitro, and adult stem cells [28]. Stem cells are also used as "seed" cells for screening and treating genetic diseases and the construction of organs in vitro. They are of great value in regenerative medicine [29]. Stem cells can be further divided into embryonic stem cells (ES cells) and somatic stem cells, depending on the stage of development [30]. Stem cells such as neural stem cells, adipose stem cells, epidermal stem cells, and mesenchymal stem cells could be found in adult organs [31]. These stem cells referred to adult stem cells which are as effective in treating various diseases [32].

In 1998, Professor Thomson [33] of the University of Wisconsin successfully obtained human embryonic stem cell lines from human fetal germ cells. It is the first study of human embryonic stem cells and has been recognized as a milestone in the study of stem cells. At the same time, Professor John Gilhart of Johns Hopkins University cultivated the first human embryonic germ cell line and established a human pluripotent stem cell line [34]. These achievements marked the initiation of regenerative medicine. In theory, ESCs could provide as a source of transplants. However, immunologic rejection to differentiated ESCs still exists [35].

2.2 The Frontier Development of Stem Cell and Regenerative Medicine

2.2.1 Use Stem Cells Derived from the Body to Regenerate Treatment

The most commonly used stem cells are hematopoietic stem cells, mesenchymal stem cells, etc. [36]. Hematopoietic stem cells are a group of adult stem cells which give rise to blood cells [37]. Researches on hematopoietic stem cells have been greatly prospered by improvement of mouse models. The application of advanced genetic techniques has established different models for various studies [38]. With the help of single-cell technology, scientists successfully analyzed the gene expression changes during the transformation from hematopoietic stem cell precursors to hematopoietic stem cells [39].

Hematopoietic stem cells are the earliest type of stem cells that have been used for clinical treatment [40–42], such as bone marrow transplantation [43]. Years of practice has made hematopoietic stem cell transplantation to be the most mature type of regenerative medicine [44].

2.2.2 Regeneration Therapy Using In Vitro Cultured Stem Cell Lines

Despite the remarkable achievements in clinical application of hematopoietic stem cells, other types of stem cells have not yet been well studied due to their rarity and the difficulty in obtaining them [45].

One resolution is to establish stem cell lines by in vitro culture techniques. The other way is to induce differentiation on ESCs [46]. In order to prevent transplant rejection, patient's own stem cells are often used [47]. Studies found that stem cells could also be obtained from somatic cells. In 2006, Professor Shinya Yamanaka invented a new way to reprogram differentiated cells back to ESCs by transferring a combination of four transcription factors (Oct4, Sox2, Klf4, and c-Myc) into these differentiated somatic cells. These ESC-like cells are named as iPSC. The discovery of iPSC makes it possible to produce large numbers of individual-specific, well-differentiated stem cells [48].

2.2.3 Small Molecule Compound-Induced Cell Reprogramming

Small molecule compounds play a key role in the development of medicine [49]. The advanced biotechnology enables rapid progressing in cellular studies [50]. There are three major issues in the reprogramming of somatic cells, pluripotent cell differentiation, transdifferentiation, and pluripotency of embryonic stem cells.

The invention of the technology to induce pluripotent stem cell enables humans to obtain autologous cell and do organ transplantation. It is a milestone in the history of regenerative medicine [51]. Compared to genetic therapy, small molecule has the advantages of no genome integration, simple operation, easy-to-controllable dose, and able to be reversed [52]. Studies have shown that small molecules can influence cell fate through epigenetic modifications, signaling pathways, and metabolic kinases. These lead to improved reprogramming efficiency and replacement of relevant transcription factors [53]. The use of small molecules to regulate the signaling pathway can also significantly affect the resonance cell reprogramming process [54]. Transdifferentiation of pluripotent cells into cells of certain lineages or cells of different lineages is an important way to obtain functional cells in vitro [55–59].

2.3 The Application of Stem Cell and Regenerative Medicine

2.3.1 Stem Cells and Diabetes Treatment

Diabetes is one of the most common debilitating diseases that threaten human health [60]. At present, about 150 million people worldwide suffer from diabetes. Both type 1 diabetes and type 2 diabetes result from absolute or relative deficiency of insulin secretion. Insulin deficiency will result in dysregulation of sugar, lipids, proteins, and electrolytes [61]. Drug treatment and long-term injection of exogenous insulin are the main treatments for diabetes. However, these methods do not cure the disease, nor do they prevent the diabetic complications. Islet transplantation could potentially cure diabetes. In 1990, Scharp et al. reported that the first human allogeneic islet cell transplantation has been successfully used to treat type 1 diabetes [62]. So far, about 1000 diabetic patients have been treated with islet cell transplantation.

Suberi et al. used human embryonic stem cells (Hes-H9) to produce cells with β-cell characteristics [63]. Assady et al. [64] and Lumelsky et al. [65] reported that human ESCs could differentiate into insulin-producing cells. Bone marrow mesenchymal cells are found to be effective in treating type 1 diabetic mice [66]. Using islet cell transplantation is a promising treatment for diabetes, although it remains in its early stage with a lot of unsolved problems.

2.3.2 Stem Cells and Cardiomyocyte Injury Treatment

Cardiovascular disease could induce ventricular wall remodeling and thus decrease cardiac function. In acute setting, restoration of blood flow is the best way to prevent damage to the heart muscle. It became more complicated if revascularization is delayed or failed, since the dead cardiac cells cannot be replaced by functional cells. For a long time, heart has been recognized as a nonregenerative organ. The emergence of regenerative medicine makes it theoretically possible to keep cardiac function from ischemic state. In a study, bone marrow stem cells were injected directly into the coronary arteries in six patients with myocardial infarction. After 10 weeks, the infarct size was reduced by nearly 1/3, and their cardiac function was improved as well [67]. The REPAIR-AMI trial (reinfusion of enriched progenitor cells and infarcted remodeling in acute myocardial infarction) was the first randomized, double-blind, clinical study that aimed to prove the efficacy of stem cell therapy in myocardial infarction [73].

2.3.3 Stem Cells and Nerve Injury Diseases

Like myocardiocytes, neuro cells are the other group of cells with extremely limited regenerative ability. Depending on the site of injury, neuro cell damage could cause tremendous disability. The most vulnerable side is the central nervous system. The repair of central nervous injury has always been a huge challenge. Stem cells replace damaged cells to secrete neurotrophic factors that can promote regeneration, protect neurons, and thus reduce secondary damage. The process of stem cell therapy includes forming bridge-guided nerve regeneration in injured areas, digesting collagenous scars, removing cell debris, regulating immune responses, and repairing spinal cord nonnervous tissue. In theory, stem cell therapy is plausible in treatment of neuro-injury. However, the human nervous system may be more complicated than we thought. How the implanted stem cell will interact with surrounding cells remains uncertain. Plus ethical challenges are daunting, making large clinical trials very difficult.

3 Discovery of circRNA

circRNA was first discovered in the RNA virus in the 1970s [68]. In 1990, scientists used two-dimensional gel electrophoresis and electron microscopy to observe the circRNA molecule in a *Saccharomyces* fungus [69]. In the following decades, some circular RNA transcripts were also found in other transcripts, such as the deletion in colorectal carcinoma gene (DCC) transcripts [70] and human Ets-1 gene (ETwenty-Six-1, Ets-1) transcript [71], sex-determining region Y (SRY) transcript [72], cytochrome P450 2C24 gene transcript [73], circular INK4 gene block antisense noncoding RNA in the INK4 locus (cANRIL) [74], etc.

In recent years, with the rapid development of bioinformatic technology, great achievements have been made in exploring circular RNA function [75]. Salzman et al. [76] discovered a good amount of circRNAs related to human gene expression. Jeck et al. [77] detected up to 25,000 circRNAs in human fibroblasts. Memczak et al. [78] identified 1950 human circRNAs, 1903 mouse circRNAs, and 724 nematode circRNAs from RNA sequencing data. Guo et al. [79] found 7112 circRNAs in 39 human cell lines. The circRNAs discovered so far can be divided into the following three categories based on their origins and constituent sequences: exonic circRNAs [76, 77, 80], intronic circular RNA (circRNA) [81], and retained-intron circRNA [82].

4 The Main Function of circRNAs

Salmena et al. [83] proposed the famous competing endogenous RNA (ceRNA) regulation hypothesis in which the biological function of ceRNA is accomplished through miRNA response element (MRE). ceRNAs have different numbers and types of MREs that can competitively bind miRNAs and reduce the inhibitory effect of miRNAs. There are multiple miRNA complementary binding sites on the circRNA, which can absorb miRNAs, just like sponges. On the other hand, circRNA lacks poly (A) tails and 5′ ends, which enables them to escape degradation [84]. As a result, even a small portion of circRNAs can inhibit a large number of miRNAs [85].

Zhang et al. [81] identified a group of circRNAs which are abundantly expressed in nucleus and can guide gene transcription. Ci-ankyrin repeat domain 52 (ci-ankrd52) is one of them. It is located near the transcription site and can affect the elongation of the RNA polymerase II complex. The positive regulator of the complex exerts a cis-regulatory effect on its maternal gene. Silent information regulator 7 (ci-sirt7) also has a similar function. The INK4/ARF site-associated long-chain noncoding RNAANRIL inhibits the transcription of the coding gene INK4/ARF by binding to the PcG complex. This site also encodes cANRIL which also has transcriptional regulatory function [74].

Some circRNAs can bind to proteins. For example, CDR1as and Sry can bind to the miRNA effector Argonaute (AGO) [86] and inhibit miRNAs-mRNA cleavage. CircRNA52 interacts with RNA polymerase II complexes and affects gene transcription [87]. In addition, Bohjanen et al. [88] designed a circRNA that specifically binds to the transactivating regulatory protein (Tat), which inhibited the expression of human immunodeficiency virus type 1 (HIV-1) gene.

Most circRNAs are present in the cytoplasm [76], suggesting that they can be loaded into the ribosome and translated into peptides. Like many linear mRNAs without a 5′ cap and 3′ poly (A) tail, circRNA lacks a valid structure to initiate translation. Some circRNAs are able to code protein. Once an internal ribosome entry site (IRES) is activated, circRNAs could be translatable. For example, the core of hepatitis D virus (HDV) contains a negative single-stranded circRNA, which encodes the related protein HDV antigen (HDAg) and plays a role in disease progression.

In human osteosarcoma cells, some circular RNAs are found to be able to code protein, although their translation efficiency is very low [89]. In addition, Talhouarne et al. [90] found that the oocyte nuclei contained a large number of circRNAs, most of which are 1000 nt in length and resistant to exonuclease RNase R. Their content is significantly higher than in the nucleus. During the development of fertilized eggs, circRNAs can be transmitted to offspring, indicating that they play an important role in RNA-mediated genetic inheritance.

As more and more ribosomal data is available, whether circRNA can be translated in other cell types or species is worthy of further study [91].

5 The Function of circRNAs in Stem Cell Pluripotency and Reprogramming

circRNAs are a new class of noncoding RNAs, which have been studied in the recent 20 years. They are covalently closed, single-stranded, and found to be prevalent in eukaryotes [71].

The development of high-throughput sequencing technology facilitated the discovery of many different circRNAs. However, few studies have reported their functions in somatic reprogramming or embryonic stem cells. Like other noncoding RNAs, circRNAs work by regulating gene expression. In addition, circRNAs also affect protein transcription by "sponging" certain microRNAs. It was found that circBIRC6 is involved in regulation of stem cell pluripotency [92]. The study reported that a specific cleavage factor, ESRP1, regulated the formation of circBIRC6. CircBIRC6 binds to miR-34a and miR-145 and regulates stem cell pluripotency [93, 94]. ESRP1 is regulated by transcription factors Oct4 and Nanog. This is the first study that reveals the functions of circRNAs in stem cell pluripotency.

5.1 How to Screen for Pluripotency Related to circRNAs

Stem cell pluripotency-related circRNAs are selected by high-throughput RNA sequencing. Differentially expressed circRNAs were analyzed the during ESCs differentiation. Pluripotency-related circRNAs are recorded in the database published in 2013 [78, 95].

In another study, researchers compared the expression changes of circRNAs before and after hESCs differentiated into embryoid bodies (EB spheres). It also measured the changes of corresponding linear mRNA. Among the differentially expressed 61 circRNAs, 11 were specifically expressed in hESCs [96]. The authors also validated that in the induced pluripotent stem cell (iPSC) system, all 11 circRNAs were enriched in iPSCs [97]. The expression of circRNA is often accompanied by the expression of the corresponding linear gene [96]. After the comparison analysis, the authors found that the expression pattern of the three circRNAs is not consistent with their corresponding linear genes. These three circRNAs are circBIRC6, circMAN1A2, and circILKAP. Further, the authors analyzed the correlation between the expression of circRNAs and the pluripotent state of stem cells using Northern blot. The results indicated that there was a strong correlation between the expression of three circRNAs, namely, circBIRC6, circCORO1C, and circMAN1A2, and the pluripotent state.

The above studies have demonstrated the correlation between the expression pattern of the three circRNAs and the pluripotency state of cells. However, it remains unknown whether these circRNAs are involved in the regulation of stem cell pluripotency [98]. Therefore, an RNA interference experiment was designed. The results showed that the positive rate of AP staining was significantly reduced after interfering with circBIRC6 and circCORO1C, but the effect after interfering with circMAN1A2 was not obvious. Differentiated status-related transcription factors and marker genes showed similar results.

5.2 circRNA, Not Linear RNA, Is Involved in the Regulation of Stem Cell Pluripotency

circRNA is formed by the processing precursors of mRNA [99]. In order to explore whether these precursors are involved, researchers designed a type of shRNAs that specifically targets linear RNAs and common exonic regions. Targeting the corresponding exon regions can simultaneously interfere with linear RNAs and circRNAs independently. The AP staining results showed that the linear RNAs interfere with BIRC6 and

CORO1C alone did not affect the pluripotent state of the cells. Q-PCR and immunofluorescence assays were performed to analyze pluripotency and differentiation-related genes. The results showed that the circRNAs are involved in the regulation of cellular pluripotency, while linear mRNAs did not directly affect the cellular pluripotency state.

Interfering with circBIRC6 and circCORO1C can affect the pluripotent state of stem cells. Overexpression of them can promote the transition to a pluripotent state [100]. Reporter gene experiments showed that the expression of these circRNAs alone is not sufficient to induce cell reprogramming. In contrast, in the classical four-factor (Oct4, Sox2, Klf4, and Myc, OSKM)-induced iPSC system, overexpression of circBIRC6 and circCORO1C can significantly promote cell reprogramming. Based on preliminary screening and functional studies, the authors claim that circBIRC6 and circCORO1C are involved in the regulation of stem cell pluripotency and overexpression of them can promote reprogramming of somatic cells.

6 Conclusions

Different from linear RNAs, circRNAs are covalently closed RNAs that regulate gene expression. circRNAs are characterized by high stability, high conservation, and tissue specificity. They can interact with miRNAs by sponging them and form a complicated complex to regulate gene transcription. Rapid development of high-throughput sequencing technology and bioinformatics provides us with great tools for the further study of circRNAs. More and more circRNAs have been discovered. We have already made great progress in associating them with disease and exploring their clinical significance. Some of them have already been put into clinical trials. However, due to the unclear circRNA biogenesis and their complicated interactions with other molecules, more studies need to be done to explore this huge but fascinating network.

Competing Financial Interests The authors declare no competing financial interests.

References

1. Mozaffarian D, Benjamin EJ, Go AS et al (2015) Heart disease and stroke statistics–2015 update: a report from the American Heart Association. Circulation 131(4):e29–e322
2. Mason C, Dunnill P (2008) A brief definition of regenerative medicine. Regen Med 3(1):1–5
3. Keshtkar S, Azarpira N, Ghahremani MH (2018) Mesenchymal stem cell-derived extracellular vesicles: novel frontiers in regenerative medicine. Stem Cell Res Ther 9(1):63
4. Daley GQ, Scadden DT (2008) Prospects for stem cell-based therapy. Cell 132(4):544–548
5. Arjmand B, Goodarzi P, Mohamadi-Jahani F et al (2017) Personalized regenerative medicine. Acta Med Iran 55(3):144–149
6. Dushnik-Levinson M, Benvenisty N (1995) Embryogenesis in vitro: study of differentiation of embryonic stem cells. Biol Neonate 67(2):77–83
7. Fox IJ, Daley GQ, Goldman SA et al (2014) Stem cell therapy. Use of differentiated pluripotent stem cells as replacement therapy for treating disease. Science 345(6199):1247391
8. Li F, Zhang L, Li W et al (2015) Circular RNA ITCH has inhibitory effect on ESCC by suppressing the Wnt/beta-catenin pathway. Oncotarget 6(8):6001–6013
9. Yang W, Du WW, Li X et al (2016) Foxo3 activity promoted by non-coding effects of circular RNA and Foxo3 pseudogene in the inhibition of tumor growth and angiogenesis. Oncogene 35(30):3919–3931
10. Houbaviy HB, Murray MF, Sharp PA (2003) Embryonic stem cell-specific MicroRNAs. Dev Cell 5(2):351–358
11. Itskovitz-Eldor J, Schuldiner M, Karsenti D et al (2000) Differentiation of human embryonic stem cells into embryoid bodies compromising the three embryonic germ layers. Mol Med 6(2):88–95
12. Zheng Q, Bao C, Guo W et al (2016) Circular RNA profiling reveals an abundant circHIPK3 that regulates cell growth by sponging multiple miRNAs. Nat Commun 7:11215
13. Spector M (2018) Biomedical materials to meet the challenges of the aging epidemic. Biomed Mater 13(3):030201
14. McMahon DS, Thorsteinsdottir H, Singer PA et al (2010) Cultivating regenerative medicine innovation in China. Regen Med 5(1):35–44
15. Mironov V, Visconti RP, Markwald RR (2004) What is regenerative medicine? Emergence of applied stem cell and developmental biology. Expert Opin Biol Ther 4(6):773–781
16. Kemp P (2006) History of regenerative medicine: looking backwards to move forwards. Regen Med 1(5):653–669
17. Buganim Y (2016) Back to basics: refined nuclear reprogramming techniques yield higher-quality stem cells. Science 352(6292):1401

18. Niwa H (2018) The principles that govern transcription factor network functions in stem cells. Development 145(6):dev157420
19. Yu J, Vodyanik MA, Smuga-Otto K et al (2007) Induced pluripotent stem cell lines derived from human somatic cells. Science 318(5858):1917–1920
20. Thrall JH (2004) Personalized medicine. Radiology 231(3):613–616
21. Murphy MJ, Shahriari N, Payette M et al (2016) Development of a curriculum in molecular diagnostics, genomics and personalized medicine for dermatology trainees. J Cutan Pathol 43(10):858–865
22. Terzic A, Nelson TJ (2010) Regenerative medicine advancing health care 2020. J Am Coll Cardiol 55(20):2254–2257
23. Das AK, Pal R (2010) Induced pluripotent stem cells (iPSCs): the emergence of a new champion in stem cell technology-driven biomedical applications. J Tissue Eng Regen Med 4(6):413–421
24. Terzic A, Nelson TJ (2013) Regenerative medicine primer. Mayo Clin Proc 88(7):766–775
25. Kordelas L, da Silva Nardi F, Wagner B, et al (2018) Elevated soluble human leukocyte antigen G levels in patients after allogeneic stem cell transplantation are associated with less severe acute and chronic graft-versus-host disease. Bone Marrow Transplant: https://doi.org/10.1038/s41409-018-0145-1
26. Safinia N, Grageda N, Scotta C et al (2018) Cell therapy in organ transplantation: our experience on the clinical translation of regulatory T cells. Front Immunol 9:354
27. Lo Monaco M, Merckx G, Ratajczak J et al (2018) Stem cells for cartilage repair: preclinical studies and insights in translational animal models and outcome measures. Stem Cells Int 2018:9079538
28. Esteban MA, Wang T, Qin B et al (2010) Vitamin C enhances the generation of mouse and human induced pluripotent stem cells. Cell Stem Cell 6(1):71–79
29. Onteniente B (2013) The multiple aspects of stroke and stem cell therapy. Curr Mol Med 13(5):821–831
30. Zhang M, Ngo J, Pirozzi F et al (2018) Highly efficient methods to obtain homogeneous dorsal neural progenitor cells from human and mouse embryonic stem cells and induced pluripotent stem cells. Stem Cell Res Ther 9(1):67
31. Evans MJ, Kaufman MH (1981) Establishment in culture of pluripotential cells from mouse embryos. Nature 292(5819):154–156
32. Alison MR, Islam S (2009) Attributes of adult stem cells. J Pathol 217(2):144–160
33. Thomson JA, Itskovitz-Eldor J, Shapiro SS et al (1998) Embryonic stem cell lines derived from human blastocysts. Science 282(5391):1145–1147
34. Shamblott MJ, Axelman J, Wang S et al (1998) Derivation of pluripotent stem cells from cultured human primordial germ cells. Proc Natl Acad Sci U S A 95(23):13726–13731
35. Fagoonee S, Hobbs RM, De Chiara L et al (2010) Generation of functional hepatocytes from mouse germ line cell-derived pluripotent stem cells in vitro. Stem Cells Dev 19(8):1183–1194
36. Fagoonee S, Pellicano R, Altruda F (2010) Stem cell world 2010: crucial questions from clinicians to biologists. Panminerva Med 52(2):95–96
37. Butler AE, Huang A, Rao PN et al (2007) Hematopoietic stem cells derived from adult donors are not a source of pancreatic beta-cells in adult nondiabetic humans. Diabetes 56(7):1810–1816
38. Alonso-Ferrero ME, van Til NP, Bartolovic K et al (2018) Enhancement of mouse hematopoietic stem/progenitor cell function via transient gene delivery using integration-deficient lentiviral vectors. Exp Hematol 57:21–29
39. Wang Y, Tian H, Cai W et al (2018) Tracking hematopoietic precursor division ex vivo in real time. Stem Cell Res Ther 9(1):16
40. Kemp CB, Knight MJ, Scharp DW et al (1973) Transplantation of isolated pancreatic islets into the portal vein of diabetic rats. Nature 244(5416):447
41. Ricordi C, Lacy PE, Finke EH et al (1988) Automated method for isolation of human pancreatic islets. Diabetes 37(4):413–420
42. Ryan EA, Paty BW, Senior PA et al (2005) Five-year follow-up after clinical islet transplantation. Diabetes 54(7):2060–2069
43. Lacotte S, Berney T, Shapiro AJ et al (2011) Immune monitoring of pancreatic islet graft: towards a better understanding, detection and treatment of harmful events. Expert Opin Biol Ther 11(1):55–66
44. Copelan EA (2006) Hematopoietic stem-cell transplantation. N Engl J Med 354(17):1813–1826
45. Corbett MS, Webster A, Hawkins R et al (2017) Innovative regenerative medicines in the EU: a better future in evidence? BMC Med 15(1):49
46. Nishiga M, Guo H, Wu JC (2018) Induced pluripotent stem cells as a biopharmaceutical factory for extracellular vesicles. Eur Heart J 39:1848. https://doi.org/10.1093/eurheartj/ehy097
47. Giacalone JC, Sharma TP, Burnight ER et al (2018) CRISPR-Cas9-based genome editing of human induced pluripotent stem cells. Curr Protoc Stem Cell Biol 44:5B 7 1–5B 7 22
48. Moldovan L, Barnard A, Gil CH et al (2017) iPSC-derived vascular cell spheroids as building blocks for scaffold-free biofabrication. Biotechnol J 12(12):1700444
49. Sangeetha KN, Sujatha S, Muthusamy VS et al (2017) Current trends in small molecule discovery targeting key cellular signaling events towards the combined management of diabetes and obesity. Bioinformation 13(12):394–399
50. Wolpaw AJ, Stockwell BR (2014) Multidimensional profiling in the investigation of small-molecule-induced cell death. Methods Enzymol 545:265–302
51. Yamanaka S, Takahashi K (2006) Induction of pluripotent stem cells from mouse fibroblast cultures. Tanpakushitsu Kakusan Koso 51(15):2346–2351
52. Yao Y, Li S, Cao J et al (2018) Development of small molecule biosensors by coupling the recognition of

the bacterial allosteric transcription factor with isothermal strand displacement amplification. Chem Commun (Camb) 54:4774. https://doi.org/10.1039/C8CC01764F
53. Li W, Li K, Wei W et al (2013) Chemical approaches to stem cell biology and therapeutics. Cell Stem Cell 13(3):270–283
54. Marson A, Foreman R, Chevalier B et al (2008) Wnt signaling promotes reprogramming of somatic cells to pluripotency. Cell Stem Cell 3(2):132–135
55. D'Amour KA, Agulnick AD, Eliazer S et al (2005) Efficient differentiation of human embryonic stem cells to definitive endoderm. Nat Biotechnol 23(12):1534–1541
56. Efe JA, Hilcove S, Kim J et al (2011) Conversion of mouse fibroblasts into cardiomyocytes using a direct reprogramming strategy. Nat Cell Biol 13(3):215–222
57. Graf T (2011) Historical origins of transdifferentiation and reprogramming. Cell Stem Cell 9(6):504–516
58. Sekiya S, Suzuki A (2011) Direct conversion of mouse fibroblasts to hepatocyte-like cells by defined factors. Nature 475(7356):390–393
59. Huang P, He Z, Ji S et al (2011) Induction of functional hepatocyte-like cells from mouse fibroblasts by defined factors. Nature 475(7356):386–389
60. Saifan C, Nasr R, Mehta S et al (2013) Diabetes insipidus: a challenging diagnosis with new drug therapies. ISRN Nephrol 2013:797620
61. Devendra D, Liu E, Eisenbarth GS (2004) Type 1 diabetes: recent developments. BMJ 328(7442):750–754
62. Scharp DW, Lacy PE, Santiago JV et al (1990) Insulin independence after islet transplantation into type I diabetic patient. Diabetes 39(4):515–518
63. Suberi M, McKeever WF (1977) Differential right hemispheric memory storage of emotional and nonemotional faces. Neuropsychologia 15(6):757–768
64. Assady S, Maor G, Amit M et al (2001) Insulin production by human embryonic stem cells. Diabetes 50(8):1691–1697
65. Lumelsky N, Blondel O, Laeng P et al (2001) Differentiation of embryonic stem cells to insulin-secreting structures similar to pancreatic islets. Science 292(5520):1389–1394
66. Urban VS, Kiss J, Kovacs J et al (2008) Mesenchymal stem cells cooperate with bone marrow cells in therapy of diabetes. Stem Cells 26(1):244–253
67. Strauer BE, Brehm M, Zeus T et al (2001) Intracoronary, human autologous stem cell transplantation for myocardial regeneration following myocardial infarction. Dtsch Med Wochenschr 126(34–35):932–938
68. Wu Q, Wang Y, Cao M et al (2012) Homology-independent discovery of replicating pathogenic circular RNAs by deep sequencing and a new computational algorithm. Proc Natl Acad Sci U S A 109(10):3938–3943
69. Matsumoto Y, Fishel R, Wickner RB (1990) Circular single-stranded RNA replicon in Saccharomyces cerevisiae. Proc Natl Acad Sci U S A 87(19):7628–7632
70. Pasman Z, Been MD, Garcia-Blanco MA (1996) Exon circularization in mammalian nuclear extracts. RNA 2(6):603–610
71. Cocquerelle C, Daubersies P, Majerus MA et al (1992) Splicing with inverted order of exons occurs proximal to large introns. EMBO J 11(3):1095–1098
72. Capel B, Swain A, Nicolis S et al (1993) Circular transcripts of the testis-determining gene Sry in adult mouse testis. Cell 73(5):1019–1030
73. Zaphiropoulos PG (1997) Exon skipping and circular RNA formation in transcripts of the human cytochrome P-450 2C18 gene in epidermis and of the rat androgen binding protein gene in testis. Mol Cell Biol 17(6):2985–2993
74. Burd CE, Jeck WR, Liu Y et al (2010) Expression of linear and novel circular forms of an INK4/ARF-associated non-coding RNA correlates with atherosclerosis risk. PLoS Genet 6(12):e1001233
75. Wang PL, Bao Y, Yee MC et al (2014) Circular RNA is expressed across the eukaryotic tree of life. PLoS One 9(6):e90859
76. Salzman J, Gawad C, Wang PL et al (2012) Circular RNAs are the predominant transcript isoform from hundreds of human genes in diverse cell types. PLoS One 7(2):e30733
77. Jeck WR, Sorrentino JA, Wang K et al (2013) Circular RNAs are abundant, conserved, and associated with ALU repeats. RNA 19(2):141–157
78. Memczak S, Jens M, Elefsinioti A et al (2013) Circular RNAs are a large class of animal RNAs with regulatory potency. Nature 495(7441):333–338
79. Guo JU, Agarwal V, Guo H et al (2014) Expanded identification and characterization of mammalian circular RNAs. Genome Biol 15(7):409
80. Danan M, Schwartz S, Edelheit S et al (2012) Transcriptome-wide discovery of circular RNAs in archaea. Nucleic Acids Res 40(7):3131–3142
81. Zhang Y, Zhang XO, Chen T et al (2013) Circular intronic long noncoding RNAs. Mol Cell 51(6):792–806
82. Salzman J, Chen RE, Olsen MN et al (2013) Cell-type specific features of circular RNA expression. PLoS Genet 9(9):e1003777
83. Salmena L, Poliseno L, Tay Y et al (2011) A ceRNA hypothesis: the Rosetta Stone of a hidden RNA language? Cell 146(3):353–358
84. Huntzinger E, Izaurralde E (2011) Gene silencing by microRNAs: contributions of translational repression and mRNA decay. Nat Rev Genet 12(2):99–110
85. Hansen TB, Wiklund ED, Bramsen JB et al (2011) miRNA-dependent gene silencing involving Ago2-mediated cleavage of a circular antisense RNA. EMBO J 30(21):4414–4422
86. Peng Y, Song X, Zheng Y et al (2017) Circular RNA profiling reveals that circCOL3A1-859267 regulate type I collagen expression in photo aged human

dermal fibroblasts. Biochem Biophys Res Commun 486(2):277–284
87. Wilusz JE (2017) Circular RNAs: unexpected outputs of many protein-coding genes. RNA Biol 14(8):1007–1017
88. Bohjanen PR, Colvin RA, Puttaraju M et al (1996) A small circular TAR RNA decoy specifically inhibits tat-activated HIV-1 transcription. Nucleic Acids Res 24(19):3733–3738
89. Bolha L, Ravnik-Glavac M, Glavac D (2017) Circular RNAs: biogenesis, function, and a role as possible Cancer biomarkers. Int J Genomics 2017:6218353
90. Talhouarne GJ, Gall JG (2014) Lariat intronic RNAs in the cytoplasm of Xenopus tropicalis oocytes. RNA 20(9):1476–1487
91. Yang Y, Wang Z (2018) Constructing GFP-based reporter to study back splicing and translation of circular RNA. Methods Mol Biol 1724:107–118
92. Yu CY, Li TC, Wu YY et al (2017) The circular RNA circBIRC6 participates in the molecular circuitry controlling human pluripotency. Nat Commun 8(1):1149
93. Choi YJ, Lin CP, Ho JJ et al (2011) miR-34 miRNAs provide a barrier for somatic cell reprogramming. Nat Cell Biol 13(11):1353–1360
94. Xu N, Papagiannakopoulos T, Pan G et al (2009) MicroRNA-145 regulates OCT4, SOX2, and KLF4 and represses pluripotency in human embryonic stem cells. Cell 137(4):647–658
95. Wu CS, Yu CY, Chuang CY et al (2014) Integrative transcriptome sequencing identifies trans-splicing events with important roles in human embryonic stem cell pluripotency. Genome Res 24(1):25–36
96. Suzuki H, Zuo Y, Wang J et al (2006) Characterization of RNase R-digested cellular RNA source that consists of lariat and circular RNAs from pre-mRNA splicing. Nucleic Acids Res 34(8):e63
97. Buganim Y, Markoulaki S, van Wietmarschen N et al (2014) The developmental potential of iPSCs is greatly influenced by reprogramming factor selection. Cell Stem Cell 15(3):295–309
98. Gabut M, Samavarchi-Tehrani P, Wang X et al (2011) An alternative splicing switch regulates embryonic stem cell pluripotency and reprogramming. Cell 147(1):132–146
99. Ashwal-Fluss R, Meyer M, Pamudurti NR et al (2014) circRNA biogenesis competes with pre-mRNA splicing. Mol Cell 56(1):55–66
100. Kawamura T, Suzuki J, Wang YV et al (2009) Linking the p53 tumour suppressor pathway to somatic cell reprogramming. Nature 460(7259):1140–1144

The Role of Circular RNAs in Cerebral Ischemic Diseases: Ischemic Stroke and Cerebral Ischemia/Reperfusion Injury

25

Jian Yang, Mengli Chen, Richard Y. Cao, Qing Li, and Fu Zhu

Abstract

Cerebral ischemic diseases including ischemic stroke and cerebral ischemia reperfusion injury can result in serious dysfunction of the brain, which leads to extremely high mortality and disability. There are no effective therapeutics for cerebral ischemic diseases to date. Circular RNAs are a kind of newly investigated noncoding RNAs. It is reported that circular RNAs are enriched in multiple organs, especially abundant in the brain, which indicates that circular RNAs may be involved in cerebral physiological and pathological processes. In this chapter, we will firstly review the pathophysiology, underlying mechanisms, and current treatments of cerebral ischemic diseases including ischemic stroke and cerebral ischemia/reperfusion injury. Secondly, the characteristics and function of circular RNAs will be outlined, and then we are going to introduce the roles circular RNAs play in human diseases. Finally, we will summarize the function of circular RNAs in cerebral ischemic diseases.

J. Yang · R. Y. Cao · Q. Li · F. Zhu (✉)
Zhongshan-Xuhui Hospital, Fudan University/Shanghai Clinical Research Center, Chinese Academy of Sciences, Shanghai, China

M. Chen
Department of Cardiology, The First Affiliated Hospital of Nanjing Medical University, Nanjing, China

Keywords

Circular RNAs · Cerebral ischemic diseases · Ischemic stroke · Cerebral ischemia/reperfusion injury

1 Introduction

The brain is the most sensitive organ to hypoxia. When cerebral blood flow is suddenly interrupted, tissues are deprived of oxygen and glucose, leading to dysfunction of brain as well as other parts of body controlled by the cerebral ischemic regions, which is called ischemic stroke or cerebral ischemia/reperfusion injury without or with restoration of blood flow and oxygen [1]. Ischemia elicits tissue anoxia which is the basis of ischemic injury and primes the tissue for subsequent reperfusion damage [2]. Ischemic stroke and cerebral ischemia/reperfusion injury can both result in serious dysfunction of the brain, which leads to extremely high mortality and disability [3, 4]. However, there are no effective therapeutics against cerebral ischemic diseases. Therefore, it is required to develop novel effective therapeutic strategies for treatment of cerebral ischemic diseases. As a special type of noncoding RNAs, circular RNAs play a vital regulatory role in RNA metabolism [5]. It is known that circular RNAs are abundant in various organs, especially highly enriched in synapses. The expression levels of certain circu-

J. Xiao (ed.), *Circular RNAs*, Advances in Experimental Medicine and Biology 1087,
https://doi.org/10.1007/978-981-13-1426-1_25

lar RNAs can be much higher than the canonical linear transcripts of their parent genes during the central nervous system development, which indicates that circular RNAs may be involved in cerebral physiological and pathological processes [6–8]. This chapter will be divided into three parts. In the first part, we will outline the pathophysiology, underlying mechanisms, and current treatments of cerebral ischemic diseases including ischemic stroke and cerebral ischemia/reperfusion injury. In the second part, we will review the characteristics and function of circular RNAs and then briefly introduce the roles circular RNAs play in human diseases. In the final part, we will focus on talking about the function of circular RNAs in cerebral ischemic diseases.

2 Cerebral Ischemic Diseases

2.1 Ischemic Stroke

2.1.1 Pathophysiology

Stroke is among the leading causes of deaths and disability worldwide, the high morbidity and mortality caused by stroke bring heavy financial and mental burden to society and family [9, 10]. Each year, 15 million people worldwide suffer from stroke. Ischemic stroke and hemorrhagic stroke are two main types recognized, and the former accounts for more than 80% of total cases [11]. Ischemic stroke is mainly caused by middle cerebral artery occlusion by blood clot or plaque formation which results in reduction of blood supply to the brain tissues, leading to insufficient supply of oxygen and glucose to the affected brain area [12]. Ischemic stroke is a dynamic process, during which neurons adjacent hypoperfused areas undergo necrosis since ischemia leads to disturbances in membrane integrity and cell stability. It has been estimated via serial MR brain imaging that each minute of ischemia leads to the death of 1.9 million neurons and the destruction of 14 billion synapses in human patients with middle cerebral artery occlusion [13]. The high mortality and disability caused by stroke makes it urgent for us to pay attention and to search for solutions. In order to achieve this, thorough understanding about the pathophysiology of stroke is necessary.

2.1.2 Underlying Mechanisms

The mechanisms of ischemic stroke are complicated. Currently, it is acknowledged that the mechanisms of ischemic stroke injury include metabolic disorders, inflammation, depolarization of penumbra, and cellular apoptosis [14–17].

Metabolic Disorders

The brain is the biggest consumer of energy in the human body and therefore very sensitive to reduction of oxygen and glucose. When ischemic stroke happens, blood supply to brain is blocked, which results in lack of oxygen and glucose, leading to dysfunction in energy metabolism and substance metabolism in the brain [14]. It has been demonstrated that combining NAD+ with NADPH can provide more neuroprotective effects in both cellular and animal models of ischemic stroke [18]. Besides, the downregulation of energy metabolism-related peroxisome proliferator-activated receptor gamma (PPARγ) gene was reported to prevent relapse of stroke and other vascular disorders in patients with stroke [19]. Apart from energy metabolism, substance metabolism is also involved. It is reported that an increasing glucose range quartile was positively associated with initial neurologic severity, and the association remained significant after dichotomization regarding glycated hemoglobin levels on admission [20]. Another research on pancreatic β-cell function and prognosis of nondiabetic patients with ischemic stroke demonstrates the association between pancreatic β-cell function and an increased risk of 12-month poor prognosis in nondiabetic patients with ischemic stroke [21].

Inflammation

Ischemic stroke can induce the activation inflammatory cascade. The activation of innate and adaptive immune system induced by ischemic stroke causes a massive migration of peripheral leukocytes into the brain, which triggers focal inflammatory responses, promotes tissue death, and deteriorates clinical outcome [22]. Increasing evidence suggests that antigen presentation cells (APC) are reduced in the periphery and increased in the ischemic brain both in rodent and human stroke; the accumulation is associated with expression of MHC class II molecules and the co-stimulatory molecules CD80 [23–26]. Besides, both CD4+ and CD8+ T cells are involved in cerebral ischemic injury [27]. Moreover, monocytes and macrophages play important roles in cerebral ischemic stroke because of their expression of the anti-inflammatory and pro-inflammatory mediators [15, 28]. Not only inflammatory cells but also inflammatory mediators participate in this pathological process. For instance, chemokines are considered pro-inflammatory because of their ability of mediating signals that induce leukocyte recruitment to damaged tissues. They are comprised of four subfamilies including CC, CXC, XC, and CX3C chemokines [29–31]. However, it has been reported that chemokines can induce both neuroprotective and deleterious effects on postischemic stroke patients, suggesting that more explorations are needed for the role of chemokines in ischemic stroke and the chemokines represent novel and potential therapeutic targets for prevention of ischemic stroke and reduction of consequent disability [32]. Previous studies have suggested that the expression of IL-6 is increased when ischemic stroke occurs and IL-6 is believed to act as a robust early marker for outcome in acute ischemic stroke [33, 34]. Collectively, inflammation cascade reaction, including activation of different types of immune cells and initiation of various inflammatory mediators, plays a vital role in the pathophysiology of cerebral ischemic stroke.

Cellular Apoptosis

Increasing evidence indicates that cellular apoptosis contributes to ischemic stroke injury. Along with neuronal necrosis, cellular apoptosis is another important form of delayed neuronal death following ischemic stroke [35, 36]. The family of caspases plays vital role in the pathway of apoptosis. Previous studies have shown that caspase-6 and caspase-8 are involved in neuronal apoptosis following ischemic stroke and inhibition of caspase-6 or caspase-8 can ameliorate the neuronal damage resulting from ischemic stroke [37]. Besides, it is reported that administration of mesencephalic astrocyte-derived neurotrophic factor (MANF) can protect against ischemic stroke-induced neuronal apoptosis by inhibiting the cleavage of caspase-3 [38]. PARP-1 is another important protein in cell death regulation, which is cleaved by caspase-3 into two fragments and acts as a marker for apoptotic cell death [39]. A report published on *Stroke* journal revealed that the high level of chelating zinc decreases brain damage and improves neurological functions via inhibition of PARP-1 [40]. In addition to neuronal apoptosis, platelet apoptosis is also critical in the pathogenesis of stroke. Platelet can induce caspase-3 gene expression when stimulated with ADP, which plays a key role in apoptosis signaling. Furthermore, compared to the control group, the expression level of platelet cytochrome-c and caspase-3 is significantly increased in stroke patients [41, 42]. Collectively, cellular apoptosis is a significant part of ischemic stroke pathophysiology.

Penumbra

The penumbra was classically described as the tissue between the ischemic core and normal area of patients suffering from ischemic stroke; and the blood flow of penumbra is too low to maintain electric activity but sufficient to preserve ion channels [17]. Since 1977 when penumbra was firstly defined, the physiologic characteristics of ischemic penumbra and the underlying mechanisms that mediate penumbra cell death have been deeply investigated [43]. The ultimate goal of neuroprotective treatments is to save penumbra. If not treated in time, infarction will develop inside penumbra due to consequent damage such as rapid depolarization, cell apoptosis, and inflammation like ischemic core [44]. Previous studies based on experimental cerebral ischemic stroke models have suggested that the depolarizations of penumbra are positively associated with the infarct volume, indicating that inhibition of peri-infarct depolarization maybe a therapeutic strategy for improving the poor outcome of ischemic stroke [45]. Besides, a reduction on protein synthesis in the penumbra is a sensitive metabolic result of cerebral ischemia. Moreover, the low blood flow in penumbra is responsible for reduction of adenosine 5′-triphosphate and failure of Na/K pumps, which results in an increase of intracellular calcium and consequently induces irreversible damage to brain tissues [46]. In a word, the penumbra plays an essential role in the pathophysiology and diagnosis of ischemic stroke.

2.1.3 Treatment

Clinical Treatment

The most effective treatment approach for ischemic stroke is to recover cerebral arteries blood flow occluded by thrombus or embolism. The approach is called reperfusion therapy that is required as soon as possible [47–49]. There are two types of strategies for recanalization of blocked vessels: chemical and mechanical ways [50]. Intravenous thrombolysis is the main type of chemical treatment for ischemic stroke. Until now, recombinant tissue plasminogen activator (rtPA) is the only FDA-approved drug for patients following ischemic stroke. The narrow therapeutic window that is from 3 h of the onset of a stroke until 4.5 h is usually a major limitation for its clinical application [51–54]. Thrombectomy is the main mechanical treatment for ischemic stroke. A tiny catheter is used to penetrate into the blocked cerebral vessel along the natural line, reach the focal lesion, and remove the thrombus out of the vessel [55]. Despite of the efficacy of recanalization, there is a chance that rtPA may increase the incidence of hemorrhage, which lead to more serious outcomes [13]. As for thrombectomy, it is performed to a comprehensive stroke center that limits to an artery with a diameter >2 mm [56, 57]. Early supportive care is another clinical treatment, which is used to meet basic demands for keeping patients alive [58]. Considering the limitation of recanalization in clinical application, novel therapeutic strategies are required for improved treatment of patients following ischemic stroke.

Potential Therapeutic Strategies

Since current treatment has a lot of limitation, efforts have been made in searching for novel therapeutic strategies. Stem cells can exert multipotency in treatment of cerebral ischemia. Bone marrow stromal cells (BMSCs) have the ability of self-renewal and differentiation into neuronal and glial lineages under certain conditions [59, 60]. Increasing evidence suggests that BMSCs transplantation can obviously improve the function and outcome of animal models of cerebral ischemic stroke at different time points via different transplantation routes [61]. More than 20% of the miRNAs alter in the ischemic brain, suggesting that miRNAs are key mediators in ischemic stroke biology [62–65]. miR-124 is mainly expressed in the central nerve system neuronal cells; and the expression of miR-124 in the brain is 100 times higher than in other organs [66]. It has been reported that microRNA-124 exhibits neuronal protective effects on rodent models of cerebral ischemic stroke via inhibiting cellular apoptosis [67]. Danshen is a Chinese medicinal herb widely used for treatment of ischemic brain and heart diseases. The most abundant and bioactive component in danshen is salvianolic acid B, which can attenuate apoptosis and inflammation in experimental stroke rats through activation of SIRT1 [68, 69]. Collectively, numerous experimental studies on cerebral ischemic stroke have provided multiple novel therapeutic strategies.

2.2 Cerebral Ischemia/Reperfusion Injury

2.2.1 Pathophysiology

Stroke is the fifth leading cause of death and a major cause of adult disability [4]. The most effective therapeutic method is immediate recovering the blood flow via recanalization of the occluded arteries. Immediate restoration of the blood supply can reduce more extensive brain tissue injury by salvaging a reversibly damage of penumbra [70, 71]. However, recanalization carries risks of further cellular necrosis and neural damage. Some patients may experience serious deterioration in the form of fatal edema or intracranial hemorrhage following thrombolysis, which is called cerebral ischemia/reperfusion injury [72]. Cerebral ischemia/reperfusion injury could be defined as degeneration of ischemic but salvageable brain tissue after reperfusion [73]. Although thrombolysis and embolectomy can restore blood flow of the infarcted brain tissue, the treatment can also bring the risk of reperfusion injury [50, 74]. In order to develop more effective and feasible strategies for reduction or minimization of cerebral reperfusion injury, we need to understand the pathophysiology of cerebral reperfusion injury.

2.2.2 Underlying Mechanisms

Reperfusion can recover blood flow and the supply of oxygen as well as many other energy materials. However, this process bears risks of worsening the original brain damage caused by ischemia through different mechanisms such as release of free radicals, activation of inflammatory cascade reaction, promotion of cellular apoptosis, calcium overload, and release of excitatory amino acids, which is much more sophisticated than ischemia alone [75–78].

Free Radicals

Free radicals are cytotoxic molecules that play vital roles in the progress of cerebral ischemia/reperfusion injury. There are two main types of free radicals: reactive oxygen species (ROS) and reactive nitrogen species (RNS) [75, 79]. Physiologically, the level of ROS generated from mitochondrial inner membrane is very low for maintenance of cellular redox homeostasis [80–82]. When cerebral ischemia reperfusion happens, blood flow and oxygen supply are interrupted, which leads to accumulation of excessive ROS, resulting in tissue oxidative damage in brain tissue suffered from ischemia reperfusion. More and more evidences have shown that ROS accumulation can disturb the signal transduction and induce lipid peroxidations in neural cells, which lead to cell death, activation of inflammation factors, and breakdown of blood-brain barrier [83–85]. On the other hand, RNS is also crucial for the pathophysiological process of cerebral ischemia/reperfusion injury. The two common kinds of RNS, nitric oxide (NO) and peroxynitrite ($ONOO^-$), are reported to participate in cerebral ischemia/reperfusion injury. NO is generated together with superoxide, and they can interact with each other and produce $ONOO^-$ at a diffusion-limited rate. $ONOO^-$ inactivates aconitase and superoxide dismutase (SOD) and mediates NO-induced damage of blood-brain barrier [86, 87]. The neurotoxic effect due to overproduction of free radicals plays a significant part in cerebral ischemia/reperfusion injury; therefore, intervention of excessive free radicals is critical for treatment of patients with cerebral ischemia.

Inflammation

The inappropriate reperfusion of ischemic tissues can produce numerous free radicals and many other messengers, which trigger a chain of inflammatory reaction including activation of inflammatory cells and release and interaction of inflammatory mediators [88]. The most important inflammatory cells involved in cerebral ischemia/reperfusion injury are leukocytes, microglia, and astrocytes [89]. The signals received from ischemic area can activate and induce leukocytes to secret inflammatory cytokines which in turn recruit more leukocytes. Accumulating of leukocytes and increasing cytokines related consequently deteriorate the vicious circle of inflammation [76]. Microglia are pro-macrophage cells in the brain which are intimately associated with inflammation in cerebral ischemia/reperfusion injury. In the first few minutes of ischemia, microglia are activated and transformed into cerebral macrophages as soon as cellular apoptosis occurs, resulting in neurotoxic effects in the brain [90]. Astrocytes can promote the differentiation and proliferation of microglia, enhance the phagocytic activity of macrophage, and induce generation of inflammatory mediators [91]. It has been demonstrated that interleukin-1β is able to reduce cerebral blood flow and increase the recruitment of neutrophils and generation of superoxide anion (O_2^-) in animal models of middle cerebral artery occlusion (MCAO) [92]. Plenty of reports have indicated that inflammatory mediators including cytokines, chemokines, and cell adhesion molecules can mediate the pathophysiology of cerebral ischemia/reperfusion injury [93–96].

Calcium Overload

Calcium is an important element in the body and takes a vital part in generation of bioelectricity and regulation of cellular function as well as metabolism. Recovery of blood flow in ischemic tissues significantly increases the level of calcium which leads to cellular injury. This phenomenon is called calcium overload [77, 97, 98]. Calcium overload-induced cytotoxic effects cause neural damage through the following ways. Firstly, the augmented free calcium activates calmodulin, which binds to free calcium and form a compound that cramp brain vessels and consequently aggravate blood and oxygen deficiency in cerebral ischemic area. Secondly, phospholipase C, phospholipase A2, and nitric oxide synthase (NOS) as well as calcium-dependent protease activated by calcium overload can catalyze the generation of oxygen free radicals. Thirdly, calcium overload enlarges the gaps between cerebral vascular endothelial cells, resulting in high permeability of blood-brain barrier and consequent cerebral edema. Finally, calcium overload can raise the production of excitatory amino acids which in turn aggravate calcium overload, inducing the death of neurons [99–102]. Collectively, the level of calcium greatly affects the cerebral pathophysiology, and this can be another therapeutic target against cerebral ischemia/reperfusion injury.

Cellular Apoptosis

Cellular apoptosis is triggered in penumbra several days after cerebral ischemia [103]. When reperfusion injury appears, the damage can be exacerbated. On one hand, deficiency of blood and oxygen stimulates the expression of the apoptosis-related genes. On the other hand, ROS, calcium overload, NO, and energy metabolism disorders will also induce neuronal apoptosis through different mechanisms. The caspases family is the initiators and performers of apoptosis in mammal which is deeply associated with cerebral ischemia/reperfusion injury [104]. Besides, several other genes related to apoptosis including Bcl-2, FAS, and P53 participate in the process of cellular apoptosis [105–107]. During cerebral ischemia reperfusion, various signaling pathways, including MAPK signaling pathway, PI3K/Akt/GSK-3β signaling pathway, JAK-STAT signaling pathway, and transcription factor NF-κB-involved signaling pathway, function in neurons and astrocytes [108–110].

2.2.3 Treatment

Clinical Treatment

As described above, the most effective treatment for cerebral ischemic diseases is restoration of ischemic area. However, restoration carries the risks of further aggravating dysfunction induced by ischemia. Therefore, inhibition of reperfusion injury is a crucial part of overall treatment for cerebral ischemia. Hypothermia therapy for neuroprotection was first proposed by Busto in 1987. It has been proved that hypothermia therapy is able to decrease the cerebral metabolic rate, decrease the production of inflammatory mediators, and inhibit the generation of free radicals and excitatory amino acids. In addition, hypothermia therapy can be utilized in patients following cardiac arrest which apparently improve the function of the brain, especially when combined with other therapeutics [111–114]. Moreover, cerebral ischemia/reperfusion injury can be treated by reducing inflammation; antioxidants are effective agents for removing free radicals and inhibiting lipid peroxidation. Aspirin is a validated antioxidative drug for brain tissue protection [23, 115]. Besides, vitamin C and vitamin E are regarded as preventive neuroprotective agents due to their radical scavenging ability [116, 117]. In addition, calcium antagonists can reduce calcium influx by blocking calcium channel and consequently alleviate brain damage [118]. However, current clinical treatments have various limitations, and advanced novel therapeutics are urgently required.

Experimental Treatment

The accumulation of glutamate in periphery area of ischemic region causes high levels of intracellular sodium and calcium, resulting in neuronal injury. Lubeluzole is the antagonist of glutamate, and pre-administration of lubeluzole can alleviate neuronal damage in cerebral ischemia models [119]. It has been reported that tamoxifen could inhibit the release of excitatory amino acids in penumbra and improve the praxeology and histopathology results after reperfusion [120]. Another research indicates that natural compounds from traditional herbal medicine can be applied in the treatment of cerebral ischemia reperfusion injury [121]. In fact, Chinese medication and therapy such as acupuncture are suggested to have unique effects on treatment of cerebral ischemia reperfusion injury [122].

3 Circular RNAs

In the past few decades, protein-coding genes and their transcripts in eukaryotes have been investigated deeply [123]. However, the recent development of high-throughput RNA sequencing has identified that 98% of the whole-genome transcripts are noncoding RNAs and the protein-coding RNAs constitute only 2% [124]. Noncoding RNAs play vital roles in gene regulation. Currently, there are various noncoding RNAs have been found including microRNAs, long noncoding RNAs, small interfering RNAs, and small nuclear RNAs. In addition, circular RNAs that recently catch our attention are suggested with surprisingly strong function in regulating gene expression [5, 125]. Due to their specific structure, uncertain function, and low abundance, circular RNAs were ignored as trash of genomes for decades [126]. Nowadays, circular RNAs regain the attention of scientists because more and more evidence has revealed their potency and prospect.

3.1 Characteristics of Circular RNAs

Circular RNAs are structurally different from the other kinds of noncoding RNAs whose head 3′ and tail 5′ ends combine covalently and form a closed loop structure [127]. In 1976, circular RNAs were first identified by electron microscopy in a research associated with RNA viruses [128]. At that time, biotechnology was not advanced enough to validate the abundant existence of circular RNAs, let alone their functions. However, emerging evidences suggest that circular RNAs are abundant in mammal and more stable than linear RNAs because of their covalently closed loop structure which enables them to be resistant to RNA exonucleases [129]. In addition, genes encoding circles in one species have great possibility of encoding circles in other species, indicating that circular RNAs are evolutionarily conserved among diverse species [130–132]. There are three types of circular RNAs, exonic circular RNAs, intronic circular RNAs, and retained-intron circular RNAs, which are transcribed from pre-mRNA sequences by RNA polymerase [133]. According to published documents, most of circular RNAs show a sophisticated tissue- and cell type-specific expression pattern. Intriguingly, compared with other organs, it seems that circular RNAs are more abundant in the brain, which gives a clue of the potential function of circular RNAs in neurology [8, 133–135].

3.2 Biological Functions of Circular RNAs

Circular RNAs have substantial impact on gene regulation. In this part, we will list some biological functions of circular RNAs as follows.

3.2.1 MicroRNA Sponges

MicroRNAs are an enriched class of noncoding RNAs and mediate regulation of mRNA transcription through microRNA response elements (MREs) [136]. Emerging studies reveal that several circular RNAs can bind to MREs, acting as competitive endogenous (ce)RNAs for microRNA-binding sites, which in turn modulate the activity of microRNAs. For instance, the sex-determining region Y (SRY) is a circular RNA found in mouse testis in 1993; and then it has been proved to be a microRNA-138 sponge by regulation microRNA-138-associated mRNA translation [135, 137, 138]. In addition to circ-SRY, cerebellar degeneration-related protein 1 (CDR1) transcript, called CIRS-7, is another circular RNA discovered in mammal, which is reported to act as a microRNA-7 sponge [139]. Besides, recent investigation suggests that circHIPK3, derived from the HIPK3 gene Exon2, can bind to 9 microRNAs with 18 potential binding sites and particularly serve as a microRNA-124 sponge [140].

3.2.2 Transcription Regulation

The distribution of circular RNAs in cells provides hints on their probable functions. According to multiple researches, intronic circular RNAs and retained-intron circular RNAs usually locate in the nucleus, while exonic RNAs are primarily situated in the cytoplasm [141]. Apart from acting as microRNA sponges, several intronic circular RNAs can exert posttranscriptional regulation on gene expression. For example, c-sirt7, derived from lariats and interacting with the Pol II complex, can downregulate the gene expression of the relevant ankyrin repeat domain 52 or sirtuin 7 [132, 142]. Moreover, some retained-intron circular RNAs can interact with the U1 component and recruit RNA polymerase II to the promoter region of genes and upregulated the expression of their target genes [143].

3.2.3 Competition with Linear Splicing

Back-splicing or splicing of pre-mRNA determined the structure of RNA produced. Back-splicing, the way how circular RNAs generate, compete with splicing of pre-mRNA, leading to decreased production of linear mRNAs [144]. It has been uncovered that the Muscleblind (MBL) gene carries sequences which are able to form a circular RNA transcript. CircMBL can compete with MBL pre-mRNA splicing, exhibiting negative regulatory effects on canonical splicing [133]. Besides, a circular RNA originated from SEPALLATA3 gene can form a R-loop and suspend the transcription. It also has an impact on recruitment of splicing factors for initiating transcription [145]. It has also been reported that a great deal of genes can play roles in biogenesis of circular RNAs; and the mechanism of splicing control resembles a switch that flips between back-splicing or linear splicing [132, 141].

3.3 Circular RNAs in Human Diseases

Besides the diverse functions of circular RNAs physiological processes, the effects of circular RNAs on pathological processes have also been revealed. The change of circular RNA expression is involved with multiple types of diseases such as cardiovascular diseases, cancers, infections, and neurological diseases.

3.3.1 Circular RNAs and Cardiovascular Diseases

CircANRIL is a circular antisense noncoding RNA in the INK4 locus which has been reported to be associated with atherosclerosis. CircANRIL can bind to pescadillo homologue 1 (PES1) and in turn impair ribosome biogenesis in vascular smooth muscle cells, which is protective for arteries [146, 147]. Besides, a recent study has revealed that the well-known circular RNA Cdr1as can also serve as a microRNA-7a sponge in myocardial cells and modulate the function of microRNA-7a in myocardial infarct injury [148]. Furthermore, circ-Foxo3 is found highly expressed in heart samples of aged patients and mice, indicating that circFoxo3 is linked with cardiac senescence [149]. Another research suggests that the heart-related circular RNA (HRCR) can regulate cardiomyocyte hypertrophy and heart failure via acting as a microRNA-223 sponge [150].

3.3.2 Circular RNAs and Cancer

Considering the role of circular RNAs in regulation of cell cycles, cell proliferation, and cellular senescence, it is reasonable that circular RNA is implicated with cancer. For instance, ciRS-7 is a ceRNA of microRNA-7 that can promote the initiation and progress of cancer by upregulating oncogenic EGRF and XIAP gene and downregulating tumor-suppressed KLF4 [151, 152]. In addition, has-circ-001988 has been proved to be involved in colorectal cancers and linked with the differentiation and perineural invasion of cancer [153]. Moreover, it has been suggested that the abundance of circular RNAs may be a potential marker for cell proliferation in breast cancer [154].

3.3.3 Circular RNAs and Preeclampsia

Emerging investigations provide hints on the role of circular RNAs in preeclampsia. Based on results from circular RNA microarray and previous studies, circular RNAs are hypothesized to make contributions to the pathogenesis of preeclampsia via functioning as microRNA sponges [155]. It has also been reported that circular RNAs in blood corpuscles can serve as predictor of preeclampsia in early stage when combined with plasma protein factor [156].

3.3.4 Circular RNAs and Neurological Diseases

Circular RNAs are extremely abundant in the brain, particularly in neuropils and dendrites, indicating the potential roles of circular RNAs in regulating synaptic function and neural plasticity [8, 134]. Cdr1as has been demonstrated to be associated with neurodegenerative diseases like Alzheimer's disease (AD). Cdr1as can increase the AD-linked targets like ubiquitin protein ligase A via acting as a microRNA-7 sponge, increasing the clearance of amyloid peptides [157, 158]. Besides, Cdr1as causes high neural expression of α-synuclein protein which is an essential component of Lewy bodies in Parkinson's disease (PD) by sequestering microRNA-7 [159, 160]. Circular RNAs play roles in memory as well. For example, circPAIP2 can increase the expression of memory-related gene PAIP2 through poly(A)-binding protein-associated signaling [161].

4 Circular RNAs and Cerebral Ischemic Diseases

As previously mentioned, the valid therapies for cerebral ischemic diseases including acute ischemic stroke and cerebral ischemia/reperfusion injury are not satisfying. Despite current advanced biological and medical technologies, prognosis of patients with cerebral ischemic diseases is poor, which urges us to look for more effective interventions. During the past few decades, noncoding RNAs are hot topics in life science. In this part, we briefly review the roles of microRNAs and long noncoding RNAs by focusing on the roles of circular RNAs, a new star in the family of noncoding RNAs, in cerebral ischemic diseases.

4.1 Noncoding RNAs in Cerebral Ischemic Diseases

4.1.1 MicroRNAs in Cerebral Ischemic Diseases

First, we talk about microRNAs. One publication has revealed that increased microRNA-129-5p levels exert protective effect against ischemia reperfusion injury via alleviating neuronal inflammation and blood-spinal cord barrier damage through regulation of high-mobility group box-1 (HMGB1) and the Toll-like receptor (TLR)-3 pathway [162]. Besides, it has been demonstrated that microRNA-130b can protect the brain from ischemic stroke by targeting water channel protein aquaporin 4 [163]. In addition, microRNA-431 also has a positive effect on rats following cerebral ischemia/reperfusion injury by modulating the Rho/Rho-kinase signaling pathway [164]. Similarly, microRNA-93 has been reported to be protective for cerebral ischemia/reperfusion via suppressing inflammation and cellular apoptosis [165].

4.1.2 Long Noncoding RNAs in Cerebral Ischemic Diseases

As to the role of long noncoding RNAs in cerebral ischemic diseases, there are plenty of corresponding documents exist. It has been reported that Malat1 inhibits endothelial cell death and inflammation and consequently protects the cerebral microvasculature and parenchyma from cerebral ischemic insults [166]. Another study indicates that as a competing endogenous RNA of microRNA-21, long noncoding RNA MEG3 is able to regulate ischemic neuronal death through targeting microRNA-21/PDCD4 signaling pathway [167]. Moreover, it has been found that the silence of long noncoding RNA RMST can apparently reduce the brain infarct area and ameliorate neurological function in mice following cerebral ischemic insult [168]. A paper published in 2017 suggested that long noncoding RNA H19 can activate autophagy, thereby inducing cerebral ischemia/reperfusion injury [169].

4.2 Circular RNAs in Cerebral Ischemic Diseases

Actually, there are only a few papers about the direct roles of circular RNAs in cerebral ischemic diseases. In this part, we will first describe these several literatures about the function of circular RNAs in cerebral ischemic diseases; and then we will briefly review the roles of circular RNAs in atherosclerosis which represents a validated risk factor for cerebral ischemia.

4.2.1 Circular RNAs and Cerebral Ischemic Diseases

Ying Bai and his team published an article on the *Journal of Neuroscience* in 2017. In this study, they used the plasma of acute ischemic stroke patients including 13 females and 13 males as well as a mouse stroke model for research. The results showed that circular RNA DLGAP4 (circDLGAP4) is increased both in the plasma of patients and in rodent models. Besides, circDLGAP4 was shown to serve as a microRNA-143 sponge that inhibits the expression of HECT domain E3 ubiquitin protein ligase 1. More importantly, they found that overexpression of circDLGAP4 can significantly reduce neurological deficits, diminish infarct area, and alleviate damage of blood-brain barrier in mouse stroke model following transient middle cerebral artery occlusion, indicating that circDLGAP4 is involved in cerebral ischemia which provide a novel therapeutic target for treatment of cerebral ischemic diseases [170]. Based on mouse model subjected to transient middle cerebral artery occlusion, a research team used circular RNA microarrays and real-time PCR to identify the circular RNAs expression pattern at different time points: 6, 12, and 24 h after reperfusion. The results suggested there are 283 circular RNAs altered compared with sham control, of which 16 candidates are altered at all three time points of reperfusion after ischemia. Furthermore, these 16 circular RNAs were validated to carry binding sites for plenty of microRNAs. According to the bioinformatics analysis, these identified circular RNAs are functionally linked with pathophysiology of stroke [171]. Another similar study was carried out based on a model of oxygen-glucose deprivation/reoxygenation (OGD/R) in HT22 cells by using circular RNA microarray for identification of changes in circular RNAs expression profiles. The results identified 2 upregulated circular RNAs and 13 downregulated circular RNAs in the OGD/R model. Among those 15 circular RNAs, the upregulation of mmu-circRNA-015947 was verified by quantitative real-time PCR and bioinformatics analysis, indicating that it can interact with several target microRNAs. In addition, mmu-circRNA-015947 may be involved in apoptosis, metabolism, and immune-related pathways, suggesting that mmu-circRNA-015947 is likely to play a role in the pathophysiology of cerebral ischemia/reperfusion injury [172].

4.2.2 Circular RNAs and Atherosclerosis

Atherosclerosis is a chronic pathological process of arterial walls which is a critical risk factor of cerebrovascular diseases [173]. The major cause of cerebral ischemic stroke is the blockage of cerebral vessels by embolism which is a common consequence of atherosclerosis [174]. It has been reported that the circular noncoding RNA ANRIL can promote the progress of atherosclerosis via aggravating the inflammation of endothelial cells and increasing the serum levels of lipids, TGs, LDL, IL-1, IL-6, MMP-9, and CRP [175]. However, another study reported a contrary result that circular noncoding RNA ANRIL can protect atherosclerosis [146]. In addition, a screening study identified that has-circ-0003575 was significantly raised in oxLDL-induced human umbilical vein endothelial cells (HUVECs) and knockdown of has-circ-0003575 was able to advance the angiogenesis, indicating its role in atheroprotection [176]. All of these observations suggest a potential role of circular RNAs in atherosclerosis-related cerebrovascular diseases such as cerebral ischemia.

5 Conclusion and Perspective

Cerebral ischemic diseases like ischemic stroke and cerebral ischemia/reperfusion injury can lead to extremely high mortality and disability due to the lack of effective interventions. Although thrombolysis and the other supportive medication can improve the prognosis of the patients, the overall high mortality and disability has not been significantly ameliorated. The major mechanisms for the onset and development of cerebral ischemic diseases include inflammation, oxidative stress, cellular apoptosis, calcium overload, and the disorders of energy metabolism. There are lots of researches regarding the treatment and underlying mechanisms of cerebral ischemic diseases, in order to uncover potential therapeutics. Among these therapeutics, the newly studied circular RNAs hold the potential to improve the quality of life of patients suffering from cerebral ischemic diseases someday. Ying Bai and his team have demonstrated that upregulation of circDLGAP4 can significantly relieve the damage of blood-brain barrier in rodent model following transient middle cerebral artery occlusion, which indicated the potential therapeutic function of circular RNAs in cerebral ischemic diseases [170]. However, current investigations about the role of circular RNAs in cerebral ischemic diseases are insufficient in both quantity and quality. In addition to searching for the existence of dysregulated circular RNAs in patients and animal models of cerebral ischemic diseases, we should focus on the functions and underlying mechanisms of circular RNAs in cerebral ischemic diseases in the future.

Competing Financial Interests The authors declare no competing financial interests.

References

1. Turley KR, Toledo-Pereyra LH, Kothari RU (2005) Molecular mechanisms in the pathogenesis and treatment of acute ischemic stroke. J Investig Surg 18(4):207–218
2. Eltzschig HK, Eckle T (2011) Ischemia and reperfusion--from mechanism to translation. Nat Med 17(11):1391–1401
3. Ma Z, Xin Z, Di W et al (2017) Melatonin and mitochondrial function during ischemia/reperfusion injury. Cell Mol Life Sci 74(21):3989–3998
4. Benjamin EJ, Virani SS, Callaway CW et al (2018) Heart disease and stroke statistics-2018 update: a report from the American Heart Association. Circulation 137:e67. https://doi.org/10.1161/CIR.0000000000000558
5. Hentze MW, Preiss T (2013) Circular RNAs: splicing's enigma variations. EMBO J 32(7):923–925
6. Xie L, Mao M, Xiong K et al (2017) Circular RNAs: a novel player in development and disease of the central nervous system. Front Cell Neurosci 11:354
7. Earls LR, Westmoreland JJ, Zakharenko SS (2014) Non-coding RNA regulation of synaptic plasticity and memory: implications for aging. Ageing Res Rev 17:34–42
8. Rybak-Wolf A, Stottmeister C, Glazar P et al (2015) Circular RNAs in the mammalian brain are highly abundant, conserved, and dynamically expressed. Mol Cell 58(5):870–885
9. Lees KR, Bath PM, Naylor AR (2000) ABC of arterial and venous disease. Secondary prevention of transient ischaemic attack and stroke. BMJ 320(7240):991–994
10. Edwards JD, Kapral MK, Fang J et al (2017) Trends in long-term mortality and morbidity in patients with no early complications after stroke and transient ischemic attack. J Stroke Cerebrovasc Dis 26(7):1641–1645
11. Grysiewicz RA, Thomas K, Pandey DK (2008) Epidemiology of ischemic and hemorrhagic stroke: incidence, prevalence, mortality, and risk factors. Neurol Clin 26(4):871–895 vii
12. Woodruff TM, Thundyil J, Tang SC et al (2011) Pathophysiology, treatment, and animal and cellular models of human ischemic stroke. Mol Neurodegener 6(1):11
13. Saver JL (2006) Time is brain–quantified. Stroke 37(1):263–266
14. Cobley JN, Fiorello ML, Bailey DM (2018) 13 reasons why the brain is susceptible to oxidative stress. Redox Biol 15:490–503
15. Jin R, Liu L, Zhang S et al (2013) Role of inflammation and its mediators in acute ischemic stroke. J Cardiovasc Transl Res 6(5):834–851
16. Fricker M, Tolkovsky AM, Borutaite V et al (2018) Neuronal cell death. Physiol Rev 98(2):813–880
17. Ramos-Cabrer P, Campos F, Sobrino T et al (2011) Targeting the ischemic penumbra. Stroke 42(1 Suppl):S7–S11

18. Huang Q, Sun M, Li M et al (2017) Combination of NAD(+) and NADPH offers greater neuroprotection in ischemic stroke models by relieving metabolic stress. Mol Neurobiol 55:6063. https://doi.org/10.1007/s12035-017-0809-7
19. Liu J, Wang LN (2015) Peroxisome proliferator-activated receptor gamma agonists for preventing recurrent stroke and other vascular events in patients with stroke or transient ischaemic attack. Cochrane Database Syst Rev 8(10):CD010693. https://doi.org/10.1002/14651858.CD010693.pub3
20. Kim YS, Kim C, Jung KH et al (2017) Range of glucose as a glycemic variability and 3-month outcome in diabetic patients with acute ischemic stroke. PLoS One 12(9):e0183894
21. Pan Y, Chen W, Jing J et al (2017) Pancreatic beta-cell function and prognosis of nondiabetic patients with ischemic stroke. Stroke 48(11):2999–3005
22. Iadecola C, Anrather J (2011) The immunology of stroke: from mechanisms to translation. Nat Med 17(7):796–808
23. Felger JC, Abe T, Kaunzner UW et al (2010) Brain dendritic cells in ischemic stroke: time course, activation state, and origin. Brain Behav Immun 24(5):724–737
24. Kostulas N, Li HL, Xiao BG et al (2002) Dendritic cells are present in ischemic brain after permanent middle cerebral artery occlusion in the rat. Stroke 33(4):1129–1134
25. Gelderblom M, Leypoldt F, Steinbach K et al (2009) Temporal and spatial dynamics of cerebral immune cell accumulation in stroke. Stroke 40(5):1849–1857
26. Yilmaz A, Fuchs T, Dietel B et al (2009) Transient decrease in circulating dendritic cell precursors after acute stroke: potential recruitment into the brain. Clin Sci (Lond) 118(2):147–157
27. Yilmaz G, Arumugam TV, Stokes KY et al (2006) Role of T lymphocytes and interferon-gamma in ischemic stroke. Circulation 113(17):2105–2112
28. Hallenbeck JM (1996) Significance of the inflammatory response in brain ischemia. Acta Neurochir Suppl 66:27–31
29. Zhai SK, Volgina VV, Sethupathi P et al (2014) Chemokine-mediated B cell trafficking during early rabbit GALT development. J Immunol 193(12):5951–5959
30. Fernandez EJ, Lolis E (2002) Structure, function, and inhibition of chemokines. Annu Rev Pharmacol Toxicol 42:469–499
31. Murphy PM (2002) International Union of Pharmacology. XXX. Update on chemokine receptor nomenclature. Pharmacol Rev 54(2):227–229
32. Mirabelli-Badenier M, Braunersreuther V, Viviani GL et al (2011) CC and CXC chemokines are pivotal mediators of cerebral injury in ischaemic stroke. Thromb Haemost 105(3):409–420
33. Herrmann O, Tarabin V, Suzuki S et al (2003) Regulation of body temperature and neuroprotection by endogenous interleukin-6 in cerebral ischemia. J Cereb Blood Flow Metab 23(4):406–415
34. Waje-Andreassen U, Krakenes J, Ulvestad E et al (2005) IL-6: an early marker for outcome in acute ischemic stroke. Acta Neurol Scand 111(6):360–365
35. Xu N, Zhang Y, Doycheva DM et al (2018) Adiponectin attenuates neuronal apoptosis induced by hypoxia-ischemia via the activation of AdipoR1/APPL1/LKB1/AMPK pathway in neonatal rats. Neuropharmacology 133:415–428
36. Lossi L, Castagna C, Merighi A (2015) Neuronal cell death: an overview of its different forms in central and peripheral neurons. Methods Mol Biol 1254:1–18
37. Shabanzadeh AP, D'Onofrio PM, Monnier PP et al (2015) Targeting caspase-6 and caspase-8 to promote neuronal survival following ischemic stroke. Cell Death Dis 6:e1967
38. Yang W, Shen Y, Chen Y et al (2014) Mesencephalic astrocyte-derived neurotrophic factor prevents neuron loss via inhibiting ischemia-induced apoptosis. J Neurol Sci 344(1–2):129–138
39. van Wijk SJ, Hageman GJ (2005) Poly(ADP-ribose) polymerase-1 mediated caspase-independent cell death after ischemia/reperfusion. Free Radic Biol Med 39(1):81–90
40. Zhao Y, Pan R, Li S et al (2014) Chelating intracellularly accumulated zinc decreased ischemic brain injury through reducing neuronal apoptotic death. Stroke 45(4):1139–1147
41. Rengan R, Ochs HD, Sweet LI et al (2000) Actin cytoskeletal function is spared, but apoptosis is increased, in WAS patient hematopoietic cells. Blood 95(4):1283–1292
42. Cevik O, Adiguzel Z, Baykal AT et al (2013) The apoptotic actions of platelets in acute ischemic stroke. Mol Biol Rep 40(12):6721–6727
43. Astrup J, Symon L, Branston NM et al (1977) Cortical evoked potential and extracellular K+ and H+ at critical levels of brain ischemia. Stroke 8(1):51–57
44. Liu R, Yuan H, Yuan F et al (2012) Neuroprotection targeting ischemic penumbra and beyond for the treatment of ischemic stroke. Neurol Res 34(4):331–337
45. Mies G, Iijima T, Hossmann KA (1993) Correlation between peri-infarct DC shifts and ischaemic neuronal damage in rat. Neuroreport 4(6):709–711
46. White SH, Brisson CD, Andrew RD (2012) Examining protection from anoxic depolarization by the drugs dibucaine and carbetapentane using whole cell recording from CA1 neurons. J Neurophysiol 107(8):2083–2095
47. Yoshie T, Ueda T, Takada T et al (2018) Effects of pretreatment cerebral blood volume and time to recanalization on clinical outcomes in endovascular Thrombectomy for acute ischemic stroke. J Stroke Cerebrovasc Dis 27:1802. https://doi.org/10.1016/j.jstrokecerebrovasdis.2018.02.009
48. White CJ, Abou-Chebl A, Cates CU et al (2011) Stroke intervention: catheter-based therapy for acute ischemic stroke. J Am Coll Cardiol 58(2):101–116

49. Powers WJ, Derdeyn CP, Biller J et al (2015) 2015 American Heart Association/American Stroke Association focused update of the 2013 guidelines for the early management of patients with acute ischemic stroke regarding endovascular treatment: a guideline for healthcare professionals from the American Heart Association/American Stroke Association. Stroke 46(10):3020–3035
50. Moussaddy A, Demchuk AM, Hill MD (2018) Thrombolytic therapies for ischemic stroke: triumphs and future challenges. Neuropharmacology 134:272. https://doi.org/10.1016/j.neuropharm.2017.11.010
51. Del Zoppo GJ, Saver JL, Jauch EC et al (2009) Expansion of the time window for treatment of acute ischemic stroke with intravenous tissue plasminogen activator: a science advisory from the American Heart Association/American Stroke Association. Stroke 40(8):2945–2948
52. Wapshott T, Blum B, Kelsey W et al (2017) Investigation of gender differences and exclusive criteria in a diabetic acute ischemic stroke population treated with recombinant tissue-type plasminogen activator (rtPA). J Vasc Interv Neurol 9(6):26–32
53. Guo Z, Yu S, Chen X et al (2018) Suppression of NLRP3 attenuates hemorrhagic transformation after delayed rtPA treatment in thromboembolic stroke rats: involvement of neutrophil recruitment. Brain Res Bull 137:229–240
54. Shere A, Goyal H (2017) Cannabis can augment thrombolytic properties of rtPA: intracranial hemorrhage in a heavy cannabis user. Am J Emerg Med 35(12):1988 e1981–1988 e1982
55. Jovin TG, Chamorro A, Cobo E et al (2015) Thrombectomy within 8 hours after symptom onset in ischemic stroke. N Engl J Med 372(24):2296–2306
56. Berkhemer OA, Fransen PS, Beumer D et al (2015) A randomized trial of intraarterial treatment for acute ischemic stroke. N Engl J Med 372(1):11–20
57. Campbell BC, Mitchell PJ, Kleinig TJ et al (2015) Endovascular therapy for ischemic stroke with perfusion-imaging selection. N Engl J Med 372(11):1009–1018
58. Shafi N, Kasner SE (2011) Treatment of acute ischemic stroke: beyond thrombolysis and supportive care. Neurotherapeutics 8(3):425–433
59. Thwaites JW, Reebye V, Mintz P et al (2012) Cellular replacement and regenerative medicine therapies in ischemic stroke. Regen Med 7(3):387–395
60. Deng W, Obrocka M, Fischer I et al (2001) In vitro differentiation of human marrow stromal cells into early progenitors of neural cells by conditions that increase intracellular cyclic AMP. Biochem Biophys Res Commun 282(1):148–152
61. Yang Z, Zhu L, Li F et al (2014) Bone marrow stromal cells as a therapeutic treatment for ischemic stroke. Neurosci Bull 30(3):524–534
62. Wu P, Zuo X, Ji A (2012) Stroke-induced microRNAs: the potential therapeutic role for stroke. Exp Ther Med 3(4):571–576
63. Bhalala OG, Srikanth M, Kessler JA (2013) The emerging roles of microRNAs in CNS injuries. Nat Rev Neurol 9(6):328–339
64. Vemuganti R (2010) The MicroRNAs and stroke: no need to be coded to be counted. Transl Stroke Res 1(3):158–160
65. Zheng HW, Wang YL, Lin JX et al (2012) Circulating MicroRNAs as potential risk biomarkers for hematoma enlargement after intracerebral hemorrhage. CNS Neurosci Ther 18(12):1003–1011
66. Mishima T, Mizuguchi Y, Kawahigashi Y et al (2007) RT-PCR-based analysis of microRNA (miR-1 and -124) expression in mouse CNS. Brain Res 1131(1):37–43
67. Sun Y, Gui H, Li Q et al (2013) MicroRNA-124 protects neurons against apoptosis in cerebral ischemic stroke. CNS Neurosci Ther 19(10):813–819
68. Wang ZS, Luo P, Dai SH et al (2013) Salvianolic acid B induces apoptosis in human glioma U87 cells through p38-mediated ROS generation. Cell Mol Neurobiol 33(7):921–928
69. Lv H, Wang L, Shen J et al (2015) Salvianolic acid B attenuates apoptosis and inflammation via SIRT1 activation in experimental stroke rats. Brain Res Bull 115:30–36
70. Li D, Shao Z, Vanden Hoek TL et al (2007) Reperfusion accelerates acute neuronal death induced by simulated ischemia. Exp Neurol 206(2):280–287
71. Jung JE, Kim GS, Chen H et al (2010) Reperfusion and neurovascular dysfunction in stroke: from basic mechanisms to potential strategies for neuroprotection. Mol Neurobiol 41(2–3):172–179
72. The NINDS t-PA Stroke Study Group (1997) Intracerebral hemorrhage after intravenous t-PA therapy for ischemic stroke. Stroke 28(11):2109–2118
73. Pan J, Konstas AA, Bateman B et al (2007) Reperfusion injury following cerebral ischemia: pathophysiology, MR imaging, and potential therapies. Neuroradiology 49(2):93–102
74. Gobin YP, Starkman S, Duckwiler GR et al (2004) MERCI 1: a phase 1 study of mechanical embolus removal in cerebral ischemia. Stroke 35(12):2848–2854
75. Nanetti L, Raffaelli F, Vignini A et al (2011) Oxidative stress in ischaemic stroke. Eur J Clin Investig 41(12):1318–1322
76. Jin R, Yang G, Li G (2010) Inflammatory mechanisms in ischemic stroke: role of inflammatory cells. J Leukoc Biol 87(5):779–789
77. Kristian T, Gido G, Kuroda S et al (1998) Calcium metabolism of focal and penumbral tissues in rats subjected to transient middle cerebral artery occlusion. Exp Brain Res 120(4):503–509

78. Wang GH, Lan R, Zhen XD et al (2014) An-Gong-Niu-Huang Wan protects against cerebral ischemia induced apoptosis in rats: up-regulation of Bcl-2 and down-regulation of Bax and caspase-3. J Ethnopharmacol 154(1):156–162
79. Allen CL, Bayraktutan U (2009) Oxidative stress and its role in the pathogenesis of ischaemic stroke. Int J Stroke 4(6):461–470
80. Circu ML, Aw TY (2010) Reactive oxygen species, cellular redox systems, and apoptosis. Free Radic Biol Med 48(6):749–762
81. Jezek P, Hlavata L (2005) Mitochondria in homeostasis of reactive oxygen species in cell, tissues, and organism. Int J Biochem Cell Biol 37(12):2478–2503
82. Lee HL, Chen CL, Yeh ST et al (2012) Biphasic modulation of the mitochondrial electron transport chain in myocardial ischemia and reperfusion. Am J Physiol Heart Circ Physiol 302(7):H1410–H1422
83. Crack PJ, Taylor JM (2005) Reactive oxygen species and the modulation of stroke. Free Radic Biol Med 38(11):1433–1444
84. Chan PH (2001) Reactive oxygen radicals in signaling and damage in the ischemic brain. J Cereb Blood Flow Metab 21(1):2–14
85. Heo JH, Han SW, Lee SK (2005) Free radicals as triggers of brain edema formation after stroke. Free Radic Biol Med 39(1):51–70
86. Jian Liu K, Rosenberg GA (2005) Matrix metalloproteinases and free radicals in cerebral ischemia. Free Radic Biol Med 39(1):71–80
87. Chen XM, Chen HS, Xu MJ et al (2013) Targeting reactive nitrogen species: a promising therapeutic strategy for cerebral ischemia-reperfusion injury. Acta Pharmacol Sin 34(1):67–77
88. Gyoneva S, Davalos D, Biswas D et al (2014) Systemic inflammation regulates microglial responses to tissue damage in vivo. Glia 62(8):1345–1360
89. Zaleska MM, Mercado ML, Chavez J et al (2009) The development of stroke therapeutics: promising mechanisms and translational challenges. Neuropharmacology 56(2):329–341
90. Ekdahl CT, Kokaia Z, Lindvall O (2009) Brain inflammation and adult neurogenesis: the dual role of microglia. Neuroscience 158(3):1021–1029
91. Gleichman AJ, Carmichael ST (2014) Astrocytic therapies for neuronal repair in stroke. Neurosci Lett 565:47–52
92. Fabian RH, Perez-Polo JR, Kent TA (2000) Electrochemical monitoring of superoxide anion production and cerebral blood flow: effect of interleukin-1 beta pretreatment in a model of focal ischemia and reperfusion. J Neurosci Res 60(6):795–803
93. Yan YP, Sailor KA, Lang BT et al (2007) Monocyte chemoattractant protein-1 plays a critical role in neuroblast migration after focal cerebral ischemia. J Cereb Blood Flow Metab 27(6):1213–1224
94. Wasserman JK, Yang H, Schlichter LC (2008) Glial responses, neuron death and lesion resolution after intracerebral hemorrhage in young vs. aged rats. Eur J Neurosci 28(7):1316–1328
95. Berti R, Williams AJ, Moffett JR et al (2002) Quantitative real-time RT-PCR analysis of inflammatory gene expression associated with ischemia-reperfusion brain injury. J Cereb Blood Flow Metab 22(9):1068–1079
96. Caso JR, Moro MA, Lorenzo P et al (2007) Involvement of IL-1beta in acute stress-induced worsening of cerebral ischaemia in rats. Eur Neuropsychopharmacol 17(9):600–607
97. Baird GS (2011) Ionized calcium. Clin Chim Acta 412(9–10):696–701
98. Zheng J, Zeng X, Wang S (2015) Calcium ion as cellular messenger. Sci China Life Sci 58(1):1–5
99. Costa C, Martella G, Picconi B et al (2006) Multiple mechanisms underlying the neuroprotective effects of antiepileptic drugs against in vitro ischemia. Stroke 37(5):1319–1326
100. Zhou X, Fan GC, Ren X et al (2007) Overexpression of histidine-rich Ca-binding protein protects against ischemia/reperfusion-induced cardiac injury. Cardiovasc Res 75(3):487–497
101. Lopez-Caamal F, Oyarzun DA, Middleton RH et al (2014) Spatial quantification of cytosolic Ca(2)(+) accumulation in nonexcitable cells: an analytical study. IEEE/ACM Trans Comput Biol Bioinform 11(3):592–603
102. Li LH, Tian XR, Hu ZP (2015) The key target of neuroprotection after the onset of ischemic stroke: secretory pathway Ca(2+)-ATPase 1. Neural Regen Res 10(8):1271–1278
103. Liu HJ, Yang JP, Wang CH et al (2009) Endoplasmic reticulum in the penumbra following middle cerebral artery occlusion in the rabbit. Neurol Sci 30(3):227–232
104. Butterick TA, Duffy CM, Lee RE et al (2014) Use of a caspase multiplexing assay to determine apoptosis in a hypothalamic cell model. J Vis Exp 16. https://doi.org/10.3791/51305(86)
105. Wawryk-Gawda E, Chylinska-Wrzos P, Lis-Sochocka M et al (2014) P53 protein in proliferation, repair and apoptosis of cells. Protoplasma 251(3):525–533
106. Feng L, Balakir R, Precht P et al (1999) Bcl-2 regulates chondrocyte morphology and aggrecan gene expression independent of caspase activation and full apoptosis. J Cell Biochem 74(4):576–586
107. Daehn IS, Varelias A, Rayner TE (2010) T-lymphocyte-induced, Fas-mediated apoptosis is associated with early keratinocyte differentiation. Exp Dermatol 19(4):372–380
108. Zhou J, Du T, Li B et al (2015) Crosstalk between MAPK/ERK and PI3K/AKT signal pathways during brain ischemia/reperfusion. ASN Neuro 7(5):175909141560246

109. Si Y, Bao H, Han L et al (2013) Dexmedetomidine protects against renal ischemia and reperfusion injury by inhibiting the JAK/STAT signaling activation. J Transl Med 11:141
110. Tabassum R, Vaibhav K, Shrivastava P et al (2015) Perillyl alcohol improves functional and histological outcomes against ischemia-reperfusion injury by attenuation of oxidative stress and repression of COX-2, NOS-2 and NF-kappaB in middle cerebral artery occlusion rats. Eur J Pharmacol 747:190–199
111. Field JM, Hazinski MF, Sayre MR et al (2010) Part 1: executive summary: 2010 American Heart Association guidelines for cardiopulmonary resuscitation and emergency cardiovascular care. Circulation 122(18 Suppl 3):S640–S656
112. International Liaison Committee on R (2005) 2005 International consensus on cardiopulmonary resuscitation and emergency cardiovascular care science with treatment recommendations. Part 5: acute coronary syndromes. Resuscitation 67(2–3):249–269
113. Hypothermia after Cardiac Arrest Study G (2002) Mild therapeutic hypothermia to improve the neurologic outcome after cardiac arrest. N Engl J Med 346(8):549–556
114. Bernard SA, Gray TW, Buist MD et al (2002) Treatment of comatose survivors of out-of-hospital cardiac arrest with induced hypothermia. N Engl J Med 346(8):557–563
115. Strbian D, Kovanen PT, Karjalainen-Lindsberg ML et al (2009) An emerging role of mast cells in cerebral ischemia and hemorrhage. Ann Med 41(6):438–450
116. Polyak E, Ostrovsky J, Peng M et al (2018) N-acetylcysteine and vitamin E rescue animal longevity and cellular oxidative stress in pre-clinical models of mitochondrial complex I disease. Mol Genet Metab 123:449. https://doi.org/10.1016/j.ymgme.2018.02.013
117. Kim JH, Kim YC, Nahm FS et al (2017) The therapeutic effect of vitamin C in an animal model of complex regional pain syndrome produced by prolonged hindpaw ischemia-reperfusion in rats. Int J Med Sci 14(1):97–101
118. Chan SL, Cipolla MJ (2017) Treatment with low dose fasudil for acute ischemic stroke in chronic hypertension. J Cereb Blood Flow Metab 37(9):3262–3270
119. Koinig H, Vornik V, Rueda C et al (2001) Lubeluzole inhibits accumulation of extracellular glutamate in the hippocampus during transient global cerebral ischemia. Brain Res 898(2):297–302
120. Zhang Y, Jin Y, Behr MJ et al (2005) Behavioral and histological neuroprotection by tamoxifen after reversible focal cerebral ischemia. Exp Neurol 196(1):41–46
121. Wu PF, Zhang Z, Wang F et al (2010) Natural compounds from traditional medicinal herbs in the treatment of cerebral ischemia/reperfusion injury. Acta Pharmacol Sin 31(12):1523–1531
122. Xia WG, Zheng CJ, Zhang X et al (2017) Effects of "nourishing liver and kidney" acupuncture therapy on expression of brain derived neurotrophic factor and synaptophysin after cerebral ischemia reperfusion in rats. J Huazhong Univ Sci Technolog Med Sci 37(2):271–278
123. Esteller M (2011) Non-coding RNAs in human disease. Nat Rev Genet 12(12):861–874
124. Consortium EP (2012) An integrated encyclopedia of DNA elements in the human genome. Nature 489(7414):57–74
125. Patil VS, Zhou R, Rana TM (2014) Gene regulation by non-coding RNAs. Crit Rev Biochem Mol Biol 49(1):16–32
126. Nigro JM, Cho KR, Fearon ER et al (1991) Scrambled exons. Cell 64(3):607–613
127. Guo JU, Agarwal V, Guo H et al (2014) Expanded identification and characterization of mammalian circular RNAs. Genome Biol 15(7):409
128. Sanger HL, Klotz G, Riesner D et al (1976) Viroids are single-stranded covalently closed circular RNA molecules existing as highly base-paired rod-like structures. Proc Natl Acad Sci U S A 73(11):3852–3856
129. Rong D, Sun H, Li Z et al (2017) An emerging function of circRNA-miRNAs-mRNA axis in human diseases. Oncotarget 8(42):73271–73281
130. Wang PL, Bao Y, Yee MC et al (2014) Circular RNA is expressed across the eukaryotic tree of life. PLoS One 9(6):e90859
131. Salzman J, Chen RE, Olsen MN et al (2013) Cell-type specific features of circular RNA expression. PLoS Genet 9(9):e1003777
132. Jeck WR, Sorrentino JA, Wang K et al (2013) Circular RNAs are abundant, conserved, and associated with ALU repeats. RNA 19(2):141–157
133. Ashwal-Fluss R, Meyer M, Pamudurti NR et al (2014) circRNA biogenesis competes with pre-mRNA splicing. Mol Cell 56(1):55–66
134. You X, Vlatkovic I, Babic A et al (2015) Neural circular RNAs are derived from synaptic genes and regulated by development and plasticity. Nat Neurosci 18(4):603–610
135. Memczak S, Jens M, Elefsinioti A et al (2013) Circular RNAs are a large class of animal RNAs with regulatory potency. Nature 495(7441):333–338
136. Bartel DP (2009) MicroRNAs: target recognition and regulatory functions. Cell 136(2):215–233
137. Capel B, Swain A, Nicolis S et al (1993) Circular transcripts of the testis-determining gene Sry in adult mouse testis. Cell 73(5):1019–1030
138. Hansen TB, Jensen TI, Clausen BH et al (2013) Natural RNA circles function as efficient microRNA sponges. Nature 495(7441):384–388
139. Xu H, Guo S, Li W et al (2015) The circular RNA Cdr1as, via miR-7 and its targets, regulates insulin transcription and secretion in islet cells. Sci Rep 5:12453

140. Zheng Q, Bao C, Guo W et al (2016) Circular RNA profiling reveals an abundant circHIPK3 that regulates cell growth by sponging multiple miRNAs. Nat Commun 7:11215
141. Salzman J, Gawad C, Wang PL et al (2012) Circular RNAs are the predominant transcript isoform from hundreds of human genes in diverse cell types. PLoS One 7(2):e30733
142. Zhang Y, Zhang XO, Chen T et al (2013) Circular intronic long noncoding RNAs. Mol Cell 51(6):792–806
143. Li Z, Huang C, Bao C et al (2015) Exon-intron circular RNAs regulate transcription in the nucleus. Nat Struct Mol Biol 22(3):256–264
144. Zhang XO, Wang HB, Zhang Y et al (2014) Complementary sequence-mediated exon circularization. Cell 159(1):134–147
145. Conn VM, Hugouvieux V, Nayak A et al (2017) A circRNA from SEPALLATA3 regulates splicing of its cognate mRNA through R-loop formation. Nat Plants 3:17053
146. Holdt LM, Stahringer A, Sass K et al (2016) Circular non-coding RNA ANRIL modulates ribosomal RNA maturation and atherosclerosis in humans. Nat Commun 7:12429
147. Burd CE, Jeck WR, Liu Y et al (2010) Expression of linear and novel circular forms of an INK4/ARF-associated non-coding RNA correlates with atherosclerosis risk. PLoS Genet 6(12):e1001233
148. Geng HH, Li R, Su YM et al (2016) The circular RNA Cdr1as promotes myocardial infarction by mediating the regulation of miR-7a on its target genes expression. PLoS One 11(3):e0151753
149. Du WW, Yang W, Chen Y et al (2017) Foxo3 circular RNA promotes cardiac senescence by modulating multiple factors associated with stress and senescence responses. Eur Heart J 38(18):1402–1412
150. Wang K, Long B, Liu F et al (2016) A circular RNA protects the heart from pathological hypertrophy and heart failure by targeting miR-223. Eur Heart J 37(33):2602–2611
151. Peng L, Yuan XQ, Li GC (2015) The emerging landscape of circular RNA ciRS-7 in cancer (review). Oncol Rep 33(6):2669–2674
152. Hansen TB, Kjems J, Damgaard CK (2013) Circular RNA and miR-7 in cancer. Cancer Res 73(18):5609–5612
153. Wang X, Zhang Y, Huang L et al (2015) Decreased expression of hsa_circ_001988 in colorectal cancer and its clinical significances. Int J Clin Exp Pathol 8(12):16020–16025
154. Nair AA, Niu N, Tang X et al (2016) Circular RNAs and their associations with breast cancer subtypes. Oncotarget 7(49):80967–80979
155. Qian Y, Lu Y, Rui C et al (2016) Potential significance of circular RNA in human placental tissue for patients with preeclampsia. Cell Physiol Biochem 39(4):1380–1390
156. Zhang YG, Yang HL, Long Y et al (2016) Circular RNA in blood corpuscles combined with plasma protein factor for early prediction of pre-eclampsia. BJOG 123(13):2113–2118
157. Lukiw WJ (2013) Circular RNA (circRNA) in Alzheimer's disease (AD). Front Genet 4:307
158. Bingol B, Sheng M (2011) Deconstruction for reconstruction: the role of proteolysis in neural plasticity and disease. Neuron 69(1):22–32
159. Junn E, Lee KW, Jeong BS et al (2009) Repression of alpha-synuclein expression and toxicity by microRNA-7. Proc Natl Acad Sci U S A 106(31):13052–13057
160. Danan M, Schwartz S, Edelheit S et al (2012) Transcriptome-wide discovery of circular RNAs in archaea. Nucleic Acids Res 40(7):3131–3142
161. Khoutorsky A, Yanagiya A, Gkogkas CG et al (2013) Control of synaptic plasticity and memory via suppression of poly(A)-binding protein. Neuron 78(2):298–311
162. Li XQ, Chen FS, Tan WF et al (2017) Elevated microRNA-129-5p level ameliorates neuroinflammation and blood-spinal cord barrier damage after ischemia-reperfusion by inhibiting HMGB1 and the TLR3-cytokine pathway. J Neuroinflammation 14(1):205
163. Zheng Y, Wang L, Chen M et al (2017) Upregulation of miR-130b protects against cerebral ischemic injury by targeting water channel protein aquaporin 4 (AQP4). Am J Transl Res 9(7):3452–3461
164. Han XR, Wen X, Wang YJ et al (2017) Protective effects of microRNA-431 against cerebral ischemia-reperfusion injury in rats by targeting the Rho/Rho-kinase signaling pathway. J Cell Physiol 233:5895. https://doi.org/10.1002/jcp.26394
165. Tian F, Yuan C, Hu L et al (2017) MicroRNA-93 inhibits inflammatory responses and cell apoptosis after cerebral ischemia reperfusion by targeting interleukin-1 receptor-associated kinase 4. Exp Ther Med 14(4):2903–2910
166. Zhang X, Tang X, Liu K et al (2017) Long noncoding RNA Malat1 regulates cerebrovascular pathologies in ischemic stroke. J Neurosci 37(7):1797–1806
167. Yan H, Rao J, Yuan J et al (2017) Long non-coding RNA MEG3 functions as a competing endogenous RNA to regulate ischemic neuronal death by targeting miR-21/PDCD4 signaling pathway. Cell Death Dis 8(12):3211
168. Hou XX, Cheng H (2018) Long non-coding RNA RMST silencing protects against middle cerebral artery occlusion (MCAO)-induced ischemic stroke. Biochem Biophys Res Commun 495(4):2602–2608
169. Wang J, Cao B, Han D et al (2017) Long non-coding RNA H19 induces cerebral ischemia reperfu-

sion injury via activation of autophagy. Aging Dis 8(1):71–84

170. Bai Y, Zhang Y, Han B et al (2018) Circular RNA DLGAP4 ameliorates ischemic stroke outcomes by targeting miR-143 to regulate endothelial-mesenchymal transition associated with blood-brain barrier integrity. J Neurosci 38(1):32–50
171. Mehta SL, Pandi G, Vemuganti R (2017) Circular RNA expression profiles Alter significantly in mouse brain after transient focal ischemia. Stroke 48(9):2541–2548
172. Lin SP, Ye S, Long Y et al (2016) Circular RNA expression alterations are involved in OGD/R--induced neuron injury. Biochem Biophys Res Commun 471(1):52–56
173. Yamauchi H, Nishii R, Higashi T et al (2011) Silent cortical neuronal damage in atherosclerotic disease of the major cerebral arteries. J Cereb Blood Flow Metab 31(3):953–961
174. Schaller B, Graf R (2004) Cerebral ischemia and reperfusion: the pathophysiologic concept as a basis for clinical therapy. J Cereb Blood Flow Metab 24(4):351–371
175. Song CL, Wang JP, Xue X et al (2017) Effect of circular ANRIL on the inflammatory response of vascular endothelial cells in a rat model of coronary atherosclerosis. Cell Physiol Biochem 42(3):1202–1212
176. Li CY, Ma L, Yu B (2017) Circular RNA hsa_circ_0003575 regulates oxLDL induced vascular endothelial cells proliferation and angiogenesis. Biomed Pharmacother 95:1514–1519

Part VII

Circular RNAs in Plants and in Archaea

CircRNAs in Plants

26

Xuelei Lai, Jérémie Bazin, Stuart Webb,
Martin Crespi, Chloe Zubieta, and Simon J. Conn

Abstract

Circular RNAs (circRNAs) are covalently closed, single-stranded transcripts that are ubiquitously expressed in all eukaryotes and even prokaryotic archaea. Although once regarded as splicing artifacts, circRNAs are a novel class of regulatory molecules with diverse biological functions, including regulation of transcription, modulation of alternative splicing, and binding of miRNAs and proteins. The majority of studies of circRNAs have been performed in animals with a focus on the biogenesis, function, and mechanistic characterization of these molecules. In contrast, the study of circRNAs in plants is just emerging. Interestingly, recent circRNA profiling studies in model plant systems show distinct features of plant circRNAs compared with those from animals, including putative roles in stress response, differences in expression patterns, and novel biogenesis mechanisms. This provides a great opportunity to broaden our knowledge of circRNAs using plant model systems, such as Arabidopsis and rice, which are ideal for phenotypic characterization and genetic studies. In this review, we summarize current knowledge of plant circRNAs, discuss their identification and biogenesis, describe potential functions, and propose future perspectives for plant circRNA study.

Keywords

circRNAs · Plants · Transcriptomics · Genome-wide profiling

Author contributed equally with all other contributors. Xuelei Lai and Jeremie Bazin

X. Lai · C. Zubieta (✉)
Laboratoire de Physiologie Cellulaire et Végétale, CNRS Univ. Grenoble Alpes, CEA, INRA, BIG Grenoble, Grenoble, France
e-mail: chloe.zubieta@cea.fr

J. Bazin · M. Crespi
Institute of Plant Sciences Paris-Saclay, IPS2, CNRS-INRA-University of Paris Sud, Paris-Diderot and Evry, University of Paris Saclay, Gif sur Yvette, France

S. Webb · S. J. Conn (✉)
Flinders Centre for Innovation in Cancer, College of Medicine & Public Health, Flinders University, Bedford Park, Australia
e-mail: simon.conn@flinders.edu.au

1 Introduction

CircRNAs are a novel type of endogenous and largely noncoding RNA. Unlike their cognate messenger RNA (mRNA), circRNAs possess a covalent bond linking their 3′ and 5′ ends and, as such, are hyperstable RNA molecules [1]. While human circRNAs were initially reported almost three decades ago [2], they were regarded as transcriptional noise derived from splicing errors [3]. However, in recent years they have received

J. Xiao (ed.), *Circular RNAs*, Advances in Experimental Medicine and Biology 1087,
https://doi.org/10.1007/978-981-13-1426-1_26

increased attention partly due to advances in next-generation sequencing techniques and highly efficient bioinformatics algorithms for their discrimination and mapping. This has unequivocally shown that circRNAs are abundant and ubiquitous among eukaryotes [4], including yeast, *C. elegans*, mouse, human and plants, and even prokaryotic archaea [5], adding a novel class of RNA molecules with diverse regulatory functions to the extensive collection of noncoding RNAs present in eukaryotes [6].

The majority of circRNAs exhibit cell-type-, tissue-, or developmental stage-specific expression patterns [7–10] and are highly regulated [11], implying their deliberate production by the cell. In addition, some circRNAs are syntenically conserved. These findings are suggestive of potential functionality of circRNAs. Indeed, emerging evidence has shown that specific circRNAs are involved in regulation of gene expression at the transcriptional and posttranscriptional levels. For example, increasing numbers of circRNAs have been shown to act as miRNA sponges, able to sequester and prevent miRNAs from binding to their corresponding target genes [9, 12], or in one particular case, stabilizing them to improve miRNA targeting [13]. Some circRNAs contain multiple binding sites for RNA-binding proteins (RBP) and were therefore proposed to act as RBP sponges [14]. More recently, several studies show that a subpopulation of circRNAs retain an internal ribosomal entry site (IRES) which enables ribosome occupancy and can result in their translation into proteins/peptides [15–18]. However, no overarching functionality has been ascribed to the majority of circRNAs, and the aforementioned functions may or may not operate in different organisms such as plants.

Compared with the extensive studies of circRNAs in animals, the systematic characterization of circRNAs in plants has only recently begun and focused primarily on model plant systems such as Arabidopsis and rice. In this review, we concentrate on what is known with respect to the profiling, biogenesis, and function of circRNAs in plants. Studies of circRNAs from other systems have been extensively reviewed elsewhere [14, 19–24] and by accompanying chapters of this book. First, we summarize profiling of circRNAs from different plant species and their distinct features. We then discuss biogenesis of plant circRNAs based on studies in animals. Finally, we propose potential functions of plant circRNAs and future perspectives of circRNA studies in plants.

2 Plant circRNA Profiling

2.1 Experimental circRNA Detection

Next-generation sequencing (NGS) has greatly expanded the diversity of detected circRNAs. A number of strategies exist to enrich for circRNAs from RNA extraction through to library preparation. CircRNAs are known to be depleted in poly(A) + libraries (mRNA-seq) because they lack poly(A) tails. In addition, by their chemical nature, they do not exhibit free termini. Consequently, intact circRNAs are unavailable for direct ligation of RNA adapters common among several RNA-seq library preparation protocols. However, they are known to be present in transcriptomic datasets produced from rRNA-depleted samples and RNA ligation-independent protocols, including total RNA-seq [25]. Fortuitously, many datasets are publicly available in NGS repositories for a number of plants species including model organisms such as Arabidopsis and various crops. The first genome-wide identification and characterization of circRNAs in plants were conducted during the last few years and exploited these publicly available total RNA-seq datasets from Arabidopsis, rice, and barley [26, 27]. As RNA splicing in plants differs from that of animals, dedicated predictive algorithms needed to be designed for plants. Thousands of circRNA were identified in each of these species, with the vast majority of them deriving from mRNA-coding exons. Strikingly, the parent genes of more than 700 exonic rice and Arabidopsis circRNAs were orthologous between the two species, suggesting a conservation of circRNAs in plants, as had been shown in mammals

[9, 27, 28]. However, most circRNAs are expressed at low levels, and detecting and quantifying them in total RNA-seq are challenging. A high sequencing depth is required to have a significant number of reads supporting their expression. An additional protocol has been developed to greatly enrich for circRNAs, using the linear ribonuclease R (RNase R) prior to rRNA depletion and RNA-seq library preparation. This approach was successfully used in plant species and yielded high-depth interrogation of circRNAs (and intron lariats and other paired RNAs) in Arabidopsis [29], rice [30], soybean [31], tomato [32], and kiwifruit [33].

2.2 Bioinformatics Detection of Circular RNA

As mentioned, exonic circRNAs are formed by back-splicing reactions which consist of a covalent ligation event between a downstream 3′ donor splice site and an upstream 5′ acceptor splice site from a linear pre-mRNA [1]. Sequencing reads that span this backsplice junction are considered as chimeric and are routinely rejected by most standard read-mapping algorithms. However, this feature can also be exploited to specifically detect exonic circular RNAs. Indeed, read fragments overlapping a known or de novo predicted splice junction with opposite orientation are likely to arise from a backsplice event. A number of tools have been used to recover such features. These tools use two basic approaches. The first one, called "pseudo-reference based" relies on the reconstruction of all possible circRNA sequences by shuffling the exon-exon junctions based on genome annotation, prior to read mapping [34]. This requires an accurate genome annotation with a thorough knowledge of all alternatively spliced mRNA isoforms. One of the available tools called KNIFE [34] has been very recently applied to publicly available and RNase R-treated RNA-seq datasets from maize and detected 85 high confidence circRNA candidates [35]. The second group of algorithms does not directly rely on prior knowledge of gene annotation. In this approach, algorithms are used to identify back-splicing junctions from the mapping information of a multiple split-read alignment. This approach, termed "segmented read approach," uses the unmapped reads from aligner programs, such as Bowtie [36], extracts their two end segments, and performs a second alignment. It then looks for cases where the two end segments are mapped within spliced exons in the opposite orientation, which indicates a putative back-splicing event. Then, depending on the software and the dataset used, the presence of a canonical splice junction, the number of reads overlapping the splice junction, and the enrichment of sequences containing the splice junction in RNase R-treated samples are used as criteria to define high-confidence circRNAs. The latter tools have been most extensively used in plants where they have led to the identification of thousands of circRNAs in multiple plant species. Most of the resources produced by these studies have been or will be implemented into web-based databases such as PlantcircBase (http://ibi.zju.edu.cn/plantcircbase/), in which users can browse over 90,000 circRNAs detected among 8 plants species (*Oryza sativa*, *Arabidopsis thaliana*, *Zea mays*, *Solanum lycopersicum*, *Glycine max*, *Gossypium hirsutum*, *Triticum aestivum*, *Hordeum vulgare*), identify those with miRNA seed sites which could act as miRNA sponges, and probe competing endogenous RNA networks (circRNA-miRNA-mRNA) [37].

2.3 Limitations and Challenges

All NGS approaches have inherent limitations, and one should use caution when interpreting this data to extrapolate common features of circRNAs. For instance, in a study reanalyzing plant RNA-seq datasets, Ye et al. [27] identified 12,037 and 6,012 circRNAs from rice root and Arabidopsis leaf, respectively. In addition to exonic circRNAs, a large number of circRNAs are derived from noncoding regions or spanning two or more genes. However, they could not validate any of these experimentally, suggesting that distinct genomic features, such as repetitive sequences or gene duplication that differ between plants and animals, decrease the specificity of prediction algorithms that were originally developed for applications in animals. The development of dedicated tools for plant circRNA calling such as PcircRNA_finder [38] has at least partially overcome such problems in the detection of plant circRNAs. In addition, depending on the tool used and the cut-off applied, the number of detected circRNAs can be very different and only partially overlap [39]. Indeed, benchmarking analysis of five existing software on the same RNA-seq datasets highlighted the large difference between tools, suggesting that several algorithms should ideally be combined to achieve reliable predictions [40]. A summary of plant circRNA profiling studies, including identification methods, total circRNAs, stress treatments, and circRNAs with potential miRNA-binding sites, is given in Table 26.1. These data highlight the conservation and distinct features of plant circRNAs from different plant species.

3 Plant circRNA Biogenesis

CircRNAs in plants and animals can arise from exons, introns, and/or intergenic regions. In animals, circularized exons are typically bracketed by long introns that contain complementary sequences such as ALU elements [28, 41] and/or micro-repeat regions sufficient to promote intron base-pairing [42, 43]. In contrast, yeast circRNAs could be generated through exon-containing lariat precursors that lacked noticeable flanking intronic secondary structure or repeating sequences [44]. Unlike the animal circRNAs, most plant circRNAs have limited repetitive and reverse complementary sequences in intronic sequences flanking exons [27], suggesting that plants might favor different mechanisms of circRNA biogenesis compared with those of animals. One such mechanism has been proposed in maize, where transposons seem to play a role in the biogenesis of circRNAs [35].

Recently, several studies have shown that RNA-binding proteins may serve as regulatory activators or inhibitors in the formation of circRNAs in some conditions, such as quaking (QKI) [11], muscleblind (MBL) [42], adenosine deaminase 1 (ADAR1) [45], FUS [46], and DHX9 [47]. Homolog proteins of these factors, while identified in plants, have not yet been studied, and their putative role in circRNA biogenesis remains to be investigated. Here we summarize these homolog proteins in Arabidopsis (Table 26.2) and assess their potential functions in circRNA biogenesis in plants based on knowledge of their counterparts from animals.

4 QKI Homolog: KH Domain–Containing Proteins in Arabidopsis

QKI regulates circRNA biogenesis during the human epithelial/mesenchymal transition [11]. It binds to the introns flanking the exon on the host RNA and homodimerizes; therefore, it could promote circRNA biogenesis by exon looping, bringing the 3′ and 5′ termini into close proximity for circularization. At the protein level, QKI contains a STAR domain responsible for RNA binding and homodimerization that is composed of a central KH domain flanked by QUA1 and QUA2 domains [48]. The two key features that are important for QKI's function as a circRNA regulator are its dimerization ability and RNA-binding activity [11]. QKI dimerization involves the dimerization domain, QUA1 [49] and the RNA-binding domain, KH [36] which, together with the QUA2 domain, are responsible for RNA-binding activity [36]. The QKI dimer preferentially binds to two A/U-rich motifs [50–52] that can be on the same or separate RNA molecules [49].

QKI homolog proteins in Arabidopsis are KH domain-containing proteins, which are RNA-binding proteins that have been shown to participate in pre-mRNA processing [53] and to be important for various domains of plant development and survival, including flowering regulation [54, 55], stress responses [56–58], as well as hormone signaling [59]. There are 26 KH domain-containing proteins in Arabidopsis, 5 of which are highly similar to QKI (hereafter

Table 26.1 Identification and characteristics of circRNAs currently identified in plant species

Plant species	Arabidopsis			Kiwifruit	Soybean	Tomato	Wheat	Maize	Rice		Barley
Tissues	Leaves	Leaves	Whole plants, roots, stems, leaves, flowers, and siliques	Leaf, root, and stem tissues	Leaf, root, and stem tissues	Fruit	Leaves of wheat seedlings	Seedling leaves	Root	Mature leaf and panicle tissues	Leaves, grains and grain transfer cells
Datasets or circRNA library prep	Reanalysis of published RNA-seq data (PRJNA218215)	RNA-seq data (GEO accession GSE43616)	rRNA depletion and RNase R treatment	rRNA depletion and RNase R treatment	rRNA depletion and RNase R treatment	rRNA depletion and RNase R treatment	CircRNA enrichment kit (cloud-seq Inc.)	RNase R-seq, 977 public maize RNA-Seq	Reanalysis of published RNA-seq data (GenBank PRJNA215013)	rRNA depletion and RNase R treatment	RNA-seq (Genbank SRA297575)
Bioinformatics program	BOWTIE2 (v2.0.5)	MapSplice (version 2.0)	BOWTIE2 (v2.0.5), find_circ program (Memczak et al., 2013)	CIRI	CIRI	CIRI	CIRI	KNIFE, circ_finder, CIRCexplorer2, CIRI	BOWTIE2 (v2.0.5)	Self-designed pipeline	CIRI
Treatment conditions	High-light	N/A	Drought, salinity and heat	Pathogen infection (*Pseudomonas syringae* pv. *actinidiae*)	N/A	Chilling	Dehydration stressed and well-watered conditions	N/A	Pi-starvation	N/A	Micronutriants such as iron and zinc
Total circRNA count (exonic/intronic/intergenic/others)	6012 (85.70%/0.015%/0.5%/13.76%)	168 (94.04%/NA/5.95%/NA)	5861 (85.14%/3.77%/11.09%/NA); unique mitochondrial and chloroplast circRNAs sere also detected	3582 (21.44%/14.55%/64.01%/NA)	5372 (46.43%/48.05%/5.55%/NA)	854 (72.01%/3.63%/24.35%/NA)	88 (6.82%/2.27%/60.22%/2.27%); 28.41% exon-intronic	5329 in total, 2804 high confident (mostly exonic)	12,037 (50.46%/4.03%/5.86%/39.65%)	2354 (57.60%/2.55%/29.74%/NA); 1.66% exon-intronic	NA
circRNAs with miRNA-binding sites	5.0% contain potential miRNA-binding sites	N/A	39 (0.67%) 9 have more than 1 different miRNA-binding sites	N/A	2134 (39.7%) circRNAs contained predicted binding sites for 92 miRNAs. Of these circRNAs, only 352 had 2 to 6 miRNA-binding sites	102 circRNAs with predicted miRNA-binding sites for 24 miRNAs	Six of the differentially expressed circRNAs have putative miRNA-binding sites (3 to 8 sites); 26 miRNAs were predicted	circRNAs have an average of 1.33 miRNA-binding sites ranging from 1 to 9	6.6% contain potential miRNA-binding sites	235 exonic circRNAs contain putative miRNA-binding sites, among which only 31 circRNAs had 2 to 6 miRNA-binding sites	N/A

(continued)

Table 26.1 (continued)

Plant species	Arabidopsis			Kiwifruit	Soybean	Tomato	Wheat	Maize	Rice		Barley
Differential expression	Temporal expression pattern and stress-dependent expression pattern; mutant (tnr-1) background expression pattern	circRNAs showed developmental-specific expression pattern in Arabidopsis leaves, e.g., circ-AT1G29965 was only expressed in the early stage of leaf growth, while circ-AT5G18590 was only detected in the mature stage of leaf growth, and circ-AT4G08300 was expressed only in the senescence process	Developmental stage- and tissue-specific expression pattern	The expression of the majority of circRNAs was tissue-, taxon-, and stage- specific pathogen infection dependent expression	Tissue-specific expression: 62.4% (484) of leaf circRNAs, 83.5% (2647) of root circRNAs and 72.2% (1563) of stem circRNAs, only 2.7% (143) of the total circRNAs were expressed in all the tissues	163 circRNAs had significant difference between the control and chilling injury group, 138 circRNAs upregulated, 25 circRNAs downregulated	62 circRNAs showed significant difference between the PEG and control treatment groups, containing 16 upregulated circRNAs and 46 downregulated circRNAs	N/A	Temporal- and stress-dependent expression pattern	N/A	Cellular level alterations across tissues and in response to micronutrients iron and zinc
Distinct features of circRNAs	A significantly positive correlation was observed for the expression profiles of some circRNAs and their parent genes	N/A	N/A	Both exonic and intronic circRNAs were significantly positively correlated to parent protein-coding genes and intronic circRNAs are a class of highly remarkable regulators the parent genes comparing to that of exonic circRNAs	4451 (82.8%) circRNAs were generated from the paralogous genes	N/A	Predicted miRNA-binding circRNAs contain more abundant miRNA-binding sites than those predicted from other plant species	The flanking intron length of the junctions of circRNAs is significantly larger than that of linear transcripts of randomly selected genes; LINE1-like elements (LLEs) and their reverse complementary pairs (LLERCPs) are significantly enriched in the flanking regions of circRNAs	27 rice exonic circRNAs were found to be differentially expressed under phosphate-sufficient and starvation conditions; a significantly positive correlation was observed for the expression profiles of some circRNAs and their parent genes	Parental genes with multiple exons are preferentially circularized; a large number of circRNA isoforms derived from alternative backsplicing	Fluctuations of circRNAs do not correlate with the levels of their parental-loci encoded linear transcripts; circRNA from mitochondria genome were detected

Proposed function of circRNAs	N/A	Potential posttranscriptional regulators in the senescence of Arabidopsis leaves	N/A	Regulate host pathogen interactions	Tissue differentiation in soybean	Play roles in the chilling injury regulation	Potential miRNA sponges	circRNAs are likely to be involved in the modulation of phenotypic variation by LLERCPs	N/A	A case study: Overexpression of Os08circ16564 reduces expression of its parental gene. Os08circ16564 contain potential binding sites of miR172, a miRNA plan crucial role in the development of the rice spikelet and floral organ)	Regulatory role in micronutrient homeostasis; regulate both nuclear and organellar gene expression
Reference	[27]	[64]	[35]	[33]	[31]	[32]	[33]	[35]	[27]	[30]	[26]

Table 26.2 Arabidopsis homologs of RBPs implicated in animal circRNA biogenesis

RBP	Binding sites/ motifs	circRNA biogenesis mechanism	Features important as circRNA biogenesis promoter/ repressor	Homolog/similar proteins in Arabidopsis	Putative/known function(s) of the homolog proteins in Arabidopsis
QKI [11]	A/U-rich motifs	Promote circRNA biogenesis by exon-looping	Dimerization, RNA binding	KH domain-containing proteins: AT4G26480, AT5G56140, AT3G08620, AT1G09660, and AT2G38610	mRNA processing, flowering regulation, stress response, and hormone signaling
MBL [42]	MBL-binding sites (expanded CUG or CCUG repeats) [88,89]		Dimerization, RNA binding	No homologs in Arabidopsis, closest matches: Zinc finger CCCH domain-containing proteins: AT2G47850, AT3G02830, AT3G06410, AT3G48440, AT5G16540, AT5G18550, and AT5G63260	Putative activity of RNA binding or nuclear acid binding
FUS [46]	FUS-binding sites		RNA-binding activity	No homologs in Arabidopsis, closest matches: Glycine-rich RNA-binding protein 3 (AT5G61030), TBP-associated factor 15 (AT1G50300), and early flowering 9 (AT5G16260)	N/A
DHX9 [47]	Inverted-repeat Alu elements	Repress circRNA biogenesis by acting as nuclear RNA resolvase	Nuclear RNA helicase activity, interact with ADAR, a co-repressor	DExH-box ATP-dependent RNA helicases: AT2G35920, AT2G01130, AT1G77030, AT1G33390, AT1G48650, and AT5G04895	ATP-dependent RNA helicase activity
ADAR1 [45]	N/A	Repress circRNA biogenesis with putative mechanisms of **(a)** compete with circRNA-promoting factors, such as MBL. **(b)** Editing or hyper-editing of introns flanking circRNAs	RNA editing activity	ADAR proteins can be found in nearly all metazoa but absent in all protozoa, yeast, and plants [90]	N/A

called Arabidopsis QKI-like proteins for simplicity) (Table 26.2) [60]. Protein sequence alignment suggests that Arabidopsis QKI-like proteins share highly conserved amino acid residues with those in the central STAR domains of human QKI and GLD-1, a *C. elegans* QKI homolog protein (Fig. 26.1). The crystal structures of both proteins in complex with RNA were resolved, and key modules/residues responsible for homodimerization (α1 and α2 of QUA1 domain) and RNA binding (α3, α4 and β3 of KH domain, and α7 of QUA2 domain) were identified [49] (Fig. 26.1). Interestingly, all of these modules/residues are highly con-

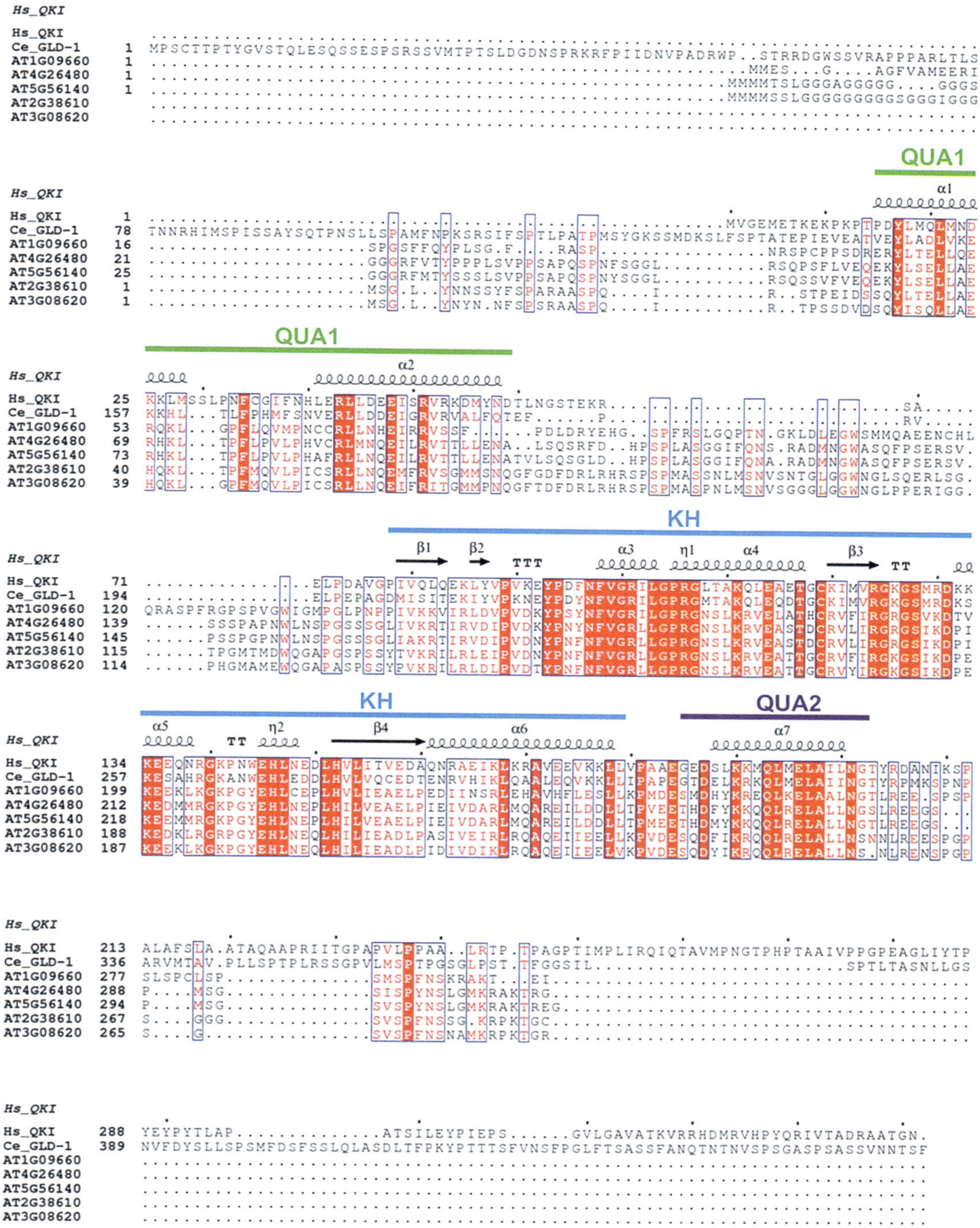

Fig. 26.1 Protein sequence alignment of *Homo sapiens* QKI (Hs_QKI), *C. elegans* GLD-1 (Ce_GLD-1), and Arabidopsis QKI-like proteins. The STAR domain of QKI and GLD-1 consists of QUA1, KH, and QUA2 subdomains, which share highly conserved residues with that of Arabidopsis QKI-like proteins. Identical residues are boxed and colored in white and red background; conserved residues are boxed and colored in red without background; secondary structure elements are derived from crystal structure of QKI (PDB: 4jvh) [49]; TT and T.T represent turns that connect defined secondary structure elements

served in the Arabidopsis QKI-like proteins (Fig. 26.1), suggesting that these proteins may share similar physiological functions with QKI, e.g., in circRNA biogenesis [11]. Although this is an appealing hypothesis, no experimental data is available yet to support or refute this putative function. Extensive studies of QKI homologs in Arabidopsis would require characterization of the RNA-binding motif in plants; mutant/overexpression lines of the Arabidopsis QKI-like proteins; biochemical activity studies, such as homodimerization and RNA-binding activity; and biophysical studies, e.g., RNA-looping ability.

5 Plant circRNA Function

A number of distinct functional mechanisms for animal circRNAs have been identified, and plant circRNAs may exhibit similar conserved functions. These include miRNA sponging, transcriptional modulation, translation of circRNAs into proteins/peptides, and altering protein function through direct protein binding. The consequences of these activities include altering cell cycle progression [61], cell proliferation [62], and cell migration [63], all of which are indispensable for plant growth and development. With this in mind, we summarize the known plant circRNA functions below, with scope for much greater functional characterization of circRNAs in plants, leveraging animal circRNA functional studies.

5.1 CircRNAs as miRNA Sponges

Based on the work in the mammalian field, circRNA function in plants has largely focused on their potential role as miRNA sponges or as acting in the miRNA pathway. Recent work by Liu et al. [64] has investigated the putative circRNA-miRNA-mRNA network based on the hypothesis that both mRNAs and circRNAs are targets of the same miRNA. Based on differentially expressed mRNAs, circRNAs, and miRNAs, an anticorrelation pattern was shown for a number of circRNA-mRNA-miRNA pairs during M- to S- and G- to M-stages, suggesting a possible role for circRNAs in leaf senescence in this network.

Initial studies have attempted to identify circRNAs from plants that may act as true miRNA sponges, yet no circRNA with high density of miRNA-binding sites as described for Sry and ciRS-7 from mammals has been shown for a plant circRNA [12]. Further identification of circRNAs from plant species may reveal the presence of circRNAs that can act as miRNA sponges with multiple miRNA-binding sites, however, to date this has not been shown. Unlike in mammals, miRNAs from plants exhibit much higher sequence complementarity to their target site in order to cleave and/or negatively regulate mRNA expression, facilitating the identification of miRNA targets [65, 66]. The availability of bioinformatics prediction tools for miRNAs and their targets in Arabidopsis and the publicly available databases such as miRTarBase (http://mirtarbase.mbc.nctu.edu.tw/) which collects validated miRNA targets will facilitate scanning of circRNAs for miRNA-binding sites [67, 68]. For a comparison of miRNA prediction tools for plants, see Srivastava, et al. [69]. These tools have predicted numerous plant circRNAs which could bind miRNAs and impact miRNA-mRNA regulatory networks; however, unlike their animal counterparts, these putative circRNA-miRNA interactions have yet to be experimentally validated. The development of new pipelines incorporating prediction software, miRNA databases, and new filtering procedures offers great promise for more robust prediction and validation of miRNA targets which will likely reveal many new connections between miRNAs and circRNAs [68].

Some of the most powerful methods for identifying protein-RNA complexes are crosslinking immune-precipitation sequencing (CLIP-seq) techniques. For example, to identify miRNA-circRNA species genome wide, AGO-CLIP techniques can be used [70]. Argonaute (AGO) proteins are part of the RNA-induced silencing complex (RISC), and this complex recognizes mRNA- or circRNA-containing sequences complementary to the miRNA. These techniques have been successfully performed in plants and are one method to help identify circRNAs that may act as miRNA sponges [68]. CLIP experiments targeting different RNA-binding proteins (RBPs) involved in circRNA activity are also easily envisaged.

5.2 CircRNAs in Stress Response

Plants, as sessile organisms, encounter various environmental stresses, such as drought, heat, salinity, cold, and pathogen infection [71]. To tackle these environmental challenges, plants have evolved various sophisticated biological pathways, in which massive gene expression reprogramming occurs. RNA molecules, such as microRNAs (miRNAs) and small interfering RNAs (siRNAs), have been shown to play critical roles in gene expression regulation under different stress conditions in plants [72, 73]. Interestingly, circRNAs have been shown to display stress-specific expression patterns in Arabidopsis in the first report of plant circRNAs [27]; therefore, one intriguing question is whether or not circRNAs also play a role in stress responses of plants or are the products of stress responses on different RNA machineries such as the spliceosome.

Studies under different stress conditions including oxidative stress, drought, and nutrient deficiency have further revealed differential expression of circRNAs which in most cases does not correlate with expression of their associated mRNA, further suggesting a functional role of circRNA in environmental and stress response. For example, 62 circRNAs were identified that exhibit differential expression under dehydration stress and are associated with photosynthesis and hormone signal pathways in wheat [74]. Moreover, a number of circRNAs in rice and barley have also been reported to respond to nutrient depletion such as phosphate, iron, and zinc [26, 27]. These studies suggest a posttranscriptional role for circRNAs; however, the mechanism of this remains to be elucidated.

5.3 CircRNA in Gene Expression Regulation: A Mechanistic Study

While circRNAs exhibit differential expression patterns in leaf senescence, photosynthesis, and stress response, a mechanistic basis for the putative role of circRNAs in these processes has not been described. To date, the only example of a mechanism for circRNA function with respect to gene regulation was postulated by Conn et al. [75]. In this seminal study, the authors describe the differential expression of floral MADS gene-derived circRNAs under different temperature conditions, demonstrate a positive correlation with their parental splice variant for a few of the circRNAs, and propose a mechanism of R-loop-mediated alternative splicing (AS), favoring generation of the circRNA and its associated mRNA construct. The *SEPALLATA3* gene gives rise to the canonical mRNA that includes all exons, an exon 6-skipped splice variant and a circRNA consisting of exon 6. Overexpression studies of SEP3 exon 6 circRNA and a control circRNA demonstrated that the exon 6 circRNA was able to alter the splicing balance, favoring its own biogenesis and the biogenesis of its associated mRNA at the cost of the canonical splice form. These studies suggest the tantalizing possibility that circRNAs may have general gene regulatory roles in AS via R-loop formation, although this remains to be proven on a wider scale. Indeed, over 60% of genes in Arabidopsis are alternatively spliced, and the putative role of circRNAs in splicing may provide a general role for these molecules. As tight control of mRNA and AS transcript abundance is a hallmark of many genes, including key regulatory genes, profiling circRNAs in plant NGS datasets, examining the potential role of specific circRNAs in modulating gene expression and AS, and determining the effects of circRNAs at the phenotypic level are crucial challenges in the field.

5.4 CircRNA as Biomarkers

CircRNAs are highly stable molecules and can be enriched in many cell types, thus making them good candidates as biomarkers [19, 23, 76, 77]. Over the last few years, many studies have shown the promising relevance of circRNAs as potential molecular markers for disease diagnosis and treatment in humans, with particular emphasis on various types of cancers [78–84]. In plants, various biomarkers have been studied and proven be valuable tools for both fundamental research and applied practices in crop breeding. For example, biomarker genes can be harnessed to assess the response of plants to varying nitrogen conditions and therefore are potential agronomic tools to monitor and optimize nitrogen fertilizer usage [85]. Another study suggested that biomarker genes can be used to identify and differentiate microbial pathogens [86]. Furthermore, metabolic biomarkers allow the prediction of crop product quality [87]. circRNAs as biomarkers in plants are still a new concept but there exists a potential utility for these molecules due to their long half-lives, resistance to common degradation pathways, and the ease and specificity of detection. Only recently, Conn et al. suggested that circRNAs could function as bona fide biomarkers of AS event for MADS box genes in Arabidopsis [75]. Hence, in future research on the function of circRNA in plants, biomarker function is an attractive topic to investigate.

6 Perspectives

CircRNAs are the most contemporary class of regulatory noncoding RNAs, and a plethora of roles for these molecules in animals is emerging. We propose leveraging this information to elucidate the functions of specific circRNAs among the thousands already identified in numerous plant species. With the ability to study whole organism function in higher eukaryotes, plants offer a perfect model to investigate the functions of these molecules. With the breadth of characterized functions for circRNAs, the cellular consequences can be recapitulated in plants. A concerted effort among laboratories will illuminate the functions and interactions of circRNAs in plant development and also offer potential parallels to new functional roles in other eukaryotes.

Acknowledgments This work was supported by Australian Research Council Future Fellowship (FT160100318 to S.C.), Action Thématique et Incitative sur Programme (ATIP)-Avenir (to C.Z.), Agence Nationale de la Recherche (project FloPiNet to C.Z. and X.L.), Grenoble Alliance for Integrated Structural Cell Biology (ANR-10-LABX-49-01 to C.Z.), and the "Laboratoire d'Excellence (LABEX)" Saclay Plant Sciences (SPS; ANR-10-LABX-40) and the ANR grant SPLISIL, France (to M.C.).

Competing Financial Interests The authors declare no competing financial interests.

References

1. Chen LL (2016) The biogenesis and emerging roles of circular RNAs. Nat Rev Mol Cell Biol 17(4):205–211
2. Nigro JM, Cho KR, Fearon ER et al (1991) Scrambled exons. Cell 64(3):607–613
3. Cocquerelle C, Mascrez B, Hetuin D et al (1993) Mis-splicing yields circular RNA molecules. FASEB J 7(1):155–160
4. Wang PL, Bao Y, Yee MC et al (2014) Circular RNA is expressed across the eukaryotic tree of life. PLoS One 9(4):e95116
5. Danan M, Schwartz S, Edelheit S et al (2012) Transcriptome-wide discovery of circular RNAs in Archaea. Nucleic Acids Res 40(7):3131–3142
6. Kopp F, Mendell JT (2018) Functional classification and experimental dissection of long noncoding RNAs. Cell 172(3):393–407
7. Salzman J, Chen RE, Olsen MN et al (2013) Cell-type specific features of circular RNA expression. PLoS Genet 9(9):e1003777
8. Salzman J, Gawad C, Wang PL et al (2012) Circular RNAs are the predominant transcript isoform from hundreds of human genes in diverse cell types. PLoS One 7(2):e30733
9. Memczak S, Jens M, Elefsinioti A et al (2013) Circular RNAs are a large class of animal RNAs with regulatory potency. Nature 495(7441):333–338
10. Gao Y, Wang J, Zhao F (2015) CIRI: an efficient and unbiased algorithm for de novo circular RNA identification. Genome Biol 16:4
11. Conn SJ, Pillman KA, Toubia J et al (2015) The RNA binding protein quaking regulates formation of circRNAs. Cell 160(6):1125–1134

12. Hansen TB, Jensen TI, Clausen BH et al (2013) Natural RNA circles function as efficient microRNA sponges. Nature 495(7441):384–388
13. Piwecka M, Glazar P, Hernandez-Miranda LR et al (2017) Loss of a mammalian circular RNA locus causes miRNA deregulation and affects brain function. Science 357(6357):eaam8526
14. Greene J, Baird AM, Brady L et al (2017) Circular RNAs: biogenesis, function and role in human diseases. Front Mol Biosci 4:38
15. Pamudurti NR, Bartok O, Jens M et al (2017) Translation of CircRNAs. Mol Cell 66(1):9–21 e27
16. Granados-Riveron JT, Aquino-Jarquin G (2016) The complexity of the translation ability of circRNAs. Biochim Biophys Acta 1859(10):1245–1251
17. Abe N, Matsumoto K, Nishihara M et al (2015) Rolling circle translation of circular RNA in living human cells. Sci Rep 5:16435
18. Yang Y, Fan X, Mao M et al (2017) Extensive translation of circular RNAs driven by N(6)-methyladenosine. Cell Res 27(5):626–641
19. Barrett SP, Salzman J (2016) Circular RNAs: analysis, expression and potential functions. Development 143(11):1838–1847
20. Lasda E, Parker R (2014) Circular RNAs: diversity of form and function. RNA 20(12):1829–1842
21. Salzman J (2016) Circular RNA expression: its potential regulation and function. Trends Genet 32(5):309–316
22. Cortés-López M, Miura P (2016) Emerging functions of circular RNAs. Yale J Biol Med 89(4):527
23. Fischer JW, Leung AK (2017) CircRNAs: a regulator of cellular stress. Crit Rev Biochem Mol Biol 52(2):220–233
24. Patop IL, Kadener S (2018) circRNAs in Cancer. Curr Opin Genet Dev 48:121–127
25. Toubia J, Conn VM, Conn SJ (2018) Don't go in circles: confounding factors in gene expression profiling. EMBO J 37(11):e97945
26. Darbani B, Noeparvar S, Borg S (2016) Identification of circular RNAs from the parental genes involved in multiple aspects of cellular metabolism in barley. Front Plant Sci 7:776
27. Ye CY, Chen L, Liu C et al (2015) Widespread noncoding circular RNAs in plants. New Phytol 208(1):88–95
28. Jeck WR, Sorrentino JA, Wang K et al (2013) Circular RNAs are abundant, conserved, and associated with ALU repeats. RNA 19(2):141–157
29. Chen G, Cui J, Wang L et al (2017) Genome-wide identification of circular RNAs in Arabidopsis thaliana. Front Plant Sci 8:1678
30. Lu T, Cui L, Zhou Y et al (2015) Transcriptome-wide investigation of circular RNAs in rice. RNA 21(12):2076–2087
31. Zhao W, Cheng Y, Zhang C et al (2017) Genome-wide identification and characterization of circular RNAs by high throughput sequencing in soybean. Sci Rep 7(1):5636
32. Zuo J, Wang Q, Zhu B et al (2016) Deciphering the roles of circRNAs on chilling injury in tomato. Biochem Biophys Res Commun 479(2):132–138
33. Wang ZP, Liu YF, Li DW et al (2017) Identification of circular RNAs in kiwifruit and their species-specific response to bacterial canker pathogen invasion. Front Plant Sci 8:413
34. Szabo L, Morey R, Palpant NJ et al (2015) Statistically based splicing detection reveals neural enrichment and tissue-specific induction of circular RNA during human fetal development. Genome Biol 16:126
35. Chen L, Zhang P, Fan Y et al (2018) Circular RNAs mediated by transposons are associated with transcriptomic and phenotypic variation in maize. New Phytol 217(3):1292–1306
36. Feracci M, Foot JN, Grellscheid SN et al (2016) Structural basis of RNA recognition and dimerization by the STAR proteins T-STAR and Sam68. Nat Commun 7:10355
37. Chu Q, Zhang X, Zhu X et al (2017) PlantcircBase: a database for plant circular RNAs. Mol Plant 10(8):1126–1128
38. Chen L, Yu Y, Zhang X et al (2016) PcircRNA_finder: a software for circRNA prediction in plants. Bioinformatics 32(22):3528–3529
39. Szabo L, Salzman J (2016) Detecting circular RNAs: bioinformatic and experimental challenges. Nat Rev Genet 17(11):679–692
40. Hansen TB, Veno MT, Damgaard CK et al (2016) Comparison of circular RNA prediction tools. Nucleic Acids Res 44(6):e58
41. Westholm JO, Miura P, Olson S et al (2014) Genome-wide analysis of Drosophila circular RNAs reveals their structural and sequence properties and age-dependent neural accumulation. Cell Rep 9(5):1966–1980
42. Ashwal-Fluss R, Meyer M, Pamudurti NR et al (2014) circRNA biogenesis competes with pre-mRNA splicing. Mol Cell 56(1):55–66
43. Liang D, Wilusz JE (2014) Short intronic repeat sequences facilitate circular RNA production. Genes Dev 28(20):2233–2247
44. Barrett SP, Wang PL, Salzman J (2015) Circular RNA biogenesis can proceed through an exon-containing lariat precursor. Elife 4:e07540
45. Rybak-Wolf A, Stottmeister C, Glazar P et al (2015) Circular RNAs in the mammalian brain are highly abundant, conserved, and dynamically expressed. Mol Cell 58(5):870–885
46. Errichelli L, Dini Modigliani S, Laneve P et al (2017) FUS affects circular RNA expression in murine embryonic stem cell-derived motor neurons. Nat Commun 8:14741
47. Aktas T, Avsar Ilik I, Maticzka D et al (2017) DHX9 suppresses RNA processing defects originating from the Alu invasion of the human genome. Nature 544(7648):115–119
48. Foot JN, Feracci M, Dominguez C (2014) Screening protein--single stranded RNA complexes by NMR

spectroscopy for structure determination. Methods 65(3):288–301
49. Teplova M, Hafner M, Teplov D et al (2013) Structure-function studies of STAR family quaking proteins bound to their in vivo RNA target sites. Genes Dev 27(8):928–940
50. Ryder SP, Williamson JR (2004) Specificity of the STAR/GSG domain protein Qk1: implications for the regulation of myelination. RNA 10(9):1449–1458
51. Ryder SP, Frater LA, Abramovitz DL et al (2004) RNA target specificity of the STAR/GSG domain post-transcriptional regulatory protein GLD-1. Nat Struct Mol Biol 11(1):20–28
52. Hall MP, Nagel RJ, Fagg WS et al (2013) Quaking and PTB control overlapping splicing regulatory networks during muscle cell differentiation. RNA 19(5):627–638
53. Cheng Y, Kato N, Wang W et al (2003) Two RNA binding proteins, HEN4 and HUA1, act in the processing of AGAMOUS pre-mRNA in Arabidopsis thaliana. Dev Cell 4(1):53–66
54. Mockler TC, Yu X, Shalitin D et al (2004) Regulation of flowering time in Arabidopsis by K homology domain proteins. Proc Natl Acad Sci USA 101(34):12759–12764
55. Rodriguez-Cazorla E, Ripoll JJ, Andujar A et al (2015) K-homology nuclear ribonucleoproteins regulate floral organ identity and determinacy in Arabidopsis. PLoS Genet 11(2):e1004983
56. Jiang J, Wang B, Shen Y et al (2013) The Arabidopsis RNA binding protein with K homology motifs, SHINY1, interacts with the C-terminal domain phosphatase-like 1 (CPL1) to repress stress-inducible gene expression. PLoS Genet 9(7):e1003625
57. Guan Q, Wen C, Zeng H et al (2013) A KH domain-containing putative RNA-binding protein is critical for heat stress-responsive gene regulation and thermotolerance in Arabidopsis. Mol Plant 6(2):386–395
58. Jeong IS, Fukudome A, Aksoy E et al (2013) Regulation of abiotic stress signalling by Arabidopsis C-terminal domain phosphatase-like 1 requires interaction with a k-homology domain-containing protein. PLoS One 8(11):e80509
59. Thatcher LF, Kamphuis LG, Hane JK et al (2015) The Arabidopsis KH-domain RNA-binding protein ESR1 functions in components of Jasmonate Signalling, unlinking growth restraint and resistance to stress. PLoS One 10(5):e0126978
60. Lorkovic ZJ, Barta A (2002) Genome analysis: RNA recognition motif (RRM) and K homology (KH) domain RNA-binding proteins from the flowering plant Arabidopsis thaliana. Nucleic Acids Res 30(3):623–635
61. Du WW, Yang W, Liu E et al (2016) Foxo3 circular RNA retards cell cycle progression via forming ternary complexes with p21 and CDK2. Nucleic Acids Res 44(6):2846–2858
62. Zheng QP, Bao CY, Guo WJ et al (2016) Circular RNA profiling reveals an abundant circHIPK3 that regulates cell growth by sponging multiple miRNAs. Nat Commun 7:11215
63. Hsiao KY, Lin YC, Gupta SK et al (2017) Noncoding effects of circular RNA CCDC66 promote Colon Cancer growth and metastasis. Cancer Res 77(9):2339–2350
64. Liu T, Zhang L, Chen G et al (2017) Identifying and characterizing the circular RNAs during the lifespan of Arabidopsis leaves. Front Plant Sci 8:1278
65. Mendes ND, Freitas AT, Sagot MF (2009) Current tools for the identification of miRNA genes and their targets. Nucleic Acids Res 37(8):2419–2433
66. Wang XJ, Reyes JL, Chua NH et al (2004) Prediction and identification of Arabidopsis thaliana microRNAs and their mRNA targets. Genome Biol 5(9):R65
67. Bulow L, Bolivar JC, Ruhe J et al (2012) 'MicroRNA Targets', a new AthaMap web-tool for genome-wide identification of miRNA targets in Arabidopsis thaliana. BioData Min 5(1):7
68. Bonnet E, He Y, Billiau K et al (2010) TAPIR, a web server for the prediction of plant microRNA targets, including target mimics. Bioinformatics 26(12):1566–1568
69. Srivastava PK, Moturu TR, Pandey P et al (2014) A comparison of performance of plant miRNA target prediction tools and the characterization of features for genome-wide target prediction. BMC Genomics 15:348
70. Chi SW, Zang JB, Mele A et al (2009) Argonaute HITS-CLIP decodes microRNA-mRNA interaction maps. Nature 460(7254):479–486
71. Suzuki N, Rivero RM, Shulaev V et al (2014) Abiotic and biotic stress combinations. New Phytol 203(1):32–43
72. Phillips JR, Dalmay T, Bartels D (2007) The role of small RNAs in abiotic stress. FEBS Lett 581(19):3592–3597
73. Khraiwesh B, Zhu JK, Zhu J (2012) Role of miRNAs and siRNAs in biotic and abiotic stress responses of plants. Biochim Biophys Acta 1819(2):137–148
74. Wang YX, Yang M, Wei SM et al (2017) Identification of circular RNAs and their targets in leaves of Triticum aestivum L under dehydration stress. Front Plant Sci 7:2024
75. Conn VM, Hugouvieux V, Nayak A et al (2017) A circRNA from SEPALLATA3 regulates splicing of its cognate mRNA through R-loop formation. Nat Plants 3(5):17053
76. Chen Y, Li C, Tan C et al (2016) Circular RNAs: a new frontier in the study of human diseases. J Med Genet 53(6):359–365
77. Huang S, Yang B, Chen BJ et al (2017) The emerging role of circular RNAs in transcriptome regulation. Genomics 109(5–6):401–407

78. Meng S, Zhou H, Feng Z et al (2017) CircRNA: functions and properties of a novel potential biomarker for cancer. Mol Cancer 16(1):94
79. Han YN, Xia SQ, Zhang YY et al (2017) Circular RNAs: a novel type of biomarker and genetic tools in cancer. Oncotarget 8(38):64551–64563
80. Zhang HD, Jiang LH, Sun DW et al (2018) CircRNA: a novel type of biomarker for cancer. Breast Cancer 25(1):1–7
81. Huang YS, Jie N, Zou KJ et al (2017) Expression profile of circular RNAs in human gastric cancer tissues. Mol Med Rep 16(3):2469–2476
82. Lu LS, Sun J, Shi PY et al (2017) Identification of circular RNAs as a promising new class of diagnostic biomarkers for human breast cancer. Oncotarget 8(27):44096–44107
83. Shao Y, Li J, Lu R et al (2017) Global circular RNA expression profile of human gastric cancer and its clinical significance. Cancer Med 6(6):1173–1180
84. Zhu X, Wang X, Wei S et al (2017) hsa_circ_0013958: a circular RNA and potential novel biomarker for lung adenocarcinoma. FEBS J 284(14):2170–2182
85. Yang XS, Wu J, Ziegler TE et al (2011) Gene expression biomarkers provide sensitive indicators of in planta nitrogen status in maize. Plant Physiol 157(4):1841–1852
86. Yan X, Gurtler J, Fratamico P et al (2011) Comprehensive approaches to molecular biomarker discovery for detection and identification of Cronobacter spp. (Enterobacter sakazakii) and Salmonella spp. Appl Environ Microbiol 77(5):1833–1843
87. Steinfath M, Strehmel N, Peters R et al (2010) Discovering plant metabolic biomarkers for phenotype prediction using an untargeted approach. Plant Biotechnol J 8(8):900–911

Circular RNAs and Plant Stress Responses

27

Celso Gaspar Litholdo Jr.
and Guilherme Cordenonsi da Fonseca

Abstract

Circular RNAs (circRNAs) are a novel class of noncoding RNAs that have been extensively explored in the past few years. The advent of new high-throughput sequencing technologies coupled with bioinformatics tools revealed the presence of these molecules in the transcriptome of a wide range of organisms. In animals, circRNAs can modulate gene expression and act as sponges of miRNAs to inhibit their activity. It has been demonstrated that they have the potential to be diagnostic biomarkers as their expression is closely associated to human diseases, such as Alzheimer and cancer. However, in plants their function remains elusive. Recently, the role of the circRNAs in plant stress responses has been studied. During the infection of *Pseudomonas syringae* in kiwifruit plants, 584 circRNAs were differentially expressed in leaf samples, and a group of them could be further associated with the stage of infection. Under phosphate deficiency conditions, 27 rice circRNAs were reported to be differentially expressed. In tomato, 163 circRNAs demonstrated chilling-responsive expression, with 102 containing miRNA-binding sites and are predicted to act as miRNA sponges. Additionally, *Arabidopsis* seedlings presented 1583 heat-specific circRNAs, and it was also reported that heat stress could increase the quantity, length, and alternative circularization events of circRNAs. Finally, wheat seedlings under dehydration stress had 62 circRNAs differentially expressed, with 6 being predicted as miRNA sponges. Although the role of plant circRNAs during the biotic and abiotic stresses is still poorly characterised, these molecules have the potential to expand the number of targets and tools in the biotechnology field.

C. G. Litholdo Jr. (✉)
Laboratoire Génome et Développement des Plantes, Centre National pour la Recherche Scientifique (CNRS), Perpignan, France
e-mail: celso.litholdo@univ-perp.fr

G. C. da Fonseca
Centro de Biotecnologia, Universidade Federal do Rio Grande do Sul, Porto Alegre, RS, Brazil

1 Introduction

1.1 Circular RNAs: Biosynthesis and Function in Plants

Circular RNAs (circRNAs) are single-stranded covalently closed loop RNA molecules that were first described in 1976, when it was demonstrated that plant viroids are composed of single-stranded RNA molecules lacking free 5′ and 3′ends [1]. Posteriorly, another group identified for the first time the existence of endogenous circRNA

J. Xiao (ed.), *Circular RNAs*, Advances in Experimental Medicine and Biology 1087,
https://doi.org/10.1007/978-981-13-1426-1_27

transcripts. Studying the candidate tumor suppressor gene (DCC), it was observed that several abnormal transcripts composed of exons were scrambled during the splicing process [2]. These molecules were found at relatively low levels in the nonpolyadenylated component of cytoplasmic RNA from human and rodent cells [2]. Over the following years, more endogenous circRNAs were discovered in eukaryotes [3, 4]; however, they were long considered transcriptional byproducts or the result of aberrant RNA splicing without a clear function [5].

Until recently, the study of circRNAs has been neglected due to the difficulties in the detection of these molecules through the most used methods of RNA analysis. Since circRNAs lack free 5′- or 3′- ends, they cannot be detected by techniques that need polyadenylated free ends, such as rapid amplification of cDNA ends (RACE) or RNAseq from poly(A)-enriched samples, which is the most commonly used strategy for transcriptome analysis [6]. However, the improvement of the high-throughput sequencing technology, particularly with the advent of the RibominusSeq technique (RNAseq from rRNA-depleted samples), coupled with the development of new bioinformatics tools, allowed the identification of hundreds of circRNAs in different organisms [7, 8]. Over the past few years, circRNAs have been identified in many plant species, including *Arabidopsis*, rice, barley, wheat, tomato, cotton, soybean, potato, kiwi, maize, and orange [9–18].

The biosynthesis of the circRNAs (Fig. 27.1) is considered to be conserved in all eukaryotes and, as protein-coding genes, is dependent of RNA polymerase II and a noncanonical splicing mechanism of the pre-messenger RNAs (pre-mRNAs), termed backsplicing [19]. The backsplicing occurs when the 5′-end of the upstream exon is linked to 3′-end of the downstream exon producing a head-to-tail splicing junction [20]. In some cases, the circularization retains introns between exons producing the exon-intron circRNAs or EIciRNAs [21]. circRNAs can be also derived from lariat introns that escape debranching and form a stable circular molecule, termed intronic circRNA or ciRNA [22].

The regulation of the circularization events is dependent by *cis* and *trans* elements during the spliceosome formation, including RNA pairing, by either repetitive elements [23] or complementary sequences in the flanking introns [24]. However, circRNAs identified in plants possessed few repetitive and reverse complementary sequences in the flanking introns, when compared to animals [9]. In *Arabidopsis*, it was identified that circRNAs have at least two different short complementary sequences that ranged from 4 to 11 nucleotides near the splice sites in more than 33% of the cases, suggesting that multiple combination of these sequences can compensate for their short length [25]. It was demonstrated in rice that more than 90% of the identified circRNAs were flanked by noncanonical splicing signals, contrasting with humans circRNAs, which most are flanked by the canonical GT/AG splicing sites [26]. Altogether, these findings indicated that plants have specific mechanisms for the regulation of the circRNAs when compared to other organisms [27].

The most striking function of the circRNA that has been discovered so far is their ability to modulate gene expression, acting as miRNA sponges [28]. MiRNA sponges are transcripts that have miRNA-binding sites and therefore are able to "sequest" miRNAs, to inhibit their activity [29]. These transcripts are also called competing endogenous RNAs in animals or target mimicry in plants [30], and they have been used as molecular biology tools to study the function of miRNAs [29, 31]. The advantage of the circRNAs as miRNA sponges is their intracellular high stability, with half-lives exceeding 48 h, whereas their linear counterparts exhibited half-lives of less than 20 h [23]. This higher stability is due to their resistance to RNA exonucleases, since circRNAs do not have 5′- or 3′- free ends [5]. However, no evidence has been shown that plant circRNAs act as miRNA sponges [27]. Several studies show that plant circRNAs have the potential to be miRNA sponges, but compared to animals, they represent a smaller proportion of the total circRNAs (around 5%) and have lesser miRNA-binding sites [9, 12, 32].

In plants, it has been demonstrated that circRNAs can negatively modulate the expression of their parental genes. In rice, the overexpression of a circRNA greatly reduces the expression

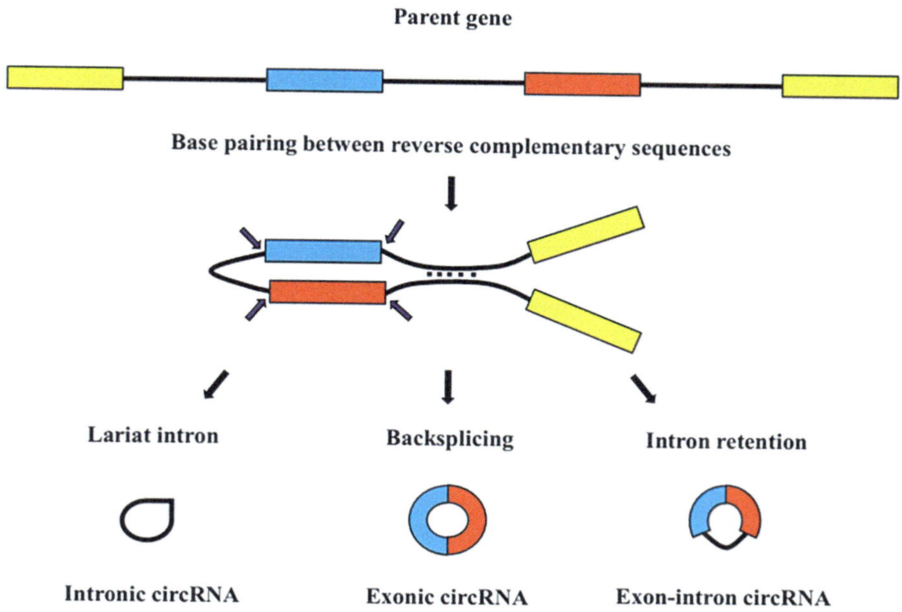

Fig. 27.1 Circular RNA biosynthesis
The parental gene of the circRNA is transcribed by RNA polymerase II. The backsplicing can occur by RNA pairing of complementary sequences in the flanking introns. The result of the backsplicing is a circular molecule produced by a head-to-tail splicing junction. The circRNA that contains only exons is termed exonic circRNA. In some cases, the circularization retains introns between exons producing the exon-intron circRNAs. When a lariat intron escapes the debranching, it can form a stable circular molecule termed intronic circRNA. The purple arrows indicate the splicing sites. The colored rectangles represent the exons, while the black lines represent the introns

levels of its parental gene in different tissues [33]. In a more recent work, a circRNA derived from exon 6 of the *SEPALLATA3* (*SEP3*) gene of *Arabidopsis thaliana* was demonstrated to bind to its cognate DNA locus through the formation of a RNA:DNA hybrid resulting in transcriptional pausing and an decrease on the abundance of its sourced gene [34]. This circRNA also increases the abundance of an exon-skipped alternative splicing variant that lacks the exon 6 driving floral homeotic phenotypes [34]. These findings provided new insights about the biological roles of the plant circRNAs.

1.2 Methods to Identify and Study circRNAs in Plants

High-throughput sequencing and bioinformatics analysis have been the most used tools to characterize and identify circRNAs. Although RibominusSeq libraries are the most commonly used method, these molecules can also be found in the traditional RNAseq libraries from poly(A)-enriched samples, however with a lower efficiency of detection [33]. Several works are also using RNAseq libraries from rRNA-depleted samples further treated with RNAse R, an enzyme that degrades linear but not circular RNAs, to improve the sensibility and decrease the number of false positives for the detection of circRNAs [15, 23, 35, 36].

The identification of circRNAs relies mainly by detection of sequence reads that map only in the backsplicing junctions, but not in the reference genome, often termed backsplicing reads or junction reads. To address that goal, several algorithms and programs were developed including CIRI [37], find_circ [7], MapSplice [38], CIRCexplorer [24], and testrealign [39]. The differences among them are the use of different read aligners, the requirement of inputs as gene annotations, and the search for non-exonic circRNAs [40]. A study compared five different algorithms and evaluated the levels of *bona fide* and false-positive circRNAs using data from libraries treated with RNase R and untreated samples [41]. These results showed that the false-positive rate ranged from 12% to 28%, but was especially high for highly expressed circRNAs, suggesting that the more reliable approach is the combination of at least two circRNA identification methods [41]. However, all these programs were designed primarily for human or animal datasets, and considering the differences in the organization of plant and mammal genomes, a software called PcircRNA_finder was developed [42]. This software was able to predict, with more high-confidence, circRNAs from rice libraries when compared to CIRCexplorer and find_circ [42].

Over the past few years, an increasingly number of studies have allowed the development of plant circRNA databases. The PlantcircBase [43] compiles the information of almost 90,000 circRNAs from eight plant species. Detailed information is available for each circRNA entry, such as ID, genomic position, annotation and experimental validation of the backsplicing site. Additionally, splicing signals of each circRNAs are provided, which is a particularly useful information, considering that most of the plant circRNAs do not have the canonical GT-AG splicing signals [26]. Another interesting feature found in this database is the prediction of networks involving circRNA-miRNA-mRNA, allowing the identification of putative miRNA sponges.

The mostly used technique to validate circRNAs from *in silico* analysis is the reverse transcription (RT)-PCR using convergent primers to detect linear transcripts, as the parental mRNA, and divergent primers to detect the circular RNA [9, 33]. Divergent primers are designed as such: the forward primer aligns in the downstream exon, and the reverse primer aligns in the upstream exon of the circRNA sequence. The reverse primer is also used in the cDNA synthesis of the circular and linear transcripts [44].

2 Potential Role of circRNAs During Plant Stress Responses

Circular RNAs are new discovered players in the RNA-mediated gene regulation, acting in several biological processes, at the transcriptional and posttranscriptional levels. In animals, circRNAs are abundant in transcriptomics data, and several has been demonstrated to participate in the occurrence of human diseases, such as cancers and Alzheimer, which creates diagnostic and therapeutic targets based on circRNAs [45, 46]. In plants, circRNA function is poorly characterized although its presence in plant transcriptomics is found widespread and highly abundant, since the first report in the plant model *Arabidopsis* [8]. We will here compile all the information available in the literature to give us hints of the role of circRNAs during plant stress responses, which is also indicated in Table 27.1.

2.1 circRNAs and Biotic Stresses

The first report identifying circRNAs under a biotic stress condition was the study of the plant-pathogen interaction between *Arabidopsis* and *Pseudomonas syringae* pv. tomato. Leaf samples were sequenced in the Illumina HiSeq 2000 platform, after removal of ribosomal and linear RNAs, by magnetic beads and RNase R, respectively, the latest a biochemical approach named CircleSeq [25]. 85 million reads per sample were obtained, and 803 circRNAs were identified using 16 distinct RNA sequencing data sets from *A. thaliana*.

The authors used the following filter criteria for circRNA identification: (1) presence of alignment of the reads with mapping quality of >10; (2) presence of either a "GT" or "AG" signal within five nucleotides of the donor or acceptor site, respectively, or a back-splice site; (3) segments from the same read must be in the same strand with a reversed order; and (4) circRNA sequence was presented in at least two *Arabidopsis* datasets [25]. 108 circRNAs out of

Table 27.1 Biotic and abiotic stresses currently known to deregulate plant circular RNAs expression

	Plant stress	Plant species	Number of differentially expressed circrnas	Reference
Biotic	Bacterial infection (*Pseudomonas syringae pv. actinidiae*)	*Actinidia sp.* (kiwi fruit)	584	[10]
	Bacterial infection (*Pseudomonas syringae pv. Tomato*)	*Arabidopsis thaliana*	803	[25]
Abiotic	Low- and high- light	*Arabidopsis thaliana*	6012	[9]
	Heating (38 °C)	*Arabidopsis thaliana*	1583	[47]
	Iron and zinc treatments	*Hordeum vulgare* (barley)	62	[13]
	Phosphate-starvation	*Oryza sativa* (rice)	27	[9]
	Chilling (0 °C)	*Solanum lycopersicum* (tomato)	163	[12]
	Dehydration (PEG)	*Triticum aestivum* (wheat)	62	[11]

The number of differentially expressed circRNAs identified during plant stress conditions is indicated

803 overlapped with the exons, showing that complementary sequences are not exclusive of introns, but found in the flanking sequences. In addition, the chloroplast genome showed a higher source of circRNA than the mitochondria; 6% of identified circRNA were found in the chloroplast compared to 1% of circRNA in mitochondria. Furthermore, the results revealed that both general and alternative splicing mechanisms generate circRNAs [25]. Although this work identified hundreds of circRNAs in *Arabidopsis* under pathogenic interaction and contributes to understanding this endogenous class of RNA, the biological function of circRNAs under biotic stresses remains to be revealed.

More recently, circRNAs were also found to respond to bacterial invasion in kiwifruit plants [10]. Leaf samples of three species belonging to the genus *Actinidia* have their ribosomal RNA removed and sequenced using the Illumina HiSeq 2000 sequencer. These samples were previously infected with *Pseudomonas syringae* pv. *actinidiae*, the causal agent of kiwi canker disease. In addition, the authors identified that circRNAs expression have tissue-specific patterns. A total of 2884 million paired-end reads were sequenced, and 2914 circRNAs from the bacterial infected samples were identified after CircRNA Identifier (CIRI) tool and manual filtering. Among these circRNAs, 64% were intergenic, 21% were exonic, and 14.55% were intronic circRNAs [10].

A total of 584 circRNAs were differentially expressed during *P. syringae* infection, and the expression of particular circRNAs was further associated with the stage of infection [10]. Moreover, exonic and intronic circRNAs showed a positive correlation with the originating protein-coding genes, and a weighted gene co-expression network analysis (WGCNA) for searching potential biological process that are associated with genes of interest identified circRNAs that were closely associated with respiratory burst, MAP kinase activity, and intracellular signal transduction. Therefore, this study represents the first report on the potential biological role of circRNAs in plant defense response against bacterial pathogen infection [10].

2.2 circRNA and Abiotic Stresses

Previous reports have shown that circRNAs are differentially expressed under abiotic stresses, such as phosphate, zinc, and iron imbalance, low and high light, chilling, and drought. However, the biological significance underlying circRNA regulatory function during these conditions remains to be elucidated.

Computational analysis, using two publicly available ribosomal RNA-depleted RNA sequencing data from different tissues and phosphate-starvation and light treatments, identified 12037 and 6012 circRNAs in *Oryza sativa* roots and *A. thaliana* leaves, respectively [9]. circRNAs were selected based on back-spliced reads, and 50.5% and 85.7% represented exonic circRNAs in rice and *Arabidopsis*, respectively. Approximately, 700 exonic circRNAs were originated from homologous genes between both plant species, suggesting a conservation in plants [9]. The authors found 27 differentially expressed circRNAs under phosphate deficiency condition in rice, 6 upregulated and 21 downregulated; moreover, several circRNAs presented positive correlation with their parental genes. These results suggest that a regulatory role between circRNAs and circRNAs-originating genes also exists in plants, and additionally, the differentially expressed circRNAs have a putative role in response to phosphate deficiency [9].

Another recent study revealed the expressional changes of circRNAs in response to nutrients [13]. This study was carried out across barley tissues and in response to iron and zinc treatment on both foliar and seed tissues. The authors used a RNAseq dataset available in the GenBank and performed sequencing using the Illumina platform HiSeq 2000 and TruSeq technology. A total of 262 putative circRNAs were identified by these approaches, and after manual filtering, 62 circRNAs were selected. The majority of selected circRNAs presented a weak and negative transcriptional correlation with the parental genes. Moreover, when comparing treatments in seeds and leaves, an opposite trend was observed. For instance, *CAX2* gene, which is an H+/cation antiporter and sequesters of calcium, iron, and zinc, had its transcription level inverted correlated with the CAX2 circRNA, in an tissue-dependent manner [13].

In addition, several identified circRNAs were derived from mitochondrial and nuclear genes previously known to be involved with cellular functions that are related to respiratory complex, ion homeostasis, intracellular protein transport, amino acid biosynthesis, transcription and translation, and hormonal signaling. These genes include *COX1* subunit of the cytochrome c oxidase, apocytochrome b, NADH dehydrogenase *NAD9*, and *CTR1*-like serine/threonine-protein kinase [13]. Interestingly, the authors also found noncoding protein transcripts as parental source of circRNAs, such as microRNA and long noncoding RNAs after iron and zinc treatments.

Another abiotic stress that has been shown to alter circRNAs expression is extreme temperature, either heat or chilling stress; therefore, circRNAs have possibly also a role on temperature-related stress responses. Low-temperature stress was studied in tomato plants, as chilled tomatoes are susceptible to develop symptoms in this condition [12]. Tomato fruits were kept under 0° C for 72 h, and rRNA-removed RNA samples were sequenced by Illumina platform. Using CIRI tools and differential expression analysis, the authors identified 3019 chilling-responsive circRNAs, with 770 unique circRNAs that were only identified under the stress condition when compared to the control sample. Stress-induced exonic circRNAs represented 72% of the total parental genes, including genes that encode to enzymes involved in, cellular redox homoeostasis network, cell wall degradation, membrane lipid peroxidation, arginine and polyamine metabolism, energy metabolism, and jasmonic acid and abscisic acid signalling. circRNAs related to genes that encode to temperature-induced transcription factors, such as the core-binding factor (CBF) and WRKY superfamily, were also identified [12]. In addition, the authors predicted 102 circRNAs that can potentially act as miRNA sponges, due to the predicted miRNA-binding site for 24 distinct mature miRNAs. Further analysis showed that 163 circRNAs presented differential expression changes between control and chilled plants. The vast majority of the deregulated circRNAs showed upregulation under chilling treatment [12].

Similar pattern of expressional changes were observed in *A. thaliana* plants under heat treatment [47]. *Arabidopsis* seedlings were treated with 38°C for 3 h, and rRNA-depleted and circRNAs-enriched RNA samples were submitted to Illumina HiSeq 4000 sequencer. The authors discovered 1599 novel circRNAs, with 1583 showing heat-specific accumulation. When comparing control samples, heat stress treatment triggered a significant increase in circRNA quantity, size, and alternative circularization events. Interestingly, different from the other reports on circRNAs and plant stress responses, the production of circRNAs was originated mainly from nuclear genes rather than chloroplast and mitochondria, and moreover, compared to chilling treatment [12], the majority of circRNAs under high temperature treatment showed a positive correlation with their parental genes [47]. Additionally, the authors predicted the miRNAs that could be sequestered by heat-induced circRNAs and built an interacting network comprising the miRNA target genes. The predicted miRNA sponge-like circRNAs can potentially influence the expression of genes related to heat stress, hydrogen peroxide, and phytohormones, by sequestering, for instance, the miR156, miR157, miR159, miR395, miR172, and miR857.

Our last herein mentioned plant stress that has shown to regulate circRNAs is drought. Wheat seedlings treated with polyethylene glycol (PEG), which artificially mimics dehydration stress were analyzed by high-throughput sequencing and real-time PCR [11]. The authors identified 62 candidate circRNAs that showed a differential expression under drought-like stress, with 16 circRNAs being upregulated and 46 downregulated at this condition. This work also predicted the sponge action of circRNAs by showing that 6 out of the 62 circRNAs have predicted miRNA-binding site that can potentially regulate 26 distinct wheat miRNAs [11]. Further analysis, using Gene Ontology (GO) and Kyoto Encyclopedia of Genes and Genomes (KEGG), demonstrated an involvement of the identified circRNAs in several cellular processes, such as photosynthesis, porphyrin and chlorophyll metabolism, oxidative phosphorylation, amino acid-related process, and phytohormone signaling.

3 Concluding Remarks

Circular RNAs are a class of noncoding RNAs that have been increasingly studied in plants during the past 5 years. In animals, the role of circRNAs in the control of transcriptional and posttranscriptional gene expression is uncovered; however, the equivalent function in plant species still lacks consistent and experimental validation. Further, the already known biological role of animal circRNAs during development and progress of diseases represents the chance to target circRNAs for therapeutic and diagnostic purposes. Despite the progress in understanding the role and action of animal circRNAs, there is little information about plant circRNAs and even less about its biological role during plant development and stress responses.

Based on high-throughput sequencing technologies and *in silico* analyses, there are indications that circRNAs-mediated gene regulation might play functional roles in plant immune response (Table 27.1), through different mechanisms by acting directly on mRNAs, or as microRNA sponges. Another recently raised question is the putative translation of exonic circRNAs into functional peptides [35]; the majority of identified plant circRNAs are derived from exons, but the existance of such peptides still demands further additional research. circRNAs represent, therefore, a new and intriguing class of noncoding RNAs still poorly explored in the field of plant science; their putative regulatory role on functional coding and noncoding transcripts associated with bacterial, temperature, nutrient starvation and toxicity, and drought resistances needs to be fully addressed. Future investigations on the biological significance of circRNAs during biotic and abiotic stresses can potentially expand the available tools and targets for applied crop biotechnology.

References

1. Sanger HL, Klotz G, Riesner D et al (1976) Viroids are single-stranded covalently closed circular RNA molecules existing as highly base-paired rod-like structures. Proc Natl Acad Sci USA 73:3852–3856
2. Nigro JM, Cho KR, Fearon ER et al (1991) Scrambled exons. Cell 64:607–613
3. Cocquerelle C, Daubersies P, Majérus MA et al (1992) Splicing with inverted order of exons occurs proximal to large introns. EMBO J 11:1095–1098

4. Capel B, Swain A, Nicolis S et al (1993) Circular transcripts of the testis-determining gene Sry in adult mouse testis. Cell 73:1019–1030
5. Cocquerelle C, Mascrez B, Hétuin D et al (1993) Mis-splicing yields circular RNA molecules. FASEB J 7:155–160
6. Jeck WR, Sharpless NE (2014) Detecting and characterizing circular RNAs. Nat Biotechnol 32:453–461
7. Memczak S, Jens M, Elefsinioti A et al (2013) Circular RNAs are a large class of animal RNAs with regulatory potency. Nature 495:333–338
8. Wang PL, Bao Y, Yee M-C et al (2014) Circular RNA is expressed across the eukaryotic tree of life. PLoS ONE 9(3):e90859
9. Ye CY, Chen L, Liu C et al (2015) Widespread noncoding circular RNAs in plants. New Phytol 208:88–95
10. Wang Z, Liu Y, Li D et al (2017) Identification of circular RNAs in Kiwifruit and their species-specific response to bacterial canker pathogen invasion. Front Plant Sci 8:413
11. Wang Y, Yang M, Wei S et al (2017) Identification of circular RNAs and their targets in leaves of Triticum aestivum L. under dehydration stress. Front Plant Sci 7:2024
12. Zuo J, Wang Q, Zhu B et al (2016) Deciphering the roles of circRNAs on chilling injury in tomato. Biochem Biophys Res Commun 479:132–138. https://doi.org/10.1016/j.bbrc.2016.07.032
13. Darbani B, Noeparvar S, Borg S (2016) Identification of circular RNAs from the parental genes involved in multiple aspects of cellular metabolism in Barley. Front Plant Sci 7:776
14. Zhao T, Wang L, Li S et al (2017) Characterization of conserved circular RNA in polyploid Gossypium species and their ancestors. FEBS Lett 591:3660–3669
15. Zhao W, Cheng Y, Zhang C et al (2017) Genome-wide identification and characterization of circular RNAs by high throughput sequencing in soybean. Sci Rep 7:5636
16. Zhou R, Zhu Y, Zhao J et al (2017) Transcriptome-wide identification and characterization of potato circular RNAs in response to Pectobacterium carotovorum subspecies brasiliense infection. Int J Mol Sci 19:E71
17. Chen L, Zhang P, Fan Y et al (2018) Circular RNAs mediated by transposons are associated with transcriptomic and phenotypic variation in maize. New Phytol 217:1292–1306
18. Zeng R-F, Zhou J-J, Hu C-G et al (2018) Transcriptome-wide identification and functional prediction of novel and flowering-related circular RNAs from trifoliate orange (Poncirus trifoliata L. Raf.). Planta 247:1191–1202
19. Ashwal-Fluss R, Meyer M, Pamudurti NR et al (2014) circRNA biogenesis competes with pre-mRNA splicing. Mol Cell 56:55–66
20. Salzman J, Gawad C, Wang PL et al (2012) Circular RNAs are the predominant transcript isoform from hundreds of human genes in diverse cell types. PLoS One 7:e30733
21. Li Z, Huang C, Bao C et al (2015) Exon-intron circular RNAs regulate transcription in the nucleus. Nat Struct Mol Biol 22:256–264
22. Zhang Y, Zhang X-O, Chen T et al (2013) Circular intronic long noncoding RNAs. Mol Cell 51:792–806
23. Jeck WR, Sorrentino JA, Wang K et al (2013) Circular RNAs are abundant, conserved, and associated with ALU repeats. RNA 19:141–157
24. Zhang X-O, Wang H-B, Zhang Y et al (2014) Complementary sequence-mediated exon circularization. Cell 159:134–147
25. Sun X, Wang L, Ding J et al (2016) Integrative analysis of Arabidopsis thaliana transcriptomics reveals intuitive splicing mechanism for circular RNA. FEBS Lett 590:3510–3516. https://doi.org/10.1002/1873-3468.12440
26. Ye C-Y, Zhang X, Chu Q et al (2017) Full-length sequence assembly reveals circular RNAs with diverse non-GT/AG splicing signals in rice. RNA Biol 14:1055–1063
27. Li Q-F, Zhang Y-C, Chen Y-Q et al (2017) Circular RNAs roll into the regulatory network of plants. Biochem Biophys Res Commun 488:382–386
28. Hansen TB, Jensen TI, Clausen BH et al (2013) Natural RNA circles function as efficient microRNA sponges. Nature 495:384–388
29. Ebert MS, Neilson JR, Sharp PA (2007) MicroRNA sponges: competitive inhibitors of small RNAs in mammalian cells. Nat Methods 4:721–726
30. Franco-Zorrilla JM, Valli A, Todesco M et al (2007) Target mimicry provides a new mechanism for regulation of microRNA activity. Nat Genet 39:1033–1037
31. Todesco M, Rubio-Somoza I, Paz-Ares J et al (2010) A collection of target mimics for comprehensive analysis of microRNA function in Arabidopsis thaliana. PLoS Genet 6:e1001031
32. Wang K, Long B, Liu F et al (2016) A circular RNA protects the heart from pathological hypertrophy and heart failure by targeting miR-223. Eur Heart J 37:2602–2611
33. Lu T, Cui L, Zhou Y et al (2015) Transcriptome-wide investigation of circular RNAs in rice. RNA 21:2076–2087
34. Conn VM, Hugouvieux V, Nayak A et al (2017) A circRNA from SEPALLATA3 regulates splicing of its cognate mRNA through R-loop formation. Nat Plants 3:17053
35. Chen G, Cui J, Wang L et al (2017) Genome-Wide Identification of Circular RNAs in Arabidopsis thaliana. Front Plant Sci 8:1678
36. Tan J, Zhou Z, Niu Y et al (2017) Identification and functional characterization of tomato CircRNAs derived from genes involved in fruit pigment accumulation. Sci Rep 7:8594
37. Gao Y, Wang J, Zhao F (2015) CIRI: an efficient and unbiased algorithm for de novo circular RNA identification. Genome Biol 16:4
38. Wang K, Singh D, Zeng Z et al (2010) MapSplice: accurate mapping of RNA-seq reads for splice junction discovery. Nucleic Acids Res 38:e178

39. Hoffmann S, Otto C, Doose G et al (2014) A multi-split mapping algorithm for circular RNA, splicing, trans-splicing and fusion detection. Genome Biol 15:R34
40. Gaffo E, Bonizzato A, Kronnie G et al (2017) CirComPara: a multi-method comparative bioinformatics pipeline to detect and study circRNAs from RNA-seq data. Non-Coding RNA 3:8
41. Hansen TB, Venø MT, Damgaard CK et al (2016) Comparison of circular RNA prediction tools. Nucleic Acids Res 44:e58
42. Chen L, Yu Y, Zhang X et al (2016) PcircRNA_finder: a software for circRNA prediction in plants. Bioinformatics 32:3528–3529
43. Chu Q, Zhang X, Zhu X et al (2017) PlantcircBase: a database for plant circular RNAs. Mol Plant 10:1126–1128
44. Kulcheski FR, Christoff AP, Margis R (2016) Circular RNAs are miRNA sponges and can be used as a new class of biomarker. J Biotechnol 238:42–51
45. Ghosal S, Das S, Sen R et al (2013) Circ2Traits: a comprehensive database for circular RNA potentially associated with disease and traits. Front Genet 4:283
46. Chen LL (2016) The biogenesis and emerging roles of circular RNAs. Nat Rev Mol Cell Biol 17:205–211
47. Pan T, Sun X, Liu Y et al (2017) Heat stress alters genome-wide profiles of circular RNAs in Arabidopsis. Plant Mol Biol 96:217–229

Part VIII

Future Prospects

28 Prospective Advances in Circular RNA Investigation

Siti Aishah Sulaiman, Nor Azian Abdul Murad, Ezanee Azlina Mohamad Hanif, Nadiah Abu, and Rahman Jamal

Abstract

circRNAs have emerged as one of the key regulators in many cellular mechanisms and pathogenesis of diseases. However, with the limited knowledge and current technologies for circRNA investigations, there are several challenges that need to be addressed for. These include challenges in understanding the regulation of circRNA biogenesis, experimental designs, and sample preparations to characterize the circRNAs in diseases as well as the bioinformatics pipelines and algorithms. In this chapter, we discussed the above challenges and possible strategies to overcome those limitations. We also addressed the differences between the existing applications and technologies to study the circRNAs in diseases. By addressing these challenges, further understanding of circRNAs roles and regulations as well as the discovery of novel circRNAs could be achieved.

Keywords

circRNA · Challenges · Identification and analysis · Future directions

1 Introduction

The emergence of circulating RNAs (circRNAs) was reported in the 1970s, where they were observed as circular structures captured by electron microscopy in the eukaryotic cytoplasm. Interest in this noncoding RNA did not last long due to the belief then that it was low in abundance. circRNAs are often assumed to be splicing noise or byproducts of splicing [1–4], and that microRNAs (miRNAs) may outcompete their activation [5]. However, circRNAs have reemerged and have become a hot topic in many areas such as agriculture [6, 7] and medicine with the discovery of their roles in gene regulation [1, 8–11]. circRNAs have been identified through advances in RNA sequencing (RNA-seq) and bioinformatics analyses, where data have been successfully pooled to establish databases for predicting circRNA downstream targets. The establishment of circRNA databases provides guidelines for the prediction and further evaluation of circRNAs to further understand their biogenesis and roles.

S. A. Sulaiman · N. A. Abdul Murad (✉)
E. A. Mohamad Hanif · N. Abu (✉) · R. Jamal
Universiti Kebangsaan Malaysia (UKM)
Medical Molecular Biology Institute (UMBI),
Kuala Lumpur, Malaysia
e-mail: nor_azian@ppukm.ukm.edu.my;
nadiah.abu@ppukm.ukm.edu.my

J. Xiao (ed.), *Circular RNAs*, Advances in Experimental Medicine and Biology 1087,
https://doi.org/10.1007/978-981-13-1426-1_28

2 Current Knowledge of circRNA Functions

2.1 Bona Fide Disease Biomarkers and Therapeutic Targets

circRNA expression is tissue specific, indicating that they harbor biological functions and can be used as biomarkers to distinguish cell origin and also tumor types [4, 12]. circRNAs can adopt a protective tumor-suppressive role and can also exhibit oncogenic properties. Currently, only a few circRNAs and their intersection molecules and/or pathways have been identified, namely, cZNF292 in glioma [13], circ-ITCH in lung cancer [14], and CDR1 (or ciRS-7) in the ischemic myocardium (MI) [15]. Circ-ITCH downregulation has been observed in esophageal squamous carcinoma (ESC). Circ-ITCH overexpression suppresses tumor growth via miR-7, miR-17, and miR-124 sponging to activate *ITCH* mRNA, which then undergoes degradation and ubiquitination by phosphorylated Dvl2, ultimately repressing the Wnt/β-catenin pathway [16]. In lung cancer, circ-ITCH expression in tumor tissues was significantly decreased compared to the adjacent non-tumor tissues (n = 78) [14]. Ectopic expression of circ-ITCH results in increased expression of the *ITCH* gene, its parental cancer-suppressive gene. This subsequently leads to inhibition of lung cancer cell proliferation. circ-ITCH is a sponge of the oncogenic miR-7 and miR-214, which are important for increasing *ITCH* expression and suppressing Wnt/β-catenin signaling activation [14]. cZNF292 is crucial for tube formation in glioma, and cZNF292 silencing reduced cell proliferation and cell cycle progression in human glioma U87MG and U251 cell lines [13]. Here, the S/G2/M phase in cell cycle progression was arrested through the Wnt/β--catenin signaling pathway and several genes, including *PRR11* (proline rich 11), *CCNA* (cyclin A), p-*CDK2* (phosphorylated cyclin-dependent kinase 2), *VEGFR1/2* (vascular endothelial growth factor 1/2), p-*VEGFR1/2*, and *EGFR* (epidermal growth factor receptor) [13].

In addition to their roles as biomarkers, circRNAs are involved in many cellular processes, including regulating the processes of cellular energy, cell death and senescence, genome stability, angiogenesis, invasion and metastasis, inflammation, and metabolism [17]. In a study of lead-induced neurotoxicity, circ-Rar1 acted as a sponge of miR-671 and promoted apoptosis by inducing caspase-8 and p38 expression [18]. Moreover, circRNAs also regulate the epithelial-to-mesenchymal transition (EMT). For example, circHIPK3 played a crucial role in miR-558 sponging, leading to inhibition of migration, invasion, and angiogenesis [19]. circHIPK3 is highly expressed in liver cancer, and its depletion has proliferative effects on liver cancer cells [20]. Likewise, circPVT1 upregulation in gastric cancer (GC) is postulated to have an effect on GC tissue proliferation, acting as a sponge of miR-125 [21]. A similar effect has been observed in other cancer cells; for example, Huh7, HCT116, and HeLa cell proliferation was significantly reduced upon knockdown of circHIPK3 [20]. circRNA profiling has yielded positive insights in the classification of circRNA biomarkers in cancer.

Expanding on the potential role of circRNAs as disease biomarkers, several studies have explored the function and implication of circRNAs in metabolic and non-cancer diseases. One such circRNA is ciRS-7, a well-identified circRNA that acts as a miR-7 a/b sponge or inhibitor in brain tissues or islet cells [22]. Hypoxia treatment led to increased expression of *Cdr1as* and miR-7a in mice with myocardial infarction (MI), causing an increase in the cardiac infarct size [15]. *Cdr1as* overexpression in mouse cardiomyocytes (MCM) promotes apoptosis and is reversed by miR-7a overexpression. Sp1 transcription factor (SP1) and poly (ADP-ribose) polymerase (PARP) were both targeted by miR-7a and induced apoptosis under hypoxia treatment [15]. Myosin VIIA and Rab-interacting protein (*Myrip*) and paired box 6 (*Pax6*) activation were observed as the downstream targets of miR-7 [23], indicating that ciRS-7 is a potential alternative diagnostic strategy in metabolic disease, namely, diabetes. However, further studies are required to rule out the functional roles of ciRS-7 and to prove whether it truly has any implications in the advancement of diabetes.

2.2 circRNAs in Agriculture

In humans, circRNA biogenesis is dependent on RNA polymerase II transcription and the back-splicing reaction of the precursor mRNA (pre-mRNA). In plants, circRNAs are localized in the nucleus or other organelles, such as the mitochondria and chloroplast, or outside the cells, such as with viroids [6]. The functional assessments of circRNAs in plants are still in their infancy. However, circRNAs may be involved in plant developmental processes and stress responses (abiotic versus biotic stress), and similar to humans, they also serve as miRNA sponges. circRNAs are tissue organelle dependent, indicating that they play a role in plant organelle development [7, 24] and can be distinguished via RNA-seq datasets. There is evidence that circRNAs also contribute to malnutrition in *Oryza sativa* (rice) and barley [25, 26]. The described regulatory pathways indicate that a number of circRNAs may be involved in cellular metabolism, hormonal signaling, intracellular protein sorting, respiration, and carbohydrate metabolism [25]. An RNA-seq dataset of *Arabidopsis thaliana* showed that circRNAs are clustered in the chloroplast, suggesting that circRNAs may play a vital role in plant photosynthesis [27]. The composition of circRNAs has been proven to be an important element in the field of plant RNAs. In-depth studies aimed at understanding the mechanism behind plant development are crucial, as they may benefit the future agriculture sector.

3 Challenges to Current circRNA Investigation

3.1 Challenges in Regulating circRNA Biogenesis

It is accepted that circRNAs act as miRNA and RNA-binding protein (RBP) sponges in addition to roles in regulating gene transcription and expression, as well as protein/peptide translators [28]. circRNAs are formed in a circular tail-less mode at the 5′ end, as also seen in the canonical linear mRNA. circRNAs are commonly configured from pre-mRNA. They are organized by the joining of the 3′ end of a downstream exon of a gene and then back-spliced to the 5′ end of the first exon, forming the covalently circularized RNA [1].

circRNA back-splicing is generated by the presence of inverted repeat sequences or Alu repeat elements in the introns flanking the exons that bring the spliced sites of exons at the 3′ end to the 5′ end of the first exon in close proximity to one another via base-pairing [12]. circRNAs are mainly derived from exonic coding regions and can appear in three different forms: exonic circRNAs (e-circRNAs), intronic circRNAs (ciRNAs), and retained-intron circRNAs (ElcircRNAs).

circRNAs have become a hot topic for research with an enormous amount of research being performed to understand their mechanisms. Solid evidence has emerged by in vivo and in vitro studies. circRNAs are regulated by multiple mechanisms and serve various functions. To date, four functional roles have been reported for circRNAs. First, circRNAs act as miRNA sponges or competing endogenous mRNAs. Introducing circRNAs may have sponging effects on multiple miRNAs, altogether abrogating miRNA binding to the target mRNA(s) [29]. Second, circRNAs bind and sequester proteins. For example, circularized *muscleblind* (*MBL1*) is highly abundant in *Drosophila* and humans [4, 30, 31]. The start site coding sequence present in circMbl and the binding sites of *MBL1* mRNA signifies the recruitment binding of *MBL1*, indicating the capacity of circRNAs to sequester proteins through *MBL1* ribosomes to initiate translation in *Drosophila* [31]. It has been suggested that circRNA protein binding can also inhibit translation by acting as a competing counterpart, as observed in cytoplasmic circ-FOXO3 (forkhead box O3), where interactions with ID1 (inhibitor of DNA binding 1, HLH protein) and E2F1 (E2F transcription factor 1) [32] deplete activation in the cytoplasm; it has also been implicated in the promotion of cell cycle arrest in NIH3T3 mouse embryo fibroblast cells [33]. The formation of circRNA–mRNA complexes plays an inhibitory role in cellular and biological functions. However, there is a crucial need for in-depth exploration of the physical interactions of circRNAs with multiple target proteins to understand their actual mechanisms and pathways. Third, circRNAs regulate mRNA splicing or transcription. ElcircRNAs are localized in the nucleus, implying that they will undergo alternative splicing and gene transcription [12]. Unlike the canonical linear mRNAs, transcripts containing introns in the nucleus will be exported from the nucleus and be degraded in the cytoplasm [34]. Intron-containing circRNAs (ElcircRNAs and ciRNAs) remain in the nucleus, indicating their involvement in mRNA transcriptional regulation [4]. Nevertheless, ciRNAs do not have defined roles or biological processes. Lastly, circRNAs can be translated. circRNA transformation by N^6-methyladenosine (m^6A) gives rise to the initiation of protein translation [35]. Translating ribosomes have been associated with circRNAs known as ribo-circRNAs via ribosome footprinting. Ribo-circRNAs harbor a conserved termination codon at which translation initiation occurs by utilizing the start codon of the target mRNA bound to membrane-associated ribosomes [3]. Important evidence from in vitro and in vivo studies has demonstrated the translational activity of circRNAs. The protein encoded by *Drosophila mbl* has been detected via mass spectrometry, indicating that the translational mechanism occurs in normal settings [3]. circRNAs have also been implicated as driver mechanisms of protein synthesis. An internal ribosome entry site (IRES) was stably integrated in a circRNA minigene to induce expression in vitro, and western blotting and flow cytometry showed that it activated a number of putative proteins. This is further strong evidence that circRNAs may also function as mRNAs to drive protein synthesis [10].

Studies on the involvement of circRNAs in various diseases are still in the infancy stage, where expression has been identified but further functional and population validation are still lacking [15, 36–39]. Some key functions of circRNAs have been identified; however, the emerging circRNA–miRNA–mRNA interactions require further validation. In addition, the studies of circRNAs in cancer have not focused on the mechanisms of initiation, progression, and metastasis. In cancers, most of the studies involved identification of circRNAs, including lung, colorectal, breast, and bladder cancers as well as glioma [4]. Currently, several circRNAs have been identified to be upregulated or downregulated in cancers. The analysis of circRNAs poses several challenges including in terms of experimental design, techniques to be used, and most importantly, the bioinformatics tools to be used. The current knowledge regarding the regulation of circRNA biogenesis is limited, which renders the studies on circRNAs more difficult.

3.2 Challenges in Experimental Design and Technique

CircRNAs can be identified using several approaches, including RNA-seq, microarray, and reverse transcription (RT)-PCR [17, 40, 41]. As circRNAs are novel products of regulated alternative splicing processes, detecting splicing through RNA-seq is the most favorable method. However, there are several limitations, where circRNA cannot be detected due to several RNA library preparation steps, including RNA purification, RNA or cDNA size selection, and RNA fragmentation. For purification, rRNA-depleted libraries is much better compared to poly(A) selected or depleted since they retain the circRNAs in a particular sample. circRNAs lack a poly(A) tail; however, poly(A) enrichment step will significantly reduce circRNAs, thus making them difficult to be detected by RNA-seq. For size selection, random priming is preferable as this method is not biased where small sizes of circRNAs can also be detected. The RNA needs to be fragmented prior to adaptor ligation or priming in order to distinguish between the small RNA and circRNAs. In addition, RNA-seq protocols can introduce technical artifacts during ligation and RT. The ligation step can produce chimeric complementary DNAs (cDNAs), thus might generate low levels of artefactual circRNAs [42]. On the other hand, reverse transcriptase can lead to extensive template-switching artifacts, where two different RNA molecules can join together and affect the discovery of novel RNA isoforms [43, 44]. RT can also lead to strand displacement, which subsequently leads to differences in circRNA quantitative measurement [45]. RT is highly dependent on good RNA quality, and most importantly, the right library preparation method should be selected [46]. Other techniques that can be used to detect circRNAs include the microarray and RT-quantitative (Q) PCR techniques [17]. Most genome-wide studies use microarray for detecting circRNAs, but it is performed in limited numbers of cancer types and their adjacent normal tissue samples. However, microarray cannot identify novel circRNAs. RT-QPCR is commonly used to validate the circRNAs identified in the discovery phase in a larger sample size. Currently, most circRNA studies have been performed retrospectively; thus conducting a large prospective clinical trial is important for confirming circRNAs as disease biomarkers. The details regarding the challenges in bioinformatics are discussed in the next section.

3.3 Challenges in Bioinformatics Analysis

CircRNA identification in single-end (SE) RNA-seq data is performed by aligning the reads to the back-splice junctions. Many available algorithms can be used to identify circRNAs from the sequence data; however, their performance differs [47]. Five circRNA prediction algorithms, i.e., circRNA_finder, CIRCexplorer, find_circ, CIRI, and MapSplice, have different false-positive and false-negative circRNA detection rates [40]. The ability of the algorithms to identify a circRNA based on the splice site distance also varies. CIRI and circRNA_finder can identify circRNAs with very low proximal splice sites (<100 bp) compared to other algorithms; however, most of the circRNAs found were false positives [40]. On the other hand, CIRCexplorer and MapSplice produce the most reliable list of circRNAs with low positive rates. However, these algorithms took 2–3 days to complete individual dataset predictions [40]. Hence, identifying circRNAs with high specificity and sensitivity using a single algorithm is challenging. The usage of more algorithms for circRNA identification is advisable [41]. More importantly, individual validation is crucial for novel circRNAs with regard to the type of algorithms used, as false-positive results are common [40]. The technical challenges to studying circRNAs are discussed in the next section.

4 Comparisons Between Current circRNA Investigation Methods

4.1 Comparison Between RNA Preparation Methods for Identification of circRNAs

As circRNAs are biologically closed loops of RNA and lack open ends, it is quite difficult to identify and validate their presence. To date, there are three main methods of screening for the presence of circRNAs in biological samples: microarray, RNA-seq, and bioinformatics analyses of readily available data (Table 28.1) [48, 49]. Nevertheless, before proceeding with any analysis, it is important to enrich the circRNA population and remove any other linear RNAs that may skew the population [24, 50]. Total cellular RNA contains a wide array of RNA species and artifacts, and to avoid false positives, samples are usually treated with RNAse R [24] to digest all other linear RNAs with >7 nucleotides, which will exclude circRNAs because they lack the required 3′ end [24]. For microarray, only one available platform is Arraystar. This platform utilizes circRNA information from six published databases [24, 51–55]. Currently, this is the most straightforward approach for identifying circRNAs from human samples because it does not require extensive bioinformatics analysis. A more comprehensive approach to identifying circRNAs is by sequencing the transcriptome population. Current RNA-seq library preparations

Table 28.1 Pros and cons of different methods to identify and validate circular RNAs

Method	Pros	Cons
Identification		
1. Microarray	Cost-effective	Only cover reported species of circular RNA
	Some targets have been wet lab-validated	Cannot be used for new discovery of circular RNAs
	Known targets, easier to validate	Human, mouse, rat targets only [78]
		Require a larger amount of starting material
		Most targets are exonic and intronics circular RNAs
2. RNA-Seq	Less starting material is required than microarray	Library preparations can be biased against circular RNAs
	Can discover new species circular RNAs	High cost
	Can cover both intronic, intergenic, and exonic circular RNAs	Bioinformatics pipeline is not well established yet [66]
3. Bioinformatics approach of readily available data	Can reuse the data for other analyses	Needs more validation
		Bioinformatics pipeline is not well established yet
		Most of the library preparation protocol does not enrich for circular RNA population [66]
		Require higher expertise in bioinformatics
Validation		
1. qRT-PCR	Cost-effective	False positives can happen
	Readily available in most labs	Can amplify linear splicing events also [60]
		Rolling RT effects [60]
2. DdPCR	Less false positives as it eliminates rolling RT effects [60]	High cost
		Requires special machine/equipment to perform
3. Northern Blot	Specific target [61]	Low sensitivity [95]
	Less false positives [47]	Laborious and time-consuming [62]
4. FISH/ISH	Less false positives	Low sensitivity [96]

vary greatly, particularly in terms of biochemical preparation [41]. RNA purification is among the first steps in RNA-seq and can greatly influence circRNA population [56]. For example, the poly (A) selection in RNA-seq library preparation can heavily deplete the circRNA population, as circRNAs do not contain poly (A) tails [41]. The size selection generally used in RNA-seq excludes nucleotides of <200 bp [24]. This would exclude smaller circRNAs (even if present) unless a small RNA-seq library kit is used [24]. Therefore, the most reliable, cost- and time-effective method for identifying circRNAs at the moment is microarray. The qRT-PCR approach is commonly used to validate the presence of circRNAs [57]. Other methods such as droplet digital PCR (ddPCR), fluorescence in situ hybridization, and northern blotting are also used [57]. It is imperative to determine the sequence of the circRNAs and identify the junction spanning 5′ and 3′ ends [58]. From there, divergent primers can be designed to flank the junction sequence and use sequencing [58, 59]. This is to avoid the primers amplifying random linear splicing events [59]. ddPCR is a new technology that enables higher sensitivity in nucleic acids quantification [60]. It is more reliable and accurate than qRT-PCR for determining the quantity of circRNAs, as it eliminates the effect of rolling RT products [60]. Northern blotting is another commonly used method for detecting circRNAs [61, 62]. Similar to qRT-PCR, probes are designed to target the junction sequence and are detected on blots. Sometimes, northern blotting is preferred because the species targeted is known, thus reducing false positives [47, 62]. Nevertheless, it remains an unpopular choice due to its time-consuming and laborious procedure

[62]. Fluorescence in situ hybridization can also be used to detect circRNAs [63, 64]. It utilizes fluorescence-coupled probes targeting the junctional sequence [63, 64]. All four methods have been proven to be able to detect circRNAs, but some are preferred over others. Table 28.1 compares the advantages and disadvantages of each method.

4.2 Comparison Between Bioinformatics Tools for Identifying circRNAs

CircRNA identification in RNA-seq data analysis was not performed until the use of splicing analysis [41, 65]. Therefore, numerous bioinformatics pipelines and algorithms have been developed for circRNA detection [51, 53, 66–75], including SE or paired-end (PE) RNA-seq data (Table 28.2). These tools or algorithms use different and unique combinations of identification strategies based on the reference genome and annotations, mapper of choice, back-splicing junction (BSJ) identification methods, and other filtering options [41, 76]. Typically, the flow of the bioinformatics analysis starts from the removal of mapped reads, followed by the mapping of splice junction sites from the unmapped reads to identify the circRNAs [76, 77]. Among these tools, the majority, except segemehl [68], use external aligners such as Bowtie [78, 79], Burrows-Wheeler Aligner (BWA) [23, 80], or Spliced Transcripts Alignment to a Reference (STAR) [81] to filter out the sequencing reads aligned to the reference genome (mapped reads) [41, 76].

To identify circRNAs from unmapped reads, two methods or algorithms are used [76]. The first method uses the split-alignment approach that focuses on BSJ reads to identify circRNAs. The algorithm is based on the fact that BSJ reads are aligned to the reference sequences in reverse, particularly to the established GU-AG splicing sequences flanking the splice sites [76]. The split-alignment approach can be gene annotation dependent, where the BSJs are identified in between annotated exons, or gene annotation independent, where the BSJs are identified from alignment to a reference genome without gene annotation. There is a slight difference between the tools (Table 28.2) in terms of executing this approach. Segemehl [68], MapSplice [69], CIRCexplorer [53], circRNA_finder [74], and double conjugated clustering (DCC) [75] create an algorithm that aligns the unmapped reads to spliced sites for detecting BSJ sequences. By contrast, UROBORUS [66] and find_circ [51] remove mapped reads and extract the first and last 20-bp anchor sequences from the unmapped reads before identifying BSJ sequences from the mapping of these anchors. CIRI [71, 73] performs local alignment using BWA-MEM to identify paired chiastic clipping (PCC) for de novo detection of BSJ sequences and filters systematically to eliminate false positives. The second method is a pseudo-reference approach that creates a pseudo-sequence database or de novo prediction of BSJs by combining the reference genome with the corresponding gene annotation to identify circRNAs (Table 28.2) [76]. Posttranscriptional exon shuffling (PTES)Finder [67] and non-co-linear transcripts (NCL)Scan [70] create putative circRNA sequences based on mapping information of the exon-to-exon junction sequences after alignment to the reference genome. However, Known and Novel IsoForm Explorer (KNIFE) [72] creates all potential exon-to-exon junction sequences from gene annotations before the alignment. Additionally, KNIFE uses mRNA forward-spliced junction reads to construct a forward-spliced sites database [72] and also for further removal of candidate reads aligned to both databases to improve accuracy.

Previous comparison studies have shown that most circRNA algorithms have little overlap between the prediction results, and no single gold-standard algorithm is sensitive and specific enough for all data [40, 77, 82, 83]. Using rRNA-depleted libraries with or without RNAse R treatment, up to 76% and 28% of false positives were detected in rRNA-depleted libraries and rRNA-depleted RNAse R–treated libraries, respectively [40]. Testing these algorithms with simulated data improved sensitivity and specificity, although there was a trade-off between sensitivity and specificity, in which the algorithms with

Table 28.2 Comparison between circRNA identification and prediction tools

Computational tools	Sequencing read types	Mappers	Splice-aware	Annotation	CircRNAs detection method	Analysis by previous studies				References
						Precision (HeLa, %)	Precision (Hs68, %)	False positives (HeLa, n)	False positives (Hs68, n)	
CIRCexplorer	PE and SE	TopHat/STAR	Yes	Yes	Split-alignment	50.09	68.54	1388	1856	[53]
circRNA_finder	PE and SE	STAR	Yes	No	Split-alignment	46.32	58.54	1597	2094	[74]
CIRI	PE and SE	BWA-MEM	No	No	Split-alignment	54.20	69.49	3210	3400	[71, 73]
DCC	PE and SE	STAR	Yes	Yes	Split-alignment	45.22	63.08	1760	2107	[75]
find_circ	SE	Bowtie2	No	No	Split-alignment	36.99	59.75	2092	2377	[51]
KNIFE	PE and SE	Bowtie, Bowtie2	No	Yes	Pseudo-reference	44.26	69.49	2055	2359	[72]
MapSplice	PE and SE	Bowtie	No	Yes	Split-alignment	54.21	76.33	1765	1854	[69]
NCLscan	PE and SE	BWA, BLAT, Novoalign	Mixed	Yes	Pseudo-reference	45.06	64.73	954	892	[70]
PTESFinder	PE and SE	Bowtie, Bowtie2	No	Yes	Pseudo-reference	35.65	63.29	2054	2474	[67]
UROBORUS	PE and SE	Bowtie, Bowtie2, TopHat	Mixed	Yes	Split-alignment	31.00	19.73	761	279	[66]
Segemehl	PE and SE	Per Se	No	No	Split-alignment	14.32	8.78	2506	3094	[68]

Summary of the 11 known circRNA identification tools including their precision analysis done by previous study [76, 77]
Note: MapSplice is a splice-aware aligner but uses Bowtie read mapper
PE paired-end, *SE* single-end

the highest specificity also had the lowest sensitivity, and vice versa [40]. Therefore, none of these tools share an exact pipeline of analysis and adapt to a combination of selection criteria and filtering.

One such difference is the implementation of gene annotation, which is often preferred for its improved accuracy and sensitivity for BSJ detection in comparison to pseudo-reference or de novo prediction [82, 83]. Annotation-dependent algorithms only detect uniquely mapped reads and require canonical splice signals [41, 76]; thus this can cause a blind spot for detecting circRNA isoforms (non-canonical signals). In other words, the pseudo-reference approach offers more flexibility for detecting non-canonical splice signals, but at the cost of high false positives. Even so, most of the non-canonical signals are small; thus many pseudo-reference algorithms choose to implement GU-AG splicing sequences as a default filter to control the false discovery rate (FDR) [41, 76], therefore potentially undermining the detection of non-canonical splice signals. Importantly, this dependency on annotation may create a blind spot in the algorithms for detecting novel circRNAs.

Another important point is the type of mapper used in the sequence mapping. find_circ [51], CIRI [71, 73], KNIFE [72], and PTESFinder [67] use the most standard reference aligners (Bowtie and BWA), which are not specifically designed to note splice sites. Excluding KNIFE [72], these algorithms implement the split-alignment approach or 20-bp anchor filters to control sensitivity, FDR, and specificity. Meanwhile, splice-aware mappers such as TopHat [84, 85], STAR [81], and Novoalign have been optimized and updated with the current reference genome, thus making them more convenient. One disadvantage is that splice-aware mappers have more restricted reporting modes, filtering, and arguments and may not be able to detect circRNAs in non-canonical signals [41, 76].

Besides that, the sensitivity and accuracy of these algorithms also depend on the recovery of unbalanced BSJ reads [41, 76]. Standard 20-bp anchor sequence filtering cannot capture the very short reads of unbalanced BSJs because there is little information for detection [41, 76]; thus algorithms such as CIRI [71, 73] can detect unbalanced BSJ reads due to its dynamic alignment algorithm, which can identify balanced BSJ sequences to control the FDR. The detection of unbalanced BSJs using balanced BSJ reads may require an additional step to improve the accuracy, especially for low-abundance circRNAs. The usage of multiple seed matching improves the recovery of unbalanced BSJs [73]. In this study [73], CIRI created two putative genomic regions from short segments corresponding to BSJs and forward-spliced sequences. Then, it divided the short segments or potential unbalanced BSJ reads into multiple short seeds and counted the rate those seeds were matched to the genomic region [73]. By comparing the matching counts and seed locations, the exact position of the unbalanced BSJs or short segments could be determined [73]. This evidence therefore indicates that CIRI is a more preferable algorithm for those who need to identify low-abundance circRNAs.

4.3 Comparison of circRNA Databases and Biological Function Tools

Besides circRNA identification tools, other computational tools have been developed to investigate the biological function of circRNAs by focusing on databases, differential expression, quantification, visualization, and length assembly. Currently, the available circRNA databases are circBase (http://www.circbase.org/) [86], circRNADb (http://reprod.njmu.edu.cn/circrnadb) [14], circ2Traits (http://gyanxet-beta.com/circdb/) [87], and CircNet (http://circnet.mbc.nctu.edu.tw/) [88]; they are all freely accessible online databases. circBase is a database of eukaryote circRNAs with evidence of their expression [86]. It supports the genome assemblies of *Homo sapiens* (hg19), *Mus musculus* (mm9), *Caenorhabditis elegans* (ce6), *Latimeria chalumnae* and *L. menadoensis* (latCha1), and *D. melanogaster* (dm3) [86]. circRNADb [14] contains 32,914 annotated e-circRNAs with information on the circRNA genetic sequences, exon splicing, IRES, open reading frames (ORF), and references [14]. circ2Traits [87] reports the associations of circRNAs and disease by estimating the likelihood of a circRNA being associated with miRNA in the disease and the network prediction of these associations between circRNAs, miRNAs, protein-coding genes, and long noncoding RNAs [87]. CircNet accounts for circRNA isoforms and incorporates a novel naming system for antisense circRNAs or ciRNAs and the location of BSJ sites on well-annotated exons [88]. CircNet also provides information on novel circRNAs, integrated miRNA–circRNA networks, expression profiles, genomic annotations, and circRNA isoform sequences [88].

A few application tools have been developed for investigating the biological functions of circRNAs. Both Sailfish-cir [39] and CircTest [75] have been used to identify circRNA differential expression or quantification. Sailfish-cir [39] estimates circRNA expression based on BSJ reads and other related reads from the data, while CircTest [75] estimates circRNA expression based on the BSJ read counts distributed in the samples. An additional algorithm in CIRI, CIRI-AS, which refers to alternative splicing, can be used for investigating the internal structure and alternative splicing of circRNAs [89]. Similarly, three other tools have been developed to characterize and visualize circRNA function and structure. Full Circular RNA Characterization from RNA-seq (FUCHS) [90] was developed to detect the known exon-skipping events within circRNAs, whereas CircPro [91] can be used to identify the protein-coding potential of circRNAs. CircView [92] can be used to aid visualization and understanding of circRNA functions, particularly regulatory elements such as miRNA response elements (MREs) and RBP binding sites. For understanding circRNA interaction with proteins and miRNAs, CircInteractome is a user-friendly, web-based tool for exploring circRNA interactions based on their potential binding sites [93]. However, more such tools are needed to further explore circRNAs, their biogenesis, and functions.

5 Suggestions on Methods to Investigate circRNAs

5.1 Suggestions for Improvement

For identifying circRNAs, technology providers should create more microarray platforms. This will help reduce bias and create more opportunities to include other types of newly discovered circRNAs. For RNA-seq, the biochemical steps in the library preparation protocol should be established and validated [41]. A proper guideline on enriching circRNAs and removing artifacts and linear splicing events should be implemented. For circRNA validation, both microarray and RNA-seq have their advantages and disadvantages, but perhaps two or three methods can be used in combination to confirm the presence of circRNAs. There should also be minimal requirements or criteria for publishing circRNA results, for example, whether the circRNAs were enriched using enzymatic reaction and how the primers/probes targeting the junctional sequences were designed and validated.

In terms of the bioinformatics pipelines for circRNA identification and prediction, there is no single perfect and accurate algorithm or analysis pipeline that has been developed so far [41, 76]. circRNA detection in RNA-seq reads requires careful consideration, as most tools depend on BSJ reads that have significant read density bias, and some of the sequence reads at the exon boundaries are usually associated with sequencing errors [2, 10, 37]. Therefore, these can lead to false-positive alignments at BSJ reads which will not be able to represent a truly expressed circRNA.

Due to the high rate of inconsistency between the bioinformatics tools discussed above, a validation approach is needed to assess the accuracy and performance of these algorithms. One such method is using simulated data to assess the sensitivity, specificity, and limitations [41] and to identify the trade-off values, false positives, and false negatives so that the user can identify the tool that best suits their needs. An example of simulated data is Benchmarker for Evaluating the Effectiveness of RNA-Seq Software (BEERS) [94], which simulates human and mouse PE RNA-seq data (Illumina) with differential expression of genes, splicing events, sequencing errors, variants, and insertions/deletions (indels). It is important to note that experiment-generated data are more complex than simulated data, and it is unknown how similar simulated data are in comparison to experimental data [41]. Following that, a study evaluating these algorithms found that CIRI [71, 73], CIRCexplorer [53], and KNIFE [72] showed more balanced performance but that their sensitivity, specificity, and cost had a few shortcomings [77]. To improve these algorithms, a recent study suggested that combining prediction algorithms and selecting the commonly identified circRNAs resulted in reduced false positives and improved the accuracy [82], particularly for CIRI, find_circ, and UROBORUS, which benefited the most. An improved algorithm that can eliminate the need for strict post-filtering is required to provide more flexibility in exploring novel circRNAs.

Rapid identification of circRNAs has revealed that the existing circRNAs are more diverse than expected. This could mean that some circRNAs may have been overlooked. Identifying these novel circRNAs is by nature difficult due to the many confounding factors, and a reference- or annotation-free approach algorithm would be able to identify such circRNAs, but at the expense of a higher FDR [41, 76]. This is due to the fact that most circRNAs are derived from novel and unannotated splice sites and are often less likely to be true positives and with reference-free algorithms having reduced accuracy of 13–27% [82]. Interestingly, the performance of CIRI, a reference-free algorithm, is comparable to that of annotation algorithms [82], suggesting that it is currently the most reliable algorithm. Therefore, developing a reference-free algorithm with greater sensitivity and accuracy is the current challenge to improving circRNA investigation.

6 Conclusions and Perspectives

It is apparent that circRNAs are now coming of age and are being rediscovered as important regulators of the cellular system. The biological functions of circRNAs include miRNA sponging, protein binding and sequestration, regulation of mRNA splicing, and transcription of circRNAs that can be translated. circRNAs are involved in many cellular processes leading to cancer and non-cancer diseases. In cancer, circRNAs are important for regulating cell death and senescence and the proliferation, migration, and invasion and inflammation processes. In non-cancer diseases, such as diabetes, circRNAs are important players in the metabolic and inflammation pathways. Nevertheless, there is still much uncharted territory when it comes to understanding the role of circRNAs. Most current circRNA investigations are generally dependent on RNA-seq data, and many experimental designs and library preparations are tailored for this purpose. Thus, this may introduce potential bias in pooling circRNA populations for identification and prediction analysis. Moreover, most if not all the algorithms that have been developed for identifying and predicting circRNA have little agreement, and their sensitivity and specificity require improvement. Given the widespread existence of circRNAs in the human genome, it is possible that current technologies in circRNA investigation may overlook other circRNA isoforms or low-abundance circRNAs that could be biologically meaningful. In conclusion, circRNAs have a promising potential as disease biomarkers and regulators; however, understanding of their biological functions and interaction with other known regulators is currently limited and needs further studies.

References

1. Haque S, Harries LW (2017) Circular RNAs (CircRNAs) in health and disease. Genes 8(12):1–17
2. Jeck WR, Sharpless NE (2014) Detecting and characterizing circular RNAs. Nat Biotechnol 32(5):453–461
3. Pamudurti NR, Bartok O, Jens M et al (2017) Translation of CircRNAs. Mol Cell 66(1):9–21 e27
4. Qu S, Liu Z, Yang X et al (2018) The emerging functions and roles of circular RNAs in cancer. Cancer Lett 414(2018):301–309
5. Tay Y, Rinn J, Pandolfi PP (2014) The multilayered complexity of ceRNA crosstalk and competition. Nature 505(7483):344–352
6. Lee SM, Kong HG, Ryu CM (2017) Are circular RNAs new kids on the block? Trends Plant Sci 22(5):357–360
7. Liu T, Zhang L, Chen G et al (2017) Identifying and characterizing the circular RNAs during the lifespan of Arabidopsis leaves. Front Plant Sci 8(July):1–9
8. Dou Y, Cha DJ, Franklin JL et al (2016) Circular RNAs are down-regulated in KRAS mutant colon cancer cells and can be transferred to exosomes. Sci Rep 6(October 2015):1–11
9. Meng S, Zhou H, Feng Z et al (2017) CircRNA: functions and properties of a novel potential biomarker for cancer. Mol Cancer 16(1):1–8
10. Wang Y, Wang Z (2015) Efficient backsplicing produces translatable circular mRNAs. RNA 21(2):172–179
11. Zhong Z, Lv M, Chen J (2016) Screening differential circular RNA expression profiles reveals the regulatory role of circTCF25-miR-103a-3p/miR-107-CDK6 pathway in bladder carcinoma. Sci Rep 6(November 2015):1–12
12. Zhang ZC, Guo XL, Li X (2018) The novel roles of circular RNAs in metabolic organs. Genes Dis 5(1):16–23
13. Yang P, Qiu Z, Jiang Y et al (2016) Silencing of cZNF292 circular RNA suppresses human glioma tube formation via the Wnt/β-catenin signaling pathway. Oncotarget 7(39):63449–63455
14. Wan L, Zhang L, Fan K et al (2016) Circular RNA-ITCH suppresses lung cancer proliferation via inhibiting the Wnt/β-Catenin pathway. Biomed Res Int 2016:1579490
15. Geng H-H, Li R, Su Y-M et al (2016) The circular RNA Cdr1as promotes myocardial infarction by mediating the regulation of miR-7a on its target genes expression. PLoS One 11(3):e0151753
16. Li F, Zhang L, Li W et al (2015) Circular RNA ITCH has inhibitory effect on ESCC by suppressing the Wnt/β-catenin pathway. Oncotarget 6(8):6001–6013
17. Kristensen LS, Hansen TB, Venø MT et al (2018) Circular RNAs in cancer: opportunities and challenges in the field. Oncogene 37(5):555–565
18. Nan A, Chen L, Zhang N et al (2017) A novel regulatory network among LncRpa, CircRar1, MiR-671 and apoptotic genes promotes lead-induced neuronal cell apoptosis. Arch Toxicol 91(4):1671–1684
19. Li Y, Zheng F, Xiao X, et al (2017) CircHIPK3 sponges miR-558 to suppress heparanase expression in bladder cancer cells. EMBO Rep:e201643581-e201643581.
20. Zheng Q, Bao C, Guo W et al (2016) Circular RNA profiling reveals an abundant circHIPK3 that regulates cell growth by sponging multiple miRNAs. Nat Commun 7:1–13
21. Chen J (2016) Circular RNA profile identifies circPVT1 as a proliferative factor and prognostic marker in gastric cancer. (December):2016–2016
22. Hansen TB, Jensen TI, Clausen BH et al (2013) Natural RNA circles function as efficient microRNA sponges. Nature 495:384
23. Wang Y, Liu J, Liu C et al (2013) MicroRNA-7 regulates the mTOR pathway and proliferation in adult pancreatic β-cells. Diabetes 62(3):887–895
24. Salzman J, Chen RE, Olsen MN et al (2013) Cell-type specific features of circular RNA expression. PLoS Genet 9(9):e1003777
25. Darbani B, Noeparvar S, Borg S (2016) Identification of circular RNAs from the parental genes involved in multiple aspects of cellular metabolism in barley. Front Plant Sci 7(June):1–8
26. Ye CY, Chen L, Liu C et al (2015) Widespread noncoding circular RNAs in plants. New Phytol 208(1):88–95
27. Sun X, Wang L, Ding J et al (2016) Integrative analysis of Arabidopsis thaliana transcriptomics reveals intuitive splicing mechanism for circular RNA. FEBS Lett 590(20):3510–3516
28. Han Y-N, Xia S-Q, Zhang Y-Y et al (2017) Circular RNAs: a novel type of biomarker and genetic tools in cancer. Oncotarget 8(38):64551–64563
29. Yang W, Du WW, Li X et al (2016) Foxo3 activity promoted by non-coding effects of circular RNA and Foxo3 pseudogene in the inhibition of tumor growth and angiogenesis. Oncogene 35(30):3919–3931
30. Ashwal-Fluss R, Meyer M, Pamudurti NR et al (2014) CircRNA biogenesis competes with pre-mRNA splicing. Mol Cell 56(1):55–66
31. Kramer MC, Liang D, Tatomer DC et al (2015) Combinatorial control of Drosophila circular RNA expression by intronic repeats, hnRNPs, and SR proteins. Genes Dev 29(20):2168–2182
32. Du WW, Yang W, Chen Y et al (2017) Foxo3 circular RNApromotes cardiac senescence by modulating multiple factors associated with stress and senescence responses. Eur Heart J 38(18):1402–1412
33. Du WW, Yang W, Liu E et al (2016) Foxo3 circular RNA retards cell cycle progression via forming ternary complexes with p21 and CDK2. Nucleic Acids Res 44(6):2846–2858
34. Braunschweig U, Barbosa-Morais NL, Pan Q et al (2014) Widespread intron retention in mammals functionally tunes transcriptomes. Genome Res 24(11):1774–1786

35. Yang Y, Fan X, Mao M et al (2017) Extensive translation of circular RNAs driven by N 6 -methyladenosine. Cell Res 27(5):626–641
36. Cui X, Niu W, Kong L et al (2016) hsa_circRNA_103636: potential novel diagnostic and therapeutic biomarker in major depressive disorder. Biomark Med 10(9):943–952
37. Guo X-Y, Chen J-N, Sun F et al (2017) circRNA_0046367 prevents Hepatoxicity of lipid peroxidation: an inhibitory role against hepatic steatosis. Oxid Med Cell Longev 2017:3960197
38. Hansen TB, Kjems J, Damgaard CK (2013) Circular RNA and miR-7 in Cancer. Cancer Res 73(18):5609–5612
39. Tang C-M, Zhang M, Huang L et al (2017) CircRNA_000203 enhances the expression of fibrosis-associated genes by derepressing targets of miR-26b-5p, Col1a2 and CTGF, in cardiac fibroblasts. Sci Rep 7:40342
40. Hansen TB, Venø MT, Damgaard CK et al (2016) Comparison of circular RNA prediction tools. Nucleic Acids Res 44(6):e58–e58
41. Szabo L, Salzman J (2016) Detecting circular RNAs: Bioinformatic and experimental challenges. Nat Rev Genet 17:679–692
42. Quail MA, Kozarewa I, Smith F et al (2008) A large genome centre's improvements to the Illumina sequencing system. Nat Methods 5(12):1005–1010
43. Houseley J, Tollervey D (2010) Apparent non-canonical trans-splicing is generated by reverse transcriptase in vitro. PLoS One 5(8):e12271
44. Luo GX, Taylor J (1990) Template switching by reverse transcriptase during DNA synthesis. J Virol 64(9):4321–4328
45. Kelleher CD, Champoux JJ (1998) Characterization of RNA strand displacement synthesis by Moloney murine leukemia virus reverse transcriptase. J Biol Chem 273(16):9976–9986
46. Panda AC, De S, Grammatikakis I et al (2017) High-purity circular RNA isolation method (RPAD) reveals vast collection of intronic circRNAs. Nucleic Acids Res 45(12):e116–e116
47. Barrett SP, Salzman J (2016) Circular RNAs: analysis, expression and potential functions. Development 143(11):1838
48. Zheng Y, Zhao F (2018) Detection and reconstruction of circular RNAs from transcriptomic data. In: Dieterich C, Papantonis A (eds) Circular RNAs: methods and protocols. Springer, New York, pp 1–8. https://doi.org/10.1007/978-1-4939-7562-4_1
49. Jakobi T, Dieterich C (2018) Deep computational circular RNA analytics from RNA-seq data. In: Dieterich C, Papantonis A (eds) Circular RNAs: methods and protocols. Springer, New York, pp 9–25. https://doi.org/10.1007/978-1-4939-7562-4_2
50. Bachmayr-Heyda A, Reiner AT, Auer K et al (2015) Correlation of circular RNA abundance with proliferation – exemplified with colorectal and ovarian cancer, idiopathic lung fibrosis, and normal human tissues. Sci Rep 5:8057
51. Memczak S, Jens M, Elefsinioti A et al (2013) Circular RNAs are a large class of animal RNAs with regulatory potency. Nature 495:333
52. Zhang Y, Zhang X-O, Chen T et al (2013) Circular intronic long noncoding RNAs. Mol Cell 51(6):792–806
53. Zhang X-O, Wang H-B, Zhang Y et al (2014) Complementary sequence-mediated exon circularization. Cell 159(1):134–147
54. Guo JU, Agarwal V, Guo H et al (2014) Expanded identification and characterization of mammalian circular RNAs. Genome Biol 15(7):409
55. Jeck WR, Sorrentino JA, Wang K et al (2013) Circular RNAs are abundant, conserved, and associated with ALU repeats. RNA 19(2):141–157
56. Cooper DA, Cortés-López M, Miura P (2018) Genome-wide circRNA profiling from RNA-seq data. In: Dieterich C, Papantonis A (eds) Circular RNAs: methods and protocols. Springer, New York, pp 27–41. https://doi.org/10.1007/978-1-4939-7562-4_3
57. Zhang Y, Liang W, Zhang P et al (2017) Circular RNAs: emerging cancer biomarkers and targets. J Exp Clin Cancer Res 36(1):152
58. Heumüller AW, Boeckel J-N (2018) Characterization and validation of circular RNA and their host gene mRNA expression using PCR. In: Dieterich C, Papantonis A (eds) Circular RNAs: methods and protocols. Springer, New York, pp 57–67. https://doi.org/10.1007/978-1-4939-7562-4_5
59. Panda AC, Abdelmohsen K, Gorospe M (2017) RT-qPCR detection of senescence-associated circular RNAs. Methods Mol Biol (Clifton, N.J.) 1534:79–87
60. Chen D-F, Zhang L-J, Tan K et al (2018) Application of droplet digital PCR in quantitative detection of the cell-free circulating circRNAs. Biotechnol Biotechnol Equip 32(1):116–123
61. Capel B, Swain A, Nicolis S et al (1993) Circular transcripts of the testis-determining gene Sry in adult mouse testis. Cell 73(5):1019–1030
62. Schneider T, Schreiner S, Preußer C et al (2018) Northern blot analysis of circular RNAs. In: Dieterich C, Papantonis A (eds) Circular RNAs: methods and protocols. Springer New York, New York, pp 119–133. https://doi.org/10.1007/978-1-4939-7562-4_10
63. Kocks C, Boltengagen A, Piwecka M et al (2018) Single-molecule fluorescence in situ hybridization (FISH) of circular RNA CDR1as. In: Dieterich C, Papantonis A (eds) Circular RNAs: methods and protocols. Springer New York, New York, pp 77–96. https://doi.org/10.1007/978-1-4939-7562-4_7
64. Zirkel A, Papantonis A (2018) Detecting circular RNAs by RNA fluorescence in situ hybridization. In: Dieterich C, Papantonis A (eds) Circular RNAs: methods and protocols. Springer, New York, pp 69–75. https://doi.org/10.1007/978-1-4939-7562-4_6
65. Huang M-S, Zhu T, Li L et al (2018) LncRNAs and CircRNAs from the same gene: masterpieces of RNA splicing. Cancer Lett 415:49–57

66. Song X, Zhang N, Han P et al (2016) Circular RNA profile in gliomas revealed by identification tool UROBORUS. Nucleic Acids Res 44(9):e87–e87
67. Izuogu OG, Alhasan AA, Alafghani HM et al (2016) PTESFinder: a computational method to identify post-transcriptional exon shuffling (PTES) events. BMC Bioinformatics 17(1):31
68. Hoffmann S, Otto C, Doose G et al (2014) A multi-split mapping algorithm for circular RNA, splicing, trans-splicing and fusion detection. Genome Biol 15(2):R34
69. Wang K, Singh D, Zeng Z et al (2010) MapSplice: accurate mapping of RNA-seq reads for splice junction discovery. Nucleic Acids Res 38(18):e178–e178
70. Chuang T-J, Wu C-S, Chen C-Y et al (2016) NCLscan: accurate identification of non-co-linear transcripts (fusion, trans-splicing and circular RNA) with a good balance between sensitivity and precision. Nucleic Acids Res 44(3):e29–e29
71. Gao Y, Wang J, Zhao F (2015) CIRI: an efficient and unbiased algorithm for de novo circular RNA identification. Genome Biol 16(1):4
72. Szabo L, Morey R, Palpant NJ et al (2015) Statistically based splicing detection reveals neural enrichment and tissue-specific induction of circular RNA during human fetal development. Genome Biol 16(1):126
73. Gao Y, Zhang J, Zhao F (2017) Circular RNA identification based on multiple seed matching. Brief Bioinform: bbx014–bbx014
74. Westholm JO, Miura P, Olson S et al (2014) Genomewide analysis of Drosophila circular RNAs reveals their structural and sequence properties and age-dependent neural accumulation. Cell Rep 9(5):1966–1980
75. Cheng J, Metge F, Dieterich C (2016) Specific identification and quantification of circular RNAs from sequencing data. Bioinformatics 32(7):1094–1096
76. Gao Y, Zhao F (2018) Computational strategies for exploring circular RNAs. Trends Genet 34:389
77. Zeng X, Lin W, Guo M et al (2017) A comprehensive overview and evaluation of circular RNA detection tools. PLoS Comput Biol 13(6):e1005420
78. Langmead B, Salzberg SL (2012) Fast gapped-read alignment with bowtie 2. Nat Methods 9(4):357–359
79. Langmead B, Trapnell C, Pop M et al (2009) Ultrafast and memory-efficient alignment of short DNA sequences to the human genome. Genome Biol 10(3):R25–R25
80. Li H, Durbin R (2009) Fast and accurate short read alignment with burrows–wheeler transform. Bioinformatics 25(14):1754–1760
81. Dobin A, Davis CA, Schlesinger F et al (2013) STAR: ultrafast universal RNA-seq aligner. Bioinformatics 29(1):15–21
82. Hansen TB (2018) Improved circRNA identification by combining prediction algorithms. Front Cell Dev Biol 6:20
83. Baruzzo G, Hayer KE, Kim EJ et al (2017) Simulation-based comprehensive benchmarking of RNA-seq aligners. Nat Methods 14(2):135–139
84. Kim D, Pertea G, Trapnell C et al (2013) TopHat2: accurate alignment of transcriptomes in the presence of insertions, deletions and gene fusions. Genome Biol 14(4):R36
85. Trapnell C, Pachter L, Salzberg SL (2009) TopHat: discovering splice junctions with RNA-Seq. Bioinformatics 25(9):1105–1111
86. Glažar P, Papavasileiou P, Rajewsky N (2014) circBase: a database for circular RNAs. RNA 20(11):1666–1670
87. Ghosal S, Das S, Sen R et al (2013) Circ2Traits: a comprehensive database for circular RNA potentially associated with disease and traits. Front Genet 4:283
88. Liu Y-C, Li J-R, Sun C-H et al (2016) CircNet: a database of circular RNAs derived from transcriptome sequencing data. Nucleic Acids Res 44(Database issue):D209–D215
89. Gao Y, Wang J, Zheng Y et al (2016) Comprehensive identification of internal structure and alternative splicing events in circular RNAs. Nat Commun 7:12060
90. Metge F, Czaja-Hasse LF, Reinhardt R et al (2017) FUCHS—towards full circular RNA characterization using RNAseq. PeerJ 5:e2934
91. Meng X, Chen Q, Zhang P et al (2017) CircPro: an integrated tool for the identification of circRNAs with protein-coding potential. Bioinformatics 33(20):3314–3316
92. Feng J, Xiang Y, Xia S, et al (2017) CircView: a visualization and exploration tool for circular RNAs. Brief Bioinform:bbx070-bbx070
93. Dudekula DB, Panda AC, Grammatikakis I et al (2016) CircInteractome: a web tool for exploring circular RNAs and their interacting proteins and microRNAs. RNA Biol 13(1):34–42
94. Grant GR, Farkas MH, Pizarro AD et al (2011) Comparative analysis of RNA-Seq alignment algorithms and the RNA-Seq unified mapper (RUM). Bioinformatics 27(18):2518–2528
95. Streit S, Michalski CW, Erkan M et al (2008) Northern blot analysis for detection and quantification of RNA in pancreatic cancer cells and tissues. Nat Protoc 4:37
96. PCvdC M, Roeland WD, PMvG R et al (1998) Sensitive mRNA detection by fluorescence in situ hybridization using horseradish peroxidase-labeled oligodeoxynucleotides and tyramide signal amplification. J Histochem Cytochem 46(11):1249–1259

The manufacturer's authorised representative in the EU is Springer Nature Customer Service Centre GmbH, Europaplatz 3, 69115 Heidelberg, Germany. If you have any concerns regarding our products, please contact ProductSafety@springernature.com

Printed and bound by CPI Group (UK) Ltd, Croydon, CR0 4YY
25/09/2026
02222841-0019